Forest Fungi Phytogeography:

Forest Fungi Phytogeography of China, North America, and Siberia and International Quarantine of Tree Pathogens

森林真菌病理地理：

美国，中国和西伯利亚以及病原真菌及食菌类

Mo Mei Chen
Jepson and University Herbaria,
University of California at Berkeley

Published by Pacific Mushroom Research and Education Center, a nonprofit corporation in California.

ISBN : 0-9723916-6-5

Pacific Mushroom Research & Education Center
P.O. Box 189326
Sacramento, CA 95818

Acknowledgments

Preparation of this book was sponsored by the Pacific Mushroom Research & Education Center, a nonprofit organization in northern California.

I would like to gratefully acknowledge the editorial and manuscript-preparation assistance of Jo-Sing Yang, Ester Kwon, Dan Norris, Wendy Helms, Xiao Zhang, Peter Werner, Adelyn Tsu and Susan Tong.

I also thank Staci Markos of the University of California at Berkeley, University and Jepson Herbarium for her encouragement.

I would also like to thank my family for their unconditional support:
My dear husband Anghe Zhang;
My daughter May and my son-in-law Bihshan Chen;
My daughter Lily and my son-in-law Shu Huang.

Forest Fungi Phytogeography:

Forest Fungi Phytogeography of China, North America, & Siberia and International Quarantine of Tree Pathogens

Preface

This book brings together, for the first time, a number of papers on plant pathogens of China, Eastern Russia and North America. These papers were based on extensive collecting and research done over a period of forty years in a variety of remote areas and thus offer an overview of the geographical and natural distribution of plant pathogens. Knowledge of phytogeography is becoming increasingly important in international trade, especially in agricultural business, at a time when devastating plant diseases can spread rapidly. Papers in this book deal not only with identification of plant pathogens, of which the fungi genus is the most important, but emphasize the importance of understanding the source, evolution and phytogeographical history of pathogens such as tree rust.

Publication of this book would not have been possible without the great help and encouragement of Berkeley/University and Jepson Herbaria especially my youth friend Jo-Sing Yang. I also received significant support from many younger colleagues at the 2001 Chinese Forest Pathologist Symposium at Huang Mountain who encouraged me to collect and publish my writings. Since this book has been put together in haste, to be ready for the IUFRO meeting, I apologize for the fact that some mistakes are inescapable. In a future edition, I intend to revise some of the materials and include more of my works.

My Background: I graduated from the Plant Protection Department of Beijing Agricultural University in the middle 1950s. Between 1955 and 1982, in addition to filling various teaching positions, I spent many summers as a member of large scale scientific expeditions in virgin forests of Northeastern China and on the Qingzhang Plateau in the Himalayas. My work focused mainly on plant pathogens, but I had a special interest in higer fungi, particularly rusts. I used methods learned from Professor Fanglan Dai in serch collecting specimens and I received valuable help from my Professors Dafu Yu and Shuqun Deng in identification and classification of collected fungi specimens.

Since 1982 I have been living in the United States and have been involved with numerous research and teaching projects, which are referred to in the following letter of recommendation written by U.C.B. Professor Emeritus Dick Parmeter. I am deeply grateful to Professor Parmeter for his help and encouragement. Through him and Professors Fields Cobb, David Wood and William J. Libby had access to U.C. Berkeley field research facilities where I was able to pursue research in Monterey Pine western gall rust and Sugar Pine rust and to investigate the life cycle of rust fungi.

For many years I have taught U.C. Extension classes on the cultivation of mushrooms for both food and medicinal purposes and I have been instrumental in introducing valuable new mushroom species into the United States. In all my work, I emphasize the importance of conservation of fungi in the natural environment.

Mo Mei Chen

To Whom It May Concern: 1 April 1992

It has been my good fortune to work with Professor Chen Mo-Mei as a professional colleague and friend for more than seven years. Her knowledge and experience in forest pathology and biogeography, her experience with international forest problems, her flair for teaching, and her boundless energy and enthusiasm have contributed greatly to our research, extension, and teaching programs. It is unfortunate that our system does not provide means to properly reward Prof. Chen for her exceptional talent and service.

Professor Chen came to us at Berkeley following her work on Dutch elm disease with Prof. E. Smalley at the University of Wisconsin. Her first work here was with Prof. F. Cobb on epidemiology of white pine blister rust in California. Her familiarity with Asian tree rusts proved valuable. Prof. Chen's subsequent work with North American tree rusts qualifies her as one of the leading authorities on international rust relationships.

As Prof. Chen was completing her work on rusts, Prof. D. Wood and I were developing a research program on interactions of pine hosts, bark beetles, and the bluestain fungi vectored by bark beetles. Prof. Chen's experience with Dutch elm disease and its associated bark beetle vectors uniquely qualified her to contribute to our studies. In connection with these studies, Prof. Chen isolated and identified large numbers of fungi from bark beetles, organized and maintained an extensive culture collection, helped develop methods to prepare inoculum and test virulence of these fungi, and she participated in data analysis and publication. Prof. Chen certainly became an authoritative integral, and valuable part of our research program. Her contributions were outstanding. Since my retirement, the bluestain work has wound down, and there is no longer financial support for Prof. Chen.

For most people, the work outlined above on Dutch elm disease, pine rusts, bluestain fungi, and other forest problems would be sufficiently satisfying, but Prof. Chen's energy, organization, and desire to excel have allowed her to make many additional contributions to science and teaching. One of the chief activities has been the facilitation of numerous exchange visits between U.S. and Chinese scientists and scholars, and she has played a key role in arranging important contacts and in developing promising programs in U.S./China cooperation. Her knowledge of geography and forestry provides special bases for planning field trips for U.S. scientists in China and Chinese scientists in the U.S. Prof. Chen is the most competent and accomplished arranger of exchanges of whom I'm aware. Her services to China and the U.S. are extremely valuable and worthy of special recognition.

Her familiarity with international aspects of forest biology and her command of Russian and Chinese literature have been of great value in assessing the dangers of importing forest materials into North America. Prof. Chen made major contributions to recent U.S. efforts to evaluate possible problems that might arise from importation of Siberian logs.

Along with all of these activities, Prof. Chen has found time to put her knowledge of higher fungi to use. She has become widely recognized in California as an authority on production of wild and commercial edible mushrooms. Her seminars and training sessions are very popular, and she is in frequent demand to provide extension programs in mushroom biology and cultivation. The current interest in exploiting forest and agricultural wastes to produce edible fungi puts Prof. Chen's knowledge and experience at a premium. Prof. Chen could easily profit from her skills, but she prefers academic pursuits to business.

Prof. Chen's knowledge of higher fungi has also been the basis for an ambitious survey and evaluation of these fungi in the Taiga forests of Alaska. This has entailed travel to Alaska, development of a large mycological herbarium, and will no doubt lead to significant publication.

I have been much impressed with Prof. Chen's contributions and accomplishments while she has been with us at the University of California. My comments provide only brief outline of her many talents, but I hope they will serve to confirm that Prof. Chen is a unique resource, a scientist and scholar with indefatigable energy, exceptional experience, special knowledge and skill, and a commitment to service. Her past contributions and her potential to accomplish much more in the future deserve recognition and reward.

Richard Parmeter, Professor of Plant Pathology, Emeritus

The Forest Biogeography of Phytopathogen in China

China is a mountainous country lying between 5 and 53 degrees north latitude. This continent includes the highest mountains on the earth. It is extraordinarily rich in flora that is very complex (according to fossil records). According to critics, it is one of the best regions in the world for researching flowering plants.

The country has 120 million hectares of forest covering 12.7% of her total land area, the total forest volume measuring 95,000 million cubic meters. China's existing forests are mainly located over the Northeastern and Southwestern part of her territory. As we look about China now, her forests can be classified into coniferous belts in the frigid and temperate zones, mixed forest belts of coniferous, deciduous, and broad leaf trees in the temperate zone, and deciduous and broad leaf forest belts in the warm temperate zone. The evergreen forest belt is in the subtropical zone, and monsoon and rainforests are in the tropical zone.

Since the 1950s, my colleagues and I conducted extensive studies on almost all the main temperate forest regions. Forest investigations of a considerable scale and with various objectives were carried out. These studies included areas investigated including the northeast, the northwest, and Tian Shan, as well as the southwest high mountain forests and the Himalayas.

China is rich in forest plants and tree species with more than 2,800 varieties of tree species. Among them are 20 genera of *pinaceae* and *taxodiacaceae* are dew redwood, (*Metasequoia glyptostroboides*); lovely garden larch, (*Pseudolarix amabilis*); Bhutan Cupress, (*Cupressus duclouxiana*); Japanese podocarpus, (*Podocarpus nagi*); and Chinese fir, (*Cunninghamia lanceolata*). Also native to China are numerous broad leaf species constituting some 260 genera including some important industrial tree species as *Paulowinia tomenfosa, Froxinum mandishurica, Juglans mandshurica, Cinnamomum camphora, Fhoebe bournel*, and *Tonna sinesis*. Important economic tree species include *Aleurites fordii, Camellia oleifera, Sapium sebiferum,* and *Rhum verhiciflua*. China's forests can be classified into seven regions:

1 TEMPERATE ZONE FOREST

The flora of the temperate zone belongs to both the mild temperate zone and the frigid zone. Some areas extend to marginal areas of subtropical and even the tropical zone. The typical families are *Fagaceae, Betulaceae, Juglandaceae* and *Salicaceae*. There is also the family of Catkin (*Camemtum*). In the high mountains there are quite a lot of Rhododendrons, *Gentianas* species. In this region there are some pathogenic genera that are the same as in North America and East Asia.

1.1 *Eurasia forest region*

Included are three forest areas: the Great Xinan Ling, the Altain shan and Tian shan.

1.1.1 *Great Xinan Ling*

The major tree species is *Larix gmelini*. It is located from 500 to 1,200 meters elevation. Dominant species are *Betula pltyphylla, Popoulous davidana* and *Quecuum mongolia*. Above 800 meters sea level are the spreading Pinus *sylvestris* var. *mongolica* and the watershed regions have a shorter tree *Pinus pumila*.

According to several investigations there are many tree species and 114 species of tree diseases.

In the cooler virgin forest, the characteristic tree diseases of the over-mature trees are Larch decay; typical pathogens are *Phellinus pini, Phaeolus schweinitzii, Polyporus sulphareus,* and *Fomes officinalis*. The native diseases are *Cronartium guercuum*, on *Pinus sylvestris* var. *mongolia*, and *Dasyscypha wiklommii* on *Larix*.

The following are three chief species of tree diseases: *Pinus sylvestris* var. *mongolica* with pine oak rusts (*Cronartium quercuum*); *Betula platyphylla* with white heart rot (*Phellinus ignirius*); *Populus davidiana* with Aspen white heart rot (*Phellinus tremulae*); and the white rot (*Fomes fomentarius*) causes many hardwood decays; But we still haven't seen the *Cronartium kamtschaticm* on the *Pinus pumila*.

1.1.2 *Altain shan*

The *Larix sibirica* occupies a dominant position. Others are *Pinus sibirica, Abies sibirica* and *Picea obovada* etc. Usually the *L. sibirica* forest is found at 1300-2600 meters elevation where the characteristic tree diseases are coniferous decay. *Phellinus pini* on the pines and larch is quite representative of the Eurasia forest region, and the *Cronatium ribicola* is found in the western fringe of China as well.

The following are the chief species of tree diseases:

a) *L. sibirica*: white pocket rot (*Phellinus pini*), and brown trunk rot of larch (*Fomes officinalis*);
b) *Pinus sibirica*: five needles pine blister rust (*Cronartium ribicola*);
c) *Picea obovata*: spruce white pocket rot (*Phellinus pini* var. *abietis*).

1.1.3 *Tian shan*

The main forest tree species are *Picea schrenkiana* and *P.* var *tianschanica*. There is spruce rust in forest humidity regions. Its characteristic forest trees' diseases are on various kinds of spruce such as the spruce needle rusts (*Chrysomyxa spp.*) and spruce cone rust (*Thekopsora areolata*).

1.2 *Asiatic desert region*

There are two forest areas, the West Central Asiatic area and the Central Asiatic area.

The main flora is desert flora; many plants from the Central Asiatic and Mediterranean areas predominate. *Haloxylon ammodendron – H. persicum*, and *Elaeagnus angustifolia* species predominate and *Populous salix*, *Ulmus* are frequently found in this region and elsewhere. Desert diseases, such as powdery mildew of *Holoxylon* and *Hedysarum*, drought rust and popular leaf rust, are representative of the region.

Diseases endemic to the area include:

a) *Haloxylon spp.*: powdery mildew of *Haloxylon, Leveillula saxaouli f. haloxyli;*
b) *Populus spp.*: Stem canker, *Vals sordida*; leaf brown spots, *Spetotis populiperda*; *Anthraccanosa, Glomerella cingulata*; white poplar leaf rust, *Melampsora rostrupii*; leaf rust, *M. pruinosae;* willow white rot, *Trametes trogii;*
c) *Elaeaganus angustifolia*: Russian olife leaf spots, *Septoria aregyraea*.

1.3 *Eurasian steppe region*

This region is commonly called the Mongolian steppe region that includes the western Inner Mongolia and northeast prairie regions with species such as *Populus*, *Salix* and *Ulmus* etc. The prairie frequently blends into a secondary forest region in which birch and aspen species are common. Today, in many places, shelterbelt planting poplar, elms and pine is conducted along with reforestation of pines, larch and other forest trees. Since drought is frequent in this region, the growth of fungi is limited. However, the stem canker is a common disease and is usually found near the wounded standing tree canker.

1.4 *Qinghai-Xi zang plateau region*

The Himalayas are the youngest mountain range on the globe having risen from the sea 40 million years ago. The region is cold and arid or semi-arid near sea level, but receives more rainfall along watercourses and in the mountainous zone. *Rododendron, Salix* and some *Hippophae* species are found along watercourses. In the dryer regions *Populus spp.* is afflicted with diseases characteristic of droughts such as larch leaf rust, Nectaria canker and drought injury in the high plateau. Typical diseases of the area includes:

a) *Salix spp.*: willow leaf rust, *Melampsora larici-campraearum*;
b) *Populus spp.*: poplar canker, *Cytospora chrysosperma*; Nectria canker, *Nectria cinnabarina*;
c) *Hippophae rhamnoides*: Seabuckthorn trunk rot, *Phellinus robustus*.

1.5 *Sino-Japan forest region*

This region is rich in forest plants and tree species that can be divided into five areas. The mixed forest belt is made up of coniferous, deciduous and broadleaf trees in the temperate zone and an evergreen forest belt in subtropical zone.

1.5.1 *Northeast China region*

The evergreen or coniferous forest originally covered the Xiao xin Ling mountains. Characteristic trees of the natural upland forest of the region are *Pinus koralensis, Picea jezoensis, Abies holophylla,* and *A. nephrolepis*. *Quercus mongolica* predominates in the eastern part of this area.

The diseases found are those typical to cold-temperate flora of virginal pine-hardwood mixed forests. It is closely related to East of North America flora such as Appalachian forest flora. The Predominant forest tree's diseases are:

a) *Pinus koraiensis*: White Pine Blister Rust (*Cronartium ribicola*); pine white pocket rote, (*Phellinus pini*); pine Cenangium canker (*Cenangium furfruceam*); Armillaria root rot, (*Armillaria mellea*);
b) *Picea spp.*: spruce white pocket rot, (*Phellinus pini* var. *abietis*);
c) *Abies spp.*: fir white rot, (*Phellinus hartigii)*;
d) *Quercus mongolica*: oak white rot, (*Polyporus dryophilus*); Schlerotina rot, (*Sclerotinia pseudotuberosa*);
e) *Fraxinus mandshurica*: ash trunk rot, (*Polyporus hispida*);
f) *Juglanda mandshurica*: walnut white rot, *Phellinus robustus*; walnut Melaconium canker, (*Melaconium juglandium*).

1.5.2 *Northern region of middle China*

The characteristic trees of the area are pines such as *Pinus tabulaeformis, Pinus tabulaeformis* var. *mukdensis, P. bungeana* and other species and oaks such as *Quercus liaotungensis, Q. acutissima, Q. dentata* and others. The region is usually followed by some native genera as *Ailanthus, Toon, Melia,*

Koelreuteria, Zizphus, Diospyros, Paulownia, Catalup, Populus and *Robinia*. *Ulmus* is growing more rapidly here than in other regions.

The climate is typically dry and cool with a rainfall of 500-800 mm annually. Characteristic diseases are powdery mildew of Ailanthus, leaf spot, *Colleatotrichum anthraenose* of *Paulownia*, and *Paulownia* and jujube witch's broom in MLO. Chief species of tree diseases are:

a) *Pinus tabulaeformis*: damping off, (*Rhizoctonia solani*);
b) *Ailanthus altissima*: powdery mildew, (*Phyllactinia corylea*);
c) *Zizyphus jujuba:* MLO, virus of jujube;
d) *Diospyros kaki*: leaf spot, (*Cercospora kaki*); leaf spot *Mycosphaerella nawae;*
e) *Populus tomentosa, P. simonii, P. euramericana*: poplar canker, (*Valsa sordida*); poplar leaf rust, (*Melamppsora larici-populina*); leaf brown spot, (*Septoria populi*); poplar gall crown, (*Agrobacterium tumefaciens*); violet root rot, (*Helicobasidium mompa*);
f) *Salix spp.*: willow canker, (*Valsa sordida*); willow trunk rot, (*Trametes trogii*);
g) *Ulmus pumila*: elm canker, (*Cytospora sp*).
h) *Paulownia elongata, P. catalpifolia, P. fortunei, P. tomentosa*: witch's broom of paulownia (MLO); scab of paulownia, *Sphaceloma paulowniae*; *Anthracnose, Collectrichum kawakamii*;
i) *Melia spp.*: leaf spot, *Cercospora subsessilis*.

1.5.3 *Central China region*

There are two big regions sometimes called Basin of China or Central China. This region extends from the warm temperate zone to the subtropical zone. Evergreens are increasingly seen and characteristic of trees of these two regions are *Pinus massoniana* and *Cunninghamia lanceolata*. This region frequently contains several economic tree species especially in silviculture, such as tea oil, (*Camellia oleosa)* (*C.sasanqua var.oleosa, Thea oleosa*) and tung oil tree (*Aleurites fordii*), lacquer, bamboo, and chestnut. Common species of disease are leaf rust of Toona (*Nyssoposa cedrelae* and *Phakopsora cheoana*) and anthrachose of tea oil. Wilt of tung oil (*Fusarium oxysporum*)and rusts on bamboo (*Stereostratum cortocioides*). Native trees are also afflicted with these same diseases. Since the area has abundant rainfall and mild temperatures, there are many species of plant pathogens. Commercially important trees are often infected. Predominant diseases are:

a) *Pinus massoniana*: pine oak rusts, *Cronartium quercuum*; Chinese pine-oak rusts, (*Cronartium sp.*); pine needle brown spot, Cercospora pine-*densiflorae*);
b) *Cunninghamia lanceolata*: Chinese fir anthracnose, *Colletotrichum sp.*; bacterial spot, *Pseudomonas cunninghumiae*; Chinese fir narrow leaf spot, *Pestalotia shiraiana*;
c) *Camellia oleifera*: Tea oil anthracnose, *Colletotrichum camelliae; Scelerotium rolfsii*; sooty blotch, *Meliola camelliae*;
d) *Aleurites fordii*: tunt oil wilt, *Mycosphaerella aleuritidis*; tunt oil wilt, *Fusarium spp.*;
e) *Bamboo spp.*: bamboo stem rust, *Puccinia corticiodes*; bamboo witch's broom, *Aciculosporium take*;
f) *Gingko biloba*: Gingko stem rot, *Macrophomina phaseoli*;
g) *Pterocarya stenoptera*: witch's broom, Chinese wingnut.

1.5.4 *South China*

Similar to the Sino-Japan forest region, this region also has distinguishing features of the subtropical and tropical zones. The tropical forest in southern China is humid and is the southern most range for many species typical of the deep tropics. There characteristic tropical and subtropical trees. The special *Lauraceae* includes 176 species belonging to 17 genera. Common tree diseases are massoniana pine leaf spot, *Anthracnose cinnamomum*, Bacteria of Olea and bacterial wilt of Horsetail heefwood (*Casuarina equisetifolia*). The principal plant diseases are as follows:

a) *Eucalyptus spp.*: eucalyptus leaf spot, *Septotis populiperda*;

b) *Pinus massoniana*: pine needle spot, *Diplodia pinea;*
c) *Olea europaea*: Olea crown gall, *Pseudomonas savastanci*;
d) *Casuarina equisetifolia*: horsetail beefwood bacterial wilt.

1.5.5 *Yunnan*

In Yunnan thc flora is more complex than in South China. Chinese fir and Massoniana pine are extremely common along with Sino-Himalayan species such as Yunnan pine. Near mountains are found several species of Keteleeria. Because most of the land lies in the high plateau, trees have adapted themselves to little rainfall and strong sunlight. Characteristic to this range are parasitic plants such as *Arcerthobium, Loranthus*, and *viscum*. Principal diseases include:

a) *Illicium velum*: truestar anise tree, anthracnose;
b) *Pinus massoniana*: pine brown spot, *Pestalotia spp.*; pine blight, *Lophodermium pinastri*;
c) *Populus spp.*: algae leaf spot, *Cephaleuros virescens*;
d) *Olea europaea*: Olea peacock spot, *Cycloconium oleaginum*;
e) Pine, spruce, fir: dwarf mistletoes, *Arcerthobium Chinese*;
f) Hardwood: *Loranthus spp.* and *Viscum spp.*

1.6 *Sino-Himalayan forest region*

There are three regions: Yunnan plateau, Heng-duan-mountain and the base of the Himalayas. The Sino-Himalayan forest is the highest plateau on earth. The plateau itself has an average elevation of 4,500 meters above sea level and mountains rise more than 3,000 meters above the surrounding planes and basins. It is the youngest mountain range having risen from the sea 40 million years ago.

This area has the richest and oldest forest flora on earth, including the temperate zone, subtropical zone and northern tropical zone. There are over 20,000 species of vegetation including 1,300 species of trees. Within the 100 families of conifers there are conifers of high economic value in eight families, 16 genera and over 40 species. The species of pathogens and fungi are also complex. As the Himalayan plateau is high in altitude, its forest has a special and complicated natural environment. The elevation of the plateau and geological changes produced numerous varieties of plants as well as species of pathogenic fungi that are closely related to those in other parts of China and the world. Most important, tree disease flora belonging to the northern temperate zone constitute about 90 percent of the total. Many of these are of the same type as found in North America and East Asia while others are unique to the Himalayas. Principal diseases are:

a) *Pinus yunnanensis, P. densata, P. Armandii*: pine white pocket rot, *Phellinus* pini; pine needle rust, *Coleosporium sp.*; *P. griffithi*: Five needle blight rust, *Cronartium ribicola*; pine needle rust, *Peridermium brevius*; *P. longifolia*: pine needle rust, *Peridermium complanatum*;
b) *Picea likiangensis*: *P. likiangensis* var. *balfouriana*, *P. likiangensis* var. *linzhiensis, P. Spinulosa P. smithiana*: spruce white pocket rot, (*Phellinus pini* var. *abietis*); spruce needle rust, (*Peridermium sinenses*); spruce cone rust, (*Thekopsora areolata*); spruce trunk rot, (*Inonotus dryadeus*);
c) *Abies georgei. A. georgei* var. *smithii, A. spectabilis*: fir trunk rot, *Phellinus hartigii*;
d) *Tsuga dumosa*: Hemlock trunk rot, *Genoderma tsugae*; Hemlock red ring rot, *Phellinus pini* var. *abietis*
e) *Quercus semicarpifolis*: oak rot, *Trametes cinnabarius*; white trunk rot, *Phellinus igniarius*; oak soft spongy white rot, *Ganoderma lucidum*;
f) *Betula spp.*: birch leaf rust, *Melampsoridium betulinum*;
g) *Castanopsis indica, C. hystrix*: trunk rot of evergreen chinkapin, *Stereum fasciatum*;
h) Hardwood: hardwood mistletoes, *Loranthus spp., Viscum spp.*
i) Coniferous: coniferous dwarf mistletoes, *Arceuthobium spp.*

2 TROPICAL ZONE

2.1 *Malaysian*

Included here are several areas: Taiwan, South China Sea, Tongking Gulf, Yunnan-Burma-Thailand areas. This region covers the coastal area from Taiwan to Hainam, extending west to the bottom of Yunnan. This area is characterized by heavy rainfall coming from the Indian Ocean and warm temperatures. Diseases are those typical to tropical areas such as tropical rust, fruit tree rot, rubber root rot and others. Principle diseases are:

a) *Castonopsis kawakamil*: brown spot, *Phyllosticta spp.*
b) *Cinnamomum camphora*: powdery mildew, *Cinnamom anthraenose; Glomerella cinnamomoni yosh*;
c) *Dalbergia odorifera*: Rosewood tar spot, *Phyllachora dalbergiicola;*
d) *Acacia confusa*: Acacia leaf rust, *Poliotelium hyalosporum*;
e) *Tectona grandis*: common teak leaf rust, *Olivea tectonae*;
f) *Elaeis guineensis:* oilpalm fruit rot
g) *Hevea brasiliensis*: rubber powdery mildew, *Oidium hevae*; rubber streak canker, *Phytophtora palmivora*; rubber root rot, *Ganoderma pseudoferreum*: rubber root rot, *Fomea noxius*; rubber root rot, *Ustulina zonata.*

Some studies of white pine blister rust flora of North America and East Asia

ABSTRACT: China and North America have similar forest rust flora. These rust flora are the most diverse in the North Temperate Zone and both are of the richest rust taxa in the world. But China is the only country in the world that includes unbroken transitional zones connecting tropical, temperate, and boreal forests. America's rust flora is characterized as North American temperate zone, and China's is characterized as a continental forest rust flora. Uredinales, (Teliomycetes) are rust fungi taxa including 14 families, 163 genera and 7000 species (Ainsworth and Bisby). Most tree and shrub rust genera belong to the Melampsoraceae family such as: *Chrysomyxa, Coleopuccinia, Coleosporium, Cronartium, Gymnosporangium, Melampsora, Melampsorella, Melampsoridium, Nyssopsora, Phakopsora, Phragmidium, Pileolaria, Puccinia, Pucciniastrum, Ravenelia, Sphaerophragmium, Thekopsora, Tranzschelia, Triphragmiopsis, Triphragmium,* and *Uromyces.* Both China and North America have these genera in common, however, the species of *Coleopuccinia sinensis* of *Cotoneaster acutifolia* is indigenous to alpine-plateau flora and *Nyssopsora cedrelae* of *Toona* sinensis is indigenous species in China. *Endocronartium harknessii* of *Pinus radiata* is in Mediterranean-Pacific rust flora in North America only.

Recent results from cladistic analyses of morphological and molecular characters have suggested that these fungi have had a longer co-evolutionary relationship with Angiosperm than Gymnosperm hosts (Vogler and Bruns, 1998). The following will present some studies of white pine blister flora of America and China. The white pine blister rust fungus (*Cronartium ribicola*) is an obligate parasite on both aceia hosts *Pinus* spp. and telia host *Ribes* spp., *Pedicularis* spp. Thirty years of research in the virginal forests and greenhouse inoculation by the author have led to a better description of the white pine blister rust flora. In North America, an eastern white pine rust (WPBR) flora (*Pinus strobes/Ribes nigrum*) and a Pacific Mediterranean WPBR flora (*Pinus lambertiana/Ribes roezlii*) are recognized. In Eastern Asia, Sino-Japan WPBR flora are found: (*Pinus koraiensis/Pedicularis resupinata, P. resupinata var. ramose, P. spicata, Ribes mandshuricum*). In China the alpine WPBR flora are found in the southwest regions:(*Pinus armandii/Ribes himalense*). Although it has been suggested that white pine blister rust is introduced in a portion of its current range, the occurrence of white pine blister rust floras in virgin forests indicates the blister rust has evolved as part of the indigenous ecosystem in each region. Scientific analysis of the biogeography of pathogens is crucial in understanding how and where pathogens become a problem. Furthermore applications of quarantines can be valuable. Future phylogenetic speciation studies of *Cronartium* and *Ribes* may shed additional light on this endeavor.

1 THE WHITE PINE BLISTER RUST FUNGUS

The white pine blister rust fungus (*Cronartium ribicola* J. C. Fischer) is a heteroecious plant pathogen, meaning it requires two different hosts to complete its life cycle. The scientific name is applied to the sexual stage, and the species epithet "ribicola" refers to the fact that the sexual stage (telia) lives on shrubs of the genus *Ribes* (currants and gooseberries) or Scrophulariaceae. Recent results from *Ribes* cladistic analyses of morphological and molecular characters show an evolutionary trend toward greater specialization among telial hosts (Ref. 20, 21, 23, 31, 35). *Cronartium ribicola* is an obligate parasite on local telia host *Ribes* or *Pedicularis.* This rust is found in the Northern Hemisphere and forms a local rust flora wherever the alternate host and white pine is prevalent.

2 THE STUDY OF WHITE PINE BLISTER RUST IN ITS NATURAL FORESTS

The author has investigated the virgin forests of North America and East Asia for thirty years. Many investigations show that in the virginal white pine forests, the rates of infection aecia phase were lower than the man-made plantations. (0.5% armandi pine, Qining, China; 1% Korea pine, Xiao xin an ling, China and 1% sugar pine, California, middle Sierra Nevada) (Ref. 4, 5, 6, 7, 8, 9, 10, 11, 12, 13, 14, 16, 26, 36).

3 THE LIFE CYCLE

The Life Cycle consists of five spore phases. The five spore phases in order consist of pycnia, aecia, uredia, telia, and basidia. (Fig. 1) They are initially caused by basidiospore infections to pine and results with the symptom of pine bark blister; after that, aeciospores transfer to alternate host and produce the tail and basidiospores. Usually for the white pine blister rust, one life cycle lasts 1-3 years. The completion of the rust life cycle is complicated by genetic and ecological factors (Ref. 3, 6, 7).

3.1 *Pycnia (Stage 0)*

The pycnia as a fluid matrix exude pycnia spores in drops of sticky masses. The pycnia is sterigmated with one-celled, pear-shaped pycniospores. (Fig. 2a, 2b). These spores each contain a single haploid nucleus pycnia and are produced from haploid hyphae. The pycniospores serve as spermatia. The pycniospores are presumed to serve as male gametes that fertilize the female's flexuous hyphae around the periphery of the pycnium. However, these functions are not known in complete detail (Ref. 6, 7, 34).

3.2 *Aecia (Stage I)*

Blister aecia break through the bark surface where the pycnia appeared previously. The peridium wall of the aecium is fragile and breaks easily when mature, (Fig.3) releasing a powdery yellow or orange mass of aeciospores with verrucose walls. The aeciospores are produced from the dikaryotic hyphae and are specialized for specific white pine transfer. The dikaryotic hyphae, produced upon germination, attack the white pine, as did the haploid hyphae. Aeciospores resist drying and can germinate after being carried by the wind for long distances. The white pine blister rust persists for a long period of time and gradually discolors, but the aeciaspores do not reinfect the pine (Ref. 6, 7, 13, 34).

3.3 *Uredia (Stage II)*

Uredia are tiny, yellow, dome-shaped powder structures scattered on the leaf. The central pore opens to release yellow obovoid urediospores each with a thick (2-3.5μm), echinulate hyaline wall. Urediospores (Fig. 4, 5) reinfect the same host species from which they are produced. The uredial stages continue the life cycle, intensifying during summer until late fall when the telia appears, sometimes replacing the uredia (Ref. 6, 7).

3.4 *Telia (Stage III)*

Telia are composed of chains teliospores joined side by side. Teliospores are long, obovoid, and light brown with a thick wall (58-43x34-27μm) (Fig. 5, 6). The teliospores are dispersed and germinate on the white pine under cool and moist conditions producing basidia that in turn give rise to basidiospores.

3.5 *Basidia and Basidiospores (Phase IV)*

The two nuclei in the mature teliospore fuse to form a single diploid nucleus. This nucleus then divides meitotically into the haploid nuclei, which forms a basidia. (Fig. 7) Four basidiospores are then developed from the basidia. The basidiospores are delicate thin walled, colorless, single cells that are released and carried by the wind to the white pine needle. Basidiospores are dispersed generally under high moisture conditions from dew, heavy fog, rain, or melting snow. Spores appear when the relative humidity in the atmosphere approaches 100 percent. The basidia die from the sun's radiation or from drying out. Basidiospores invade between the cells of the new tissue. Nutrients are obtained from living cells of the specific *Ribes* alternated species without invading the cytoplasm. The metabolically active *Ribes or Pedicularis* cells allow a certain amount of nutrient to diffuse out to the specialized feeder hyphae called haustoria, which penetrate the cell wall but not the plasma membrane (Ref. 9, 10, 13).

4 CHARACTERISTICS OF WHITE PINE BRISTLE RUST IN NORTH AMERICA AND EAST ASIA

4.1 *North America Appalachians White pine blister rust flora*

White pine name & characteristics: The eastern white pine flora (*Pinus strobus*) show a widespread population with natural ranges from the very north eastern tips including Newfoundland, spreading to the U.S. states around the Great Lakes and reaching southward through the Appalachian mountains to the north tips of that state of Georgia.

The characteristics of the White Pine Bristle Rust in North America and East Asia are classified as the following five floras:

	White Pine Bristle Rust Flora	*Pinus. SP / Ribes. SP*	*Cronartium ribicola. f. SP.*
N A	Appalachians WPBR	*Pinus strobus/Ribes nigrum*	*C. ribicola. f. nigrum*
	Pacific Mediterranean WPBR	*Pinus lambertiana/Ribes roezlii*	*C. ribicola. f. roezlii*
E A	Sino-Japan WPBR	*Pinus koraiensis/ Pedicularis resupinata*	*C. ribicola. f. pedicularis*
	Alpine WPBR	*Pinus armandii/Ribes glaciale*	*C. ribicola. f. glaciale*
	Sino-Himalayan plateau WPBR	*Pinus griffithii/ Ribes himalense*	*C. ribicola. f. himalense*

(Ref. 1, 4, 5, 17, 19, 25, 26, 27, 28, 30, 32, 37, 38)

Latitude/Longitude: Located in areas North latitude 35°-50°, West longitude 50°-95°

Alternate host species name: *Ribes nigrum* along with *Pinus strobus* is very susceptible to the rust. Eastern North America has seven *Ribes* species implicated in blister rust: *R. americanum, R. odoratum R. oxyacanthoides R. glandulossum, R.rotundifolium, R.cynosbati, R. hirtellum.* Their relative susceptibility to *C. ribicola* is moderate but among the seven, *R. americanum* is the highest. The most virulent and widely distributed rust is on the *Ribes nigrum* that has been raised for trade into many cultivars. There are numerous varieties badly mixed in trade cultivars making it difficult to classify the descendents. *Ribes nigrum,* commonly known as black currants, is much cultivated in central and northern Europe (Ref. 17, 29).

Life Cycle Character: Many problems are caused by the cultivars of *Ribes nigrum* resulting in the wide varieties of the white pine blister rust fungus (Ref. 23, 29).

Systematic biological specializations and taxonomic treatment: The systematic biological specializations and taxonomic treatment are as follows.
Formae speciales : *Cronartium. ribicola.* f. SP.
Class: Basidiomycotina
Order: Uredinales
Family: Malampsoraceae
Genus: Cronartium
Species: C. ribicola
Formae speciales:
C. ribicola f. SP. nigrum F.

Comparison with other WPBR flora: The telia and basidia do not follow the whole distribution of *Pinus Strobus* because the rust is limited to the colder and wetter areas stopping its appearance as the distribution moves south. The fungus sexual phase host- *Ribes nigrum* (Sect. Coreosma) are apparently not a monophyletic group, because it is mixed up by many susceptible cultivars (Ref. 19, 20, 23).

4.2 *North American Pacific Mediterranean White Pine Bristle Rust Flora*

White pine name & characteristics: The sugar pine (*Pinus labertiana)* ranges from western Oregon through the North Coast Range of California, the Sierra Nevada, and the mountains of southern California, with outlines in the Santa Lucia Mountain of west central California and the Sierra San Pedro Martir of northern Baja California.

Latitude/Longitude: North latitude 29°-45°/ West longitude 115°-124°.

Alternate host species name: *Ribes roezli* is prevalent in the sugar pine areas especially in the middle of the Sierra Nevada Mountains. *Ribes roezli* is commonly known as the Sierra gooseberry is the most abundant and commonly associated with the California sugar pines. It is also highly susceptible to the blister rust and cannot be easily suppressed because it can regenerate from seed stored for long periods of time. *Ribes nevadense* (the Sierra currant) is also widely distributed and hard to control because it has layering habits and tolerance to suppression by other vegetation. *Ribes cereum* is the squaw currant and is not easily infected by blister rust. Under specific conditions it is able to produce significant amounts of pine-infecting sporidia.

Life Cycle Character:

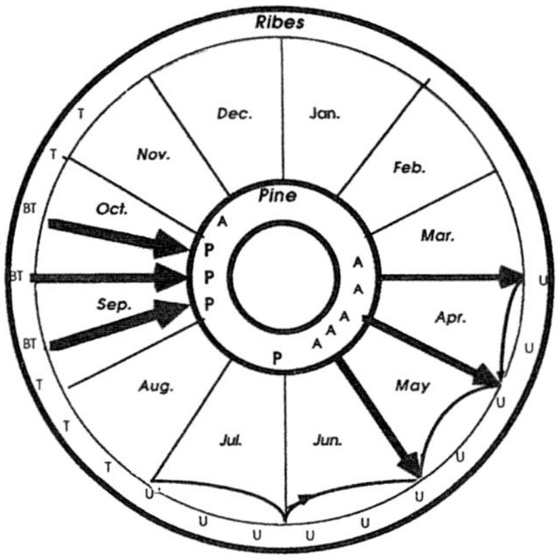

Fig. 8. Life cycle of Cronartium ribicola in the central Sirra Nevevada: Periods of production of the five spore states on Ribes sp. (outer circle) and Sugar pine (inner circle).
A – Aeciospore, B- Basidiospore, P – Pycniospore, T – Teliospore, U - Urediospore

This is a typical Pacific white pine blister flora. The WPBR does not grow through southern California because the hotter sun and dry weather limit their distribution.

Sugar pine rust life-cycle length is longer in Mediterranean weather: aeciospores last longer by 4 months. Basidiospores are massively produced because the weather is cold, cool and wet: between 10°C-20°C, with 90% precipitation. (Fig. 8)

The North American Pacific Mediterranean WPBR is characterized by summer drought and cool wet winters. Annual precipitation ranges between about 1000 mm at the central Sierra Nevada Mountain. Summer temperatures are warm, with July mean values of 20°C-30°C, and daily temperatures above 35°C are not uncommon. Winters average 10°C-12°C. Frosts are infrequent in all coast areas but the higher elevations can be high to relatively humidity favorable for teliospore germinations. The effect of dew in maintaining rust should be distinguished from the influence of soil moisture. Locations with higher than average soil moisture, such as long streams and around springs, seeps, and the edges of meadows have long been recognized as areas of high rust hazard. Moist soil favors Ribes regeneration and infection patterns from a single source of sporidia in the sugar pine region are almost always irregular (Ref. 6, 7, 24).

Systematic biological specializations and taxonomic treatment:

Formae speciales: *Cronartium. ribicola. f.* SP. Species*: C. ribicola*

Taxonomic treatment: *Formae speciales: C. ribicola. f. SP. roezli*

Comparison with other WPBR flora: Pacific temperature and moisture conditions have found this WPBR flora very favorable in the Pacific Mediterranean. The southern limit of WBPR is thought to be the result of climatic factors. Field observations and climactic data indicate that the rust thrives only in the central Sierra Nevada where the climate is relatively cool and moist. The rust is apparently absent from the mountains of southern California and adjacent northern Baja California (where both sugar pine and *Ribes* occur as potential hosts), probably because the climate in this region is relatively hot and dry (Ref. 23, 24).

4.3 *East Asia/ Sino-Japan white pine blister rust flora*

White pine name & characteristics: The East Asia Sino-Japan WPBR flora is located in China's highest latitude. *Pinus koraiensis,* a Korean pine ranges from Korea to eastern Manchuria with outliers on the Japanese islands of Honshu and Shikoku (Ref. 16, 17).

Latitude/Longitude: The Korean pine distribution of geographical area can be found at North latitude 40°45'-49°20' East longitude124° 45'-134° in the HeilongJiang, JiLing and Liaoning provinces.

Alternate host species name: Previously many green house inoculation test show these alternate host species *Pedicularis resupinata, P. resupinata var. ramose , P. spicata,* and *Ribes mandshuricum* as major telia hosts.

Life Cycle and spore characteristics: An EST test shows 16 white pine blister rust aeciospores strains collections and the results were not differentiated. Sino-Japan white pine blister rust variations were low. Their life cycle is longer than most other flora (Ref. 15). The rust morphological characters: (Ref. 14).

Pycnia	oblate-flat, in cortex, pycniospore are hyaline, pear form and 1.8-2.4x2.4-4.2μm in measure
Aecia	globoid swellings of branches and trunks, forming orange-yellow blisters bursting though the crevices of the bark, Aeciospore globoid-obovoid, have a well-developed smooth area and annulate projections with flat apices. The aeciospore measures14.4-28.8x22.8-33.6μm and optimal temperature for germination is 19-21°C
Uredia	hypophyllous, scattered or subgregarious, yellow, punciform. Urediospores are obovid, echinulate and measures 13.1-20.6x15.6-30 μm.
Telia	Only at exactly 16°C do teliospores form. Teliopsores measure at 3.6-13.5x36-59.1μm.
Basidia	At 20°C basidia and basidiospores form. Also, at 10-18 °C, they are able to infect five-needled pines. Basidiospores measure 10-12μm.

Therefore only in cold, cool, and wet conditions can the alternate host's teliospores germinate. The teliospores must be in humid conditions for 24-48 hours before they can germinate and produce basidia.

Fig. 9. WPBR Life Cycle (*Pinus koraiensis*)

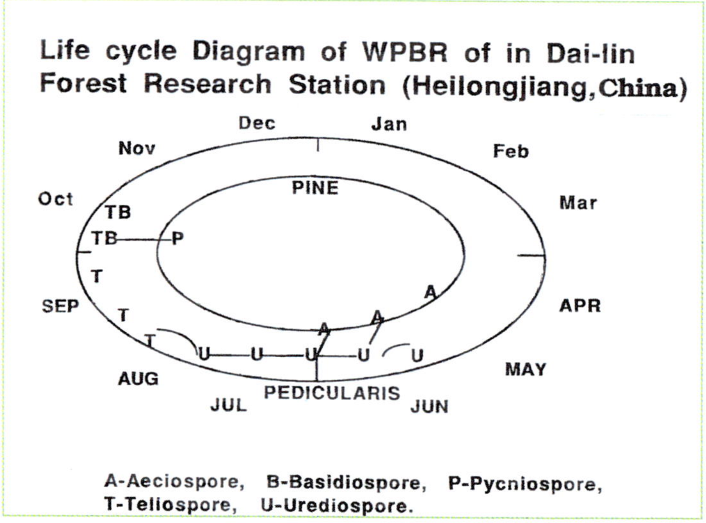

Comparison with other WPBR flora: This is the only rust flora in which the alternate hosts include *Pedicularis.*

4.5 *East Asian/ Alpine white pine blister rust flora*

White pine name & characteristics: This area is mainly occupied by *Pinus armandi* in southwest of China. The studies on the various populations of an 8-year-old juvenile forest of *P. armandi* show significant differences among different provenances. There were notable linear correlations between characteristics and latitudes in their original places. The two varieties were classified by latitudes of their origin, Yunnan-Guizhou plateau (YGP) and Qinling-Dabashan mountain (QDM). The Yunnan- Guizhou plateau grows faster, the biomass yield is higher and the needles are denser and longer than the Qinling-Dabashan Mountain. The main reason is the extreme temperatures of their origins (Ref. 5, 12, 38, 39, 22).

Latitude/Longitude: This flora is located at North latitude 23° 30'- 36°30' and East longitude 88°50'-113°.

Alternate host species name: Mainly in west of Qinling-Dabashan mountain is the species of *Ribes glacialeand va. Laciniatum,* (Ref. 38) and the three species in Central Qinling mountain are *R. emodense 2000 meters, R.glaciale 2600 meters and R. tenue 2000 meters* (Ref. 37).

Life Cycle Character:

Fig. 10. Life Cycle of Armandi Pine /Ribes Rust

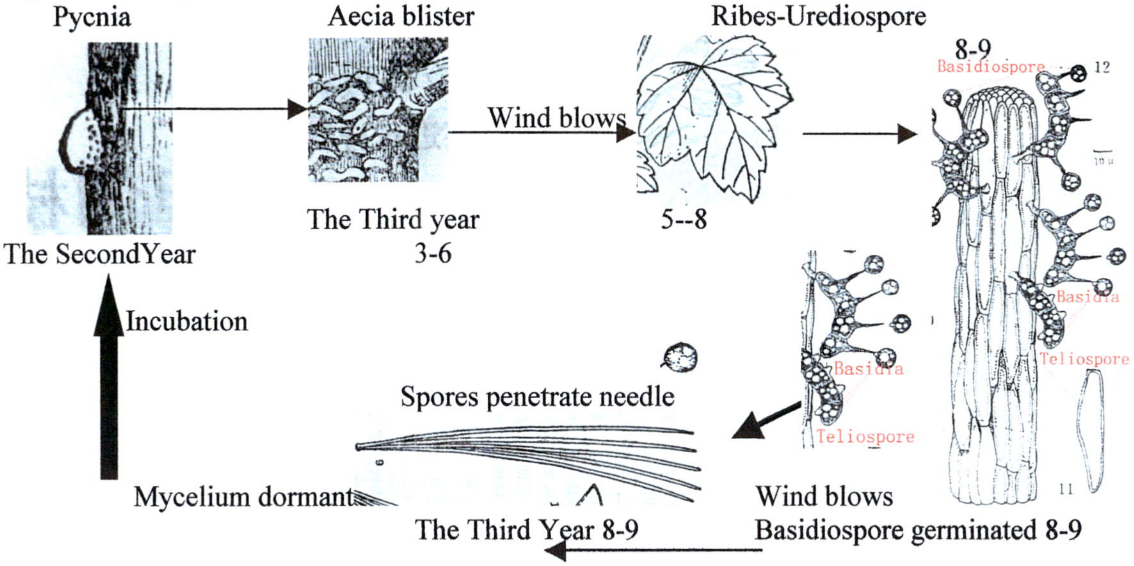

Comparison with other WPBR flora: This flora belongs to WPBR subalpine-alpine region and from this point reaches the Sino-Himalayas plateau.

This is a continental phenotype consisting of two types, Sichuan-Yunnan area and Chi ling mountain (Ref. 5, 38).

4.6 *Sino Himalayan plateau WPBR Flora*

White pine name & characteristics: During the subsequent Quaternary period, the alternating glacial and interglacial period that occurred in the Northern hemisphere complicated the migration, mixture, and differentiation of forest plants of the elevated plateau, resulting in the rich and variegated tree species there. The major white pine is *Pinus griffithii* located at Jilong and yadong, Xizhang between 1600-3500 meters elevation, general forest composition consists of *Pinus roxburghii, picea smithiana, Quercus tungmaigensis, cornus capitata, Rhododendron arboreum* etc. species are at 1700-4400 meters elevation. In addition to this white pine there can be found other white pine blister rust flora such as *Pinus armandi.*

Latitude/Longitude: North latitude 25°-35°/East longitude 70°-100° (Ref. 16, 17, 37).

Alternate host species name: There are two plateau Ribes species, *Ribes himalense* from Royle range 2600-3800 meters and *Ribes orientale* ranges from 2150-4900 meters, *R. takare* 2650-4000 meters (Ref. 37).

Life Cycle Character: The different climatic conditions of the Sino-Himalayas make a very rich tree rust flora. The first found are uredia and telia rusts in *Ribes* leaf at 3200meters, Yadong and lozha, during the 1976 expedition research at *Pinus griffithii* forests region (Ref. 11, 37).

E. Formae speciales : *Cronartium. ribicola*. f. SP.

Species: C. ribicola   *Formae speciales : C. ribicola. f. SP. Himalayas*

Comparison with other WPBR flora: The High plateau rust flora extended and linked from the alpine rust flora, but elevation and high radiation may cause the rust ecological and genetic life cycle to more completion. Therefore, Both these trends have a fundamental bearing on comparative phylogenetic study of rust white pine blister rust fungus and their hosts. For practical purposes, however, information on the host-pathogen relationship is needed for breeding rust resistant cultivars and hybrids.

5 REFERENCES

1. Berger, A. 1924. A Taxonomic Review of Currants and Gooseberries.
2. Bingham, R. T. 1972. Taxonomy, crossability, and relative blister rust resistance of 5 needled white pines. Biology of Rust Resistance in Forest Trees. *USDA For. Serv. Misc.Publ.* 1221. pp. 271-278.
3. Bureau of Entomology and Plant Quarintine. 1944. What is this Pine Blister Rust? (Repoter)
4. Chagala, E. M. 1996. Inheritance and linkage of allozymes in Pinus strobus L. *Silvae Genetica.* 45(4): 181-187.
5. Cao, Z.M. & Li, Z.Q. 1999. Rust Fungi of Qinling Mountains. China Forestry Publishing House.
6. Chen, M-M & F.W. Cobb, Jr. 1985. White pine blister rust in young sugar pine plantations in the mid-elevation Sierra Nevada. *Phytopathology* 75(11): 1367.
7. Chen, M-M & F.W. Cobb, Jr. 1988. Epidemiology of white pine blister rust in young sugar pine plantations in the central Sierra Nevada of California. Beijing International Symposium of Plant Pathology. 3.
8. Chen, M-M. 1974. Identification of shade tree and planting tree pathogens in central Inner Mongolia. Inner Mongolia Agriculture College, Huhehaote. 8.
9. Chen, M-M. 1982. Review of epidemiological history of blister rust (Cronartium ribicola) in North America and East Asia. *J. Chinese Forest Tree Quar.* 107.
10. Chen, M-M. 1985. Analysis of the distribution of forest pathogens in the United States and China: two case studies. *Phytopathology* 75(11):1322.
11. Chen, M-M. 1989. A study of the rust flora of Sino-Himalayan forests. *Canadian Journal of Botany.* 67(3): 827-833.
12. Chen, M-M and a team of forest pathologists. 1955. Report of the Western Sichuan Forest (450,000 ha) diseases (Cronartium ribicola) on Pinus armandii). The Survey and Research of Forestry team Reporters. 52.
13. Chen, M-M., et al. 1956. Investigations of Dailing Experimental Forestry Station, Heilongjiang, on forest insects and diseases (e.g., Cronartium ribicola). The Survey and Research of Forestry team Reporters. 44.
14. Chen, M-M., et al.1956. Investigation of red pine (Pinus koraiensis) diseases (e.g., Cronartium ribicola) in Langxiang County, Heilongjiang. The Survey and Research of Forestry team Reporters. 21.
15. Cheng, D.S. et.al. 1997. The studied of differential on EST of the pines stem rust's pathogens within species and in species. 1997. Scientia Silval Sincai 330-336.
16. Cheng, W.C. 1987. The Silvicultural Technology for mainly tree species of China. Chinese Forestry Press. pp. 35-42.
17. Critchfield, W.B. & Little Jr., E.L. 1966. Geographic Distribution of the Pines of the World.
18. Farr, D.F., Bills, G.F., Chamuris, G.P., & Rossman, A.Y. 1989. Fungi on Plants and Products In United States.
19. Hamelin, R.C., Beaulieu, J. & Plourde, A. 1996. Genetic diversity in populations of Cronartium ribicola in plantations and natural stands of Pinus strobus. (Annual Meeting of the American Phytopathological Society, North Central Division Indianapolis, Indiana, USA July 27-31, 1996). *Phytopathology.* 86 (1 SUPPL.): S88.
20. Hamelin, R.C. 1998. Molecular epidemiology of white pine blister rust. Proc. First IUFRO Rusts Forest Trees WP Conference, 2-7 Aug. 1998, Saariselkä 255-259.
21. Hiratsuka, Y. 1988. Cytology, taxonomy and nomenclature of autoecious pine stem rusts. First IUFRO Rusts Forest Trees WP Conference, 2-7 Aug. 1998, Saariselkä 31-35.

22. Hu, X.J. 1990. Studies on the character variations of the juvenile forest of Pinus Armandi in provenance tests. Scientia Silvae Sinicae Vol. 26, No. 4.
23. Hummer, E.K. & Sniezko, R. 2000. History of the Origin and Dispersal of White Pine Blister Rust. *HortTechnology* 10(3): 514-570.
24. Kimmey, J.W. & Wagener, W.W. 1961. Spread of White Pine Blister Rust From Ribes to Sugar Pine in California and Oregon. Techical Bulletin No. 1251.
25. Li, P.S. & Yong, Z.Z. 1994. Conference on National Parks and Protected Areas (CNPPA/IUCN) Greenness Filling in East Asia. Chinese Environmental Sciences Press.
26. Li, W.H., Han, Y.F. & Chen, M.M. 1985. The Forest of Tibet. Science Press. 375.
27. Liston, A., Robinson, W.A., Piñero, D., & Alvarez-Buylla, E.R. 1999. Phylogenetics of Pinus (Pinaceae) Based on Nuclear Ribosomal DNA Internal Transcribed Spacer Region Sequences. *Mol. Phylogenet. Evol.* 11:95-109.
28. Liu, S.J. & Chen, M.M., et al. 1982. Studies on Cronartium. J. Chinese Ministry of Forestry, Beijing (translated from Roger Peterson's English paper to Chinese), published by the Tottori Mycological Institute. *The Japan Tokyo Research Center.* Totter, Japan August 1973. pp. 203-223.
29. Maloy, O.C. 1997. White Pine Blister Rust Control in North America: A Case History. *Annu. Rev. Phytopathol.* 35: 87-109.
30. Mesler, M.R. & Sawyer, J.O. 1993. Grossulariaceae. In: Hickman J. C. (ed) The Jepson Manual: Higher Plants of California. *University of California Press, Berkeley.* pp. 676-680.
31. Messinger, W., Hummer, K. & Liston, A. 1999. Ribes (Grossulariaceae) phylogeny as indicated by restriction-site polymorphisms of PCR-amplified chloroplast DNA.
32. Raven, P.H. & Axelrod, D.I. 1978. Origin and relationships of the California flora. *Univ.Publ. Bot.* 72: 1-134.
33. Sinclair, W.A., Lyon, H.H. & Johnson, W.T. 1987. Stem and Cone Rusts of Pine Disease of Trees and Shrubs. pp. 270-273.
34. Teng, S.C. 1996. Fungi of China. Mycotaxon, LTD. 226-233
35. Vogler, D.R. & Bruns, T.D. 1998. Phylogenetic relationships among the pine stem rust fungi (Cronartium and Peridermium spp.). *Mycologia* 90(2): 244-257.
36. Wang, Y.Z. & Chen, M.M., et al. 1981. Rust fungi of western China. Acta Mycologia Sinica 20(1): 16-28.
37. Wu, C.Y. 1985. Flora XIZANGICA Vol. 2. Science Press. pp. 534-537.
38. Yang, Y.P. & Li, C.B. et 86 editors. 1990. Sichuan Forests Chinese Forestry Press. pp. 128-132.
39. Yang Z.Z. 1994. The studies of armandi pine blister rust on the techniques of pest control and quarantine. The Reporter of General Station of Sichuan province Forests Diseases and Insects Control and Quarantine.

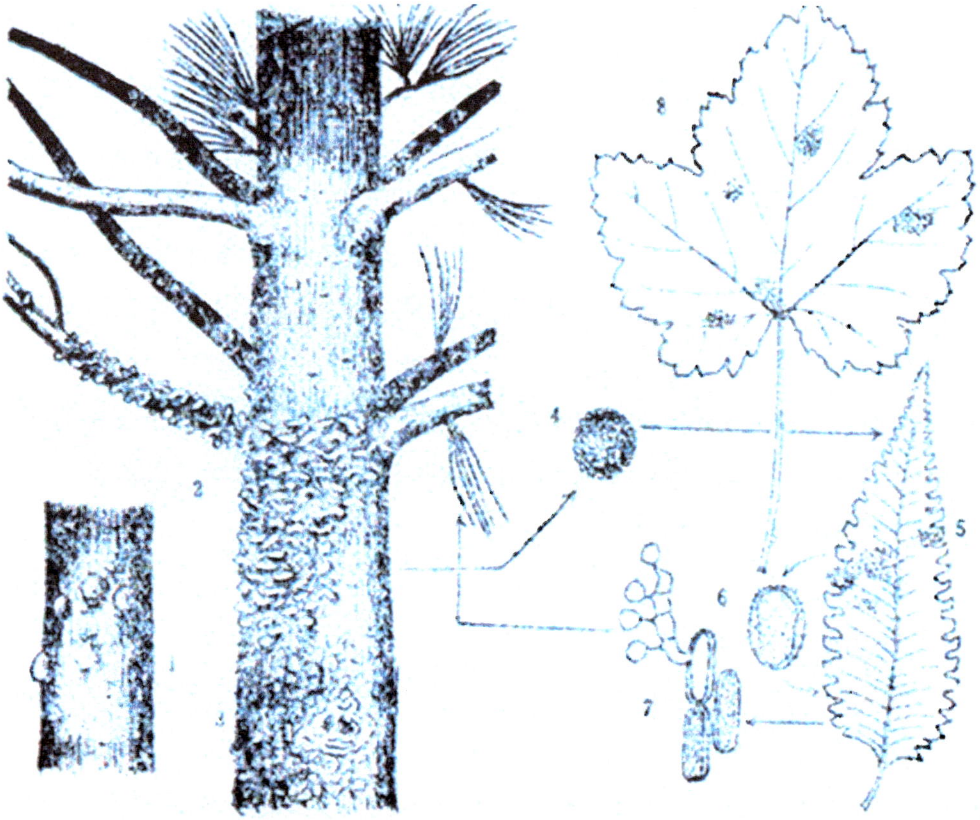

Fig 1. Red Pine Blister Rust. 1. Pycnia, 2. Blister, 3. Old bark, 4. Aecia, 5. Pedicularis and Telia, 6. Uredia, 7. Telia, basidia, and basidiospores, 8. Ribes-Telia.

Fig 2a. Pycnia as fluid

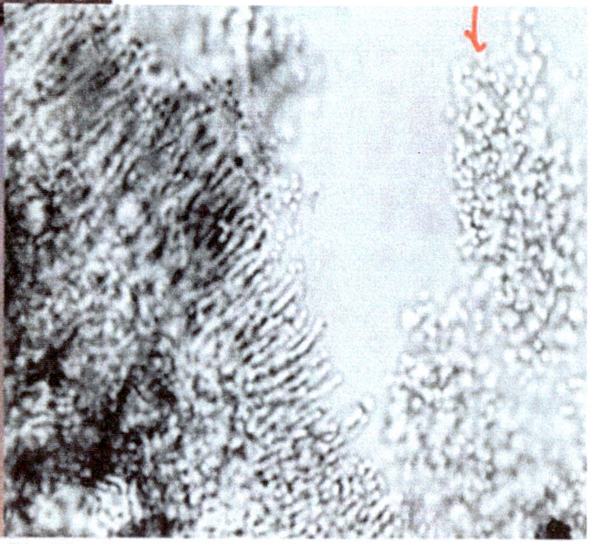

Fig 2b. Pycnia and pycniospores

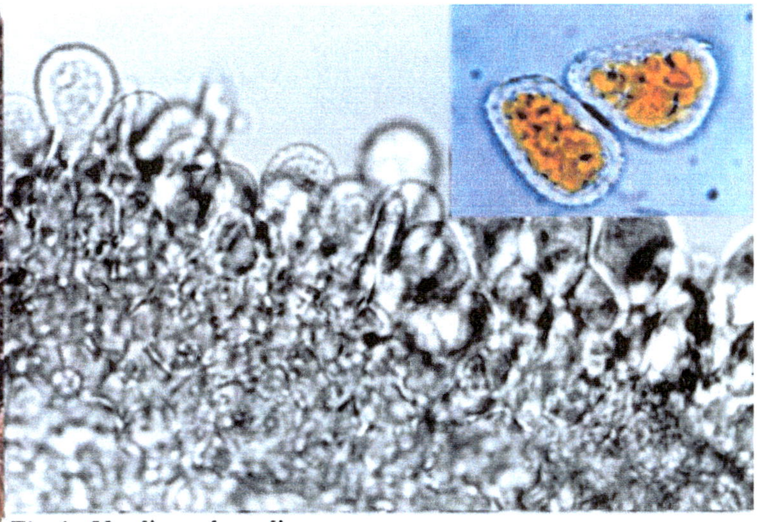

Fig 4. Uredia and urediospores.

Fig 5. Ribes with telia.

Fig 3. Blister Aecia break in bark

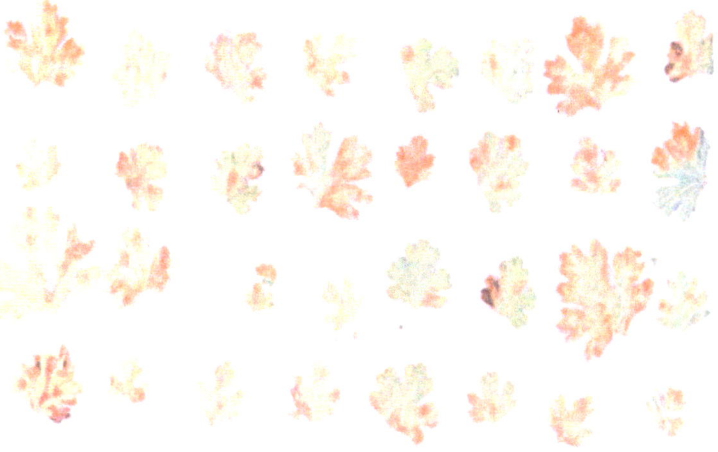

Fig 6. Telia naturally infected Ribes leaves

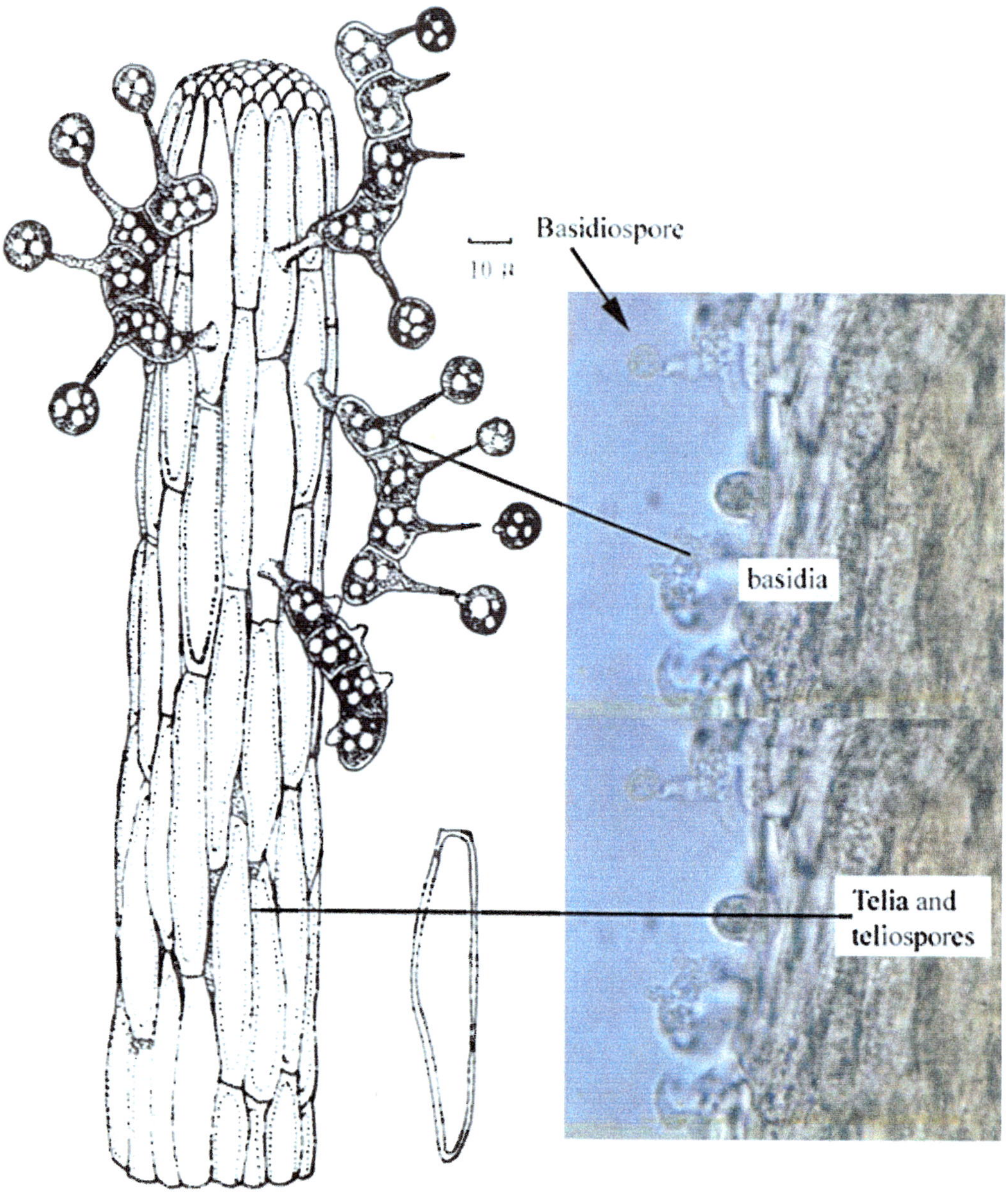

Fig. 7 Teliaspores and Basidiospore
(by L.P. Shao)

1979年1期 林 业 科 学 SCIENTIA SILVAE SINCIE 1979 Vol. 1

西藏森林植物锈菌区系初报

谌谟美

（ 中国林业科学研究院林业科学研究所 ）

锈菌广泛分布在各类森林植物群落之中，不少锈菌能引起森林病害，有的甚至是难以防治的毁灭性病害，如五针松疱锈病（ *Cronartium ribicola* Fischer ）等。每种锈菌的产生，发展成灾害都以一定的生态环境为条件，从大范围来看，它的发生、发展都有一定的地理区域的规律性。为了预防森林锈病的发生，需要研究锈菌的区域类型，找出锈菌、寄主植物和生态环境之间的联系，即锈菌区系。植物地理学是研究锈菌区系的基础，因为所有的锈菌目（ *Uredinales* ）都是专性寄生菌，在石炭纪蕨类植物上已寄生有锈菌，地质历史上锈菌就与有花植物有着密切的关系，近代的锈菌是长期适应寄主不断进化的结果。

锈菌不同于其他真菌，它的生活史中产生一定顺序、多种形式的孢子，并选择固定植物"科"为寄主，多数分别寄生在亲缘关系较远的两科植物上。

青藏高原的原始森林，受人为干扰较少，具有研究锈菌自然历史和生物区系类型的条件。分析这个地区锈菌区系特征，不仅在理论上探讨高原隆起的自然历史对微生物界的影响，而且在实践上找出各区系中森林植物和锈菌的相互关系，籍以设计防止锈病发生的措施。

1975—1976年作者考察了西藏森林植物病害和真菌。考察地区东起江达，西至日喀则西侧的吉隆，北至昌都北部，南达中喜马拉雅山和东喜马拉雅山主脉以南的聂拉木、吉隆、亚东、洛扎等峡谷区。调查区域位于青藏高原南部和东南部，约于北纬27°—31°，东经85°—98°和海拔5,000米以下的地区。地质上属于喜马拉雅造山体系，是地质时代上最新隆起的高原，地壳新构造运动十分活跃，现在山体仍在不断上升中。地势南低北高，河谷深切，气候变化大。东喜马拉雅南北两翼和中喜马拉雅南翼分布着大面积各种类型的原始森林，往往在一个数十公里的深山峡谷里，随海拔由高到低的变化，分布了从永久积雪带—高山流石滩植被带—高山草甸—高山灌丛—高山疏林—亚高山暗针叶林—山地针阔混交林—亚热带常绿阔叶林以至山地热带雨林的完整而丰富的垂直带谱。成为世界上最丰富和古老的温带，亚热带至热带北缘的植物区系的一部分，它不仅为高等植物研究而且为微生物区系研究提供了世界罕见的天然实验室。

锈菌调查是和森林植被调查紧密结合进行的[4]，在山地亚热带、温带、高山灌丛等 10个植被类型和20余种主要建群种林地（ 和宜林地 ）设调查线、调查样点、标准地，采集标本，并对其寄主、转主植物，生态环境，分布区域特征作了记载。

本文承中国科学院微生物研究所王云章教授及锈菌室同志热情指导。安徽农学院陈礼琢同志，内蒙农牧学院袁秀英同志参加部分工作，在此一并表示感谢。

过去对西藏森林真菌了解的较少，1868 年法国人达维德（ David ）和里米（ Remy ）在昌都零星采集真菌标本[2]，1915 年奥地利人莫扎特（ Mazzetti ）在澜沧江上游山区内采了一些真菌[2]，1955 年波尼在喜马拉雅采集的锈菌标本统计结果有 28 种[9] 。近年来，中国科学院青藏科学考察队在考察过程中注意了对各类真菌的采集。现在将锈菌采集结果和国内几个区域采集的种类作了比较，证明这里不仅高等植物资源丰富，而且也是我国锈菌种类比较多的地区之一（ 表 1 ）。本区锈菌经过初步整理鉴定约有 20 属，70 余种，寄生于 30 个科的植物上。其中属于无柄锈菌科 5 属 12 种。属于柄锈菌科 11 属 28 种，属于半知锈菌类 4 属 31 种，还有柄锈菌属（ *Puccinia* ）3 个新种、多孢锈菌属（ *Phragmidium* ）二个新种和 2 个新变种（ 新种、新变种将另文发表 ）。表 2 是几个新种存在的森林类型。

表 1　西藏和其他几个地区锈菌采集

地区 \ 锈菌数量	属	种
安徽黄山[1]	7	20
新　疆[1]	5	34
陕　西[1]	12	55
尼泊尔—喜马拉雅	13	34
西　藏	20	70

表 2　几个新种所在的森林类型

森　林　类　型	锈　　菌	寄　主　植　物	海拔（ 米 ）
林芝云杉林 *Picea likiangensis*（ *Fr.* ）*Pritz var. linzhiensis Cheng et L.K.Fu*	*Phragmidium*	*Rubus alexerius Focke.*	2,900
麦吊云杉林 *Picea brachytyla*（ *Fr.* ）*Pritz. var. complanata*（ *Mast.* ）*Pritz*	*Phragmidium*	*Rubus Sp.*	3,100
高山松林 *Pinus densata Mast.*	*Puccinia*	*Clematis nontana Buch.*	3,200
急尖长苞冷杉林 *Abies georgei Oti. Var. Smithii*	*Puccinia*	*Polygonum Campanulatum Hook.*	3,300
喜马拉雅冷杉林 *Abies Spectabilis*（ *D.Don.* ）*Spach*	*Puccinia*	*Morina alba Hand. Mazz.*	4,100

吴征谥等以前人的工作为基础，结合最近在青藏科学考察中收集到的资料，详细分析对比了西藏植物区系地理成份和各地优势植被的区系组成，将西藏森林植物区系划为中国—喜马拉雅亚区[3]。该亚区属泛北植物区。中国—喜马拉雅亚区下分横断山脉地区和东喜马拉雅地区，它和中国一日本亚区共同成为东亚植物区系核心部分[3]。这一亚区中不仅有许多特有的古老科、属，而且由于喜马拉雅的隆起运动速度快，延续时间持久，所以植物种类不断变化，并分化产生出新的类群。发生学观点认为，这个亚区既可能是温带区系的起源地，也是一些现代科属发生的中心。

一、中国—喜马拉雅锈菌区系类型

整个中国一喜马拉雅亚区是处于高海拔、低纬度，这里气候受高耸的喜马拉雅屏障和来自孟加拉湾、缅甸暖湿气流的显著影响，各地气候千差万别。在本区域范围内森林植物类型发生和分布，取决于一系列自然综合因素（海拔、地形、地势、土壤、气候等），不同的水热条件的差异直接影响锈菌孢子萌发，并间接地（如物候期）影响锈菌分布。经过比较可以看出（表 3），在一定范围内水热条件相同时则森林植被类型和锈菌区系类型基本趋于一致。根据不同的水热条件可以将锈菌区系划分为如下若干类型，即高原型、温带型和亚热带型。

表 3　中国—喜马拉雅锈菌区系类型的水热指标

锈菌区系类型	森林植物类型	海拔（米）	年平均温度（□）	温暖指数*	最冷月平均气温（□）	最热月平均气温（□）	绝对最低气温（□）	无霜期（日）	相对湿度（%）	降水量（毫米）
亚热带型	山地亚热带林	1,100—2,500	11—18	77—160	3—12	18—25	0—-12	200—290	60—80	＞800
温 带 型	山地温带林	1,800—3,600	4—15	25—110	-8—0	12—21	-22—-6	110—240	50—80	500—900
高 原 型	亚高山寒温带材 亚高山灌丛草原	3,000—4,000	3—8	26—63	-8—-1	10—15	-17—-25	80—160	＜60	300—500

*温暖指数系指温度大于 50℃的月平均温度，减去 5℃后的积算值。

高原型的锈菌主要分布在喜马拉雅北翼，高原面南侧边缘，以拉萨、日喀则为代表地区，年平均温度 8℃左右，年降水量 450 毫米左右，气候温凉、干燥，太阳辐射强，锈菌种类较少，栅锈菌属（*Melampsora*）是高寒、干旱地区常见属，它可分布到 4,000 米左右的阔叶乔灌木上。有的是高原型的代表种类，如角落叶松栅锈菌（*Melampsora lirici-capraearumKleb.*）其夏孢子世代危害拉萨、江孜和日喀则地区的主要造林树种—左旋柳、竹柳等，柳锈病发病株率高达 85—95%；该区系内青稞条锈病（*Puccinia glumarum*）发生严重，由于海拔高、气候温凉，形成高原小麦、青稞等锈病流行的重要原因。此类锈菌也分布在亚高山寒温带的云杉、冷杉林内，引起许多建群种（如丽江云杉、林芝云杉、川西云杉、麦吊云杉、紫果云杉）的球果和针叶发生锈病。

温带型锈菌区系一般分布在东喜马拉雅北翼，念青唐古拉山脉以南，工布江达、朗县以东以及喜马拉雅山南坡的峡谷地区，其中包括吉隆县的小吉隆、江村，聂拉木的樟木、立新乡，亚东县和阿桑桥，洛扎县的拉康、申格热、波密县的樟木、倾多、易贡、东久等地。以波密、林芝为代表的地区年平均气温为 8.4—11.4℃，年降水量为 650—960 毫米，海拔 1,800—3,600 米之间，气候以温暖湿润为主要特征。这里以山地亚热带—温带松林为主，主要建群种有乔松、云南松、高山松、华山松和小片的长叶松，植物种类丰富，是中国—喜马拉雅区系中锈菌种类最多的地区，其中有 66%的锈菌分布于本区。这里有些北温带常见的锈菌如鞘锈菌属（*Coleosporium*），柱锈属（*Cronartium*），它们的锈孢子世代寄生在松科植物上，造成松树的病害；另一些北温带常见种如茶　生柱锈菌（*Cronartium ribicola Fischer*），桦长栅锈菌（*Melampsoridium betulinum* Desm. Kleb.）、凤毛菊鞘锈菌（*Coleosporium saussureae* Thum.），这些锈菌主要寄生于世界各温带植物的属、种上。上述锈菌约有半数以上寄生在西藏高原所特有的大科植物上（如蔷薇科、毛茛科、杜鹃科、菊科、禾本科、杨柳科、松柏科等），它不仅反映了喜马拉雅亚区山的特点，而且，进一步证明中国—喜拉雅亚区基本上属于温带性质。

亚热带型的锈菌大都分布在喜马拉雅山南翼低海拔地区（1,100—2,500 米以下），如下察偶、巴安通年平均温度为 15℃左右，年降雨量在 1,000 毫米以上，这里有些是喜高温、高湿的亚热带锈菌成分，常见有寄生于木兰属（*Indigofera*）上的伞锈菌属（*Ravenelia*），蛇葡萄属（*Ampelopsis*）上的层锈菌属（*Phakopsora*），但总起来看，这里已是热带、亚热带锈菌成份的北部边缘了。

二、中国—喜马拉雅锈菌区系特点的探讨

由于本地区生态地理条件的复杂多样和青藏高原隆起的自然历史的影响，本区系内的植物和锈菌区系成份的迁移、分化、融合现象十分复杂，经过初步分析，对本区锈菌区系归纳以下一些特点。

（一）新的类群的发生地：青藏高原由于第三纪末以来一直到整个第四纪，处于地壳构造强烈上升活动之中，对植物类群的形成产生了较大的影响，十九世纪法国人迪拉威 J M —Delabay ）在云南高原地区所采的一类锈菌，经帕杜依拉（ N. Patouillard ）1889 年鉴定为新属-鞘柄锈菌属（ *Coleopuccinia* ）我们认为它是中国-喜马拉雅所产生的新类群代表之一。其模式种-中国鞘柄锈菌（ *Coleopuccinia sinesis* Pat ）图 a 已在波密（ 海拔 2,700 米 ）察偶古琴（ 海拔 3,700 米 ）采到，已知鞘柄锈菌属最北分布到甘肃南部，但大多在云南西北和西藏的东南部，它很可能是在喜马拉雅抬升过程中在这一地区发生的，然后向四周分化。此外，在本区还发现了一种古老的锈菌，这种活化石的锈菌—明痂锈菌（ *Hyalopsora* ）的夏孢子世代都是寄生在蕨类植物上的，它的出现是否可以认为：虽然喜马拉雅经过第四纪全球性的气候变冷，而许多微生物并未绝迹。这里由于地形、气候的复杂而成为古老类型的天然避难所。发生于第三纪的较原始的类型如鞘锈科（ *Coleosporiaceae* ）、金锈科（ *Chrysomyxaceae* ）和无柄锈菌科（ *Melampsoraceae* ）直到现在仍然占有一定的比例。中国-喜马拉雅锈菌区系的特点是不仅有古老成份，而且还有不少新种和新变种（ 大约 10%左右 ）由于本区独特的自然地理和自然历史过程，锈菌在高原长期特定的生态环境适应中产生了种间分化现象，一些锈菌产生了和原种不同的新的变异类型，如兰果七筋姑（ *Clintoniaudensis* Trautv. et Mey. ）上的柄锈菌冬孢子比原种长 10 — 12 微米，顶孢乳突比原种大得多（ 图版 I—1 ）另一种寄生在亚洲积雪草。（ *Centella asiatica* (L.) Urban ）上的柄锈菌它的冬孢子为椭圆形，相近似的原种冬孢子为长椭圆形或棍棒形，以上这些形态上的差异无法归于原种之中。在统计本区锈菌的寄主时有些锈菌寄主“种”属于首次记载，约占全部锈菌的 46%，有些植物“属”如五味子属（ *Schisandra* ）黄花木属（ *Piptanthus* ）冷水花属（ *Pilea* ）积雪草属（ *Centella* ）　苓草属（ *Morina* ）上的锈菌尚属于第一次记载。

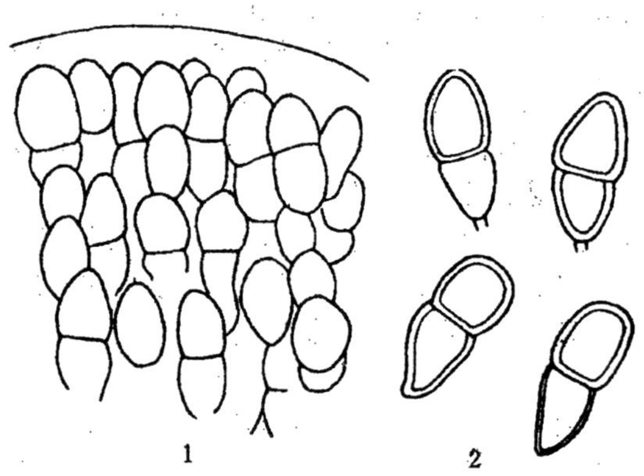

图 a　旬子木上 ***Coleopuccina sinensis*** 的冬孢子和冬孢子堆

1　冬孢子 500×　2　冬孢子堆 400

（二）区系多歧性：中国-喜马拉雅锈菌虽为一独特区系，但它和世界的种类仍有不少联系，其中属于世界性分布的种类约占 24%，如高梁锈菌属（ *Puccinia sorghi* Schw. ）龙胆柄锈菌（ *Puccinia gentianae* Mart ）栎柱锈菌（ *Cronartium quercuum* Berk ）等；属于亚洲温带种类较多如紫菀鞘锈菌（ *Coleosporium asterum*（ Diet.)Syd. ）亚洲花孢锈菌（ *Nysopsora asiatica* Lutj. ）当归柄锈菌(*Puccinia angelicae* Fuck.)；还有东亚和北美之间间断分布的种如香根芹锈孢锈菌(*Aecidium osmorrhizae* Peck)、黄水枝柄锈菌(*Puccinia tiarella* D Don) 等；资料分析说明本区锈菌中喜马拉雅山与日本相同的属、种最多，在全部种类中约有 50%

和日本的锈菌相同见表4，有些在世界上认为仅产于日本的种类如日本明痂锈菌(*Hyalopsora Japonica* Dicle)(图版I—2)。这次在西藏洛扎拉康已发现。还有寄生于菝葜属(*Smilax*)上的伊藤囊孢锈菌(*Blastospora itoana* Togashi et Onuma)(图版I—3)，第一次在我国喜马拉雅山发现，其形态特征与日本种相近。喜马拉雅的锈菌一直分布到日本，说明中国—日本植物区系和中国—喜马拉雅区系可能是有共同的起源。更多的资料表明本区主要的森林植物锈菌是和四川西部、云南西北部的森林植物锈菌同属一个区系，如寄生于四川木里丽江云杉(*Picea likiangensis* (Franch.) Pritz)上的杉李盖痂锈菌(*Thekopsora areolata* (Fr.) Magn.),

表4　　中国–喜马拉雅锈菌属和其他区域比较

锈　菌　属　名	主产地	中国—喜马拉雅	中国内陆地区	尼泊尔—喜马拉雅	印度	日本	苏联	北美	奥地利	墨西哥
明痂锈菌属(*Hyalopsora*)	北温带	+	−	−	−	+	+	+	+	−
盖痂锈菌属(*Thekopsora*)	北温带	+	+	−	−	+	+	+	+	+
长栅锈菌属(*Melampsoridium*)	北温带	+	+	−	+	+	+	+	+	−
柱锈菌属(*Cronartium*)	广　布	+	+	+	+	+	+	+	+	−
金锈菌属(*Chrysomyxa*)	北温带	+	+	+	−	+	+	+	+	−
鞘锈菌属(*Coleosporium*)	广　布	+	+	+	+	+	+	+	+	+
栅锈菌属(*Melampsora*)	北温带	+	+	+	+	+	+	+	+	−
层锈菌属(*Phakopsora*)	热　带	+	+	+	+	+	+	+	−	−
(*Monosporidium*)	印度、中国	+	+	−	+	−	−	−	−	−
单孢锈菌属(*Uromyces*)	广　布	+	+	+	+	+	+	+	+	+
柄锈菌属(*Puccinia*)	广　布	+	+	+	+	+	+	+	+	+
胶锈菌属(*Gymnosporangium*)	北温带	+	+	+	+	+	+	+	+	+
花孢锈菌属(*Nyssopsora*)	广　布	+	+	−	−	+	+	+	+	−
多孢锈菌属(*Phragmidium*)	温　带	+	+	+	+	+	+	+	+	+
伞锈菌属(*Ravenelia*)	热　带	+	+	+	+	+	−	+	−	+
囊孢锈菌属(*Blastospora*)	中国、日本	+	−	−	−	+	−	−	−	−
鞘柄锈菌属(*Colepuccinia*)	中国、日本	+	+	−	−	+	−	−	−	−
锈孢锈菌属(*Aecidium*)	广　布	+	+	−	+	+	+	+	+	+
被包锈菌属(*Peridermium*)	广　布	+	+	−	+	+	+	+	+	+
裸孢锈菌属(*Caeoma*)	广　布	+	+	−	+	+	+	+	+	+
夏孢锈菌属(*Uredo*)	广　布	+	+	−	+	+	+	+	+	+

＋ 有分布，－ 无分布

它可以一直向西发展到邻近的不丹边界森林里，寄生于麦吊云杉(*Picea* brachytula (Fr.) *Pritz.var Complanata* (Mast.) Pritz.)上，再向西则由另一个锈菌种更替了它。本区西侧的锈菌不少是印度北部西拉姆 *Silam* 地区相近的种、属 ，如寄生在乔松、长叶松针叶上的鞘锈菌(*coleosporium*)的一些种和印度西拉姆为同一区系范围。此外、洛扎、吉隆、陈塘等地还有不少锈菌是印度、尼泊尔、锡金的成份。在亚东县4,100米的冷杉林下的白花 苓草(*Morina alba* Hand. Mszz)(喜马拉雅和地中海间断分布种)上发现了柄锈菌的一个新种(图版 I—4)这个痕迹可能是喜马拉雅和古地中海之间有着自然地理的联系。

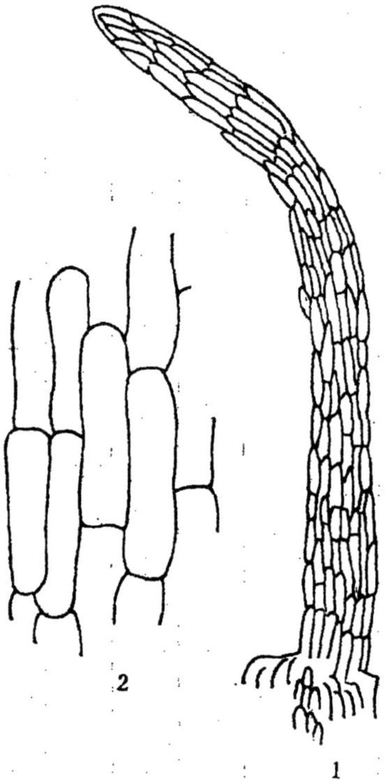

图b 茶 子上的 *Cronartium ribicola* 的冬孢子

1 冬孢子柱 200× 2.冬孢子 650×

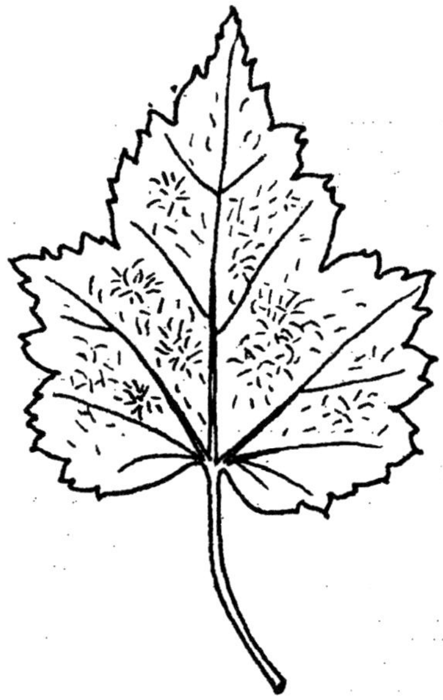

图c 茶 子叶上冬孢子柱(原大)

（三）**潜在锈病发源地**：本区系锈菌中有不少是森林的重要病菌，如寄生在长叶松、乔松上的鞘锈菌，寄生在云杉上的金锈菌，寄生于云杉球果上的盖痂锈菌，还有能引起杨、柳病害的栅锈菌。但是，目前在原始森林中锈菌在未形成灾害，锈病发病率多数仅 3—5%左右，尤其混交林的发病率更低，一般对森林生长危害并不大，如果一旦原始森林破坏，生态系统发生变化，或是人工营造了大片纯林，许多草木、灌木上寄生的转主锈菌就可能成为锈病流行的重要因素。值得注意是五针松类和茶 子(*Ribes*)、马先蒿(*Pedicularis*)之间的锈菌寄生关系，在这次考察温带型锈菌区系中发现了五针松疱锈病(*Cronartium ribicola*)的冬孢子世代图 b、c 乔松(*Pinus griffithii* Mc cle)华山松(*Pinus armandi* Fr)是五针松疱锈病(*Cronartium ribicola Fischer*)的锈子世代的转主寄主，而茶 子灌木广泛分布在本区系的山地亚热带一温带林的乔松、高山松、高山栎、华山松的林地、宜林地内。今后在相同海拔水平带如在西藏察偶、波密、林芝 2,700—2,900 米处营造华山松人工林，在吉隆、亚东、聂拉丁、拉康、错那、洛瑜、波密、易贡、墨脱的 1,100—3,300 米海拔处营造乔松林时，应注意清除茶 子灌木防治五针松疱锈病发生，严防扩展成灾。根据中国一喜马拉雅锈菌考察资料列出的本区系范围内森林植物锈菌寄主和转主植物关系表(见表 5)，对了解本区系内植物、锈菌的种群之间的结构关系和深入研究锈病生活史的参考。

表 5　中国—喜马拉雅森林植物锈菌寄主和转主植物

锈菌属名	转主或同主寄生	锈孢子世代	冬、夏孢子世代
明痂锈菌属(*Hyalopsora*)	转主寄生	*Abies* 冷杉属	*Phymatopsis* 密网蕨属
盖痂锈菌属(*Thekopsora*)	转主寄生	*Picea* 云杉属	*Prunus* 樱桃属
长栅锈菌属(*Melampsoridium*)	转主寄生	*Larix* 落叶松属	*Betula* 白桦属
柱锈菌属(*Cronartium*)	转主寄生	*Pinus* 松属	*Ribes* 茶　子属 *Quercus* 栎属
金锈菌属(*Chrysomyxa*)	转主寄生	*Picea* 云杉属	*Rhododendron* 杜鹃属
鞘锈菌属(*Coleosporium*)	转主寄生	*Pinus* 松属	*Heteropappus* 狗哇花属 *Saussurea* 款冬属 *Ligularia* 橐吾属
栅锈菌属(*Melampsora*)	转主寄生	*Larix* 落叶松属	*Populus* 杨属 *Salix* 柳属
层锈菌属(*Phakopsora*)	不明		*Ampelopsis* 蛇葡萄属
单孢锈菌属(*Uromyces*)	转主寄生 同主寄生		*Leguminosae* 豆科
柄锈菌属(*Puccinia*)	转主寄生 或同主寄生		*Gramineae* 禾本科
胶锈菌属(*Gymnosporangium*)	转主寄生	*Rosaceae* 蔷薇科	*Cupressaceae* 柏科
花孢锈菌属(*Nyssopsora*)	同主寄生		*Araliaceae* 五加科
多孢锈菌属(*Phragmidium*)	同主寄生		*Rosaceae* 蔷薇科
伞锈菌属(*Ravenelia*)	同主寄生		*Papilionaceae* 蝶形花科
囊孢锈菌属(*Blastospora*)	同主寄生		*Smilacaceae* 菝葜科

为了预防锈菌酿成生态灾害，需要注意克服人工造林的盲目性，应当根据植物锈菌区系，锈菌和寄主植物关系，从森林植物群落的结构组成，设计出能控制毁灭性森林病害发生的生态系统，这是今后开发利用本区森林资源过程中需要研究的重要课题。

参　考　文　献

[1] 王云章，1949：中国锈菌索引，1—155。

[2] 戴芳澜，1978：外人在华采集真菌史考。

[3] 武素功，1978：试论西藏植物区系分区。

[4] 李文华、韩裕丰，1977：西藏的森林，中国林业科学，第 4 期，4。

[5] G.C.Ainsworth, Fredcrick K.Sparrow and Alfred S.Sussman, 1973: The fungi, Academic press. INC (London) LTD.247-279

[6] Cummins，G. B.,1959: Illustrated genera of rust fungi, burgess, Minneapolis, Minnesota.

[7] Saccardo, P.A., 1888: Sylloge Fungroum VII, IX. Berolini Fratres Borntraeger.

[8] K.G.Mukerji, R.C.Juneja, 1962-1972: Fungi of India, Emkay publications Delhi.

[9] Balfour-Browne F.L., 1955: Some Himalayan fungi, bull. Brit. Mus (Nat. List) Bot., I, 189-218.

PRELIMINARY ANALYSIS OF THE RUST FLORA IN THE FOREST OF TIBET

Zhan Mo-mei
(*Institute of Forestry, Chinese Academy of Forest Science*)

In the period from 1975-1976, the author made pathological and fungal investigations in forest in a region located below an altitude of 5,000 meters in Tibet, where a complete and comprehensive spectrum of perpendicularly distributed forest was seen.

Here is provided a very rare natural laboratory in the world, in which there exist very good conditions for the study or the ecological aspect of the rust. After a preliminary study the rust flora investigated contains 20 genera and we nominate it the China-Himalayas rust flora.

This rust flora is classified into 3 types in according to conditions or water and temperature: 1) the plateau type. 2) The temperate zone type. 3) The sub-tropical zone type. Each type may be characterized by any species belonging to it. The analysis that has been made shows that the region investigated has the following features: It is a place of origin for a new rust type, the rust flora in this region has affinities to those in other regions, thus it is possible for this rust flora to constitute a potential source of forest disease.

The rust flora of Sino-Himalayan Forests

ABSTRACT: From 1975-1980, an investigation of rusts of the Sino-Himalayan flora was undertaken by an interdisciplinary research team of the Academia Sinica. The Sino-Himalayan plateau (Xizang) is a varied and unique physicogeographical region with combined effects of high altitude, vast expanse, and latitudinal position. A special category of montane (1000 - 6000 m) and its forest rust flora is proposed. This preliminary study of the hosts and their rusts includes 110 species and 26 genera in five families of rusts. According to biogeographical and ecological analysis, the Sino-Himalayan rust flora can be characterized as north temperate. This flora is classified into three zones: (*i*) the plateau zone; (*ii*) the temperate zone; and (*iii*) the subtropical zone. A distinct group of rust genera and species is found in each zone. The main genera and species recognized as Sino-Himalayan are also cosmopolitan in north temperate plateau regions. Sino-Himalaya rust flora mainly is a branch of East Asia flora and has close relationships with adjacent flora as well as with North American flora. This flora constitutes a potential source of rust diseases for forests in this and other regions.

1 INTRODUCTION

The Sino-Himalayan plateau (Xizang), 27-31° north latitude and 5-98° west longitude, is the largest, youngest, and highest topographic plateau on earth. The summit of the highest peak is 8842 m, and elevation commonly ranges from 1100 to 6000 m (Fig. 1). Located in the southwest of China, Xizang is sometimes called "the roof of the world." It is one of the few remaining places on earth that have not yet been fully explored (Hiratsuka 1984). In recent years, this area has been investigated by an interdisciplinary research expedition of the Academia Sinica (Fig. 2). The multidisciplinary scientific team consisted of 400 people, including geologists, geographers, biologists, foresters, mycologists, and other specialists, who concentrated on each representative altitude of the ecosystem. We focused on the effect of altitude on the mycota of the plateau. There is now a new understanding of the flora of the Sino-Himalaya. A special category of "high plateau zonation" has been proposed (Chang 1981). The vegetation and rust flora have resulted from the geologically recent massive uplift of the plateau.

Rust fungi can only thrive within host tissues. Their evolutionary tendencies must therefore be studied in connection with those of their host plants. The Xizang area provides a unique, relatively undisturbed, natural laboratory for the study of evolutionary and ecological aspects of rust fungi (Savile 1976; Chen 1979; Wang 1981).

According to preliminary investigations on the Sino-Himalayan plateau, there are over 1,498 species of trees and shrubs spread through more than 104 families and 363 genera. Among the forest plants are conifers of high economic value belonging to seven families, 15 genera, and over 46 species. In addition, there are 15 endemic species of major forest trees (Li et al. 1985; Li and Han 1981). Seventy-six families 271 genera, and 880 species of fungi have been recorded (Chen et al. 1980; Durrieu 1984; Balfour-Browne 1955).

2 METHODS

Rust fungi were gathered randomly within 1- to 3-ha zones along with a vertical transect and a contour line in forested area, and species and genera were identified by zone. The rust flora correlations were analyzed along with both information from the origin and evolution of Sino-Himalayan flora and basic environmental data (Wu and Lin 1981; Wu 1987; Li et al. 1985). The Himalayan rust flora was compared and contrasted with rust flora of other regions. We also determined the potential threat of rust diseases in this region and made recommendations for management of rust diseases in Sino-Himalayan reforestation areas (Li et al. 1985).

3 RESULTS

Eight hundred and sixty specimens of the rust flora from 18 forest population groups were collected. We described a representative vertical transect extending from 1000 to 5000 m in the Yarlung-Zangpo River region (Table 1), showing the forest habitats according to altitude and the rust and host genera present. After a study of the taxonomical literature of the rusts (Hiratsuka and Sato 1982; Cummins and Hiratsuka 1983; Savile 1976; Gallegos and Cummins 1981), a five-family classification was proposed. Preliminary study shows there are 26 genera and 110 species of rusts. Of these, six genera and 7 species are in Pucciniastraceae; five genera and 26 species are in Melampsoraceae; six genera and 48 species are in Pucciniaceae; one genus and five species are in Phragmidiaceae; and three genera and 3 species are in Raveneliaceae. The remaining 21 species are rust fungi for which the telial state is not known.

3.1 *Pucciniastraceae*

Hyalopsora japonica Diet.; *Pucciniastrum potentillae* Komarov; *Thekopsora areolata* (Fr.) Magn.; *Thekopsora sparsa* (Wint.) Magn.; *Melampsoridium betulinum* (Desm.) Kleb.

3.2 *Melampsoraceae*

Chrysomyxa expansa Diet.; *Stilbechrysomyxa himalensis* (Barcl.) Chen; *Stilbechrysomyxa stilbae* (Wang, Chen et Guo) Chen; *C. ledi* (Alb. et Schw.) de Bary var. *rhododendri* (DC.) Savile; *Cronartium flaccidum* (Alb. et Schw.) Wint.; *C. himalayense* Bagchee; *C. quercuum* Miyabe; *C. ribicola* J.C. Fisch.; *Coleosporium asterum* (Diet.) Syd.; *C. brevius* Barcl. (journ. Asiat. Soc. Bengal. 59:102. 1890); *C. clematidis* (Thüm.) Barcl.; *C. complanatum* Barcl. (Ured. Simla 101, t.v.f. 1.3:8. 1891); *C. geranii* Pat.; *C. heteropappi* (P. Henn.) Tranz.; *C. ligulariae* Thüm.; *C. lonicerae* Wang et Wei; *C. pedicularidis* Tai; *C. perillae* Syd.; *C. saussureae* Thüm.; *C. zangmui* Wang et Wei; *Melampsora coleosporioides* Diet.; *M. euphoribiaedulcis*, Otth; *M. larici-capraearum* Kleb,; *M. larici-tremulae* Kleb,; *M. ribesii-purpureae* Kleb,; *M. salicis-viminalis* Wang et Guo; *M. stellerae* Teich,; *Phakopsora ampelopsidis* Diet. et Syd.

3.3 *Pucciniaceae*

Blastospora itoana Togashi et Onuma; *Coleopuccinia sinensis* Pat.; *Gymnosporangium clavariaeforme* (Jacq.) DC.; G. jponicum Syd,; *G. nipponicum* Yamada; *Nyssopsora koelreuteriae* (Syd.) Tranz.; *Puccinia angelicae* Fuck.; *P. angelicae* P. Henn. var. *centellae* Chen; *P. asteris* Duby; *P. atrofusca* (Dudl. et Thomson) Holw.; *P. bolleyana* Sacc.; *P. caricis* Rebent.; *P. caricis-brunneae* Diet.; *P. chelonopsidis* Balfour-Br4owne; *P. circaceae* Pers.; *P. clintoniaeundensis* var. *tibetica* Chen; *P. cnici-oleracei* Pers. ex Desm.; *P. coronata* Cda.; *P. deyeuxiae*-scabrescentis Wang et Wei; P. *elymi* Westend.; *P. gentianae* (Str.) Roehl.; *P. thalaniae* Arth. et Holw.; *P. helianthi* Schw.; *P. heterocoloris* Chen; *P.*

heucherae (Schw.) Diet.; *P. himalensis* (Barcl.) Diet.; *P. kusanoi* Diet.; *P. lakanensis* Chen et Wei; *P. leveillei* Mont.; *P. menthae* Pers.; *P. nepalensis* Barcl. et Diet.; *P. paludosa* Plowright; *P. osmorrhizae* C & P.; *P. poaenemoralis* Otth; *P. recondita* Rob. ex Desm.; *P. silvatica* Schröter; *P. smilacis* Schw.; *P. sorghi* Schw.; *P. striiformis* West.; *P. taylorii* Balfour-Browne; *P. vaginatae* Juel *P. violae-reniformis* Wang et Wei; *P. yokogurae* P. Henn; *Uromyces appendiculatus* (Pers.) Unger; *U. hedysari-obscuri* (DC.) Lév.; *U. lapponicus* Lagerh.; *U. polygoniavicularis* (Pers.) Karst.; *U. viciae-fabae* (Pers.) Schroet; *Puccinia akebiae* Wang et Wei.

3.4 *Phragmidiaceae*

Phragmidium rosae-rugosae Kasai; *Phr. Rubi* (Pers.) Winter; *Phr. Shengezhense* Chen et Chen; *Phr. Zamonense* Chen et Chen; *Phr. Potentillae* (Pers.) Karst.

3.5 *Raveneliaceae*

Ravenelia indigoferae Tranz.; *Tranzschelia pruni-spinosae* (Pers.) Diet.; *Pileolaria klugkistiana* (Diet). Diet.

3.6 *Uredinales imperfecti*

Aecidium akebiae P. Henn.; *A, cimicifugatum* Schw.; *A. hyrangeae* Pat.; *A. mori* Barcl.; *A. paeoniae* Komarov; A. pusatillae Tranz.; A. rhododendri Barcl.; A. senecionis-scandentis Sawada; A. sino-rhododendri Wilson; Caeoma makinoi Kusano; Monosporidium andrachnis Barcl.; Peridermium ephedrae Cke.; Per. sinensis Wang et Guo; Per yunshae Wang et Guo; Poestelia levis (Crowell) Kern; R. sikangensis (petr.) Joerstad; Uredo clemensiae (Arth. et Cumm.) Hiratsuka f.; U. myriactidis Sumdaram; U. piptanthus Chen; U. pseudocystopteridis Wang et Wei; U. rhododendronis Wang et Guo.

4 DISCUSSION

The rust flora of the Sino-Himalaya represents a series of ecological types ranging from high altitude to low altitude, from north slope to south slope, and from cold temperature to subtropical, even tropical, zonation. The physical environment is very divergent and complex, but the horizontal zonation is for the most part closely correlated with the vertical zonation (Zheng et al. 1981; Chen 1981; Chang 1981). In some areas a zone may be as narrow as 10 miles (1 mile = 1.6 km) wide while in other places, such as on the plateau, the border of a zone may be hundreds of kilometers apart.

4.1 *Classification into ecological types*

The rust flora in this region can be classified into three ecological types, distinguished according to ambient temperatures (Table 2): (*i*) the plateau zone, (*ii*) the temperate zone, and (*iii*) the subtropical zone.

4.1.1 *(i) The plateau zone*

Areas of this type extend from plateau, high mountain meadows or scrub (6000 m) to the conifer forests of southern slopes below 4300 m, which form a nearly endless expanse of forest. The stretch from 3000 to 4300 m has an average annual temperature of about 3-8°C and annual rainfall of 450 mm. The plateau is somewhat cooler and drier than the other horizontal zones (Zheng et al. 1981). Forty-two percent of the total rust species were found in the plateau zone. Representative of the type of rusts here is the willow rust (*Melampsora larici-capraearum* Kelb.) on *Salix paraplesia, S. oxycarpa,* and *S. longistaminea* (3000-4100 m) (Li et al. 1985; Chen 1979). Some of the rusts are distributed in higher altitudes, such as *Puccinia gentianae* (at 3900-4400 m) and *Gymnosporangium nipponicum* (3700-3900 m). In the spruce and fir forests in the plateau zone, diseases of cones and needles of forest species are found, including *Picea-Prunus* rusts (*Thekopsora areolata* (Fr.) Magn.), spruce cone rust (*T. sparsa* (Wint.) Magn.), and spruce witch's-broom rust, *Peridermium yunshae* (Wang et Guo) (Wang et al. 1980). The evidence shows that, during the great geological and climatic transformation of the late Tertiary and particularly the Quaternary, glaciations and atmospheric circulation changed (Wu and Lin 1981; Xu. 1981; Shi and Li 1981; Chen 1979); these changes contributed to the evolution of the flora and the rusts. After passing through this period of harsh natural selection, some new types emerged, such as the high-altitude rust genera *Coleopuccinia* (Fig. 3), *Chrysomyxa*, and *Stilbechrysomyxa* (Tai 1947, 1948; Hiratsuka 1935; Shigeru and Hiratsuka 1984; Chen 1984), which may be considered endemic rusts of Sino-Himalayan flora.

Since the glacial period, *Rhododendron* has occupied a conspicuous position in the plateaus' floristic and vegetational formation. There are over 170 species. It stands first on the list of the largest genera of the Himalayan angiosperms (Ming and Fang 1979; Fang and Ming 1981; Shi and Li 1981). This area has a very rich *Rhododendron-Chrysomyxa* association (Hiratsuka 1935) within which *Rhododendron* leaf rust (*Stilbechrysomyxa stilbae*) Chen gen. Nov.; (Chen 1984) (3500-4700 m) probably arose. With its stilbumlike hyphal stalk at the base of the telial head (Fig. 4), which may be due to the changed conditions which were the result of the Himalayan uplift, it represents a new endemic genus with characteristics adapted to the cold aridity and high plateau radiation (Wang 1981; Chen 1979; Wu 1987). J.M. Dalavey collected a rust species in the plateau of Yunnan, which was later placed by N. Patouillard in a new genus, *Coleopuccinia* (Tai 1948; Chen 1979; Patouillard 1889). The weather found on the plateau, especially the mild temperatures at around 3000-3300 m, favors the heavy occurrence of wheat and plateau barley rust (*Puccinia stiiformis* West.) (Chen et al. 1983).

4.1.2 *(ii) The temperate zone*

The forest of mixed needle-bearing and broad-leafed trees gradually becomes more dense toward the lower elevations of the Himalayas, but the upper part is contiguous with the plateau zone in some areas. The region between 1800 and 3600 m above sea level supports a rich mixed forest of conifers and broad-leafed trees, consisting mainly of tall Chinese hemlock (*Tsuga dumosa*) several pines, the common oak (*Quercus semicarpifolia*), *Fraxinus*, *Acer*, *Betula*, *Ilex*, *Prunus*, and *Acanthopanax*, etc. Variety increases with lower altitude. With increasing humidity, and annual rainfall of 500-900 mm, and temperature increase averaging 8.4°C annually, rust fungi thrive. Some of the representative rusts typically found in this northern temperate zone are species of *Coleosporium, Cronartium, Melampsora, Gymnosporangium, Phragmidium, Chrysomyxa,* and *Melampsoridium*. Blister rust (*Cronartium ribicola*) of five-needle pines (*Pinus griffithii* and *P. armandi*), which has *Ribes* as its alternate host, is prominent. This rust flora is thus closely related to the North American and partly also to Indian flora (see Tble 3), such as the long-needle pine rust (*Coleosporium complanatum* Barcl.) on the blue pine (*Pinus roxbourghii*) and needle rust (*Coleosporium brevius* Barcl.) on Himalayan pine (*Pinus griffithii*).

More than half of the rust species are parasitic on members of the Rosaceae, Ericaceae, Compositae, Gramineae, Salicaceae, and Pinaceae. Nto only do plants belonging to such large families in the temperate zone manifest the characteristics of the Himalayan flora, but their rust parasites also are further evidence that the Himalayan fungal types are a reflection of the host plant types in a temperature zone.

4.1.3 *(iii) The subtropical zone*

The evergreen broad-leafed forest is found at 1100-2500 m above sea level. Here the annual temperature averages about 15°C and the annual rainfall totals above 1000 mm. Conifers are replaced by broad-leafed trees, mainly evergreens. The valleys lie shrouded in mist in summer. Subtropical zone rusts, such as the *Ravenelia* of *Indigofera*, *Phakopsora* of *Ampelopsis*, and *Puccinia akebiae* of *Akebia* of the monsoon region on the southern slope of the east Himalayas, are adapted to warm and damp conditions. In some area, it is overcast, misty, and moist all the year round. The representative families belong to Fagaceae, Anacardiaceae, Lauraceae, etc., which develop well on the slope at about 1100-2200 m (Chen 1981; Li and Zheng 1981). These plants are often inhabited by *Pileolaria klugkistiana* on *Rhus*, *Aecidium mori* on *Morus*, and several species of *Coleosporium* on cosmopolitan grasses, but fewer tropical rust species than temperate-zone species are present because, on the whole, this area (the northern slope of the plateau) may be considered the northern fringe of the subtropical zone.

4.2 *Relation to other rust flora*

Due to the complexity of the biogeography of this area, as influenced by its geophysical history, the biological history of the rust flora is extremely complicated. However, our preliminary study of this region revealed the following characteristics. (*i*) Eighty-five percent belong to large cosmopolitan families dominated by north-temperate species. This is the basic character of Sino-Himalayan forest rust flora. (*ii*) The remaining genera and species are newly described and endemic. Approximately forty-six percent of the host-rust combinations found were new records, and we collected more than 17 new species, such as *Uredo piptanthus* on *Piptanthus concolor* (2800 m), *Puccinia heterocoloris* on *Morina alba* (4100 m), *Phragmidium shengezhense* on *Rubus* (3100 m), and *Puccinia chintoniae-udensis* Bud. var. *tibetica* on *Clintonia udensis* (3380 m) (Durrieu 1979). (*iii*) Even though this region itself has specific geologic and climatic characteristics, the rust flora can be related to rust flora elsewhere (Table 3). Some species are distributed in a disjunctive manner between East Asia and North America, for example *Puccinia osmorrhizae* on Osmorrhiza and a few other *Puccinia* species. Many rusts are in north temperate zone environments similar to regions in North America and East Asia. Other types, such as the *Monosporidium*, are found only in the Sino-Himalayan plateau and India. These are more the south slope of Himalaya types, such as *Puccinia nepalensis* on *Rumex*, and *Pileolaria klugkistiana* on *Rhus chinensi*, while *Blastospora* and *Phragmidium hyalopsora* occur in the Sino-Himalayan plateau and Japan. Specimens on the same genera (*Coleopuccinia, Stilbechrysomyxa*) have also been described in cold arid areas in Gansu, China, and in Japan (Tai 1948; Chen 1984). The evidence suggests that these are traditional types endemic to high-altitude environments, and that their evolutionary adaptation to these conditions is ongoing. (*iv*) The rust flora may include potentially important pathogens. Many species of these rusts in the Sino-Himalayas are now known, such as poplar leaf rust (*Melampsora* spp.), spruce cone rust (*Thekopsora* spp.), willow rust (*Melampsora larici-capraerum*), pine needle rust (*Coleosporium* spp.), plateau barley, stem rust (*P. graminis*), etc., but so far they do not pose a danger to trees in the plateau and temperate zones.

Because it is a region with unique forest ecosystems, the Sino-Himalaya is a virgin forest treasure. Inside the forests there exist multiple biological communities. Fungi are small, living in close association with their hosts, and often indiscernible but they contribute to the ecological equilibrium of the forest itself. Once the equilibrium is interrupted by improper planting, certain fungi, such as *C. ribicola*, may seriously threaten *Pinus griffithii*, *Pinus armandi*, and other five-needles pines. At present teliospores of *Cronartium ribicola* are already found on *Ribes* of the temperate zone in the Sino-Himalayan area.

When the ecological equilibrium is maintained in the forests of the Sino-Himalaya, the health of the forest is maintained and no problems develop. If, however, resources are misused for human consumption and the equilibrium is disrupted (mostly by injudicious logging operations and forest operations), undesirable effects will be felt in the surrounding areas. If management can avoid disrupting the present ecological equilibrium of the Sino-Himalayan forests, there should be no reason to fear potential outbreaks of disease from pathogens that are already part of the ecosystem. If managed properly the Sino-Himalayan forest will continue to provide a balanced and healthy environment, which is desirable from the human point of view.

5 ACKNOWLEDGEMENTS

I gratefully acknowledge Prof. J. R. Parmeter, D. B. O. Savile, and Dr. G. Durrieu for their significant contribution in reviewing this paper, and Chris Hodgkinson for typing the manuscript.

6 REFERENCES

Balfour-Browne, F.L. 1955. Some Himalayan fungi. Bull. Br. Mus. (Nat. Hist.), **1**: 189 218.

Chang, D.H.S. 1981. The vegetation zonation of the Tibetan plateau. The mountain Research and Development. Vol. 1, No. 1. pp. 29-48.

Chen, M.M. 1979. Preliminary analysis of rust flora in the forest of Tibet. Linye Kexue, **79**(1): 1-8.

Chen, M.M., Chen, L.Z., G.L., Han, S.J., Wang, Y.Z., Wei, S.X. 1983. Teliomycetes. *In* Anonymous Fungi of Xizang. Sinica Press. pp. 32-61.

----- 1984. A new genus *Stilbechrysomyxa* Chen gen. Nov. of Chrysomyxaceae on *Rhododendron*. Linye Kexue, 00: 267-271.

Chen, W. 1981. The main types of forest vegetation and their distribution in Xizang. Proceedings of a Symposium on Quinghai-Xizang plateau (Beijing, China). Geological and ecological studies of Qinghai-Xizang plateau. Vol. II. pp. 1947-1950.

Durrieu, G. 1979. Uredinales du Népal. Crypto. Mycol. **1**: 33-68.

Cummins, G. B., & Hiratsuka, Y. 1983. Illustrated genera of rust fungi. American Phytopathological Society, St. Paul.

----- 1984. Biogeography of Uredinales in the Central Himalayas. Rep. Tottori Mycol. Inst. (Japan), **22**: 148-157.

Fang, R.Z., and Ming, T. 1981. Differentiation and development of genus *Rhododendron* during the process of the uplift of Quinghai-Xizang plateau. Proceedings of a Symposium on Qinghai-Xizang plateau. Vol. II. pp. 1289-1293.

Gallegos, H.M.L., and Cummins, G. B. 1981. Uredinales (royas) de Mexico. Instituto Nacional de Investigaciones Agricolas (INIA), Sinaloa, México.

Hiratsuka, N. 1935. A contribution to the knowledge of high mountains in Japan. Mem. Tottori Agric. Coll. **3**: 125-247.

----- 1984. Taxonomy of rust fungi—past, present, and future. Rep. Tottori Mycol. Inst. (Japan),**22**: 48-56.

Hiratsuka, Y., and Sato, S. 1982. Morphology and taxonomy of rust fungi. *In* The rust fungi. *Edited by* K. J. Scott and A. K. Chakravarty. Academic Press, London, New York. pp. 1-36.

Li, J.J., & Zheng, B.X. 1981. The monsoon maritime glaciers in the southeastern of Xizang. Geological and ecological studies of Qinghai-Xizang plateau. Proceedings of a Symposium on Zinghai-Xizang plateau. Vol. II. 1599-1604.

Li, W.H. & Han, Y.F. 1981. Fundamental characteristics of the forests in Xizang in relation to the uplift of the plateau. Geological and ecological studies of Qinghai-Xizang plateau. Proceedings of a Symposium on Qinghai-Xizang plateau. Vol. II. pp. 1937-1945.

Li, W.H., Han, Y.F., Chen, M.M., Xu, Y.J., Li, Y.J., Chen, H.Z. & Lo, J.C. 1985. Forests of Xizang. Science Press, Princeton, NJ.

Ming, T.J. & Fang, R.C. 1979. On the origin and geographic distribution of genus *Rhododendron.* Yunnan Zhiwa Yanjiu, **1**: 17-27.

Patouillard, N. 1889. Tab. Anal. N. 697.

Savile, D.B.O. 1976. Evolution of the rust fungi Uredinales as reflected by their ecological problems. Evol. Biol. **9**: 137-207.

Shi, Y.F., & Li, J.J. 1981. Batura glacier of Karakorum Mountains, an example of the complex type glacier. Geological and ecological studies of Qinghai-Xizang plateau. Proceedings of a Symposium on Qinghai-Xizang plateau. Vol. II. pp. 1619-1623.

Shigeru, K., & Hiratsuka, N. 1984. A preliminary list of rust fungi collected in the northern Japan Alps. Rep. Tottori Mycol. Inst. **22**: 1-8.

Tai, F.L. 1947. Uredinales of western China. Farlowia, **3**: 95-139.

----- 1948. A new species of *Coleopuccinia* with remarks on genus. Acta Agric. Sin. **Q**:979-1003.

Wang, Y.Z. 1981. On the evolution of some rust fungi in western China. Geological and ecological studies of the Qinghai-Xizang plateau. Vol. II. pp. 1161-1166.

Wang, Y.Z., Chen, M.M., Han, S.J., Wei, S.X., & Guo, L. 1980. New rust fungi from western China. Acta Microbiol. Sin. **20**(1): 16-28.

Wu, C.Y. (Editor). 1987. Flora Xizangica. Vol. 5 Science Press, Princeton, NJ.

Wu, X, & Lin, Z.Y. 1981. A preliminary analysis of climatic change during the historical time. Geological and ecological studies of the Qinghai-Xizang plateau. Vol. II. pp. 1219-1224.

Xu, R. 1981. Vegetational changes in the past and uplift of Qinghai-Xizang plateau. Geological and ecological studies of the Qinghai-Xizang plateau. Vol. I. pp.139-148.

Zheng, D., Zhang, Y.Z., & Yang, Q.Y. 1981. Physical-geographical differentiation of the Qinghai-Xizang plateau. Geological and ecological studies of the Qinghai Xizang plateau. Proceedings of a Symposium on Qinghai-Xizang plateau. Vol.II. pp.1851-1855.

(Taken from Canadian Journal of Botany, Vol. 67, 1989. pp. 827-833.)

Table 1. Representative rust fungi in vertical belts of the forest of the Yarlung-Zangpo River region, and their hosts

Altitude (m)	Forest belt (species)
5000 m	Snow line
	Puccinia heterocoloris (Morina)
	Stilbechrysomyxa (Rhododendron)
4000 m	
	Coleopuccinia (Cotoneaster)
	Melampsora (Salix)
	Gymnosporangium (Sabina)
	Nyssopsora (Aralia)
	Thekopsora (Picea)
	Chrysomyxa (Picea)
3000 m	
	Coleosporium (Pinus, Saussurea,Elsholtzia, Aster, Lonicera)
	Cronartium (Pinus, Ribes)
	Phakopsora (Ampelopsis)
2000 m	
	Ravenelia (Indigofera)
	Pileolaria (Toxicodendron)
	Aecidium (Morus)
1000 m	
	Broad-leafed forest

Table 2. Pattern of the distribution of rust flora in the Sino-Himalayas in relation to hydrothermal conditions

Type of rust flora	Type of forest	Elevation (m)	Mean annual temp. (°C)	Warmth index (°C)	Mean temperature (°C)		Average minimum (°C)	No. of frost free days	Relative humidity (%)	Mean annual precipitation (mm)
					Warmest	Coldest				
Plateau type	High mountain scrub, subalpine coniferous forest	3000-6000	3-8	26-63	10-15	-8 to -1	-17 to -25	80-160	60	300-500
Temperate zone	Mixed coniferous and broad leafed forest	1800-3600	4-15	25-110	12-21	-8 to 0	-6 to 22	110-240	50-80	500-900
Subtropical zone	Subtropical evergreen forest	1100-2500	11-18	77-160	18-25	3-12	0 to -12	200-290	60-80	1000+

Table 3. Comparison between the rust flora of the Sino-Himalayas and those of other regions

Genera of rust	Region of origin	Sino Hima-layas	Inland China	Nepal Hima-layas	India	Japan	North America	Russia	Austria	Mexico
Hyalopsora	North temperate zone	+	-	+	+	+	+	+	+	+
Thekopsora	North temperate zone	+	+	-	-	+	+	+	+	+
Melampsoridium	North temperate zone	+	+	-	+	+	+	+	+	+
Cronartium	Cosmopolitan	+	+	+	+	+	+	+	+	+
Chrysomyxa	North temperate zone	+	+	+	-	+	+	+	+	-
Stilbechrysomyxa	Sino-Himalaya	+	-	-	-	+	-	-	-	-
Coleosporium	Cosmopolitan	+	+	+	+	+	+	+	+	+
Melampsora	North temperate zone	+	+	+	+	+	+	+	+	+
Phakopsora	Tropical zone	+	+	+	+	+	+	+	-	+
Monosporidium	Sino-Himalaya	+	+	-	+	-	-	-	-	-
Uromyces	Cosmopolitan	+	+	+	+	+	+	+	+	+
Puccinia	Cosmopolitan	+	+	+	+	+	+	+	+	+
Gymnosporangium	North temperate zone	+	+	+	+	+	+	+	+	+
Nyssopsora	Cosmopolitan	+	+	-	-	+	+	+	+	-
Phragmidium	North temperate zone	+	+	+	+	+	+	+	+	+
Ravenelia	Tropical zone	+	+	+	+	+	+	-	-	+
Bastospora	Sino-Himalaya	+	-	-	-	-	-	-	-	-
Coleopuccinia	Sino-Himalaya	+	+	-	-	+	-	-	-	-
Aecidium	Cosmopolitan	+	+	-	+	+	+	+	+	+
Peridermium	Cosmopolitan	+	+	-	+	+	+	+	+	+
Caeoma	Cosmopolitan	+	+	-	+	+	+	+	+	+
Uredo	Cosmopolitan	+	+	-	+	+	+	+	+	+

1979年6月 植 物 病 理 学 报 Acta Phytopathology Sinica June 1979

中国—喜马拉雅区系中的一些真菌新种和新记录

谌谟美

（中国林业科学研究院林业研究所）

提　　要

泛北极植物区中的中国一喜马拉雅亚区是一个具有独特的自然生态系统的地区， 由于自然历史演变的结果，它有着明显的植物分布垂直带谱，森林植物和真菌资源也异常丰富， 本文将分期报道本地区的一些新种、珍贵树种病原真菌和药用真菌以及我国内陆首次发现的稀有真菌。本期报道本区系中几个新种和新记录，新种是： 1、拉康柄锈菌 *Puccina lakanensis* sp.nov.; 2、申格热多孢锈菌 *Phragmidium shengezieense* sp. nov.；3、扎木多孢锈菌 *Phragmidium zamonense* sp.nov.,新记录是：伊藤囊孢锈菌 *Blastospora Itoana Togashi* et Onuma。新种有拉丁文和汉文描述，并在每种下讨论了该种和它相近种之区别，新记录有汉文描述。

中国一喜马拉雅亚区属于泛北极植物区系，地理范围属于东喜马拉雅山和中喜马拉雅山，约位于北纬85—102度、东经27—30度之间。本区曾经历一个漫长和曲折的地质发展过程，大地貌单元有喜马拉雅山地、喜马拉雅主脉、喜马拉雅北翼湖盆和雅鲁藏布江谷地等，地质历史演变的结果，出现了独特的现代自然地理条件。本区不仅是一些温带植物和菌类科、属的发源地和古老的植物及菌类的避难所，同时由于地形、地势有着明显的垂直变化，随着海拔急剧升高，这个自然环境中的气候、生物等也随之发生由低到高的明显变化，常再一个数十公里的深山峡谷里，可以由热带、亚热带经过温带直到高寒带的垂直带谱，气候变化异常明显。在喜马拉雅山北和高原面上，由于高耸的喜马拉雅山的屏障作用，降雨显著减少，甚至主脊以北出现雨影区，气候特征是晴天多、辐射强, 日照长，夜雨多等；然而在喜马拉雅山的南翼地区，由于与印度的阿萨姆相毗连, 印度洋暖湿气流由南长驱而入，造成山南降水丰富,峡谷中生长着繁多的植物，茂密的原始森林。本区森林和真菌分布按地貌和水分条件可分为以下几个地区。

1　**藏南高山峡谷潮湿针阔混交林。**它包括“西藏的江南”——察偶和“西藏的西双

承安徽农学院陈礼琢同志、微生物所魏淑霞同志协助鉴定标本，北京农业大学董元同志协助描绘插图，王云章先生给予热情指导，在此一并表示感谢。

版纳”——墨脱，喜马拉雅山南翼的吉隆、亚东、樟木、拉康等地。常见的针叶林有云南松、高山松、华山松、丽江云杉、麦吊云杉、乌蒙冷杉、云南铁杉等。阔叶树种也比较丰富，其中以蔷薇科和杜鹃花科种类最多。由于雨量充沛，真菌种类尤为丰富，常绿阔叶林内常见有大型盘菌、大型革菌，屡见一尺宽的猴头菌和满地长着毒蘑菇和食用蘑菇、鬼笔菌、马鞍菌、锈菌和各种灵芝菌，多孔菌和巨大球形伏苓菌核等。

2 **雅鲁藏布江中游山地湿润针叶林。**它包括东久、易贡、波密、林芝、米林、朗县等地。常见的森林树种是丽江云杉，商山松，高山栎等。这些地区的真菌种类虽不及上述森林里多，但它特有种类产量仍很可观，比如波密的灵芝菌在过伐的高山栎林内比比皆是，还有蘑菇、多孔菌、地星、子囊菌等。

3 **横断山脉干热河谷半湿润块状针叶林。**它包括念青唐古拉山以南和澜沧江中上游的昌都、丁青、类乌齐地区。该地区主要树种是川西云杉和园柏，由于南来降水量的递减，所以真菌种类显著减少，有寄生于柏树上的胶锈菌，寄生于云杉林的多孔菌。本区出产经济真菌——冬虫夏草[*Cordyceps sienesis* (Berk.) Sacc.]多产于海拔3500米—4500米的杜鹃灌丛里。

4 **西藏中部半干旱灌丛草原少林带。**如拉萨、江孜和“世界的日光城”—日喀则。由于降水少，蒸发量大以及太阳辐射的毒害作用，因此真菌种类比较简单，大型真菌极少，多为腐生于枝、干、叶上的子囊菌和半知菌，还有一些危害林木的锈菌和多孔菌。

我们在上述各类森林中重点采集和鉴定了锈菌和多孔菌，本文将连续发表一些真菌新种和新记录，新种以拉丁文和汉文描述并讨论其和相近种的区别。此外，还描述一些珍贵用材树种的病原真菌和药用真菌以及我国首次发现的稀有真菌。

拉康柄锈菌

Puccinia lakanensis Chen & Wang sp. nov 图 1.

Soris teleutosporiferis hypophyllis, pulverulentis, atro-brunneis, globosis vel ellipticis, 2-4mm. daim.; teliosporis ellipiticis, late-ellipticis, brunneis, 35-52×26-31μ. apice rotundatis, basi rotundatis; poris germinativis in parte apicali cellularum superarum dispositis; membranis uniformibus 3.8-7.5μ crassis medio non vel parum consrrictis; pedicello hyalino 50-78μ longo.

Hab in follis Clematidis Tibet: Loza hsien, IX, 3, 1975, M M. Chen K-39 (Typus)

冬孢子堆著生于叶的背面，初期为寄主表皮组织所复盖，后期露出，粉状，暗褐色，圆形或椭圆形，直径2—4毫米；冬孢子椭圆形、宽椭圆形，褐色，35-52×26-31微米，顶端和基部圆形，芽孔在上面细胞的顶端，横隔膜不缢缩或稍缢缩，壁均匀，厚3.8-7.5微米；柄无色，长达50-78微米。

本菌生寄于铁线莲*Clematis* sp. 叶上。1975年9月3日，采集于海拔3200米的西藏洛扎县拉康地区。标本编号，K—39(模式)

寄生于*Clematis*上的*Puccinia*种中，本菌因冬孢子长度与*Puccinia wattiana*有些相近，但它们的冬孢子宽度完全不同，*puccinia wattiana*冬孢宽为16-22(少数26)微米，而*Puccinia lakanensis*宽度为26-31微米，特别是它们的胞壁厚度很不相同，*Puccinia lakanensis*冬孢子四周壁特别宽(3.8—7.5微米)，根据冬孢子宽度及胞壁厚度的特征*Puccinia lakanensis*可以定为一个新种。

申格热多孢锈菌

Phragmidium Shengezieense Chen & Chen sp. nov. 图 2

Soris uredosporiferis hypophyllis: uredosporis ellipticis, ovatis, hyalino 13-19×12-15米, episporio,1.5-2.6米crasso, echinulatis, Soris teleutosporiferis hypophyllis, pulverulentis, nigrifactis, 0.3—1mm. daim.; teliosporis cylindricis,atro-brummeis 78—125×20—30米,4—

8spetatis, medio non vel parum constrictis, poris germinativis in parte apicali cellularum superarum dispositis; episporio 2米crasso; pedicello 90米longo.

Hab.in follis Rubi sp. Tibet；Loza hsien, IX，1，1975，M M Chen kan—24(Typus).

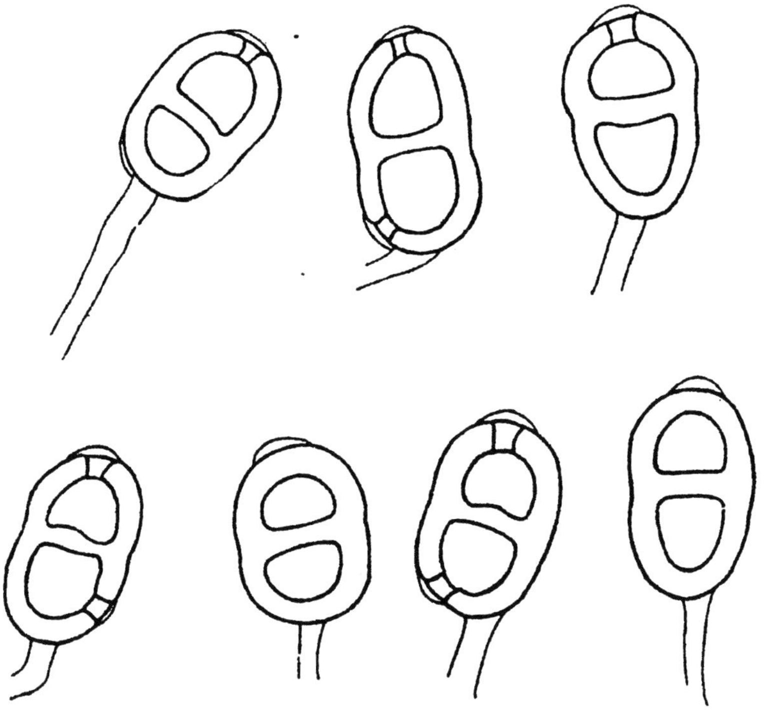

图 1. *Clematis* 叶上之 *Puccinia lakanensis* 的冬孢子 375 倍

夏孢子堆着生于叶的背面；夏孢子椭圆形、卵圆形，无色，13—19×12—15 微米，壁厚 1.5—2.6 微米，密生细疣；冬孢子堆着生于叶的背面，粉状，灰黑色，直径 0.3—1 毫米；冬孢子长圆柱形，暗褐色 78—125×20—30 微米，4—8 个隔膜，隔膜不缢缩或稍缢缩，芽孔在上面细胞的顶部，壁厚 2 微米，平滑；柄长 90 微米，柄无色。上部淡褐色。

寄生于悬钩子(*Rubus* sp.)叶上。1975 年 9 月 1 日，采集于海拔 3100 米的西藏洛扎县申格热地区。标本编号 Kan-24(模式)

在寄生于 *Rubus* 上的 *Phragmidium* 种中，大多数冬孢子细胞数目较多的都有一个共同特征，就是冬胞子壁有疣，本菌冬孢子壁光滑无疣，冬孢子数最多是 *Phragmidium grlseum* 它的冬孢子隔膜 1—6 个，冬孢子长为宽 30—89×18—32 微米，而 *Phragmidium shengezieense* 上的隔膜多数是 4—8 个，冬孢子长宽为 79—125×18—25 微米，已知所有寄生在 Rubus 上的冬孢子壁光滑无疣的 *Phragmidium* 的冬孢子长度都没有本菌这么长，因此本菌可分为一个独立的种。

扎木多孢锈菌

Phragmidium zamonense Chen & Chen sp. Nov 图 3

Soris uredosporiferis hypophyllis,pulverulentis, aurantiacis, 0.4 — 0.5mm. diam.;

uredoporis globosis, ovoideis, 21—24×15—20米,luteis, echinulatis, episporio 1.5米 crasso; Soris teleutosporiferis hypophyllis; teleutosporis oblong-ellipticis vel clavatis, 38-70×20-20米, 1-3 septatis, cinnamomeobrunneis, episporio 1.5米 crasso; poris germinativis; pedicello persistenti 78米longo, hyaline.

Hab. in follies Rubi alexeterius Focke. Tibet pome hsien, VII，28. 1976, M. M. Chen po-55 (Typus).

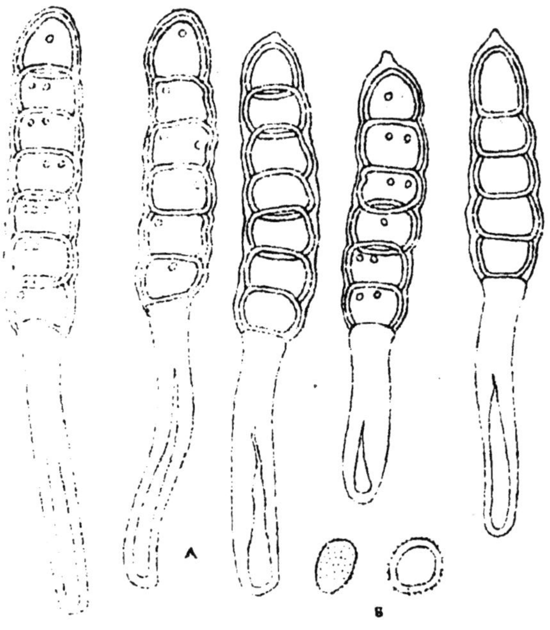

图 2. 悬钩子叶上之 *Phragmidium shengezieense* 的冬孢子和夏孢子

A、冬孢子 500 倍　　B、夏孢子 250 倍

夏孢子堆着生于叶的背面，粉状，桔黄色，直径 0.4—0.5 毫米；夏孢子球形，卵圆形，21—24×15—20 微米，黄色，具刺，壁厚 1.5 微米；冬孢子堆叶背生；冬孢子长椭圆形或棍棒形，39—76×18—25 微米，具 1—3 个隔膜。隔膜处缢缩，顶部圆形或圆锥形，基部圆形，肉桂色，壁光滑，厚 1.5 微米，每个细胞有 2—3 个芽孔；柄无色，宽 23 微米，长可达 78 微米。

寄生于波密悬钩子(*Rubus alexeterius Focke*)叶上。1976 年 7 月 28 日，采集于海拔 2700 米的西藏波密县。标本编号波—55。(模式)

在寄生于 *Rubus* 上的 *Phragmidium* 种中，本菌因夏孢子比较小，特别是冬孢子隔膜 1—3 个，普通 2 个而与 *Phragmidiumrubi-fraxinifolii* 有些相似：但两者还有其他许多不同之处，如 Phragmidium rubi-fraxinifolii 冬孢子长度长得多，宽度狭，乳突小，柄短得多，*Phragmidium zamonese* 则不是如此。根据冬孢子壁光滑无疣的 *Phragmidium* 种中，夏孢子多数为椭圆形、长椭圆形、棍棒形，而且个体比较大，本菌与上述特征完全不同，夏孢子卵圆形或近球形，而且个体比较小，因此，*Phragmidium zamonese* 应系一个独立种。

伊藤囊孢锈菌 *Blastospora Itoana* Togashi et Onuma 图 4

夏孢子堆生于叶背面，病斑为多角形，一般直径 1—1.5 毫米，以后相互聚合为白粉状的角斑；夏孢子圆形，具稀疏小刺，淡色，20.3×23 微米；冬孢子堆和夏孢子堆长在一起，冬孢子球形或近球形，内含物桔黄色，平滑，23—41×20—38 微米，壁无色，厚 1.5 微米，冬孢子具短柄，常长出柱状担子，担子无色，长达 98 微米，宽 24.5 微米。

寄生于防己叶菝葜(Smilax *menispermoides* A.DC)叶上，1976 年 9 月 13 日，采集于海拔 3500 米的西藏波密县，波墨公路。标本号波—74。

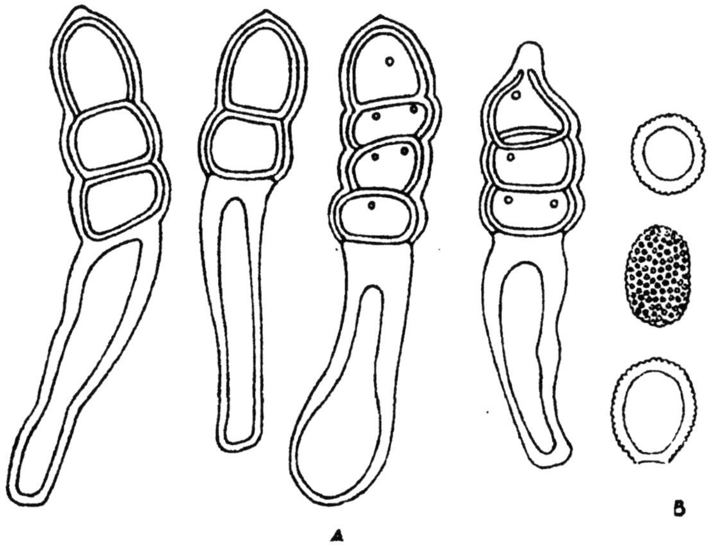

图 3　悬钩子叶上之 *Phragmidium zamonense* 的冬孢子和夏孢子

A、冬孢子 333 倍　　　B、夏孢子 333 倍

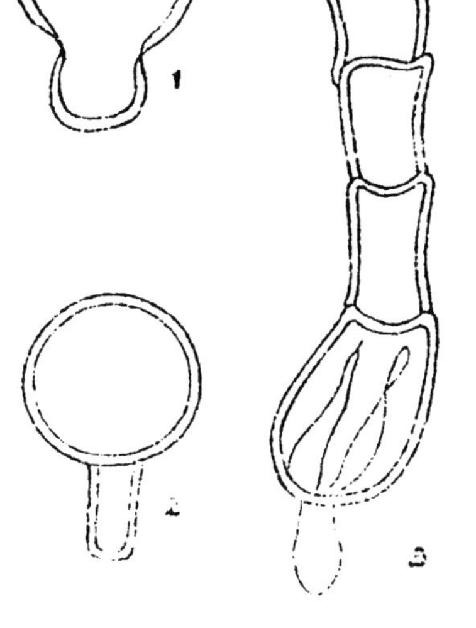

图 4. 菝 葜 叶 上 之 Blastospora Itoana 的冬孢子和担子

1. 冬孢子萌芽 333 倍
2. 冬孢子 333 倍
3. 冬孢子和担子 333 倍

参考文献

[1] 王云章, 中国锈菌索引, 1949: 科学出版社。

[2] 戴芳澜, 中国真菌总汇, 1979: 科学出版社。

[3] 伊藤诚哉，1950：日本菌类志 株式会社养贤堂，第二卷，第三号，35—45。

[4] Saccardo, P. a 1888 Sylloge fungroum VII: 742-750. IX 315-316. XI 207.

[5] Balfour- Browne F. L. 1955: Some Himalayan fungi bull. brit. Mus. (Nat.list.) Bot. I 189-218.

[6] Arthur, J. C. (1934) Manual of the rust in United Stated and Canada. 1962 ed Hafuer, New York.

[7] Wilson, M. and D. M. Henderson (1966) British rustfungl Cambridge Univ. perss. Lodon and New York.

[8] G. Durrieu, 1973: les champignons parasites des vegetaux Dans L. Himalaya du Nepal et le centre Afghanistum.

[9] Sydow, 1902-24: Monographia Urdinearumvol. Pucciniceae.

[10] 谌谟美，1979：西藏森林植物锈菌区系初报，林业科学，15（1）1—8 页。

NEW SPECIES AND NEW RECORD OF FUNGI IN THE CHINA-HIMALAYA FLORA

Chen M. M.

(Institute of Forestry protection Chinese Acad. Forest. Science Beijing, China)

ABSTRACT

Some new species and a record of Uredinales in the China-Himalaya flora are reported. The new Species are *Phragmidium zamonensis* Chen & sp. Nov., *Phragmidium shengezieense* Chenet Chen sp. Nov. Anew record is *Blastospora Itoana* Togashi et Onuma. Both Latin and Chinese diagnosis are provided for new species.

1982年 3月 植 物 病 理 学 报 Acta phytopathology Sinica March 1982

中国—喜马拉雅区系中的一些真菌新种和新记录(续)

谌谟美

(中国林业科学研究院林业研究所病理室)

提 要

本文系“中国—喜马拉雅区系中的一些真菌新种和新记录”（ 1979年)的继文，本期报导本区系中新种黄花木夏孢锈菌 *Uredo piptanthus* sp.nov.和西藏柄锈菌 *Puccinia tibetica* sp.nov.;高原特有种和古化石种中国鞘柄锈菌 *Coleopuccinia sinensis* Pat.、日本明痂锈菌 *Hyalopsora japonia* Diet 及珍贵树种、药用植物病原真菌 *Peridermium complanatum* Berc.、乔松鞘锈菌 *Peridermium brevius* Berc.、麻黄色被锈菌 *Peridermium ephedrae* CKE 当归柄锈菌 *Puccinia angilicae* Fuck.、最后描述了我国西部地区重要病原一角落叶松锈菌 *Melampsora laricicaraearum* Kleb.和木蓝锈菌 *Raveneia indigoferae* Trans.。

黄花木夏孢锈菌等[2][10]真菌资源是中国—喜马拉雅区系中所特有菌类，本区为原始森林地带，在垂直带谱上的每一类森林群落均有其特有真菌类群[7][11]，在真菌的形态和生态上也有许多特点比如胞壁比较厚，孢子堆具覆盖，着色较深以及孢子器官的包被比较大等等，这些反映了高原气候如晴天多、辐射强、日照长、夜雨多等的特征[11]。

黄花木夏孢锈 *Uredo Piptanthus* Chen sp nov [图版 I、1]

Soris uredosporiferis hypophullis, pulvervulentis brnneo-flabibus 550-630米; Uredosporis globosis flavor-brunneis,20-26×17-20米,poris germinativis conspcuus,episporio 2-3米 crasso, verricoloso uncinatus.

Hab. in follies piptanthus concolor Harr. Tibet; yadong hsien, VIII, 15, 1975, M.M. Chenia-71 (Typus)。

夏孢子堆叶背生，初期为寄主的表皮所覆盖，后期露出，粉状，褐黄色，550～630微米，夏孢子近球形，黄褐色，20～26×17～20 微米，具明显芽孔，具疣，疣顶具弯钩，基底周围凹下，壁厚 2.3 微米[9]。

寄生于喜马拉雅黄花木(*Piptanthum concolor* Harr.)的叶上，西藏亚东县，海拔 2800 米，1975年 8 月 15 日，样本编号 ia-7l(模式)。

过去在黄花木属 *Piptanthus* 上未报道过锈菌，其夏孢子疣顶端呈钩状，其疣著生处为一明显凹下部份，这些与相近种所不同，而且喜马拉雅黄花木(*Piptanthum concolor* Harr.)为一仅有种．

西藏柄锈菌 *Puccinia tibetica* Chen sp. Nov. [图版 I、2]

Urediis hypophyllis, pllide flavis, ellipsoideis, longiellipsoidies; 26-34 ×17-21 米，tellis hypophyllis, aggregates, aggregates, 0.2-0.6mm din., pulverulentis, artobrunneis; teliosporis oblongis vel rotundatis, 35-65×15-20米 membranis 1.5米 crassia, ad apic em 7-14米 crassis, pedicellis 64米, longis, hyalinis vel pallidis.

Host et Hab.: Clintonia udensis Trauto et Mey., Xizang: Luozna Shengere, alt. 3380 m, Ⅸ. Ⅰ. 1975, Chen Mo-mei Kan-20(HOLOTYPUS).

夏孢子堆叶背生，夏孢子淡黄色，椭圆形，26～34×17～21 微米，具刺；冬孢子堆叶背生，集生，直径 0.2～0.6 毫米，粉状，深褐色，冬孢子长椭圆形或棍棒形，黄褐色 35～65×15～20 微米，两端狭圆或圆形一端斜平形，膜薄 1.5 微米，顶部有的厚达 7～14 微米，柄无色或淡色，长达 64 微米[12][4]。

寄主与产地：兰果七筋姑(*Clinotonia udensis* Mey.)西藏：洛扎申格热，海拔 3380 米，1975 年 9 月 1 日。

本种的冬孢子虽比较接近于 *Puccinia clintoniae-udensis* Bud., 但这份标本具夏孢子，前者无夏孢子阶段，这份标本的冬孢子比较长，可达 35～65×15～20 微米，而 *Puccinia clintoniae-udensis* 的冬孢子长仅 32～48×17～21，尤其顶端乳突状膜厚达 7～14 微米，而 *Puccinia clintoniae-udensis* 仅为 3～6 微米，因此本种可由变种上升为一个新种。

中国鞘柄锈菌 Coleopuccinia sinensis Pat[6]。[图版 I、3]

冬孢子堆群聚或分散，生于叶背，圆形，直径 0.3～0.4 毫米；冬孢子 2 室椭圆形，顶端圆，下端略尖或圆，横隔膜处不缢缩，23～38×12～17 微米，壁厚 1 微米。

寄生于灰旬子（*Cotoneaster acutifolia*）叶上，1976 年 9 月 13 日，采集于海拔 3800 米的西藏察隅县古琴和波密县波墨公路，标本号波－117，波－77，察－32，察－31。

日本明痂锈菌 Hyalopsora japonica Diet[1] [图版 K、4]

夏孢子堆生于叶两面，大多数生于叶背面，疏松聚集，有时密集成片，桔黄红色，干后变浅黄色，病斑直径 387—628 微米；夏孢子形状多样多角形、带角圆锥形、带角卵形等，内含物微带浅黄色或淡色，26～42×21～31 微米，孢壁厚度不均匀，无色，顶壁厚 3.4～7.2 微米，侧壁厚 2.9 微米。

寄生于弯弓密网蕨（*Phymatopsis malacoudon*）叶上，1975 年 9 月 1 日，采集于海拔 3300 米的西藏洛扎县拉康，标本号 Kan—22。

长叶松鞘锈菌 *Peridermium complanatum* Berc.[1][3] [图版Ⅱ、5]

锈孢子器针叶两面生，黄白色疱状，扁平，愈合或单生，3～10×2～5 毫米，锈孢子椭圆形，矩形，近圆形或长卵圆形，密生大疣，20～31×17～23 微米，壁厚 2.9 微米；包被细胞 32～44×29～38 微米，表面具密布粗大疣，外壁 8.7～12×12 微米，内壁 8.7 微米。

寄生于西藏长叶松（*Pinus roxbour ghii*）叶上；1975 年 6 月 20 日，采集于西藏吉隆县江村，海拔 2260 米，标本号 gi—37。

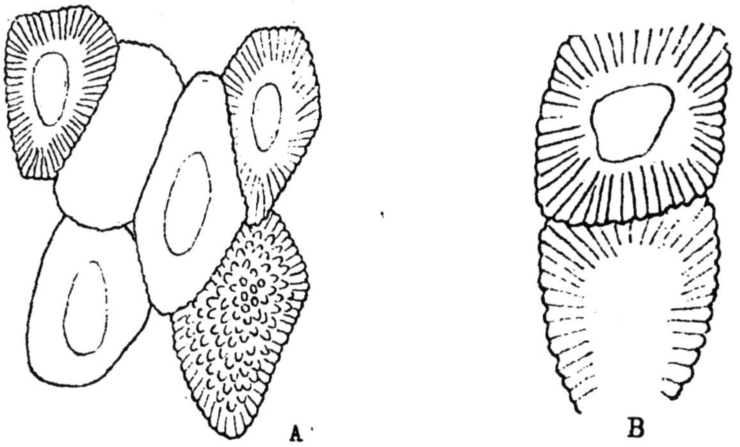

图 1 乔松鞘锈菌 A 锈孢子（800）倍；B 锈孢子器包被细胞（1000）倍

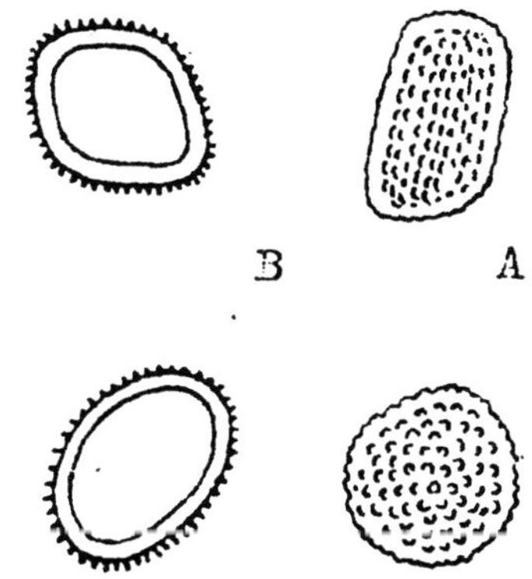

图 2 麻黄包被锈菌 A 锈子器包被细胞（1000）倍 B 锈孢子（1000）倍

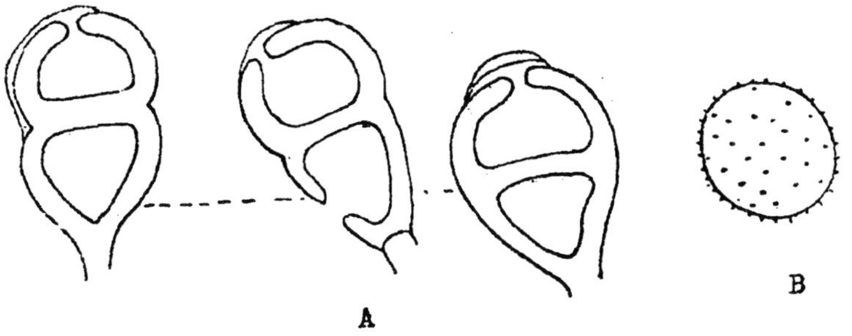

图 3 当归柄锈菌：A 冬孢子 830 倍，B 夏孢子 830 倍

乔松鞘锈菌 *Pericermium brevius* Berc.[1][3] [图 1]

锈孢子器生于针叶两面，桔黄色，疱状，0.12～0.2×0.1～0.12 毫米；锈孢子器包被细胞 35～41×23～35 微米，表面密生粗疣，胞壁具横纹，外壁 8.7～11.6 微米，内壁 5.8～3.7 微米；锈孢子椭圆形至长圆形，橙黄色，25～27×17～24 微米，壁厚 2.5 微米，密生粗疣，胞壁具横纹。

寄生于西藏乔松 (*Pinus griffithii*) 叶上，1975 年 6 月 15 日，采集于海拔 2900 米的西藏吉隆县的小吉隆，标本号 gi—9，gi—23，gi—47(2260 米)。

麻黄包被锈菌 *Peridermium ephedrae* CKE[1][8] [图 2]

锈孢子器分布于整个茎上，呈坛状，黄色，直径 420～883 微米，包被细胞矩形，23～35×13～17 微米；锈孢子成串排列，卵圆形，圆形或不规则形，棱角明显，具细密疣，19～26×15～20 微米。

寄生于麻黄(*Ephedra saxatilis*)茎上，1975 年 7 月 2 日，采集于海拔 3400 米的西藏吉隆县托当，标本号 gi—88。

当归柄锈菌 *Puccinia angilicae* Fuck[1][8]. [图 3]

冬孢子堆和夏孢子堆散生于叶背，常数个或数十个粉状孢子堆聚，冬孢子初具白色薄膜，后破裂露出暗褐色粉状孢子堆大小为 778×650 微米；冬孢子双细胞锈褐色，32～49×17～26 微米，柄长 17～46 微米，侧壁厚 2.9 微米；夏孢子堆浅桔黄色，具密疣，23～37×23～26 微米。

密生于当归(*Angelica* sp)叶上，1975 年 9 月 23 日，采集于海拔 3000 米的西藏朗县昌格，标本号 Lan—32。

角落叶松锈菌 *Melampsora larici-caraearum* Kleb. [图 4]

本种特征为夏孢子堆初生于嫩芽上，后期布满叶两面，粉点，桔黄色，椭圆形，圆形，直径 440～620 微米；夏孢子圆形，近圆形，椭圆形，内含物桔黄色，15～30×12～16 微米[11]。

寄生于左旋柳(Salix paraplesia)，竹柳 S Oxycapus 和 S. longistunina 等叶上，1975 年 5 月 25 日，采集于海拔 2800 米至 3200 米的西藏吉隆县、亚东县、拉康、波密、易贡、日喀则市，标本号 R—lo，ia—45。

木兰伞锈菌 *Ravenelia indigoferae* Tranz.[1][5][8] [图 5]

夏孢子堆叶两面生，土黄色粉斑，散生，少数密集，直径 0.5～1 毫米；夏孢子近球形，浅肉桂色，具密疣，20～26×17～23 微米，壁厚 1.5～2 微米，发芽孔 8 个，散生，侧丝棍棒状——头状，浅黄褐色，顶壁厚 4.9 微米，下端色浅；冬孢子堆头状体，扁圆形，反而凹陷，栗褐色，径宽 80～100 微米，沿直径具 4～6 个孢子，单个孢子直径 21～23 微米，孢子六角形，头状体下侧紧贴有无色泡囊附属物，泡囊体呈宽椭圆形，无色，头状体无色或成束的小短柄。

寄生于木兰（ *Indogogera* sp. ）叶上，1976年8月30日，采集于西藏下察隅，海拔1700米，标本号察一25。

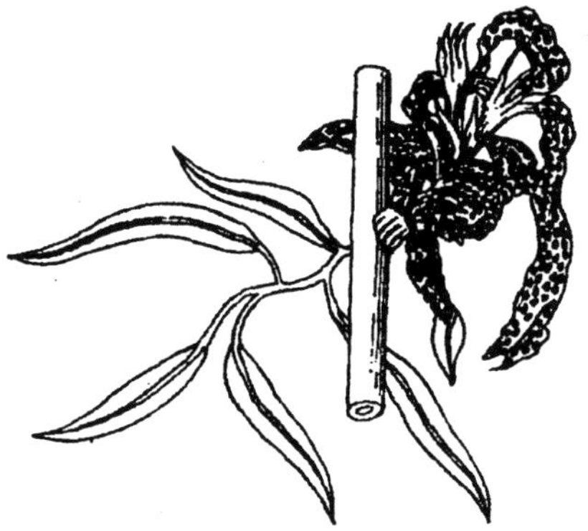

图4 角落叶松锈菌：叶上的夏孢子堆（为原物2/3）

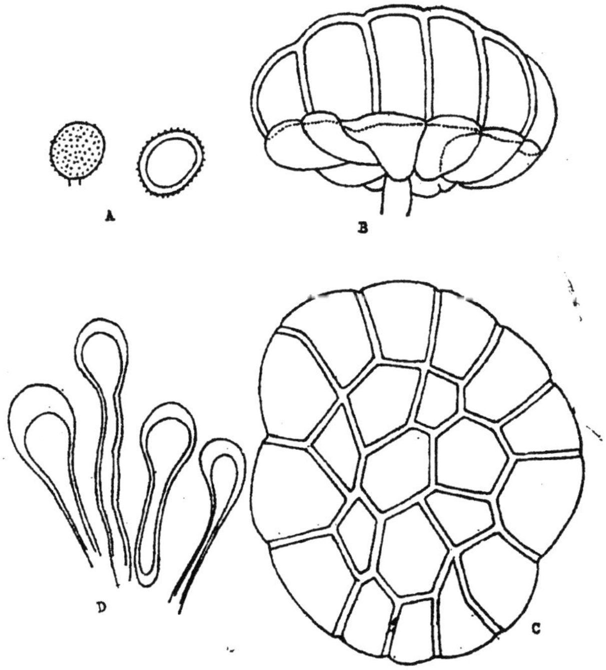

图5 木兰伞锈菌 A夏孢子500倍：B冬孢子堆500倍：
C冬孢子堆（正面）500倍；D侧丝500倍

参考文献

[1] Saccardo, P. A. 1888: Sylloge Fungrcum VII. IX. XI XII, XIII XV, XXI, XIV, XXIII, 1882~1931.

[2] Balfour. Browne F. L. 1955: Some Himalayan fungi bull. Brit. Mus (Nat. list) Bot I 189～218.

[3] Peterson,R. S. (1967). The peridermium species on the Pine stems. Bull. Torrey Bot. Club. 94:511～542.

[4] Wilson, M., and D. M. Henderson. (1966) "British Rust Fungi" Cambridge Univ. Press, London and New York.

[5] G . Durrieu les champignons paraites Des vegetaux Dans L Himalaya du Nepal et Le centre Afghanistan 1973.

[6] 戴芳澜， A new species of Coleopuccinia with remarks on the genus (Coleopuccinia Kumningensis) acta Agriculture 1:97～103, 1948.

[7] Naohide Hiratsuka 1935: A Contribution to the knowledge of the rust-flora in the alpine regions of high mountains in Japan.

[8] Sydow, 1902-24: Monographia Uredinearum (I—IV) Vol. III Pucciniaceae Vol. IV: Uredinales imperfectae

[9] Wilson, M. 1919—1921 Some fungi from Tibet notes Bot. Gard. Eninburgh 12:261-263 P1.169.

[10] Hawksworth, D. L.: Mycologist's Hand-book. Western printing Services Ltd, Bristol. Britain, 1974.

[11] 谌谟美，1979，西藏森林植物锈菌区系初报，林业科学 79—1。

[12] 王云章等，1980，中国西部锈菌，微生物学报，[20]1。

NEW SPECIES AND NEW RECORD OF FUNGI IN THE CHINA-HIMALAYA FLORA

Chen Momei

(Institute of Forestry, Chinese Academy of Forest Science)

Some new species and new records of Uredinales are reported on forest trees. They are ①. *Uredo piptanthus sp.* nov.; ②. *Puccinia tibetica sp.* nov.; ③. *Coleopuccinia sinensis* Pat.; ④. *Hyalopsora japonica* Diet.; ⑤. *Peridermium complanatum* Berc. ⑥. *Peridermium brevius* Berc. *Peridermium ephedrae* CKE ⑦. *Puccinia angiliae* Fuck. ⑧. *Melampsora larici-caraearum* Kleb. and ⑨. *Ravenelia indigoferae* Trans. Discriptions of the two new species *Uredo piptanthus* and *Puccinia tibetica* are written in Chinese with Latin diagnosis and those of the remaining species in Chinese.

A

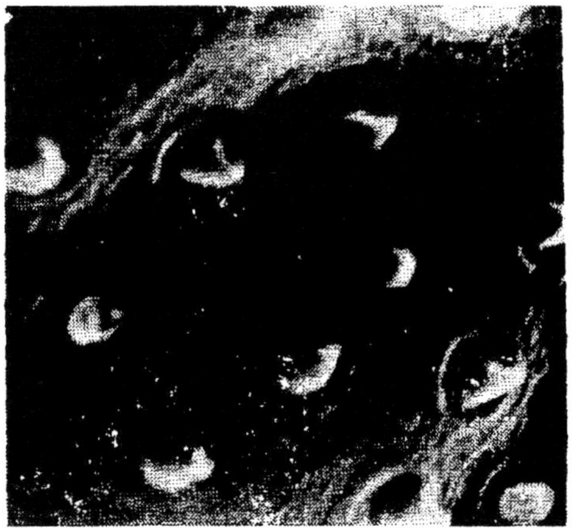

B

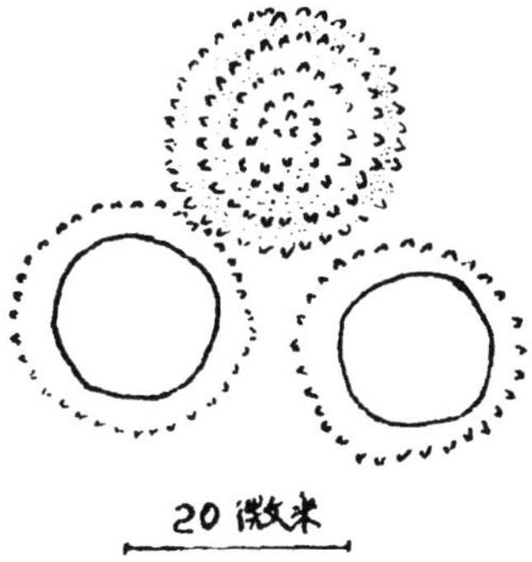

C

ABC 图 1. 黄花木夏孢锈菌　A. 夏孢子堆（ 电镜 660 倍 ）
B. 夏孢子壁疣放大（ 电镜 6600 倍 ）　C. 夏孢子

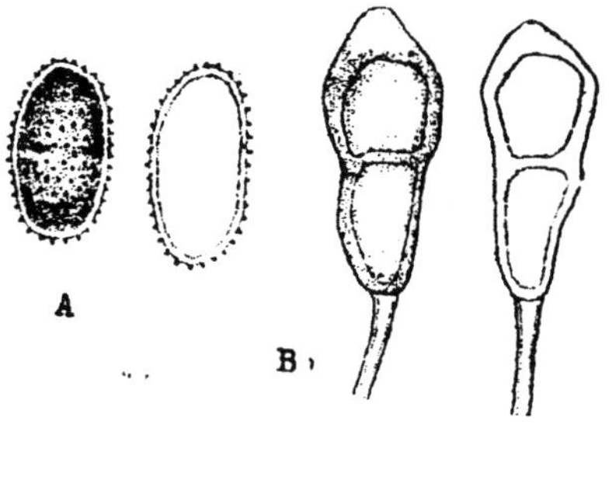

图 3.　A. 夏孢子（ 900 倍 ）
B. 冬孢子（ 900 倍 ）

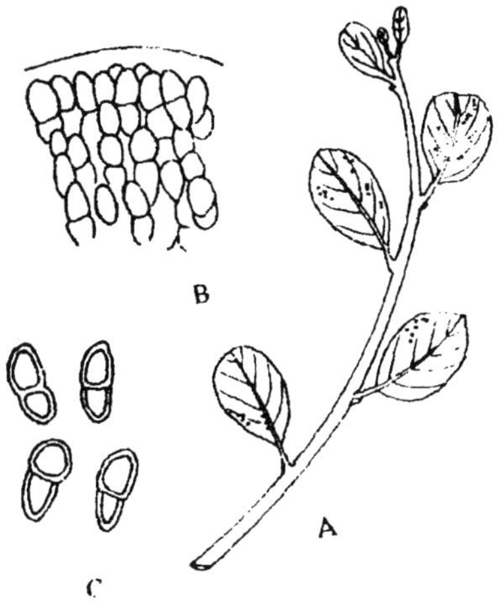

图 3.　中国鞘柄锈菌　A. 叶上冬孢子堆斑点
B. 冬孢子堆（ 130 倍 ）　C. 冬孢子（ 130 倍 ）

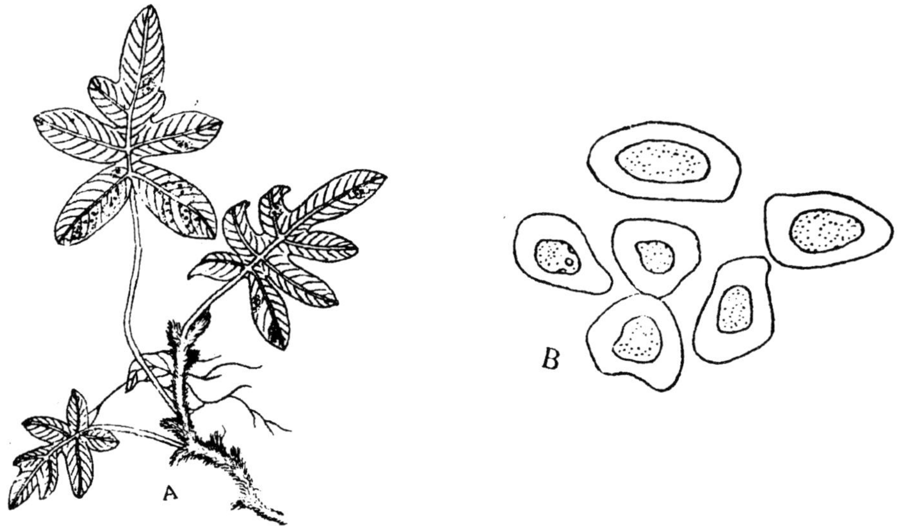

图 4. 日本明痂锈菌 A. 叶面上的夏孢子堆斑点（为原物之 2/3） B. 夏孢子（500 倍）

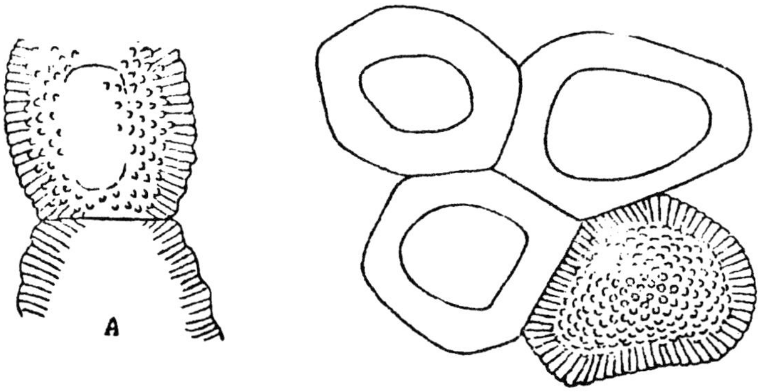

图 5. 长叶松鞘锈菌 A. 锈孢子器包被细胞 830 倍 B. 锈孢子（800 倍）

承董元同志描绘部分插图，魏江春同志校审拉丁文在此表示衷心感谢。

微 生 物 学 报 Acta Microbiologica Sinica 20(1):16-28, 1980

中国西部锈菌新种

王云章　韩树金　魏淑霞　郭林　　　　　谌谟美
(中国科学院微生物研究所，北京)　　(中国林业科学研究院林研所，北京)

1975—1978 年中国科学院综合考察委员会组织科学考察队，先后分赴西藏和新疆托木尔峰考察。真菌科学工作者臧穆、谌谟美、宗毓臣、廖银璋、卯晓岚、文华安、孙述霄分别参加考察。采得很多真菌标本，主要为伞菌、多孔菌、锈菌、小型真菌和土壤定居和暂居真菌，已写成考察报告另文发表，此文报告锈菌新种 15 种。其中在西藏地区采得 14 种，它们是 *Chrysomyxa stilbae* 寄生在杜鹃(*Rhododendron fulvum*)上；*Coleosporium zangmui* 寄生在齿冠草(*Myriactis nepalensis*)上；*Coleosporium lonicerae* 寄生在兰果忍冬变种(*Lonicera caerulea var.*)上；*Melampsora salicis-viminalis* 寄生在蒿柳(*Salix viminalis)*上；*Puccinia akebiae* 寄生在木通(*Akcbia quinata*)上；*Puccinia violae-reniformis* 寄生在肾叶堇菜(*Viola reniformis*)上；*Puccinia deyeuxiae-scabrescentis* 寄生在糙野青茅(*Deyeuxia scabrescens*)上；*Puccinia centellae* 寄生在积雪草(*Centella asiatica*)上；*Puccinia clintoniae-udensis Bub. var. tibetica* 寄生在西藏　七筋姑(*Clintonia udensis*)上；*Puccinia heterocoloris* 寄生在白花　苓草(*Morina alba*)上；*Peridermium yunshae* 和 *P*　*sinenses* 分别寄生在多种云杉(*Picea spp.*)上；*Uredo rhododen dronis* 寄生在照山白(*Rhododendron micranthum*)上；*Uredo pseudocystopteridis* 寄生在假冷蕨(*Pseudocystopteris tibetica*)上；而在新疆地区采得一种为 *Puccinia erigerontis-elongatae* 寄生在飞蓬(*Erigeron elongatus*)上。以上 15 种模式标本分别保藏在中国科学院微生物研究所标本室和中国林业科学研究院林业研究所标本室。

1.束丝梗金粘锈 新种 图 1—(3)

Chrysomyxa stilbae Wang, Chen et Guo sp. Nov.

Teliis hypophyllis, subhyalinis,0.3 —　0.5mm diam, aggregates, aurantiacis, maculis flavidis-brunneis insidentibus; capitulis teliosporarum sphaericis vel subsphaericis in mucum involutes, basis in pedicellum stilbiforme decurrentibus; pedicello 0.5-1mm alto, 0.2-0.4mm lato; teliosporis unicellulis, ellipsoidis, oblongis,15 — 43×8 — 15μm, pallide flavis,episporio 1μm crasso,ex　sporis 4—8，　plerumque 6 in catenas longitudinares 101—240μm longis compositis.

Host et Hab.: Rhododendron fulvum Bulf. Et W. W Sm., Xizang: Changdu, alt. 3600m, VI. 2. 1976, Zong Yu-chen et Liao Yin-zhang 226 (HOLOTYPUS).

冬孢子堆生叶下，橙黄色，半透明，0.3—0.5 毫米，群生，叶上呈黄褐色斑点，冬孢子结合为头状，外有粘液层，基部结为菌丝束状梗，长 0.5—1 毫米，大多为 0.2—0.4 毫米，冬孢子一室，椭圆形，长圆形，15—43×8—15 微米，浅黄色，膜厚 1 微米，由 4—8 个多数为 6

本文于 1979 年 10 月 8 日收到。部分新种标本由本院昆明植物研究所臧穆同志提供，由简荔、王碧莲同志协助描绘插图，特致谢意。

个组合为串，串长101—240微米。

寄主及产地：黄花杜鹃(*Rhododendron fulvum* Balf. f. et W. W. Sm.)，西藏：昌都 向达，1976 VI 2，海拔3600米，宗毓臣、廖银璋226(HMAS 38654主模式)。

Chrysomyxa stilbae 较接近 *Chr. succinea* (Sacc.) Tranz 冬孢子堆的形状相同，头状体下都具有菌丝束梗，但 *Chr stilbae* 冬孢子较长，15—43微米，冬孢子串长101—204微米，*Chr succinea* 冬孢子长14—27微米，冬孢子的大小区别较大。Balfour—Browne曾报道过西藏杜鹃(*Rhododendron vellerum* Hutch.)上冬孢子堆具有菌丝束梗可达2毫米长，定为 *Chr himalense*. 但此种冬孢子长22—28微米，孢子串长90—120微米，显然与 *Chr.stilbae* 在孢子大小上也有很大区别。

2 臧氏鞘锈 新种 图1—(5)

Coleosporium zangmui Wang et Wei sp. Nov.

Urediis hypophyllis, aggregates vel sparsis, rotundatis, subrotundatis, 0.5mm,diam., pulverulentis, flavor-aurantiacis; urediosporis subglobosis, ovatis,ellipsoideis vel pallide flavis,20—25×18—22μm，episporio 1μm crasso, poro germinationis obsolete vel nullis.

Tiliis hypophyllis, aggregates vel sparsis , rotundatis vel sub-rotundatis, ca 1mm dia. Sordide flavis; teliosporis ellipsoidis, cylindrico –clavatis, pallide flavidis 66—86×13—18μm, ad apicem valde incrassates (20—33μm), basi plerumque leniter attenuates.

Host et Hab: *Myriactis nepalensis* Less., Xizang: Jigung, IX. 5. 1976, Zang Mu 732 (HOLOTYPUS).

夏孢子堆生在叶下面，集生或散生，圆形，近圆形，直径0.5毫米左右，粉状，黄白色；夏孢子近球形，卵圆形，宽椭圆形，淡黄色，20—25×18—22微米，膜厚1微米，疣粗而长约1.5微米左右，芽孔不清。

冬孢子堆生叶下面，集生或散生，圆形，近圆形，1毫米左右(痂状)，土黄色；冬孢子棍棒状，淡土黄色，66－86×13—18微米，顶平厚达20—33微米，下圆或窄。

寄主及产地：齿冠草(*Myriactis nepalensis* Less)，西藏：吉贡，1976 IX 5，臧穆732(HMAS 38656主模式)。

3 忍冬鞘锈 新种 图1—(4)

Coleosporium lonicerae Wang et S. X. Wei sp. Nov.

Urediis hypophyllis, sparsis, rotundatis vel subrotundatis, pulverulentis, brnneo-flavibus vel interdum pallideflaviscentibus; urediosporis subsphaerosis, ovoides, ellipsoids, pallide-flavibus, 20—28×16—20μm., epispario ca 1μm crasso, verrucis 1.5μm longis ex parte sublevis ex sublevis ex parte poris germinationis obscures.

Teliis hypophyllis, aggregatis vel sparsis, minutis, epidermide fectis, plani vel parum convexi, eeracei, sordide flavis dein castaneo-brunneis; teliosporis cylin draceis, pallide-flavis, 50—96X15 25 rn, rotundatis, ad apicem 12 37 jim incrassa tis, episporio 1 jim erasso.

Host et Hab.: *Lonicera caerulca* L. var., Xizang Bomi, alt. 3900 m, VII. 23. 1976, Zong Yu-chen et Liao Yin-zhang 369 (HOLOTYPUS).

夏孢子堆生在叶下面，散生，圆形，近圆形，粉状，褐黄色，有时带白色；夏孢子近球形，卵形，椭圆形，淡黄色，20—28×16—20微米，膜厚约1微米，有疣，疣约1.5微米，粗而长，常常有部分膜光，芽孔不清。

冬孢子堆生叶下面，集生或散生，小形，在表皮下，腊质，平或略凸起，从暗黄色到栗褐色；冬孢子柱状，淡黄色，50—96×15—25 微米，两端圆形，顶厚 12—37 微米，膜薄约 1 微米。

寄主及产地：兰果忍冬变种(*Lonicera coerulea* L. var.)，西藏：波密林场，海拔 2900 米，1976 VII 23，宗毓臣，廖银璋 369(HMAS 38655 主模式)。

在 *Lonicera* 上只报告过 *Aplopsora* 属，这个属的夏孢子膜是刺而不是疣，冬孢子无色，顶不厚，与我们这份标本上的冬、夏孢子都不相符。

4 蒿柳栅锈 新种 图 1—(2)

Melampsora salicis-viminalis Wang et Quo sp. nov.

Urediis amphigenis, epidermide tectis minutis, 0.2 0.5mm diam., aurantiacis spare vel confluentibus; paraphysibus capitatis, pallide flavis 25 61X14 20 p,m, membrana 2 3 im crassa, ad ap 6 im erassa urediosporis ovatis, oblongo- ovatis vel ellipsoidS 15 25 x 10 14 lAm, pallide fiavis, echinulatis.

Teliis amphigenis, subepidermieis rotundatis, 0.2 — 1mm diam. Flavo-brunneis vel atro-brunneis, crustaceis, sparcis, confluentibus; teliosparis cylindri- cis, utraque rotundati.s vel kniter attennatis 43—71×8—12μm flavis, ad apicem fusco-flavis, non incrassatis, episporio 1μm crasso.

Host et Hab.: *Salix viminalis* L., Xizang Zuogong alt. 3800m, IX. 4. 1976 Zong Yu-chen et Liao Yin-zhang 521a (HOLOTYPUS).

夏孢子堆生叶两面，小圆形，0.2—0.5 毫米，桔黄色，痂状，散生或汇合；隔丝头状，淡黄色，25—61×14—20 微米，膜厚 2—3 微米，顶部有的增厚达 6 微米，夏孢子卵 形，长卵形或长椭圆形，15—25×10—14 微米，淡黄色，膜厚 1—1.5 微米，有细刺。

冬孢子堆生叶两面，表皮下，圆形，0.2—1 毫米，黄褐色或黑褐色，痂状，散生或汇合；冬孢子圆柱形，两端钝圆或稍细，43— 71×8—12 微米，黄色，顶部色深，膜厚 1 微米。

寄主及产地：蒿柳(*Salix viminalis* L.)，西藏：左贡，海拔 3800 米，1976 IX 4，宗毓臣、廖银璋 521a(HMAS 38658 主模式)。

此种接近 *M. larici-epitea* Kleb. 但是隔丝顶部厚 6 微米，夏孢子膜薄 1—1.5 微 米，冬孢子圆柱形，长而窄，(43—71×8—12 微米)而 *M. larici-epitea* Kleb. 隔丝 顶部有的可厚达 10 微米，夏孢子膜厚 1.5—3 微米，冬孢子较短粗，20—58×6.5—14 微米，由此而定为新种。

5 木通双胞锈 新种 图 1—(6)

Puccinia akebiae Wang et Wei sp. nov.

Teliis hypophyllis, petioli, caulicolis, aggregatis vel sparsis, rotundatis, pulveru lentis brunneis ve atro-brunneis; te sporis ellipsoideis, subelavatis, pallide flavescentibus, 53—75×10—2μm, apice rotundatis vel conico-rotundatis, non inerassatis, basi rotundatis vel attenuatis, medio levibus. episporlo tenu 1μm, constrictis, pedicello hyalino, 30—118μm longo, 2.5μm lato, deciduo germinatione post maturitem non dorrientibus.

Host et Hab.: *Akebia quinata* (Thunb.) Decne. Xizang: Bomi, alt. 2300 m VIII. 3. 1976, Zong Yu-chen et Liao Yin-zhang 422 (HOLOTYPTIS) alt. 2200 m 418 (PARATYPUS).

冬孢子堆生叶下面，叶柄及茎上，集生或散生，圆形，粉状，褐色或黑色；冬孢子棍棒状，淡黄色，53—75×10—12 微米，顶端圆形或圆锥形，下端圆或窄，分节处稍 缩，膜厚 1 微米，顶不厚，柄无色，长 30—118 微米，宽 2.5—4 微米，易脱落。孢子成 熟后即萌发。

寄主及产地：木通 [*Akebia quinata* (Thunb.) Decne.]，西藏：波密野贡，海拔 2300 米，1976. VIll. 3，宗毓臣、廖银璋 422 (HMAS 38661 主模式)。海拔 2200 米，1976. VIll 3，宗毓臣、廖银璋 418 (HMAS 38666 副模式)。

过去在木通上只报道过春孢子时期， 而冬孢子时期是第一次见到。

6 **肾叶堇菜柄锈** 新种 图 2—(7)

Puccinia violae-reniformis Wang et Wei sp. nov.

Urethis amphigenis, p hypophyllis rarius epiphyllis, sparsis vel aggregatis, minutis, pulveralentis, flavobrunneis; urediosporis subsphaericis, ellipsoideis, irregularibus, 25—33×21—28 rim, episporio 3—5μm scuiptura inconspicua verisimiliter aculeata poro germinationis 2, aquatuorem praeditio.

Teliis amphigenis, pleuromque hypo phyllis, sparsis wi aggregatis, minutis, pulverulentis, atris teliosporis el]ipsoidis, rubro-brunneis 30—43×20—27μm, utra pie rotundatis vel basi attenuatis, medio constrictis, apiee non cruss episporio 2.5—3.5μm crasso verruculo ad cellulam superain conspicua, pedicello hyalino brevi, decidue.

Host et Hab.: *Viola reniformis* Wall., Xizang Yadong Tanggepu, alt. 3320m VIII. 10. 1975 Zong Yu-chen 79 (HOLOTYPUS).

夏孢子堆生叶两面，以下面为主，上面稀少，散生或集生，小形，粉状，黄褐色，夏孢子近球形，椭圆形，不规则，25—33×21—28 微米，膜厚 3—5 微米，花纹不明显，少数孢子似有刺，芽孔 2 个，腰生。

冬孢子堆生叶两面，叶下面多，散生或集生，小形，粉状，黑色；冬孢子椭圆形，赤褐色，30—43×20—27 微米，两端圆或下窄，分节处稍缩，膜厚 2.5—3.5 微米，顶不厚，有沈，上室较明显，柄无色，短，易脱落。

寄主及产地：肾叶堇菜 *Viola reniformis* Wall. ）西藏：亚东唐喀铺，海拔 3320 米，1975 VIII. 10，宗毓臣 79 (HMAS 38664 主模式)。

此菌的冬孢子与其他生在本属寄主上两种近似种的区别见下表。

	夏孢子	膜厚	花纹	冬孢子	膜厚
P. aegroides Cumm.	23—34×14—21 微米	2 微米	刺	24—35×15—20 微米	1.5 微米
P. vjolae DC	21—26×17—23 微米		刺	20—35×15—20 微米	
P. violae-reniformis	25—33×21—28 微米	3—5 微米	不明显	30—43×20—27 微米	2.5—3.5 微米

7 **飞蓬双胞锈** 新种 图 2—(8)

Puccinia erigerontis-elongatae Wang et Han sp. nov.

Teliis hypophyllis, interdum amphi genis, sine maculis, rotundatis, sparsis vel confluentibus, pulverulentis, atro-brun- neis; teliosporis ellipsoideis, levibus 33—39×18—20μm, 2.7—5.0μm, apice 2.6—5.0μm incrassatis, leniter attenuatis, medio constrictis, pedicello hyalino, deciduo.

Host et flab.: *Erigeron elongatus* Ledeb., Xinjiang Tuo Mu Er Feng, alt. 3100m, VII. 18. 1977, Mao Xiao-lan et Wen Hua-an 81 (HOLOTYPUS).

冬孢子堆生叶下面，偶而也生上面或茎上，圆形常愈合，黑褐色，粉状，冬孢子椭圆形，顶端较厚，分节处明显紧缩，33—39×18—20 微米，顶厚 2.60—5.10 微米，柄无色常脱落。

寄主及产地飞蓬 *Erigeron elongatus* Ledeb. ）新疆：托木尔峰，海拔 3100 米，1977 VII 18，卯晓岚、文华安 81(HMAS 38663 主模式)。

此种与 *P. Dovrensis* Blytt. 虽然都寄生在菊科 *Erigeron* 属植物上，但它的最大不同点是孢子没有粗大的瘤，孢子顶端没有透明的乳头状突起，分节处紧缩也较明显，故定为新种。

8 糙野青茅双胞锈 新种 图 2—(9)

Puccinia deyeuxiae-scabrescentis Wang et Wei sp. nov.

Urediis epiphyllis, pleurumque in greges laxe despositis, ellipsoideis, oblong ellipsoideis 0.2—2mm diam., pulverulentis, flavor-brunneis; urediosporis globosis, subglobosis, pallede brunneo-flavidis, 18—25×18—23μm, epispono 1.5μm erasso, echinulato, poris germinationis numerosis, laxis praeditis.

Teliis pleururnque epiphyllis, totum folium oceupantibus, saepe etiam in vaginis culmisque evolutis, maculis nullis insidentibus, rotundatis vel subrotundatis, pulverulentis, atro-brunneis; teliosporis ellipsoideis vel oblongo-ellipsoideis.

Host et Hab. *Deyeuxia scabrescens* Griseb., Xizang: Changdu, alt. 3300m, IX. 22. 1976, Zong Yu-chen et Liao Yin-zhang 529 (HOLOTYPUS).

夏孢子堆生叶上面，多密布，椭圆形，长椭圆形，0.2—2 毫米，粉状，黄褐色；夏孢子球形、近球形、谈揭黄色，18—25×18—23 微米，膜厚 1.5 微米，细刺，芽孔多数，散生。

冬孢子堆以叶上为主，密布整个叶面，也生叶下面和叶鞘上，无叶斑，圆形，近圆形，粉状，黑褐色；冬袍子棒状，椭圆形，黄褐色，顶部色深，30—65×17—25 微米。 两端圆或顶部钝，下端窄，分节处紧缩，顶厚 5—10 微米，膜厚 1—1.5 微米，柄无色，可到 88 微米长，不脱落。

寄主及产地：糙野青茅(*Deyeuxia scabrescens* Griseb. ）西藏：昌都，海拔 3300 米，1976，IX 22，宗毓臣、廖银璋 529 (HMAS 38662 主模式)。

此号标本与生在 *Deyeuxia* sp.上的 *Puccinia changtuensis* Wang 比较，夏孢子比它小(27—31×27—30 微米)，膜比它薄 (2.5—5 微米)；冬孢子顶不尖(13—22 微米)。

9 积雪草柄锈 新种 图 3—(13)

Puccinia centellae Chen sp. nov Urediis hypophyllis; uredosporis ovoideis, 19—32×17—23μ, flavis, poris germinationis 2—3 in cellulis in medio dispositis, episporio 3—4μ crasso. Telils hypophylhs, arto-brunneis, 0.5 mm diam., teliosporis ellipticis 26—41×19—29μ longis apice rotundatis, basi rotundatis vel acuminatis, medionin constrictis, castaneo-brunneis, episporio 2.0—4.6μ crasso poris germinationis in cellulis superioribus vel inferioribus, pedicellis hyalinis, brevi.

Host et Hab.: *Centella asiaticci* (L.) Urban., Xizang: Yadong, alt. 2750 m, VIII. 14. 1975, Chen Mo-mei ia-100 (HOLOTYPUS) alt. 2800m, VIII. 29. 1975 Chen Mo-mel kan-30 (PARATYPUS).

夏孢子堆叶背生；夏孢子卵圆形，19—32×17—23 微米，淡黄色，腰部具发芽孔 2—3 个，膜厚3—4微米；冬孢子堆叶背生，深褐色，圆形，直径0.5毫米；冬孢子椭圆形或卵圆形，栗褐色，26—41×19—29微米，有微刺，顶端圆形，基部稍窄，中间隔膜无缢缩，膜厚2.0—4.6微米，芽孔在上面细胞的顶端部或者略靠下部，柄无色，长达29微米，易碎。

寄主及产地：亚洲积雪草[*Centella asiaticci* (L.) Urban.]，西藏：亚东，海拔2750米，1975 VIll 14，谌谟美ia-100(主模式)，海拔2800米，1975 VIll 29，谌谟美Kan-30(副模式)。

本种与*Puccinia angelicae*相近，但P angelicae冬孢子为长椭圆形或棍棒形，30—50×16—26微米，而本种冬孢子较短粗，卵圆形或椭圆形，26—41×19—29微米，同时本种胞膜也比较厚2.0—4.6微米，*P. angelicae*膜厚仅1.5—2.0微米，在积雪草属(*Centella*)尚未报道过锈菌，本种为第一次报道。

10 **西藏七筋始柄锈** 新变种 图3—(14)

Pucdnia cli ntoniae-udensis Bub. var. tibetica Chen var. nov.

Urediis hypophyllis, pallide flavis, ellipsoideis, longi-ellipsoidies; 26—34×17—21μ, tellis hypophyflis, aggregatis, 0.2—0.6mm diam., pulverulentis, arto brurineis; teliosporis oblongis vel rotunda- tis, 35—65×15—20μ membranis 1.5μ crassia, ad apicem 7—14μ crassis, pedicellis 64μ, longis, hyalinis vel pallidis.

Host et Hab.: *Clintonia udensis* Trauto et Mey., Xizang Luozha Shengere, alt. 3380 m, IX. 1. 1975, Chen Mo-mci Kan-20 (HOLOTYPUS).

夏孢子堆叶背生，夏孢子淡黄色，椭圆形或长椭圆形，26—34×17—21微米，具刺；冬孢子堆叶背生，集生，直径0.2—0.6毫米，粉状，深褐色，冬孢子长椭圆形或棍棒形，黄褐色35—65×15—20微米，两端狭或圆形或一端斜平形，膜薄1.5微米，顶部有的厚达7—14微米，柄无色或淡色，长达64微米。

寄主及产地：兰果七筋姑(*Clinotonia udensis* Trauto et Mey.)，西藏：洛扎申格热，海拔3380米，1975 IX 1。

本种比较接近*Puccinia clintomiaeudensis Bub.*，但这份标本的冬孢子比较长，可达35—65×15—20微米，而*Puccinia clintoniae-udensis*的冬孢子长为32—48×17—21，尤其顶端乳突状膜厚达7—14微米，而*Puccinia clintoniae-udensis*仅为3—6微米，本种与*Puccinia clintoniae-udensis*为同一种寄生植物兰果七筋姑，我们认为可以定为一个新变种。

11 **异色柄锈** 新种 图3—(15)

Puccinia heterocoloris Chen sp. nov.

Teiiis epiphyllis, tunicatis, cinnamo-meis, rotundatis 0.5 1 mm. diam., telio-sporis ellipticis, 36—45×21—23μ, cellulis superioribus burnneis, apice papillatis, inferioribus pallidis vel hyalinis, basi plerumqe leniter attenuatis, cellutis, duabus faciliter separatis, midio constrietis, pedicellis hyalinis 64 longis.

Host et flab. Morina alba Hand. Mazz., Xizang Yadong dongala. alt. 4100m VIII. 14. 1975, Chen Mo-mei, ia-69 (HOLOTYPTJS).

冬孢子堆叶面生，具膜质盖，肉桂色，圆形，直径0.5—1毫米；冬孢子长椭圆形，36—45×21—23微米，上面细胞褐色，顶部具乳突，下面细胞淡色或无色，向下变窄，两个细胞易分离，中间隔膜处显著缢缩，柄无色，长64微米。

寄主及产地：白花　苓草(*Morina alba* Hand　-Mazz　)，西藏：亚东东嘎拉，海拔4100米，1975　VIII　14。

过去在　苓草属(*Morina*)上未报道过锈菌，在已知 *Puccinia* 属未见过冬孢子的上面细胞和下面细胞颜色不同，而且胞膜厚度也不一致，可定为一个新种。

12　云杉被胞锈　新种　图 2—(10)

Peridermium yunshae Wang et Guo sp. nov.

Pycnidiis amphigenis, p hypophyllis, punctiformibus, sparsis, flu merosis, flavo-brunneis.

Aecidils hypophyllis, hemi-sphaerosis subroso-flavibus cellubis peridii irregula-ribus, 33—58×18—30 μm, subflavo-brun neis epispano 2—3μm, crasso, pariete exteriore subverrucoloso vel levi, pariete interiore verrucoso; aecidiosporis diversis, ellipsoides, ovoides. subsphaerosis vel ovoide-oblongis, pallide-flavibus 23—3×15—25μm, epispono ca 1μm crasso, verrucoloso.

Host et flab.: Picea sp., Xizang: Yadong Tanggepu, alt. 3300 m, VIII. 10. 1975, Zong Yu-chen et Liao Yin-chang 80 (HOLOTYPUS).

性子器生叶两面，主要在叶下，点状，　密集(系统发生的)，黄褐色。

春孢子器生叶上，半球形或半椭圆形，浅粉黄色，包被细胞不规则形，33—58×18—30微米，浅黄褐色，内外壁厚为2—3 微米，内壁有长疣，外壁有稀而小的疣。春孢子形状变化较大，椭圆形，卵形，近球形，长卵形，23—43×(—13)15—25 微米，浅黄色，膜厚 1 微米左右，疣长 1—2 微　米。

寄主及产地：云杉属之一种(*Picea* sp　)，西藏：亚东唐喀铺，海拔 3300 米，1975　VIII　10，宗毓臣、廖银璋 80(HMAS　38660 主模式)。

Peridermium yunshae 引起“丛枝病”，使受害叶子变成黑褐色，它较接近 Chrysomyxa arctostaphyli Diet　但 *Chr. arctostaphyli* 的春孢子器为疱状，春孢子为 23—35×16—25 微米，无色，疣稀而长，而 *Peridermium yunshae* 的春孢子器为半球形，半椭圆形，春孢子较大，23—43×(—13) 15—25 微米，浅黄色，疣密而短，两者显然不同。

13　中国被胞锈　新种　图 2—(11)

Peridermium sinenses Wang et Guo sp. nov.

Aecidiis amphigenis, pleurumque epiphyllis, oblongis, saepe aggregatis 6mm longis; peridio albo lacerato, cellulis peridii irregulariter oblongis, 58—80×23—40μm; aecidiosporis ellipsoidei, subsphaerosis vel sphaerosis, 40μm, aurantio-flavibus, episporio hyalino, 4—7.5μm crasso, dense verrucoso.

Host et flab.: *Picea brachytylci* (Franch.) Pritz, Xizang: Leiwuqi, alt. 3800—4200m, VIII, 1976, Zang Mu 574 (HOLOTYPUS).

Picea balfonriana Rehd. Et Wils., Sichuan: Jun Chuan, alt. 3000m, VIII. 23. 1977, Sichuan Lin Ke Suo $\frac{13}{1-2}$ (PARATYPUS)

Picea purpurea Mast., Sichuan: Hong Wei Ju 501 Chang, VII. 13. 1978, Chen Shou-chang 81 (PARATYPTUS).

春孢子器在当年生针叶两面，上面多，长形隆起，常汇集，长可达 0.6 厘米，包被膜质，白色，横向破裂、包被细胞不规则长形，58—80×23—40 微米。春孢子广椭圆形，近球形，球形，33—53×28—40 微米，内 含物桔黄色，膜无色 4—7.5 微米，疣密而短。

寄主及产地：麦吊衫[*Picea brachytylci* (Franch.) Pritz]，西藏：类乌齐，海拔 3800 一 4200 米，1976 VIII. I8，臧穆 574 (HMAS 38659 主模式)。

川西云杉[*Picea balfonriana* Rehd. Et Wils.], 四川：金川，海拔 3000 米，1977 VIll x，四川林科所 $\frac{13}{1-2}$ 号(HMAS 37741 副模式)。

紫果云杉（ *Picea purpurea* Mast. ） 四川（ 红卫局 501 场 ） 1978. VII. 13，陈守常 81 号（ HMAS 38665 副模式 ）

Chrysomyxa ledi de Bary 的春孢子小(20—34×15—23 微米)，膜薄(1.5—2.5 微米) *Chrysomyxa ledicola* Lagerh 的春孢子大(27—55×22—40 微米)，疣长(1.5—3.2 微米)。而 *Peridermium sinenses* 的春孢子大(33—53×28—40 微米)，膜厚(4—7.5 微米)疣短。是与 C. ledi, C. ledicola 不同的，由此定新种。

14 **照山白夏孢锈** 新种 图 2—(12)

Uredo rhododendronis Guo sp. nov.

Urediis hypophyllis, sparsis vel gatis, flavescentibus, 0.3 — 0.5mm diam., uredioporis oblongo-ovoidis, cylindrincis, ellipsoidis, parum subsphaerisis, leniter angularis, 26—40(-48)×13—19μm，epispono 1—1.5 dense verrucoloso utrinque 3—7μm incrassato.

Host et Hab.: *Rhododendron micranthum* Turcz., Xizang: Zuo Gong, alt. 4400m, VIII. 31. 1976, Zong Yu-chen et Liao Yin-zhang 502 (HOLOTYPUS).

夏孢子堆生叶下，疱状突起，散生或集生，浅黄色，O.3—0.5 毫米，夏孢子形状变化较大，长卵形，圆柱形，椭圆形，少数近球形，略带棱角，20—40(-48)×13—19 微米，膜厚 1—1.5 微米，两端增厚达 3—7 微米，近无色，密疣。

寄主及产地：照山白(*Rhododendron micranthum* Turcz.)，西藏：左贡，海拔 4400 米，1976 VIII，31，宗毓臣，廖银璋 502(HMAS 37834 主模式)。

Chrysomyxa rhododendri de Bary [*Chrysomyxa ledi* (A&S)]var *rhododendri* (DC) Savile 的夏孢子一般为椭圆形或近球形，略小，26—35×19—26 微米，两端不增厚，而西藏这号标本的夏孢子形状变化较大，孢子长形较多，20—40(-48)×13—19 微米，两端多增厚，HMAS 22107[*Chrysomyxa rhododendri* (DC) de Bary]。北京百花山黄安坨采的夏孢子形状是近球形，椭圆形，孢子大小为 20—24×15×20 微米，西藏的标本与它区别是较大的。

15 **假冷蕨夏孢锈** 新种 图 1—(1)

Uredo pseudocystopteridis Wang et Wei sp. nov.

Urediis hypophyllis, saepe secus nevos hospitis vel per aequaliter distributis, flavescentibus, rotundatis, pulverulentis, paraphysibus nullis; folia subtus urediosporis adhuc ignotis; amphisporis subsphaericis, ellipsoideis, piriformibus, saepe angulatis, flavovirentibus, 25—40×18—27μm, epispono ca 2.5μm crassato, subinde hinc inde usque ad 5μm incrassato, poris germinationis numerosis sparsis praeditis.

Host et Hab. *Pseudocystopteris tibetica* Ching., Xizang Bomi, alt. 2900m, VII. 16. 1976, Zong Yu-chen et Liao Yin-chang 314 (HOLOTYPUS).

夏孢子堆生叶下面，茎上，沿叶脉或密布整个叶面，圆形，近圆形，粉状，淡黄色；无隔丝；休眠夏孢子不规则有棱角椭圆形，近圆形，洋梨形，淡绿黄色，25—40×18—27微米，膜厚2 5微米左右，角处可到5微米，花纹不明显，芽孔多数散生。

寄主及产地：西藏假冷蕨(*Pseudocystopteris tibetica* Ching.)西藏：波密林场，海拔2900米，1976. VII. 16，宗毓臣、廖银璋314 (HMAS 38657主模式)。

这个种与*H. polypodii* (Diet.) Magnus有点相似，但*H. polypodii*的夏孢子生叶两面，夏孢子和休眠夏孢子同时存在，而这号西藏标本生叶下和茎上，叶下很严重但叶上未见孢子堆，也未见夏孢子，只有休眠夏孢子。

参 考 文 献

[1] Arthur, J. C. & G. B. Cummins: *Mycologia*, **25**: 397—406, 1933.

[2] Arthur, J C. & G. B. Cummins: Manual of the Rusts in the United States and Canada. Hafner Publ. Co. New York, 1962.

[3] Balfour-Browne, F: L.: *Bull. Brit. Mus.* (Nat. Hist) *Bot.*, **1**:189—218, 1955.

[4] Cummins, G. B.: The Rust Fungi of Cereals, Grasses and Bamboos. Springer-Verlag, New York, 1971.

[5] Hawksworth, D.L.: Mycologist's Hand book. Western Printing Services Ltd, Bristol. Britain, 1074.

[6] Hiratsuka, N. & S. Sato: *Trans. Mycol.* Soc. Japan, **10**(i):14—18, 1969.

[7] Ito, S.: *Mycological Plorci of Japan*, 2(2, 3) :1938, 1950.

[8] Mundkur, B. B.: Fungi of India. Supplement, T. Manager of publication, Delhi, India 1938

[9] Peterson, I S.: *Mycologia* 53:427—431, 1962

[10] Savile, D. B. 0.: *Can. J. Res.*, e28:318, 1950.

[11] Savile, D. B. 0.: *Can .1. Bot.*, 33:. 487—406, 1955.

[12] Sydow, P. et H.: Monographia. Tlredinear urn. **I**: 1004.

[13] Tai F. L.: *Farlowia*, 3(1) 95—139, 1947.

[14] Wang, Y. C.: 植物分类学报, **10**(4): 291—299, 1065.

[15] Wilson. M. & B. MBritish Rust Fungi, Cambridge Univ. Press, 1966.

NEW RUST FUNGI FROM WESTERN CHINA

Wang Yun-chan Han Shu-jing Wei Shu-xia Quo Lin

(Institute of Microbiology, Academia Sinica)

Chen Mo-mei

(Institute of Forestry, Chinese Academy of Forest Science)

In 1975-1978 the Commission of Integrated Survey of Natural Resources, Academia Sinica, organized two comprehensive scientific expedition trips to Xizang (Tibet) and Xinchang (Chinese Turkestan). Mycological workers participated the expeditions. Many specimens of fungi were collected and lists of fungi identified from these regions have been prepared and will be published elsewhere. Here in this paper are reported 15 new species of rust fungi collected from these regions. Among these new rusts some are evidently distinguished from those found in coastal regions.

Chrysomyxastibae Wang, Chen et Guo n. sp. E.g., has its telia very much alike those of authentic *Chrysomyxa.* It has, nevertheless, a stibum-like hyphae-bundle stalk somewhat like a sporophore at the base of the telial head. The hyphae-bundle stalk which measured from 0.5 to 1mm raises the telial head above the leaf surface. Balfour-Browne in a paper on Himalayan Fungi stated: "The teleutosori of *Chrysomyxa* spp. on Asiatic Rhododendron species form a series showing a gradual difference in size from the small, sessile *C. dietelii,* through the somewhat larger, still sessile, *C. taghishae* and the shortly stalked forms of *C. himalense,* from Simla and Nepal, to the large, longstalked Tibetian forms (Taylor 4324) of this latter species. Experimental work can only decide whether these forms do, in fact, represent distinct species or merely growth responses the different species of host plant." On reexamination of a specimen from Qinghai (HMAS 24398) named by the senior author as *C. rhodolendri* deposited in this Herbarium, the present authors found that it has also a stilbum-like hyphae bundle beneath the telial head. This Qinghai specimen coincident with *C. succinea* (Sacc.) Tranz., is therefore verified. The present new species from Xizang is further distinguished from other *Chrysomyxa* mentioned above in having larger teliospores and longer teliospore-chains.

As regard to the naming of the new fern rust, ***Uredo pseudocystopteridis,*** the authors have seriously considered Article 59 of the International Code of Botanical Nomenclature (1972). Having followed strictly this article, the authors have denominated it to the form genus Uredo. Otherwise, it might have been named under the generic name *Hyalopsora,* since this fern rust does have colored uredinial spores, readily distinguishable from the other two genera *Uredinopsis* and *Milesina* on ferns.

Spore state such as uredinial in the genus *Hyalopsora* which can be used to ascribe the organism to correct genus of perfect state should be legitimated and generic name of perfect state could be used in ease of the specimen not provided with telial state. Of course, it should be preferable to make a proposal to the Nomenclatural Committee of the International Botanical Congress and the International Mycological Congress for decision in the coming meeting.

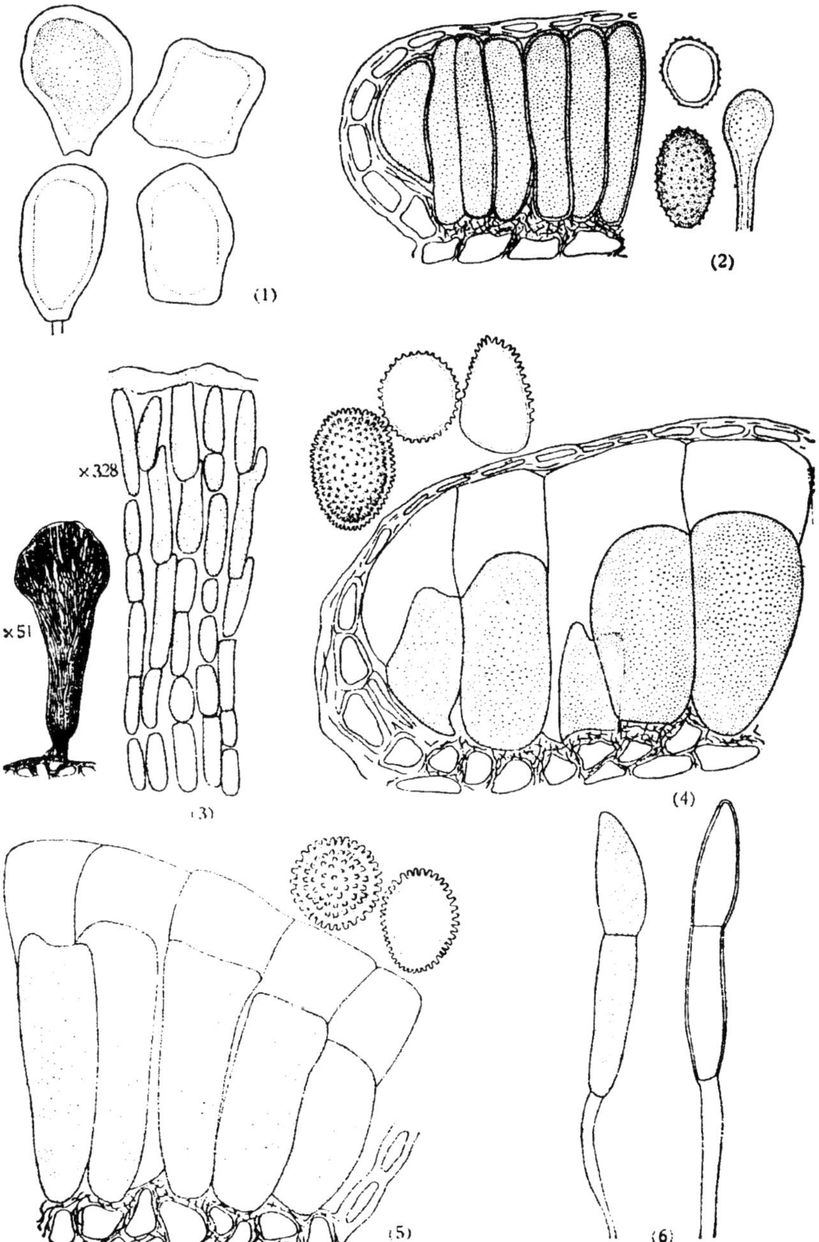

图 1

(1) *Uredo pseudocystopteridis* Wang et Wei
(2) *Melampsora salicis-viminalis* Wang et Guo
(3) *Chrysomyxa stilbae* Wang, Chen et Guuo
(4) *Coleosporium lonicerae* Wang et Wei
(5) *Coleosporium zangmui* Wang et Wei
(6) *Puccinia akebiae* Wang et Wei

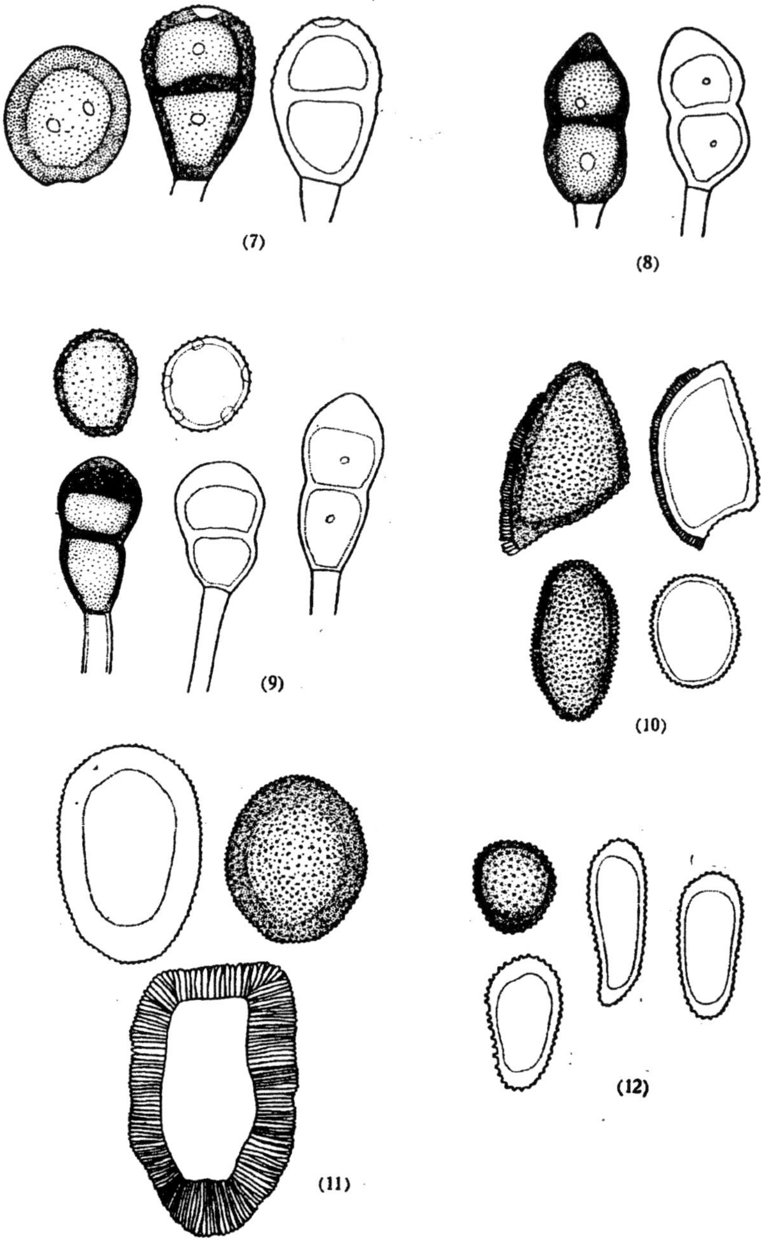

图2

(7) *Puccinia violae-reniformis* Wang et Wei
(8) *Puccinia erigerontis-elongatae* Wang et Han
(9) *Puccinia deyeuxiae-scabrescentis* Wang et Wei
(10) *Peridermium yunshae* Wang et Guo
(11) *Peridermium sinenses* Wang et Guo
(12) *Uredo rhododendronis* Wang et Guo

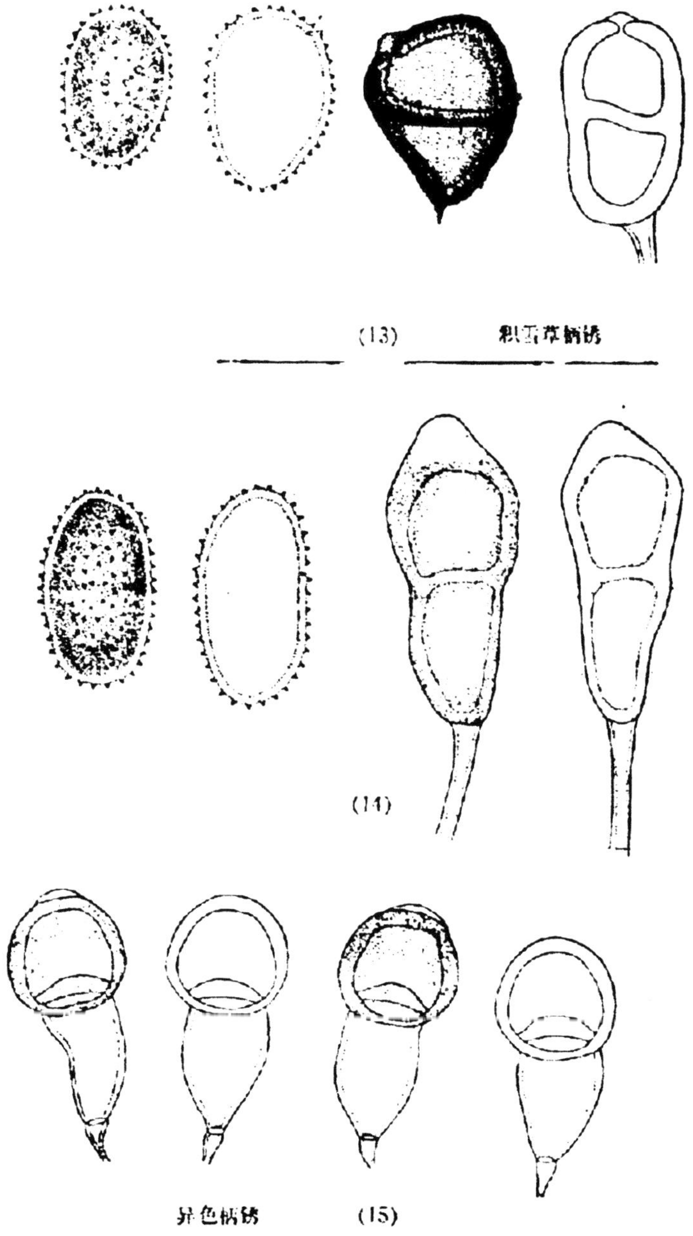

图 3

(13) *Puccinia centellae* Chen
(14) *Puccinia clintoniae-udensis Bub. vat. tibetica* Chen
(15) *Puccinia heterocoloris* Chen

The Polyporaceae Flora of Sino-Himalayas

The Polyporaceae is an advanced type of fungi. There are many large specimens of fungi which hymenophore is characterized by the presence of many pores of different shapes. They favour mild temperature and moisture.

Mycologists only knew little about the Himalayan pore fungi in the past. Fragmentary species were collected by Berkely, Balfour-Browne, and the Botanical Expeditions of University of Tokyo and the others. Table 1 shows various collections of pore fungi of the Himalayas.

In China the systematic study of forest pore fungi nearby Qinghai-Xizang Plateau and the Himalayas was made by S. C. Teng. Recently between 1974 and 1976 we joined Scientific Expedition of Qinghai-Xizang Plateau in areas 1,800-4,200 m above the sea level and collected four hundred specimens of more than 100 species of pore fungi found on the Pinus, Abies, Picea, Tusga, Quercus, etc. (Table 2) in the forests from quite different vegetation zones.

We identified and studied about one hundred species of pre fungi belonging to 21 genera. This study was based on the materials of paleo-forestry and the analysis of the population, composition and distribution of pore fungi of different species.

When the mountains were not too high, the Himalayan area might be rather warm and humid, covered with luxuriant broad-leaved forests. Here perhaps flourished with many pore fungi of ancient type, as the mountain continued upheaval and the temperature dropped down for the broad-leaved trees with pore fungi became more resistant to cold and dry weather.

Our findings have shown that there is a close relation between pre fungi and their hosts: the Phellinus pini var. abietis is inevitable on Picea (Pl. I, fig.1), the Ganoderma tsugae on Tsuga (P1. I, fig. 2), and the Piptoporus betulinus on Betula (P1. I, fig. 3).

The evolution of pore fungi from the lower to higher level may be divided into three types:

1. That of Hirshiporus etc, which is affiliated to the Hydanaceae constituting about 9%.
2. Typical pore fungi, such as Polyporus, Fomes constituting about 60%.
3. Those undergoing the transitional stage from pore fungi to pleat fungi; these constitute about 31%.

From this we can see that typical pore fungi form the majority. This is because with increasing aridity the old types of pore fungi are greatly decreasing and new types increasing.

Below the Himalayas the alpine zone is a completely different natural spectrum cut by deep south-north canyons. The south face of the Himalayas from the foothill to the forest line elevates so steeply that within a distance of only 20 to 40 kilometres its relative altitude is 1,000 to 5,000 metres above sea level. The high altitude of the Parlung Zangbo River valley clearly reveals the vertical change of the natural spectrum; the plants and the pore fungi at different altitude many be different in the types of forests and pore fungi. The forests and pore fungi below the subalpine zone are shown in Table 2.

From Table 2 we can see that the subalpine shrubberies and cushion plants (4,000-5,000 m) have no pore fungi; the fir forest (3,700-4,000 m) Phellinus hartigii (P1. I, fig. 4), the spruce forest (2,900-3,700 m) Phellinus pini var. abietis, the mixed pine and oak forest (2,700-3,700 m) --- the Phellinus pini grown in the pine forest and the Trametes cinnabarinus (P1. I, fig. 5), Ganoderma lucidum (P1. I, fig. 6) etc. in the oak trees; in the deciduous broad-leaved forests, streum spp. Are inevitable on the cornus trees.

Most importantly the pore fungi flora belonging to the N. temperate zone constitute about 90% of the total. As regards their geographical composition, many of them are of the type in areas which are closed to N. America and E. Asia, while other types (Table 3) are located only in the Himalayas and Japan.

The Pyrrhoderma adamantium (Berk.) and Daedaleopsis Purpurea (Cooke) are located only in the Himalaya and Japan.

The pore fungi flora of the Sino-Himalayas can be found in the following regions:

1. In the high mountains and gorges of S. Xizang the typical pore fungi of this kind are Polyporus cinnabarinus, Ganoderma oroflavum (P1. I, fig. 7). Ganoderma valesicum (P1. I, fig. 8), Poria cocos (P1. I, fig. 9), etc.
2. In the middle and lower reaches of the Yarlung Zangbo River the typical pore fungi of this kind are Inonotus dryadeus, Fomes fomentarius (P1. I, fig. 10), etc.
3. In the Hengduan Mountains region, the typical pore fungi of this kind are Phellinus pini var. abietis, etc.
4. In Central Xizang the typical pore fungi of this kind are Fomes robustus and Hirschioporus Pargamenus (P1. I, fig. 11).

In the studies of the ecology of Ganoderma we found six species varying from the frigid zone to the subtropical zone.

Table 4 shows the ecological environment of Ganoderma lucidum, which, as this table shows us, sometimes grows in clusters of hundreds of colonies; this is peculiar only to the Sino-Himalayas (P1. I, fig. 12).

REFERENCES

Berkeley, M. J. 1850-1854. Decades of fungi--- Himalayan fungi. Hooker's Journal of Botany, 2-6.
Balfour-Browne, E.L. 1955. F. L., Some Himalayan Fungi. Bull. Brit. Mus. Botany, 1 (7), 187.
Rokuya Imazeki. 1963. Fungi, Univ. of Tokyo, 611-626.

Taken from :
Proceedings of Symposium on Qinghai-Xizang (Tibet) Plateau (Bejing, China)
Geological and ecological studies of Qinghai-Xizang Plateau, vol II.
Science Press, Beijing 1981. Gordon and Breach, Science Publishers, Inc. New York 1981. pp 1173-1178.

Table 1. Pore fungi collected from the Sino-Himalayas

Collectors	Year of Collection	Altitude (m)	No. of genera	No. of species	Region of Collection
Balfour-Brown, E. L. (France)	1947 1952	2,500-3,900	11	20	Nepal; Nu nigaoa, Ohothas Lulokho; S. E. Xizang; Bomi
The Botanical Expedition of University of Tokyo (Japan)	1960 1963	1,600-2,600	14	21	Sikkim; Bakkim; Darjeeling; Baman, Takdah, Rayang
Qinghai-Xizang Plateau Scientific Investigation Team, Academia Sinica	1974-1976	1,800-4,200	21	102	S. E. Xizang:Qamdo, Gyirong, Zayu, etc.

Table 2. Pore fungi grown in the vertical forest spectrum of Parlung Zanggo River

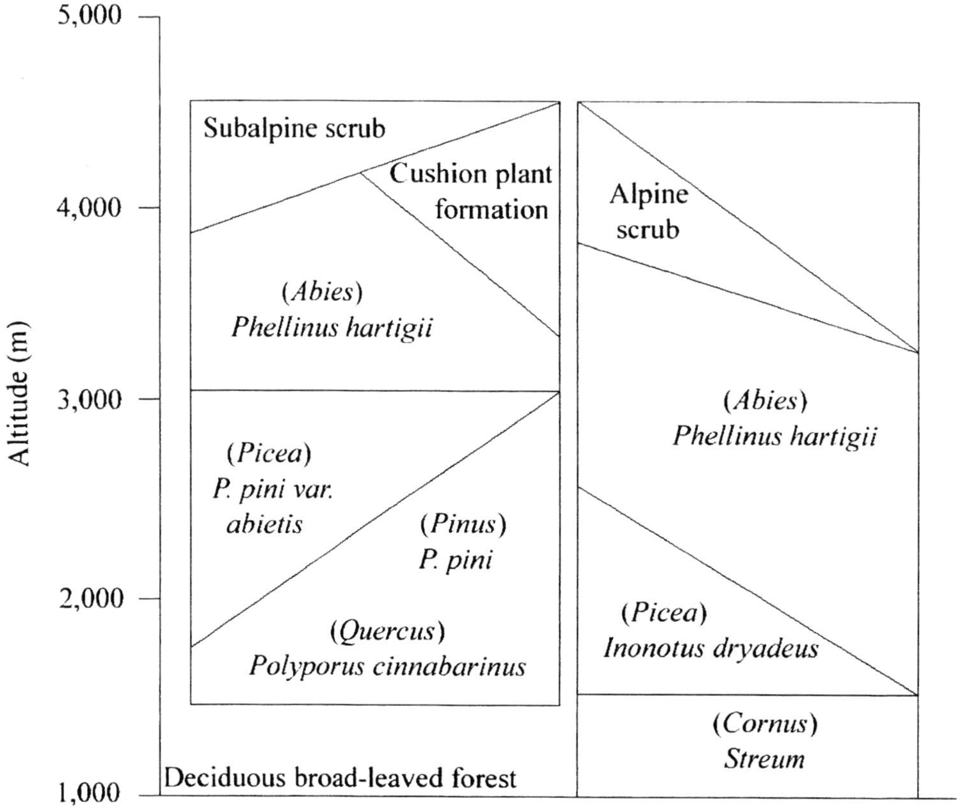

Table 3. Comparison between the common pore fungi of Sino-Himalayas and those of other regions

Pore fungi nam	Place of origin	Sino-Himalaya	Sino-Japan	N. America	Europe	India
Phellinus pini	Cosmopolotan	+	+	+	+	+
Ganoderma lucidum	N. temperate zo	+	+	+	+	+
Polyporus volvatus	N. temperate zo	+	+	+	+	-
Inonotus dryadeus	N. temperate zo	+	+	+	+	-
Pehllinus hartigi	N. temperate zo	+	+	+	+	-
Ganoderma oroflavum	Subtropical zon	+	+	-	-	+
Pyrrhoderma adamantium	Japan & Himala	+	+	-	-	-
Daedaleopsis purpurea	Japan & Himala	+	+	-	-	-

1. *Phellinus pini* var. *abietis*

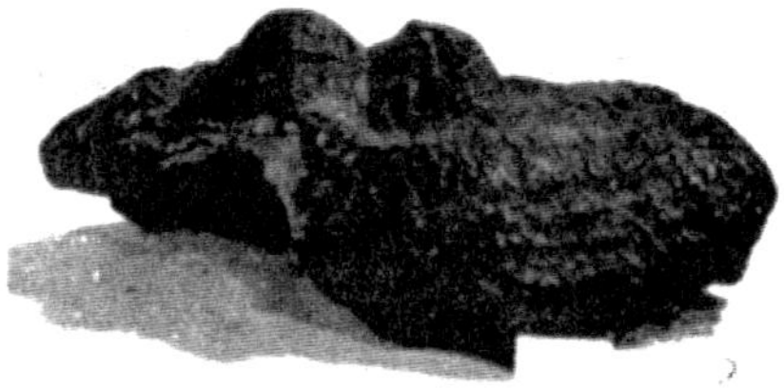

2. *Ganoderma tsugae*

3. *Piptoporus betulinus*

4. *Phellinus hartigii*

5. *Trametes cinnabarius*

6. *Ganoderma lucidum*

7. *Ganoderma oroflavum*

8. *Ganoderma volesicum*

9. *Poria cocos*

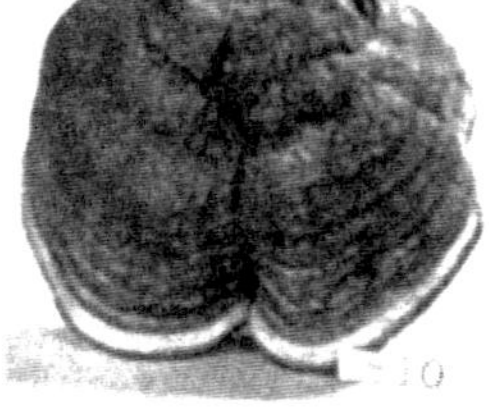

10. *Fomes fomentarius*

11. *Hirschioporus pargamenus*

12. Clusters of *Ganoderma lucidum*

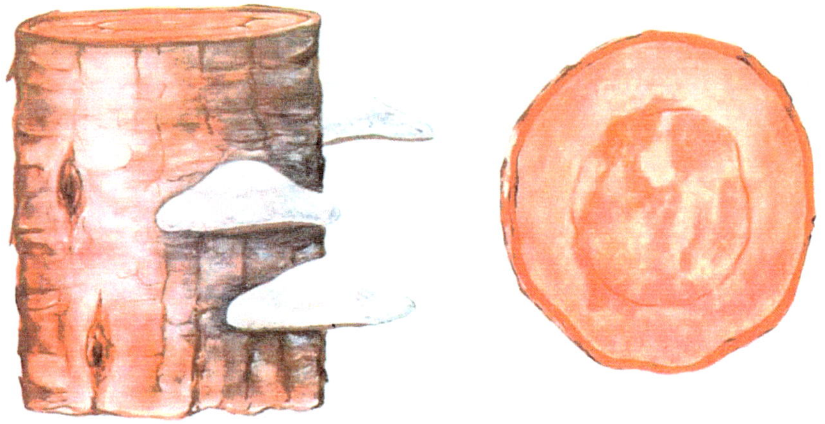

Piptoporus betulinus

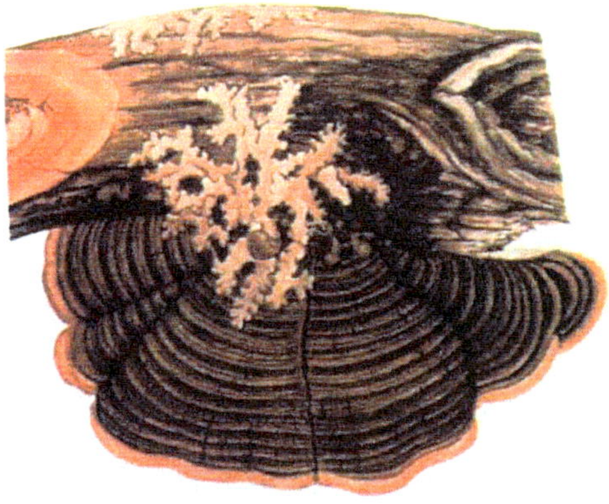

Phellinus pini var. abietis

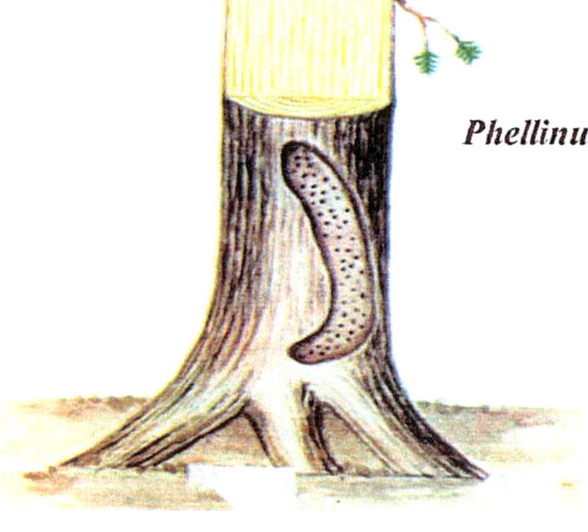

Phellinus hartigii

1984年 8月 林 业 科 学 SCIENTIA SILVAE SINICAE Aug. 1984

金锈科(Chrysomyxaceae)的一新属

——束梗锈属(*Stilbechrysomyxa* Chen gen.nov.)

谌谟美

(中国林业科学研究院林业科学研究所)

自 1840 年 Unger 创立金锈属 Chrysomyxa，之后，1890 年 Berkeley 记载了 Chrysomyxa himalense Bercley;1939 年 Tranzschel 记载了 Ch. succinea Tranzschel;1980 年王云章、谌谟美等记载了 Ch. stilbae Wang, Chen et Guo,至今该属共记载有 20 余种，笔者根据 1975 年在西藏昌都所采的 Ch. stilbae 标本鉴定，并与此属进行了详细的对比研究，认为上述 3 种锈菌与原金锈属各种基本特征不同，这 3 种菌的冬孢子堆都不是扁平垫状，也不半埋生于寄主组织中，而是以较长的束丝梗把冬孢子堆托举至寄主植物体外，而且形成胶质的头状冬孢子堆，因而另立新属束梗锈属 Stilbechrysomyxa Chen gen nov 来容纳该 3 种锈菌。文中还讨论了束梗锈属的分布及演化问题。

束梗锈属　新属

Stilbechrysomyxa Chen gen.nov. (=Chrysomyxa Unger Beit. Z. Vergl. Pathologie p.24, 1840)

Type sp.: *Stilbechrysomyxa stilbae* Chen

Soris teleutosporiferis captubiis, gelatinous, aurantiacis, subhylinis, globusis, vel subglobusis, stilbeus, hyaline vel aurantius;

Teliosporis laterali coacervatis, flavor-brunneoli cylindraceis vvel clavatis.

冬孢子堆头状体，胶质，橙黄色，半透明；圆形或半圆形，具菌丝束梗。无色或橙黄色；孢子单细胞，串生，黄褐色、长椭圆形或棍棒形(图 1、2、图版 I)。

束梗锈属为高山和亚高山地区特有属、共有 3 种，分布在中国(喜马拉雅地区和台湾)，苏联和日本。

束梗锈属与金锈菌届的区别，是冬孢子堆非埋生，而以束丝梗将冬孢子堆托举至寄主外，且带胶质呈头状，其区别如下：

冬孢子堆为垫状，基部不具菌丝束梗——*Chrysomyxa*。

冬孢子堆非垫状，基部具菌丝束梗——*Stilbechrysomyxa*。

本属 *Stilbechrysomyxa* 现有 3 种的主要区别如下：

感谢邵力平教授修改文稿。日本平冢直秀博士赠送标本及热情指导。

1. 冬孢子堆基部的菌丝束梗短，为 0.3 毫米——*Stilbechrysomyxa succinea*(图 2B)。
2. 冬孢子堆基部的菌丝束梗长度为 2 毫米 *Stil. Himalense.*
3. 冬孢子堆基部菌丝束梗长度为 0.7—1.1 毫米 *Stil. Stilbae*(图 2A，图版 I—1)。

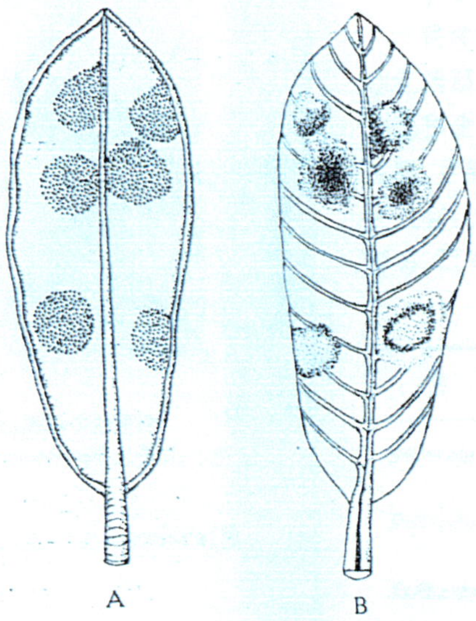

图 1 *Stilbechrysomyxa stilbae*
A.杜鹃叶背之冬孢子群（原大）B.杜鹃叶面斑点（原大）

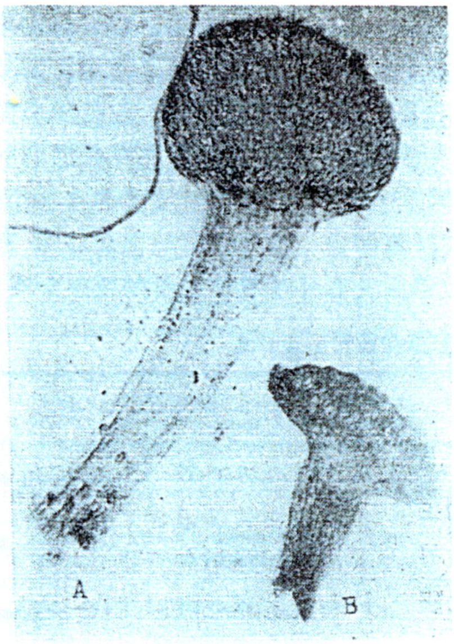

图 2
A. *Stilbechrysomyxa stilbae* 冬孢子堆；
B. *Stilbechrysomyxa succinea* 冬孢子堆

1. 喜马拉雅锈菌 Stilbechrysomyxa himalense (Bercl.) Chen Comb, nov.-Chrysomyxa himalense Bercl., in Sci. Mem. Med. Off. Army. Ind. N:83, tt I, 2 (1890).

分布：中国的西藏东南，Kongbo. Molo. 3150 米，寄生于 *Rhododendron vellereum* Hutch; 1838.5. 19. Ludlo, Shcrriff & Taylor 4234. 尼泊尔的 Chankheli 与 Darma 之间，3300 米，寄生于 *Rhododendron* sp. 1952. 5. 20., Polunin, Sykes & Williams 4140。

2 琥珀束梗锈菌 Stilbechrysomyxa succinea (Tranz.) Chen Comb. Nov.-*Chrysomyxa succine*a (Sacc.) Tranzschel, conspect, Ured. URSS. P. 314 1939; Hiratsuka, f., 1944,1960; Kuprevicz et Tranzschel, 1958。

分布：欧洲，苏联（勘察加岛，萨哈林岛）日本（千岛、北海道、本州中部）和中国（青海、台湾）之亚高山地带。II、III 寄生于 *Rhododendron aureum* Georgi; *Rhododendron brachycar pum* D. Don; *Rhododondron brachycarpum* var. roseum Koidz.; *Rhododendron metternichii* Sieb. et Zucc. Var. *pentamerum* Maxim.; *Rhododendron morii* Hayata; *Rhododendron pseudo-chrysanthum* Hayata。

3 束梗锈菌 stilbechrysomyxa stilbae Chen Chen Comb. Nov.-*Chrysomyxa stilbae* Wang Chen et Guo in Acta Microbiologica Sinica, Vol. 20 (1); 16—28, 1980, *Rhododendron campanulatum* D. Don. Changdu, 3900 米 1976. 6. 2. M. M. Chen, Chang—26 (Para Typus)

分布：中国西藏昌都 3900 米，寄生于川西云杉 (*Picea likiangensis* (Fr.) Pritz. Var. balfouriana) 林下的 *Rhododendron fulvum* Balf. F. et W. W. Sm. 和 *Rhododendron campanulatum* D. Don. 上。

关于束梗锈菌属分布与演化的讨论。

许多例子证明可用真菌的研究来确定其寄主发展史，同时寄主植物的地史资料又是研究锈菌发生演化的基础。假定杜鹃与金锈菌是相伴演化的，青藏高原大幅度的隆起使杜鹃和金锈菌同样起着剧烈地分化作用。杜鹃是典型的高山植物，为北温带分布式样的大属，主要分布中心在我国的西南地区，缅甸的北部以及向西延至东喜马拉雅。南美和非洲没有杜鹃分布，所以没有此菌的分布。金锈菌科完全按照它们的寄主杜鹃分布的格局分布，已知世界产金锈菌的地区几乎均分布北温带（ 中国、日本、苏联 ）尤其在我国（ 云南、四川、西藏等高原地带 ）最为集中，其中锈菌的冬孢子堆具束丝梗的特殊形态的种类均集中分布

三种束梗锈菌的地理分布

菌　　名	寄　主　植　物	菌丝束梗长度（　毫　米　）	海　　拔（ 米 ）	地　　区
Stilbechrysomyxa succinea	Rhododendron morii	短　柄 0.3	亚 高 山	中国、日本
Stilbechry. stilbae	Rh. Fulvum Rh. Campanulatum	长　柄 0.7— 1.1	3900—4100 米 高　原	中国西藏，青海
Stilbechry. himalense	Rh.vellereum	长　柄 2.0	3150—3300 米 高　原	中国西藏东南部，尼泊尔

与青藏高原。据地貌、沉积和动植物化石资料推测，当时康滇古陆地区气候比较暖湿，由于造山和冰川运动形成的垂直分布的明显化和相互隔离的地貌，南北走向的山脉阻隔着印度洋暖湿气流，使这里成为一些原始类型锈菌的天然“避难所”，保存了原来暖湿条件下生长的冬孢子堆为半埋生状态的金锈菌 *Chrysomyxa*。但是由于山脉不断地隆起，引起了高原环境的剧烈变化，由暖湿气候转向寒冷、干旱的趋势。这些变化很可能使导致金锈属 *Chrysomyxa* 中的一些类型，由较原始的具半埋生垫状冬孢子堆向被菌丝束梗托起具胶质的冬孢子堆演化的因素，这说明，束梗锈菌 *Stilbechrysomyxa* 是在高原（ 高山和亚高山 ）条件下，近代演化发生的类型。因此，将其定为新属，亦属必要。

(*Stilbechrysomyxa* Chen gen. nov.)

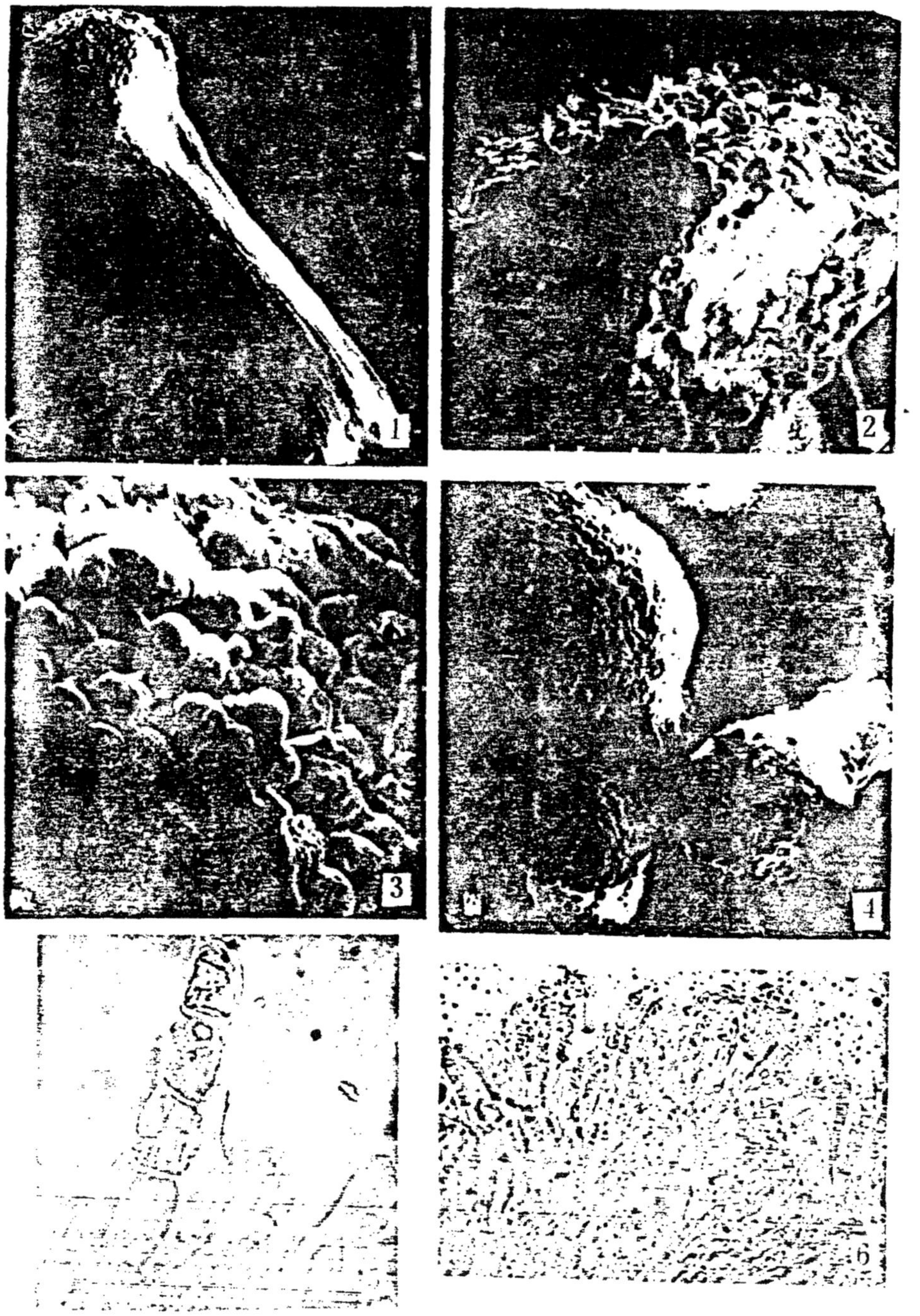

Stilbechryson stilbae 1. 单个具菌丝束梗之冬孢子堆，90×; 2.冬孢子堆头状体放大，375×; 3.带胶质冬孢子堆放大，750×; 4.三个冬孢子堆头状体，120×; 5.冬孢子串生单细胞，2250×; 6.冬孢子串生，600×。

The forest diseases and insects of the Tibetan Plateau and their integrated pest management

PREFACE: The Tibetan (Qinghai-Xizang) Plateau, known as the "world's roof," contains a treasury of information for the natural sciences. Yet, over the long years in this plateau, research was approached by only a few scientists, and it remained enshrouded in mystery. Not content with what has been known, scientists both in China and abroad have long cherished possibility of bringing this hidden treasury of information to light. A serious effort was started in China in the early 1950's to study the plateau, and in the years that followed, seven large-scale scientific expeditions to Tibet were organized by Academia Sinica, with more than 1,400 scientific workers participating. The effort was amply rewarded. Not only were many significant findings made, but research on a wide range of disciplines was done. As a result, knowledge about the plateau attained a new level.

In 1975 and 1976, the forests and economic forest diseases of the main forest distribution areas of Tibet were systematically researched. The region of research was located in the natural and man-made forest regions of the southern and southeastern parts of the Tibetan Plateau, the south and north slopes of the east side of the Himalayan Mountain range, the east side of the Nian-qing-tang-gu-la mountain range, the north side of the Heng-duan mountain range, and along the Ya-lu-zang-bu, Nu, Lan-cang, Jin-Sha, and Cha-yu River banks. Approximately between north latitude 27 – 31 degrees and east longitude 85 – 98 degrees, the research area reached Jiang-da in the east, Ji-long in the west, north of Chang-du in the north, and Cha-yu, Ya-dong, Nie-la-mu in the south.

The research on the forest diseases was combined with the research on the forests. It also collaborated with fields such as agricultural diseases, plant pathology, zoology, entomology, mycology, botany, meteorology, soil science, and natural geology. By using the point and surface method, following the horizontal and vertical distribution of plants for field research, and setting up sample areas in representative forests, the ecological and geographical characteristics of the forests and diseases were recorded and samples of fungi, insects, and parasitic plants that affect the forests were systematically collected. The total journey of the research followed a line of about 5,000 kilometers, of which the research on the forest disease was 231 kilometers, with 124 sample areas. The investigations were done in 26 main natural forest ecotypes, 62 plantation tree ecotypes, and 31 nurseries, fruit gardens, and tea plantations. Occurrence, distribution, identification, and development of the diseases that affect these Tibet forest regions were determined. A few prevention and cure experiments were carried out with the help of local production units. More than 850 diseases and fungal specimens and 120 insect specimens were collected. Identification of them was made either in the field or at the Chinese Academy of Forestry, Chinese Academiy of Sciences and Beijing Agricultural University, with involvement of many experts in forest pathology, microbiology, and entomological laboratories.

After the field work, the information collected was sorted and primarily analyzed (1977 – 1980). Besides specific analysis of the diseases that occur in each forest type and ecosystem, the distribution, occurrence, and regeneration of the Tibet forest region as a whole was also analyzed, dividing it into different floral and geographical areas (map VI-1-1). The characteristics of the forest diseases of each

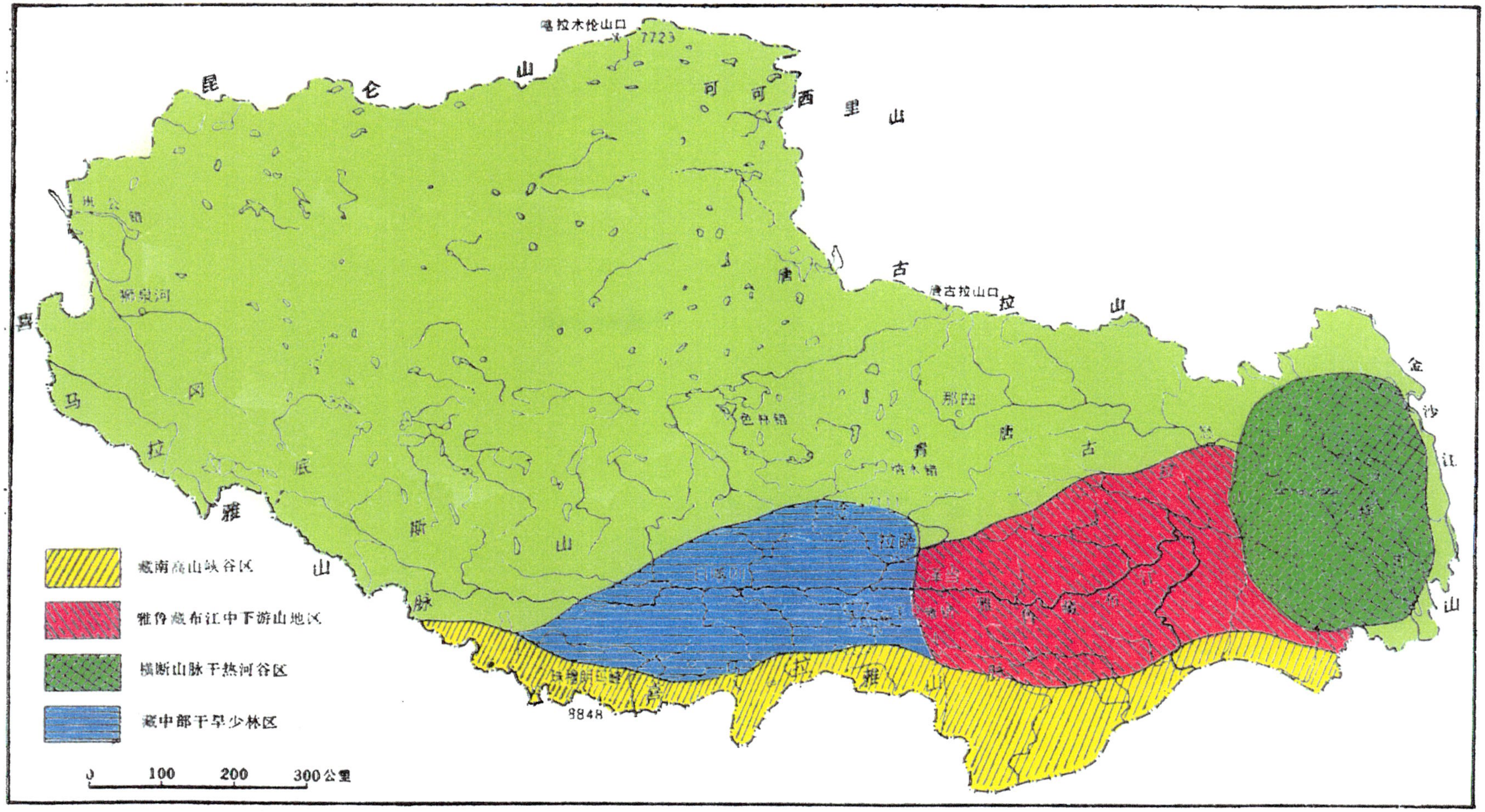

图 VI-1-1 西藏森林病虫害地理分区图

forest type

and region were described, and ways of prevention were suggested. On this basis, comparative analysis of the occurrences and effects of the main forest diseases in Tibet and similar forests in other main forest regions of China were performed. Also, the connection to similar flora of the earth was examined, therefore producing theories and support for the control of the spreading of forest diseases. According to the research, because of the unusual natural environment of the Tibetan Plateau, not only do old species survive here, new diseases and pests also occur here which are closely related to the forest diseases of the flora of China as well as some other regions of the world. At the same time, even though there are many types and variations of forest diseases in Tibet, under the condition that the ecological balance of the virgin forests are kept intact, none have caused major damage, showing the healthy characteristic of this forest ecosystem. The breaking of this balance can cause the spread of the diseases. Therefore, forest disease and insect management were suggested to be best approached by integrated forest management and especially prevention of pest introduction from nurseries and plantations. Finally, an index of the main diseases and pests of the Tibet forests were listed.

1 DESCRIPTION OF TIBET FOREST DISEASES

1.1 *Diseases of the Pine Forest*

Distributed in Tibet are Yunnan pine (Pinus yunnanensis Franch.), high mountain pine (P. densata Mast.), Himalayan pine (P. griffithii McClelland), Armand pine (P. armandii Franch.), Tibet long-leaved pine (P. roxbourghii Sarg.), and Tibet white-bark pine (Pinus gerardiana Wall.).

Because the ecosystems of the forests are different, the distribution of the diseases is different too. In the dry-heat ecotype, Tibet white-bark pine diseases are rare, while Himalaya pine and Armand pine have many diseases and fungi often seen in the north temperate zones. Automatically, the pine species differentiate into hard pines and soft pines, and the rust disease types and distribution are different too. The pine forest diseases and pest problems of the Tibet area are fundamentally similar to the pine tree diseases of other areas of China. Here we present descriptions of some representatively important diseases and insects, such as decays, stem rusts, needle rusts, and bark beetles.

1.1.1 *Pine Decay Disease*

The main pine wood decay fungus that attacks pine trees is white pocket rot (*Phellinus pini* (Thore ex Fr.) Ames) (Picture VI-1-1). It is a commonly seen polyporaceaous fungus of the north temperate zone pine and larch forests, and in this area mainly occurs in high mountain pine [common here] forests and Yunnan pine forests. Because Armand pine, high mountain pine [low here, see last comment] and Tibet long-leaved pine occur as younger trees, this disease is rarely seen on this pine species in this region. White pocket rot, also called pine stem decay, it is a disease of pine forests worldwide. Because it causes substantial loss of strength in the tree's heart wood tissue; it is a very important forest disease. The distribution and incidence of infection depend on the age and humidity of the pine tree's ecosystem. Yunnan pine usually starts decaying in the V age level of lumber table (Table VI-1-1). The decay rate of high mountain pine and Yunnan pine in the XI age level of lumber table can reach 5 – 6 %. White pocket rot has a wide range in its temperature requirement. It can survive in low temperature (-60 degrees Celsius) ecosystems, but it's humidity requirement is very specific. Comparing Yunnan pine, Red pine (*P. koraiensis* Sieb. Et Zucc.), and Chinese pine (*P. tabulaeformis* Carr.) of the same age, higher humidity Yunnan pine and Red pine types are much more susceptible than Chinese pine types. This is because only when the comparative forest humidity level is maintained above 80% can the basidiospores germinate and infect through wounds. The order of infected white pocket rot occurrence and forest humidity factor of the Tibetan pine forests (from highest to lowest) is: Yunnan pine, high mountain pine, Armand pine, Himalayan pine and Tibet long-leaved pine, Tibet white-bark pine. Because the humidity is low for Tibet white-bark pine forest ecostystem, white pocket rot was very rarely found there.

Table VI-1-1. Three pine's white pocket rot rate of Tibet

Tree Type	Number of Trees Investigated	Decay Rate At Each Age Level (%)				
		III (60-80 yrs)	IV (81-100 yrs)	V (101-120 yrs)	VI (121-140 yrs)	XI (200+ yrs)
Yunnan Pine	98	0	0.03	0.3	0.3	6.1
Tibet Long-leaved Pine	268	0	0	0	0	< 0.5
High Mountain Pine	539	0	0	1.4	1	5

White porcket rot initially causes about ground decay, but the decay may extend 3 –5 meters below the ground-level in advanced stages in old forest ecosystems. The decay rate of high mountain pine forests is the highest, because its habitat is at high altitude and it often suffers from cold damage. Also, most high mountain pine forests have reached over-maturation. Decay in over-mature pines is also serious in flat riverbank areas (Table VI-1-2). It is less severe in slope areas because those areas are dryer.

Table VI-1-2. Comparison of decay rate of high mountain pine on slopes and near riverbanks

Geography	Soil moisture	Altitude (m)	Age (years)	Number Investigated	Decay Rate (%)
Slopes above 28 Degrees	Dry	3200	160 - 180	335	28.5
Flat Riverbanks	Damp	3200	200	174	60

Other wood decay fungi that are parasites of the Tibet area pine forests include *Fomitopsis pinicola* (Sw. ex Fr.) Karst., *Polystictus pergamenus* Fr. (Picture VI-1-2), *Coriolus versicolor* (L. ex Fr.) Quel., and *Gloeophyllum saepiarium* (Wulf. Ex Fr.) Karst. Medical fungus (*Cryptoporus volvatus* (Peck.) Hubb. can often be seen on the trunk of the trees of the Yunnan pine forests after a fire. Conifer trunk fungus (*Phaeolus schweinitzii* (Fr.) Pat. grows on the bottom trunk part of the Armand pine trees. The saprophytic fungi in Himalayan pine forests includes *Gloeophyllum subferrugineum* (Berk.) Bond. & Sing., *G. saepiarium* (Wulf. Ex Fr.) Karst., *Trametes abietina* Pilat, and *T. cervina* (Schw.) Bres. Common Decay Tibet long-leaved pine forests include *Fom decay. pinicola* (SW. ex Fr.) Karst., and *G. saepiarium* (Wulf. ex Fr.) Karst. etc.

Besides wood decay pathogens, there are also some fungi in the pine forests that are considered to be favorable for medical purposes. One is fu-ling (*Poria cocos* Fr.Wolf (picture VI-1-3)). It is a specialty of the Yunnan pine forests, and has effects such as benefits for spleen and urine, and for calming nerves or relive uneasiness. Cultivation of fu-ling is preformed in pine forests that have cool temperatures and high altitudes. The Yunnan pine of the Cha-yu region of Tibet is one of the regions that has a high fu-ling production as a medicinal forest product, and optimum production per tree can be around 25 kilograms. Table VI-1-3 shows the comparison of the fu-ling produced by each type of tree.

Table VI-1-3. Comparison of fu-ling production of a few types of pine trees

Distribution Area	Tree Type	Altitude (m)	Production (kg per tree)	Sclerotiform Diameter (cm)
Cha-yu, Tibet	Yunnan Pine	2400	25	30 – 40
Hu-bei	Horse-tail Pine	400	5	10
An-hui	Huang-shan Pine	1000	5	10

The fu-ling sclerotiform grows on the roots of pine trees, and can also decay the roots. In cultivation we should change the way it is collected to achieve the goal of preserving the tree as well as collecting the fu-ling so as to protect the natural resources.

1.1.2 *Pine Stem Rust Diseases*

Tibet pine forest rust diseases are typical of the rust flora of the north temperate zone. They are generally distributed on the north side of the East Himalayas, south of the Lian-qing-tang-gu-la Mountain range, The Gong-bu-jiang-da County and Lang County, and the goerge of the south slope of the Sino-Himalayas (Tiben, China). This includes small Jilong of Jilong county, Jiangchun, Zhang-mu of Nie-la-mu, Lixin goerges, Za-mu, Qin-duo, Yi-gong, and Dong-jiu of Bo-mi county. Bo-mi Lin-zhi Counties are representative of those areas, with an annual average temperature of 8.4 – 11.4 degrees Celsius, annual rainfall of 650 – 960 mm, altitude of 1800 – 3600 meters, and warm, humid weather. Here, there are mainly mountainous semitropical – temperate zone pine forests. The main dominant trees include Himalayan pine, Yunnan pine, high mountain pine, Armand pine, and small long-leaved pine. It has very diverse plant types, and is the region with the most forest rust fungi types in the Sino-Himalaya flora. There are some rust fungi which are commonly seen in the north temperate zone, such as *Coleosporium* and *Cronartium*. Their aecial stages are pathogens of Pinus generation after generation that aecia are produced perennially from the same infection. Some other commonly seen north temperate zone rusts include white pine blister rust (*Cronartium ribicola* Fischer), birch leaf rust (*Melampsoridium betulinum* (Desm.) Kleb.), and coniferous needle rust (*Coleosporium saussureae* Tnum). More than half of the host plants of the above rust fungi and the autoecious plants (of a rust fungus:completing its life cycle on one host also can be called heteroecious) are in large families common in the north temperate zone. They are widely distributed in the Tibetan Plateau (such as *Rosaceae, Ranunculaceae, Cuculidae, Compositae, Gramineae, Salicaceae, Coniferae*). This not only reflects characteristics of the Himalayas, but also proves that the Chinese Himalayas are typical of the north temperate zone, an area naturally prone to the spread of rust diseases.
Note: The above political divisions, mountain ranges, and rivers are named by Pinyin (The official Chinese spelling approved by the United Nations)

Table VI-1-4. Pine stem rust diseases of the Tibet area

Disease	Host Plant		Alternate Host
	Five Needles Per Bundle	Two, Three Needles Per Bundle	
White Pine Blister Rust	Armand Pine		Ribes *
	Himalayan Pine *		Ribes*
Pine-Oak Rust		Yunnan Pine *	High Mountain Oak
Pine- Peony Rust		Same As Above	Peony

* Rust fungi found in Tibet area

Pine stem rusts seriously affect the pine forests. According to the rust fungi flora research of the Tibetan Plateau, the three main pine stem rust diseases of our country have all been found in Tibet (Table VI-1-4). Of these the most harmful is white pine blister rust, which affects Himalayan pine and Armamd pine the most. There are also pine-oak (*C. quercuum* (Berk) Miyabe) and pine-peony (*C. flaccidum* (Alb. et Schw.) Wint.) rusts which infect hard pine trees. White pine blister rust is a fatal disease of five-needle pines. According to recent research, white pine blister rust occurs in Armand pine in the southwest region, and might have jeopardized the planting of five needle pines on a large scale. The seriousness of white pine blister rust must be emphasized because Himalayan and Armand pine are the main forestation trees of the southwest region, and are also the most widely distributed valuable tree types of the Himalayas. To prevent the occurrences of white pine blister rust, we first need to know the geographical distribution of the five needle pine type and the ecological factors of Ribes and Pedicularis, so as to be able to mark out epidemic areas and protection areas, and to carry out reasonable forestation plans. There are 11 types of five needle pine species in China (Table VI-1-5). Most of the five needle pine forests contain the rust alternate hosts as indigenous species, and the rusts are present (Table VI-1-5).

Table VI-1-5. The distribution of epidemic diseases in our country

Tree Types	Distribution Areas	Affected Areas	Degree of Spread *
Red Pine (*P. koraiensis*)	Ji, Hei (Liao)	Hei, Liao, Ji	+++ (small trees 30 – 70%)
Xinjiang Five Needle Pine (*P. sibirica*)	Xinjiang (A-er-tai)	Xinjiang	++
Yan Pine (*P. pumila*)	Ji, Hei	0	0
Armand Pine (*P. armandi*)	Shan, Jin, Yu, E, Chuan, Gan, Qing, Tibet, Dian (Gan),Qian	E, Chuan, Shan, Dian, Qian	++
Taiwan Armand Pine (*P. armandi var. mastersiana*)	Taiwan	0	0
Da-bie-shan Five Needle Pine (*P. dabeshanensis*)	Southwest of Wan, East of E	0	0
Hainan Five Needle Pine (*P. fenzeliana*)	Hainan, Gui, Gui (Xiang)	0	0
Himalayan Pine (*P. griffithii*)	Tibet, South and North of Dian	Tibet	+
North America Eastern White Pine (*P. strobus*)	Xiong-yue, Lu-shun, Nanjing, Beijing Planted	0	+++
Taiwan Five Needle Pine (*P. morrisonicola*)	Taiwan	0	0
Huanan Five Needle Pine (*P. kwangtungensis*)	Yue, Gui, Xiang, Qian	0	0
Five Needle Pine (*P. wangii*)	Dian Planted	0	0

- Degree of Spread: + 5–10% individually distributed; ++ 10-40% medially distributed; +++ above 40% widely distributed (planted tree types in parentheses).

According to published reports, white pine blister rust mostly occurs on pine trees 1 –3 years after forestation. Before the aecial stage matures, the infected tree has no obvious symptoms. During autumn, in areas with the autochthon alternate hosts, telia grow on the underside of the leaves in the shape of short hairs. The teliospore geminate and produce masses of very thin, walled basidiospores. Basidiospores are dispersed by wind or flowing air and land on the needle surface. Infections grow through the needles and into the bark tissue of young pine trees. This process normally requires 1 – 2 years. The infected area of the bark starts to swell, and, at the end of the next autumn and beginning of winter, aecial peridia forms in the swollen area. The next spring, blisters form where the pycnia were, and the peridium breaks and emits many orange colored aecispores. Once the blister rust canker encircles the sick tree, the tree starts to die. Aeciospores are carried by wind to infect Ribes leaves. The orange-colored summer spores stage, Urediospores, are produced on the lower surface of the leaves. During the summer, they are regenerated continuously. When the weather cools in autumn, telia are produced from the old uredia. The Teliospores is a sexual stage. Each teliospore can produce basidiospores. The basidiospores infect the pine again, thereby, completing the life cycle.

Because of favorable natural ecosystem factors, and the adaptability and the good health of the young Himalayan pines, naturally regenerated young Himalayan pine trees in Tibet have a low infection rate. But because this area has warm - cool temperature and the humidity is suited for the occurrences of the basidiospores, a reasonable plan needs to be designed when planting five needle pine forests. Here are a few suggested ways to manage white pine blister rust:

1. Favor Armand pines in Cha-yu, Bo-mi, and Lin-zhi at altitudes of 2700 – 2900 meters; when planting Himalayan pines in Ji-long, Ya-dong, Nie-la-mu, La-kang, Cuo-na, Luo-yu, Bo-mi, Yi-gong, and Mo-tuo at altitudes of 1100 – 3300 meters, plants such as Ribes and other alternate host plants should be carefully removed.

2. Once white pine blister rust has been discovered, the infected area should immediately be marked out and managed to eliminate the disease source.

3. Mark out Himalayan pine and Armand pine distribution areas and protection areas, and disallow any other five needle pines to be planted in these areas.

4. Prevention and cure experiments have shown that, in infected areas, eliminate infected trees to destroy the disease source, and apply a bark paint (bu to fen) before the maturation of the pycnia (has some preventative effects).

Two needle and three needle stem rust diseases (pine-oak rust and pine-peony rust) have been found on Yunnan pine trees in the Tibet area. The former is a commonly-seen disease of the pine forests of China, and is most widely distributed in Massonia pine (*Pinus massoniana* Lamb.), Mongolia pine (*P. sylvestris* var. *mongolica* Litv.), and Yunnan pine forests, which are seriously infected. According to research, Yunnan pine forests at a high altitude in this area have a lower infection rate compared to other pine forests in China (Table VI-1-6).

Table VI-1-6. The pine-oak rusts of a few pine tree types

Host Plant	Region	Altitude (m)	Alternate Host	No. of Trees Investigated	Infection Rate (%)
Massonia Pine (P. massoniana)	Jiangsu	50	Quercus variabills Blume		30
Mongolia Pine (P. sylvestris var. mongolica)	Heilongjiang	800	Q. mongolica	815	42
Yunnan Pine	Yunnan	1400	High mountain oak	165	10 - 30
Yunnan Pine	Tibet	2400		98	5

Figure VI-1-6

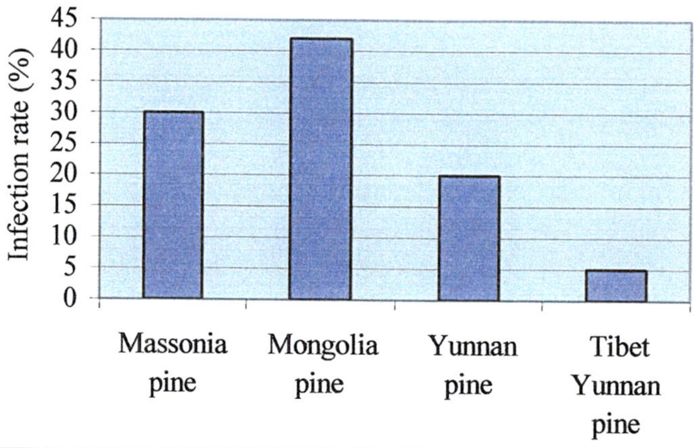

Pine-oak rust is an obligate parasite. The alternate host is high-mountain oak (*Quercus semicarpifolia* Smith). According to observation in the Tibet Yunnan pine forests, the gall rust of the Yunnan pine forms a gluey substance in pyenia in early spring. In the beginning of May, orange-colored aecia with a thin, cerebroid membrane develop on the gall surface. About 10 days later, the aecial membrane breaks, and the aeciospores spread and infect the lower surface of high-mountain oak leaves forming uredia in the shape of yellow dust. After August, dark brown telia shaped like short hairsform on the leaves. Round or ellipse shaped tumors form on the infected pine tree branch or stem. A big gall can reach 70 – 80 cm, with a rough surface and craked bark. Inside, it is filled with pitch. Serious gall rust can cause tree top decline or death of the whole tree.

Pine-oak rust is a common disease of the mixed mountain forests of pine and oak. Usually, natural forests are not seriously damaged, but key source of basidiospores for future pine plantations. According to research in the Tibet area, the occurrence of the disease is related to the density and age of the forest. When forest density degree is less than 0.4 or the trees are over-mature and weak, the pine-oak rust occurrence clearly increases. For prevention and control refer to white pine blister rust.

Pine-peony rust (*C. flaccidum*) is only found in rare cases in this area. it causes a fusiform-shaped gall on the trunk. For prevention and cure refer to white pine blister rust.

1.1.3 *Pine Needle Rust Diseases*

Needle rust of pines is a main disease of young pine tree forests. Four types have been found in this area. The uredial and telial stages of these rusts often occur on *Composita, Labiatae, Ranunculaceae*, and grasses. (Table VI-1-7).

Table VI-1-7. Pine needle rust diseases of tibet

Name of Disease	Host Plant	Alternate Host	No. of Trees Investigated	Infection Index (%)
Tibet Long-leaved Pine Needle Rust (*Peridermium complantum*)	Tibet Long-leaved Pine * (*Pinus roxbourghii*)	Not Sure	268	23
Pine Needle Rust (*Coleosporium ligulariae*)	Himalayan Pine (*P. griffithii*)	*Ligalaria* *	140	32 Plant
C. saussureae	Same As Above	*Saussurea* *	281	30 Plant
Yunnan Pine Needle Rust (*C. asterum*)	Yunnan Pine (*P. yunnanensis*)	*Aster* *	196	10 Plant
Himalayan Pine Needle Rust (*P. brevius*)	Himalayan Pine * (*P. griffithii*)	Not Sure	813	47

* Confirmed in Tibet

There are five pine needle rust diseases known in other parts of China. Most often seen are Mongolia pine needle rust, Chinese pine needle rust, Yunnan pine needle rust, and Himalayan pine needle rust. The seriousness of the infection depends on the plant density of the plant or the humidity of the micro-environment (Table VI-1-8).

Table VI-1-8. A few pine needle rust diseases

Name Of Disease	Weather Zone	Warmth Index	Rainfall (mm)	No. of Trees Investi-gated	Infection Index (%)
Mongolia Pine Needle Rust (Coleosporum pulsatillae)	Cold temperate zone (Da-xing-an-ling Geng River)	26 – 63	300 - 500	451	65
Himalayan Pine Leaf Rust (Peridermium brevius)	Temperate Zone (Small Jilong, Jilong County)	25 –110	500 - 900	813	47
Tibet Long-leaved Pine Leaf Rust (P. complanatum)	Semitropical Zone (Jiang village, Jilong county)	77 – 160	> 800	268	23
Chinese Pine Leaf Rust Disease (*C. campanula* f. sp. *adenophorae*)	Temperate Zone (Qinling, Shanxi)	25 – 90	600 - 1000	117	25
Yunnan Pine Leaf Rust Disease (*C. asterum*)	Temperate Zone (Lijiang, Yunnan)	Annual Avg. Temp. + 12 – 16 Degrees Celsius	1000	345	30

From the table, it can be seen that each weather zone has its own unique needle rust type. The fungus typically requires ecological factors that are similar to those of its aecial and telial host plants. For instance, *Peridermium complanatum* (picture VI-1-4) and Tibet long-leaved pine trees are both distributed in the semitropical zone, as opposed to *C. pulsatillae*, which requires a humid and cool type of micro-environment

Research results from Tibet long-leaved pine forests and their diseases in the Jilong area (Table VI-1-9) show that Tibet long-leaved pines grow healthily, their trunks are straight and whole, and they rarely have flaws or rusts. However, when extablishing a young forest of Tibet long-leaved pines, we still need to pay attention to the distribution and effects of the long-leaved pine leaf rust disease.

Table VI-1-9. Ecological factors associated with long-leaved pine leaf rust in Tibet

Eco-Type of Forest[1]	Altitude (m)	Slope	Composition	Age (years)	Height (m)	Diameter (cm)
G/ TLEP Forests	2260	NE 33 degrees	6 Long-leaved pine, 4 Himalayan pine	46	16	32
TLEP/AB Forests	2300	Southwest 30 degrees	10 Long-leaved pine	60	28	42

Density	M^3/hectare	No. Investigated	Infection Rate (%)	Rate of Spore Spread	
				Shrub	Herbaceous
0.5	200	270	40	Sol[1]	Cop[3]
0.5	600	178	5	Cop[3]	Sol[1]

Notes: 1. G/TLEP=Grass-Tibet long-leaved pine forest; TLEP/AB=Tibet long-leaved pine-arrow bamboo forest; 2. Vegetation density: Sol→ Cop; 1-3=Lower vegetation density---→ higher vegetation density Sol: Vegetation density less covered; Col 1-Cop2-Cop3: Vegetation more high density; Sp: middle;

After initial research, the rredial or telial stages of Tibet long-leaved pine needle rust have not been found. The aecial stage mainly infects needles of young trees. The infection rate of the needles of older trees can reach about 50%. The infection rate of the young trees is related to vegetation cover. The Grass-Tibet long-leaved pine forests have less vegetation cover (spread is Sol), and are humid, helpful to the spread and germination of the aeciospores, so they have a higher infection rate. On the other hand, Tibet long-leaved pine forests are densely populated with bamboo (spread is Cop[3]), preventing the aeciospores from spreading, so they have a lower infection rate.

Research on Himalayan pine forest diseases (Table VI-1-10) show that Himalayan pine needle rust is widespread in Himalayan pine forests, and Himalayan pine has the highest infection rate of the five-needle pines. Needle rust mainly depends on air movement and vegetation density. Forests with low tree density have more air movement and high humidity, and consequently have a high infection rate. On the other hand, forest types with dense shrubs have a lower infection rate. In the spread of needle rust, the effect of reduced aeciospore dispersal is very clear.

Table VI-1-10. Qiao pine needle rusts under each forest type

Forest Type	Alti-tude (m)	Composi-tion	Age (year)	Height (m)	Dia-meter (cm)	Class of Forest Ssoil
Less-Grass/Himalayan Pine Forest	2620	10 Himalayan Pine	40	23	21	Above Ic
Shrub-Grass-Himalayan Pine Forest	2600	10 Himalayan Pine	49	30	25	Ia
Xian He Grass/ Himalayan Pine Forest	2700	10 Himalayan Pine	57 30	31	61 12	Above Ic
Himalayan Pine Natural Young Forest	2760	10 Himalayan Pine	5	12		Ie

Forest Type	Density	M³/ hectare	No. Investi-gated	Infection Rate (%)	Rate of Spore Spread	
					Shrub	Herbaceous
Less-Grass/Himalayan Pine Forest	0.8	500		47	Sol[1]	Cop³
Shrub-Grass Himalayan Pine Forest	0.8	426	140	25	Sol	Cop³
Xian He Grass/Himalayan Pine forest	0.7	405	215	40	Cop³	Sol
Himalayan Pine Natural Young Forest	1.0		201	52	Sp	Cop³

The following prevention methods should be used against pine needle rust s.

1. Research on the types and distribution of the host and alternate host plants of every kind of needle rust disease.
2. Before planting a young plantation, pay attention to removing the alternate host plants of the rusts that infect Himalayan pine, Yunnan pine, and Tibet long-leaved pine s.
3. In areas of young forests that are seriously affected, spray 65% Zineb (Shi 1979 P. 445) and 70% thiophanate methyl (Shi, 1979 P.445)

Reference for Fungicide: D. Z. Shi 1979. Dictionary of English-Chinese Agricultural Pestcides

1.1.4 *Mistletoe*

Mistletoe (*Arceuthobium*) is a high phanerogam plant that is a parasite on pine trees and cedar trees in China. It belongs to the family *Loranthaceae* and sub-family *Viscocideae*. *Arceuthobium* is a small shrub with yellowish or brownish, fragile, joined stems, and with the segments glabrous and often +/- four-angled. Leaves are Decussate and reduced to connate scales. Flowers are solitary or have several forms. The berry is fleshly, compressed, and dehiscing on a short recurved pedicel. Infected branches are deformed and swollen, often become dry and rotten.

There are three types of mistletoe known in China, and they are parasitic on almost all of the pine and cedar species of Tibet. In the Plateau region, this disease occurs only on pine and cedar trees. In this area, host trees include high-mountain pine, Himalayan pine, Tibet long-leaved pine, and Armand pine. (Table VI-1-11) (Picture VI-1-6).

Table VI-1-11. A few mistletoe plants of Tibet

Mistletoe Plant	Characteristic of shape	Host Plant	Altitude (m)	No. Investigated	Parasite Rate (%)	Distribution Area
Arceuthobium chinese	Fruit has short stem, plant faded yellow-green colored	Himalayan Pine	2500 - 3200	1167	43	Small Jilong, Jilong county, Ya-dong, La-kang, Luo-zha county
A. chinese	Same As Above	Tibet Long-leaved Pine	2000 - 2400	268	20	Jiang village, Jilong county
A. pini	Fruit has no pedicel, plant brown colored, scarce branches,1-2 merous, fruit fusiform-shaped	High-mountain Pine	2800 – 3700	335	61	Lang county, Jia-cha, Mi-lin
A.oxycedri	Fruit has no stem, plant brown colored, dense branches, 3 – 4 merous, branch thick, fruit ellipse-shaped	High-mountain Juniper	2500 - 3000	126	34	La-kang, Luo-zha county

Of the infected forests, the high-mountain pine forest with a high altitude has the highest infection rate. From the view of flora, high-mountain pine mistletoe and long-leaved pine oil fir mistletoe are parasitic plants belonging to two different ecosystems. The former is mountain type, mostly distributed in temperate zone forest regions of the middle and downstream Ya-lu-zang-bu River. The latter is closer to the semitropical zone, mostly distributed in the lower part of the valley region of south Tibet. (We found Mistletoe most in conifers specie)

1.1.5 *Fallen-needle disease*

Fallen-needle disease is caused by *Lophodermium pinastri.* Himalayan pine and Tibet long-leaved pine are most commonly affected in this area. The infection rate of Himalayan pine is 52%. Long-leaved spruce infection rate is 72%. This disease is a pathogen of many kinds of pine needles. It is known that Chinese pine, Armand pine, Massonia pine, Japanese red pine, black pine, white-bark pine (*P. bungeana* Zucc. ex Endl.), Mongolian pine, and Yunnan pine of China all can be infected. The ventilation of the pine forest is closely related to the breakout of the disease. The obvious characteristic of this disease is the needles turn red-brownish like they are burnt. The fallen-needle disease is dormant in the winter in fallen leaves, the next spring when rainfall is abundant and the forest is humid, the ascospores are blown to and infect through the stomata of the needles. According to research in the Himalayan pine forest of the canyon south of Tibet, the disease often breaks out after about a 2 month incubation period. Because the Himalayan pine forests are warm and humid, this disease occurs often. To make man-made pine forests, attention should be paid to the prevention of the fallen-needle disease. In the nursery or man-made young forests, eliminate the source of the disease by spraying lime and sulfur or a Bordeaux mixture on the young seeds and trees.

1.1.6 *Pine Forest Insects*

According to research results, large-scale leaf-eating pest problems have not been found in the natural pine forests of this area. Problems such as bark beetles have been found in each forest region because of fire, denudation, and grazing. For instance, there is the southern pine beetle (*Dendroctonus frontalis*) that attacks high-mountain pine, Yunnan pine, and Himalayan pine. The Yunnan bark beetle (*Dendroctonus sp.)* also attacks the high-mountain pine. Both attack the thick part of the bark and the trunk of the tree. According to observations in the southeast regions of Tibet, adults appear approximately between the end of May and June. There is also the spruce hairy bark beetle that affects the thin part of the bark, often seen on the treetop. This is a wide spread type of bark beetle of pines of the Himalayas, affecting high-mountain pine, Yunnan pine, Aramand pine, and often seen in the valley forest areas of Jilong. There is also De-chang root bark beetle (*Hylestes techangensis* Tsai et Hwang) that attacks Himalayan pine and Tibet fir. It mainly aggressively attacks the thick part of the bark and the tree trunk areas. Its body is elongated and nearly black. It is distributed in Lijiang, Yunnan and the Yunnan pine forests of Chayu (2300 meters), and Tibet. Also, there is the largest stem bark beetle (*Hylurgops major Eggers*) that is easily confused for this bark beetle's character. Its main characteristics are a thick and strong body shape, near brown, unevenly sized dots on the pronotum, a gall on the sunken area of the proala crustacea, and little hair on the proala crustacea. These attack the thick bark areas too, but they have different shapes from the large stem bark beetle.

The following pests also exist in the natural forests: Tibet aphis (*Cinara tibetapini* Zhang) of the Himalayan pine (Jilong) and high mountain pine aphis (*C. paxilla* Zhang) of the high-mountain pine. These suck the juice inside the needles, causing the needles to blight.

There are also Chinese pine needle scale (*Matsucoccus sinensis* Chen) (Jilong 3300 meters) and pine white powder scale (*Crisicoccus pini*) (Yadong 3100 meters) of the Himalayan pine forests. Both are specific to the high mountain pine forests. These possess characteristics such as high adaptability to the Plateau and high reproduction ability. They should be carefully quarantined to prevent future breakouts in forest regeneration.

1.2 *Fir Forest Diseases and Insects*

Tibet fir forests possess many kinds of trees and vast resources, but the decay rate of fir forests is high, and decay can be serious. This becomes a clear problem for managing and utilizing fir forests. According to research in the Tibet fir forests, large-scale leaf-eating pest problems usually occur in some areas as an addition to these diseases.

2.1 *Fir Decay Disease*

According to research on the decay rates of a few types of fir, when the tree age is 160 – 200 years, the decay rate is commonly 40 – 80 % (Table VI-1-12), exemplifying the high decay rate of the fir of China. The decay rate of the Min River fir (*Abies faxoniana* Rehd. et Wils. the Maerkang region of Sichuan is the highest, that of the stinky fir (*A. nephrolepis* (Trautv.) Maxim.of Heilongjiang is second. The fir forests in the Tibet area are lower than the northeast and southwest regions, but the overall decay rate is still high. The decay rate of the fir forests of Tibet is influenced by the geography and weather. For instance, the *A. georgei* fir forests or Tibet fir (A. spectabilis (D.Don) forests have lower decay rates because of the flat landscape and the good ventilation, but they are still around 40%. Some areas such as the long bud fir (*A. georgei* Orr.) forests of the Sejila Mountains of Linzhi, and the Tibet fir forests of small Jilong of Jilong, have decay rates up to 50 – 60 %.

Table VI-1-12. The decay rates of a few fir forests of China

Tree Type	Altitude (m)	Distribution	Age (years)	No. Investigated	No. Decayed (%)
Abies nephrolepis	900	Zhangguang-chailin-Dahailin of Heilongjiang	160-180	200	59
A. faxoniana	3950	Maerkang of Sichuan	161-180	200	75-82
A. georgei	3700	Muli of Sichuan	161-180	376	34-41
A. georgei var. smithii	3700	Linzhi, Bomi of Tibet	160-200	537	40
A. spectabilis	3800	Jilong of Tibet	160-200	380	50-60

Comparative analysis of some information shows that the decay types and pathogens types of the Tibet fir forests are roughly the same as that of the fir forests of west Sichuan and the northwest region of Yunnan (Table VI-1-13).

Table VI-1-13. Main decay fungi of the fir forests

Types of Decay	Abies faxoniana	A. georgei	A. georgei var. smithii	A. spectabilis	A. nephrolepis
Hidden Wound Decay	+++	+++	++	++	++
Phellinus pini var. *abietis*	++	+	++	+	+
Ph. hartigii	++	++	+	+	++
Inonotus dryadeus	++	++	++	+	
Fomes conatus	+	+	+	+	
Phaeolus schweinitzii	+	+	+	+	+
Laetiporus sulfurous	+	+	+	+	+

Note: + rarely distributed; ++ moderately distributed; +++ commonly distributed

The decay type is often hidden wound decay (it's a symptom) In the later stage of the decay, the trees often do not form fruiting bodies. The decay types can be distinguished into white beehive-shaped decay(*Phellinus pini* var. *abietis)* white sponge-shaped decay, tree trunk wound decay and root and trunk decay *Phaeolus schweinitzii*. Of these, tree trunk wound decay and white sponge-shaped decay occur most often. The main pathogens that cause fir forest decay problems include spruce white pocket rot (*Phellinus pini* (Thore ex Fr.) Ames var. *abietis* Karst.) and fir sapwood rot (*Phel. hartigii* (karst.) Bourd et Galz.). Fir fiber pore rot (*Inonotus dryadeus* (Pers. Ex Fr.) Murr.) is the main wood decay (Photo VI-1-7) that causes the most economic damage. Conifer root rot (*Heterobasidion annosus* (fr.) Bref.) (Photo VI-1-8) often causes stem decay.

The amount of decay in fir is related to the age of the forest, the older the forest, the worse the decay. The amount of decay is also related to the type of the forest, of which azalea-fir forests and fern-fir forests have the highest decay rates. Under most circumstances, it is also related to the growth of the trees, usually with high altitude and low class soil sites having a lower decay rate (Table VI-1-14), and with low altitude and better class soil sites having a lower decay rate. Even old forests, if the standard soil condition is good, have a comparatively lower decay rate. For instance, the long bud fir of the Bomi area of Tibet are around 300 years old (this area has an annual average temperature of 35 degrees Celsius and annual rainfall above 1000 mm), diameter is around 100 cm, and tree height around 40 meters. More than 90% of the sample trees in the research had no decay, showing that because of the good soil site, the trees grow big and healthy, which greatly raised the trees' resistance to decay. This situation is commonly seen in the Daguaiwanju area of the Yaluzangbu River. Overall, the decay rates of the fir forests in Tibet and the southwest region are higher compared to that of other types of forests. The main reason is fir is yin tree (This is Chinese forester's concept, which was related the ecological factors), the texture is soft and weak, the bark is thin, there is no normal resin duct beneath the wood, and the wood contains low amounts of tannins to resist decay. Another reason is the roots of fir trees are often infected by root rot (*H. annosus*) and cold-resistant fungi such as *Polyporus schweinitzii* (photo VI-1-13) and *L. sulphureus*, causing many trees to fall or break during strong winds. On these dead trees, many decay fungi grow and spread into the healthy forest. The ecosystem of the high mountains is also good for the spread of the decay disease. Fir is a tree type near timberline, situated at an altitude of 3500-4400 meters, the temperature is cold in the high mountains and fluctuates a lot, often causing frost on the trees, and numerous wounds make it possible for the spread of the hidden decay. In the summer, the humidity of the fir forests is moderate (relative humidity 70-80%), providing good conditions for spore germination and penetration.

Table VI-1-14. Decay rates of the fir forests of Tibet

Region	Forest Type	Altitude (m)	Tree Type	Age (years)	(Diameter) (cm)	Height (m)	Main Fungus	No. Investigated	Decay Rate (%)
Lielamu	Shrub Forests	2890	Tibet Fir	230	62	32	*Phellinus pini* var. *abietis*	370	30
Yadong	Azalea-Fir Forests	4100	Same As Above	120	20	8	Same As Above	240	62
Jilong	Fern-Fir Forests	3500	Same As Above	230	39	19	Same As Above	590	61
Bomi	Fir Forests	3300	Long Bud Fir	250	40	36	*Inonotus dryadeus*	514	21
Linzhi	Fir Forest	3400	Same As Above	250	32	24	*Phellinus pini* var. *abietis*	120	40

Polyporaceaous fungi of the fir forests also include *Polyporus frondosus* (Licks.) Fr., *Polystictus xanthopus* Fr., *Gloeophyllum subferrugineus* (Berk.) Bond. et Sing., and *Polyporus brumalis* (Pers.) Fr. In addition, some useful fungi are parasites of the fir, such as big wood ear which is edible; branched monkey head mushroom (*Hericium coralloides* (Scop. ex Fr.) Pers. Ex Gray) which can be used for medical purposes such as benefiting internal organs (liver, kidney, spleen, heart, lungs), helping digestion, and nutrition. It is also a top grade treat of the banquet cuisine.

1.2.2 *Takahashi lasiocampoid (Selenephera lunigera Esper)*

Takahashi lasiocampoid occurs around the Pulagou River of Yuren, Bomi County in Tibet. In the fir forests with 3600-3900 meters altitude, according to field investigations and individual reports, the pest problem possibly occurred during the summer of 1972. The larvae eat the needles from May to

September. After repeated occurrences, this can cause large fir forests to decay and wither. From afar, the trees seem to be burned. After the summer of 1976, some affected fir trees seemed to grow green needles.

Research was done to evaluate the effect of defoliation on decay of fir. To set up sample plots in the major occurrence area, healthy trees and infected trees were categorized into the following groups: healthy trees – green needles, no pest, or recovered from pest; weak trees – Takahashi lasiocampoid are eating the leaves, some leaves are gone, but tree is still alive; decaying trees – pest problem on a large scale, most leaves gone, tree is almost dead; decayed trees – all needles eaten by Takahashi lasiocampoid, tree is already dead. After the trees were categorized according to the above criteria, the number of trees in each category was counted. The results of this research are shown in Table VI-1-15.

Sample plots of the azalea-fir forests on the south slope have some (10%) trees with green leaves, but most are decayed trees. Using 10 X 10 cm^2 as a sample space, the density of the pests on the trees was 12 larva. In the field research of 1200 m^2, we found 17 Ashmead and ichneumons, ichneumon-flies, longtail wasps (*Ichneumonidae)*, and saw 2 predator birds.

Long bud fir (*Abies georgi* Orr var. *smithii* (Viguie et Gaussen) Cheng et L. K. Fu.) is the main tree type of the cold-warm zone forests. Near timberline is basically pure fir forest. Mixed in are some Linzhi spruce, big-fruit spruce, aspen and southwest birch. This area is characterized by moderate and humid temperature, rainfall of 500-700 mm, and annual average temperature of 3-8 degrees Celsius. The hottest monthly average temperature is around 12 degrees Celsius, and the absolute lowest temperature can reach – 25 degrees Celsius. There are around 200 days with a temperature of 0 degrees Celsius, with frost throughout the year, and sometimes there is snow even during June. Long bud fir is one of the most cold-resistant tree types of the Tibet dark conifer forests. Because of the drought (annual rainfall 60% less than normal) in this area in 1971-1972, people reported the bad growth of grass, preventing grazing. This kind of drought leads to the large-scale occurrence of leaf-eating pests. According to field research results from 1974, 1976, and 1978, the pest entered the maturation period from 1973-1974. In 1974, more than 50% of the fir trees had been affected and had no signs of re-growth of leaves. Because of enough rainfall in 1975, grass flourished and fir forests were humid, causing an increase in the number of trees affected. After this, the pest problem seemed to drop slightly. In 1976 research of the azalea-fir forests of, 16-30% of the decayed trees that were completely affected had grown new leaves by September 1976. According to observations in 1978, the outbreak cycle had almost ended. This kind of self-regulation ability is often seen in nature.

Table VI-1-15. Takahashi lasiocampoid research

Forest Type	Altitude (m)	Tree Species Composition	Height (m)	Diameter (cm)	Age (years)	Density	No. Investigated
Lichen-Fir	3700	10 Fir	26	30	270	0.8	350
Azalea-Fir	3750	10 Fir	22	32	200	0.7	140

Forest type	Category of Affected Tree %			
	Healthy	Weak	Decaying	Decayed
Lichen-Fir	0	70	10	20
Azalea-Fir	10	10	10	70

Listed below are suggestions for managing affected fir forests:

Construct roads for transportation of the affected trees. Before the road is finished, organize citizens to chop down and peel off the bark of dead trees to prevent bark beetles and wood borers from laying eggs on them. The trees can also be chopped into pieces or firewood and stored on the side of the road. Although the pest problem has disappeared now, we still need to research the occurrence pattern, prevention, and the use of natural predators of the fir moth. Because the humidity of the fir forests is helpful to the occurrence of disease, we can conduct research on natural contagious insect diseases to kill the pests. If the pest breaks out again, airplanes should be used to apply pesticides to kill the pests.

1.2.3 *Other Pests of the Fir Forest*

Other than the above large-scale pest problems, the following pests also occur in the fir forests. The hairy southern pine beetle (*Polygrapnus major* Stebbing) of Tibet fir trees is found in the Pulan region, and pine mealybug (*Crisicoccus pini* Kuwnua) of the Todang fir is found in Jilong. To summarize, although pest problems are not common in fir forests, we should still pay attention to serious pests such as pine mealybug that can spread the attack.

1.3 *Diseases of the Spruce Forests*

The main species of the Tibet spruce forests include Li jiang spruce (*Picea Likiangensis* (Franc.) Pritz.), western Sichuan spruce (*Picea. likiangensis* var. *balfourian* (Rehd. et Wils.) Cheng ex Hu), and Linzhi spruce (*Picea. likiangensis* var. *linzhiensis* Cheng et L.K.Fu). Diseases include spruce decay, spruce leaf-rust disease, and spruce cone rust disease. There are also pest problems such as spruce bark beetle and spruce wood borers.

1.3.1 *Spruce Decay Disease*

There are many decay fungi of the spruce forest, mainly spruce white-decay disease, root white-decay disease, *Armillariella mellea, Fomes connatus, Phaeolus schweinitzii, Trametes abietina,* and *Polystictus pergamenus* (Table VI-1-16). Spruce white-decay disease and *Trametes abietina* are widely distributed. Root white-decay disease and *Armillariella mellea* cause root decay and are harmful to the regeneration of the forests.

Spruce white-pocket rot (*Phellinus pini* (Thore ex Fr.) Ames var. *abietis* Karst.) (Photo VI-1-16) is the most widely distributed stem wood decay fungus of the north temperate zone dark conifer forests, affecting spruces, firs, and hemlocks. Previously it was put in the same category as *P. pini*, but it is now clear that the range of hosts and the shape of the fungi are very different (Table VI-1-17). The shape of the fruiting structure of *P. pini var. abietis* is different from *P. pini* in that its fruit body grows flatly on the side branches, sometimes can reach 30-40 cm, the size of the fruit body is big, and the edge of the cap is thin.

Table VI-1-16. Main decay fungi of spruce forests

Fungus Type	Part Affected	Decay Extent	Distribution	Seriousness
Phellinus pini var. abietis	Heartwood of Trunk	Almost Whole Tree Rot	++	++
Heterhasidion annosum	Roots	Cause Death of Young Trees	+	+++
Armillariella mellea	Sapwood of Roots	Same As Above	++	+++
Fomes connatus	Heartwood of Trunk	Trunk	++	++
Phaeolus schweinitzii	Heartwood of Trunk	Trunk	+	++
Trametes abietina	Heartwood, Sapwood	Almost All Attacked	++	++
Polystictus pergamenus	Same As Above	Same As Above	++	++

Distribution: + rare; ++ moderate.
Seriousness: + decayed volume is 5-10 % of whole tree volume; ++ 11-40 %; +++ above 40 %.

Table VI-1-17. Comparison of white-decay of pine and white-decay of spruce

Name of Disease	Host	Characteristics of Fruit Body			Type of Decay
		Fruit Body	Cap	Lateral	
Pine white pocket rot	*Pinus, Larix*	Horse hoof Shaped	Marginal Dull	Absent	White Beehive-shaped Decay
Spruce white pocket rot	*Picea, Abies, Tsuga*	Horse hoof Shaped - Flat	Marginal Thin	Absent or Spread, flat	Same As Above

The decay rate in the spruce forests of China (Table VI-1-18) is second only to that of the fir forests, and is a serious problem affecting the utilization of the resources and regeneration of the forests. The decay rate differs with the forest flora. According to published reports, the decay rate of the spruce forests distributed in Eurasian forest flora and China-Japan forest flora is higher, and that of the spruce of Eurasian prairie flora is lower. This is related to the geography of the forests. The difference between the decay rate of each tree type is usually closely related to the ecological condition and the age of the forest. Among similar ecological conditions, the older the forest, the higher the decay rate. But in some areas of Tibet, because of the high altitude and the warmer temperature caused by the plateau, in areas with very good ventilation, old spruce trees do not have high decay rates.

Table VI-1-18. Decay rate in a few spruce forests of China

Tree Type	Altitude (m)	Distribution	Age (years)	No. Investigated	No. Decayed (%)	Main Decay Fungus
Northeast Spruce (*Picea jezoensis* var. *microsperma)*	500	Peony River, Heilongjiang	> 180	294	24	*Phellinus pini* var. *abietis*
Li Jiang Spruce (*P. likiangensis*)	3200	Muli, Sichuan	240-280	402	43	*P. pini* var. *abietis* *Armillariella mellea*
Western Sichuan Spruce (*P. likiangensis* var. *balfouriana)*	4000	Changdu, Tibet	180-200	582	34	*P. pini* var. *abietis* *Heterobasidion annosum*
Linzhi Spruce (*P. likiangensis* var. *linzhiensis*)	3500	Bomi, Tibet	> 300	1006	14	*P. pini* var. *abietis*
Xinjiang Xueling Spruce (*P. schrenkiana)*	1200	Yili Valley, Xinjiang	240-260	100	28	*P. pini* var. *abietis* *Armillariella mellea*

Table VI-1-19 Comparison between the decay of western Sichuan spruce and Li River spruce

Spruce Type	Altitude (m)	Sunlight Hours	Sunlight Percent (%)	Avg. Rainfall (mm)	Biomass (m^3)	Age (years)	Density	No. Investigated	No. Decayed (%)	Wood Production Rate (%)
Western Sichuan	4110	2200	51	400	610	240	0.6	302	31	58
Li Jiang Spruce	3300	1900	48	600	526	240	0.6	371	47	43

Table VI-1-20 Decay rate comparison of same-aged spruce forests

Spruce Type	Altitude (m)	Distribution	Age (years)	Avg. Rainfall (mm)	No. Decayed (%)
Li Jiang Spruce	3200	Muli, Sichuan	260-280	1000	50
Linzhi Spruce	3500	Bomi, Tibet	260-280	> 1000	0.2
Xinjiang Xueling Spruce	2000	Tianshan, Xinjiang	260-280	700	32.5

According to research, the decay rate of the Tibet spruce forests depends on the age and the ecological condition of the forest. The Li River spruce forest group can be arranged according to the decay rate of each forest, from the highest to the lowest: Li Jiang spruce – western Sichuan spruce – Linzhi spruce (Table VI-1-18). According to much research, the decay rate of the Li Jiang spruce forest is the highest of the spruce forests of China. Its IX age-level decay rate is 21.7%, and the XII age-level and above decay rate is 40-50%. Li Jiang spruce forests have a high decay rate because the trees are over-mature, which allows the decay fungus to grow the fastest, especially in humid conditions where the decay rate is even faster. The spruce forests of western Sichuan are located in high mountains. The forests are characterized by cold, drought, and low annual rainfall. Compared to Li Jiang spruce forests the weather is more arid (Table VI-1-19). Also because of the high altitude of the western Sichuan spruce forests (the highest of similar spruce forests, 3400-4200 meters), the intense sunlight radiation may have a certain restraining effect on the growth of the pore fungi. This is why the decay rate of western Sichuan spruce forests is lower than that of the Li Jiang spruce forests and the economic wood production rate is higher.

The Linzhi spruce forests are mainly distributed west of Linzhi. Most grow in areas with good ventilation along the middle and lower parts of Yaluzangbu River, such as Linzhi, Milin, Bomi, Cuona, Lang County, Longzi, and north of the Himalayas, 2700-3400 meters high on the east side of Nianqingtanggula. This is especially the case for the Linzhi forest in the region of the big turn of the Yaluzangbu River because of the effect of the warm and humid air current of the Indian Ocean. There, the trees are large, healthy, and strong, characterized by high storage volume and long life. The decay rate is 2-3 times less than that of the western Sichuan spruce and the Li Jiang spruce. Under its special geography and weather, the Linzhi spruce forest not only has enormous production power, it also has unusual resistance to decay diseases. For instance, spot sample area research on 14 Linzhi spruce trees 310-410 years old showed that the trunks grow full, straight, (Fig. *Picea linzhiensis* (Cheng et L. K. Fu) Cheng et L. K. F, splendidly, and had no decay. This phenomenon proves that decay is not wholly dependent on age, because forests under different ecological conditions have different mature ages. Maybe because of characteristics of the high mountain regions such as the cold weather of the growing season of the trees and the warmer temperature of the winter, the mature age of the trees is lengthened and the decay age of the Linzhi spruce is delayed. The decay rate is clearly lower. It is clear from the comparison of the decay rate of different spruce forests of the same age (Table VI-1-20) that, when comparing forests of the same age (260-280 years), Linzhi spruce has a decay rate that's 1/100 of other spruces. This phenomenon is a scientific question of silvics and plant pathology to be further discussed. It shows that under certain ecological conditions, production can be increased and the resistance against decay diseases can be raised.

Other pore fungi that are saproparasites of the Linzhi spruce forests include *Fomitopsis pinicola* (Sw. ex Fr.) Karst., *Polyporus resinosus* Fr., *P. brumalis* (Pers.) Fr., and *Fomitopsis roseus* (Alb. et Schw. Fr.) Cke. Parasites of the western Sichuan spruce forests include *F. pinicola* (Sw. ex Fr.) Karst., *Hircshioporus fusco-violaceus* (Schrad ex Fr.) Donk, and *Trametes cervina* (Schw.) Bres.

1.3.2 *Spruce Rust Diseases*

Chrysomyxa rust fungi, parasites of spruce needles, and Thekopsora rust fungi, parasites of spruce cones, are the main rust fungi of the spruce forests of this area. Much research shows that the rust fungi of the spruce forests of this area belong to the same category as the forest plant rust fungi of western Sichuan and northwest Yunnan.

1.3.2.1 *Spruce Cone Rust Diseases*

Spruce cone rust (*Thekopsora areolata* Fr. Magn.) is a parasite of Li Jiang Spruce (*Picea likiangsis*) in Muli, Sichuan. It can spread into the Bhutan border forests to the west and infect Maidiao spruce (*Picea braonytyla*). Because of the evolution and special natural geography, the species differentiation occurred in rust fungi under the constant specific ecological environment of the high mountains. New varieties of spruce cone rust fungi also appeared, different from the original species, *Thekopsora areolata*. Cone rust occurrence is only 3-5 %, and mixed forests have lower infection rates. Under ecological balance, the disease usually has little effect on forests, but research on regeneration of young trees shows that if virgin forests are destroyed and the ecosystem changed, or if large pure forests are man-made, many parasitic rust fungi may become important.

Table VI-1-21. Spruce cone rust disease

Spruce Type	Disease Name	Alternate Host	No. Investigated	Cone Infection Rate (%)	Natural Renewal
Long-leaved Spruce	-	-	140	Not Found	Not Good
Spruce (*Picea asperata*)	Spruce Cone Rust *Thekopsora areolata*	Prunus	100	18-49	Not Good
Li Jiang Spruce	Same As Above	Same As Above	100	20-30	Not Good
Purple-seed Spruce (*Picea purpurea*)	Same As Above	Same As Above	100	3-11	Not Good
Linzhi Spruce	Same As Above	Not Clear	1006	10	Not Good
Western Sichuan Spruce	T. sparasa	Not Clear	563	21	Not Good
Tibet Spruce	Same As Above	Not Clear	600	17	Not Good
Fish-scale Spruce	T. areolata	Prunus	100	9-42	Not Good

Spruce cone rust disease is a kind of cone disease caused by *Thekopsora*. It is the most common cone disease of the spruce forests (Photo VI-1-9). Other than long-leaved spruce, which hasn't been affected by this disease, the spruce forests of Tibet, including western Sichuan spruce, Linzhi spruce, and Tibet spruce, all have been affected to different degrees (Table VI-1-21).

According to observations in the Tibet Bomi and Linzhi spruce forests, small rice-shaped dark brown sporangia that are filled with orange spores grow on infected cone scales. When the infected cone scales open (Photo VI-1-10, VI-1-11), and the seeds inside the cone are found to be all replaced by spores. Linzhi spruce cone rust is different from the cone rust disease of western Sichuan and Tibet spruce. Through an electricaron microscope, the characteristics of the surface of the spores can be observed (Table VI-1-22). One is completely covered by warts, and half of the surface of the other is smooth. (Photo VI-1-12).

Table VI-1-22. Comparison of characteristics of two rust fungi spore surfaces

Cone Rust Fungus	Size of Spore	Surface Characteristics
Thekopsora areolata	29-33 X 15-23	Dense warts like scrub brush, half smooth
Thekopsora sp.	26-32 X 22-29	Dense warts like scrub brush

Table VI-1-23. Needle rust infection rate of spruces in China

Region	Tree Type	Disease Name	No. Investigated	Autoecious or Heteroecious	Infection Rate (%)
Xinjiang	Xueling Spruce	*Chrysomyxa deformans*		Autoecious	50
Heilongjiang	Fish-scale Spruce	-	-	-	-
Sichuan	Spruce	*Chrysomyxa ledi*	250	Heteroecous	30 (man-made forest)
Tibet	Long-leaved Spruce	*Chrysomyxa piceae*	140	Not Clear	15
Tibet	Tibet Spruce	*Chrysomyxa piceae*	60	Not Clear	10-15

According to observations on the spruce cone rust diseases of Jilong, Yadong, and Bomi, the occurrence of the disease is closely related to the forest ecosystem. When the forest is sparse, humidity is high, the forest stands are over-mature, and the growth is weak, the infection rate of the cones is high. Otherwise, it is low.

The prevention of spruce cone rust diseases should involve both the administration of forests and collection of seeds. No infected cones should be picked for seeds. Also, infected cones should be burned to destroy the source of disease. If possible, herbicides can be used to destroy the alternate host plants of the seed collection area to lessen the disease.

1.3.2.2 *Spruce Needle Rust Diseases*

This is a young spruce tree disease caused by *Chrysomyxa*. According to the distribution of the spruce needle rust disease of China (Table VI-1-23), the disease caused by *Chrysomyxa* requires good ventilation of the spruce forest. Thus, the infection rate of the long-leaved spruce and Tibet spruce forests of the valley region south of Tibet is the highest, and that of the Li Jiang spruce forests is clearly lower. The spruce forests of Sichuan and Xinjiang are also affected by spruce needle rust, but the drought-resistant spruce of northeast and central China are very rarely infected. Spruce needle rust caused by *Chrysomyxa* is the main disease of young man-made spruce forests of Tibet. According to research on the spruce forests of western Sichuan, young man-made forests often severely lose needles after infection, causing the growth of young trees to drop by 30%. In plantations, after the young trees are heavily infected, the growth is affected and the quality of the trees drops. The top shoots of young trees curl up (Photo VI-1-

14, VI-1-15). Orange aecia grow in rows on the needles. Infected leaves are spicate-shaped. Seriously infected needles are brown colored and withered. The alternate host of this disease in Tibet still needs to be researched. This disease affects the long-leaved spruce forests of Small Jilong, Jilong County, and the Tibet spruce forests of Yadong County the most. The occurrence and distribution of the disease is related to the age of the forest. Middle- and young-aged forests are the most seriously affected, while mature forests are rarely infected.

To take care of the newly planted trees after cutting down the old spruce forests in the Tibet area, one should pay attention to the occurrence and known relationship of host and alternate hosts. Prevention of the needle rust disease should be by destruction of alternate hosts and weeds, and picking the infected needles to be burned.

1.3.3 *Pests of the Spruce Forests and Wood Flaw*

Research shows that Linzhi spruce, Tibet spruce, and long-leaved spruce have few pest problems, but western Sichuan spruce have more types of pests and are more seriously affected. The most common are wood flaws and moth.

Western Sichuan spruce has wood flaws because there are many dead knot, decayed knot, and wood borers-grind holes, which cause the quality of the wood to drop. For instance, small volume logs (200 and 250 mm wide) of western Sichuan spruce have high quality I level wood, 80% is low quality III-IV level wood. Only 20% of large volume logs is I level (Table VI-1-24). Because of the high altitude (can reach above 4000 meters) of the forests of western Sichuan and the sparsely distributed trees, under natural pruning, branches are bad and have many dead knots. The dead knots have an average diameter of 7-8 cm. In a 1-meter-long piece of wood there are on average 5-7 dead knots.

Other than too many knots, the wood flaws of western Sichuan spruce also include holes created by spruce wood borers (*Tetropium oreinum* Gahan) (Photo VI-1-17) that can seriously affect the quality of the wood. On average, a 1-meter-long piece of wood has 35 hatch holes, and can reach 95. Observations at the wood storage ranch in Changdu lumber farm by forestry research stations show that the adults lay their eggs on bark-included wood. After hatching, the larva grinds its way inside the wood. The hole is flat elliptical, about 4 X 6 mm in size, 3-4 cm deep (can reach 10 cm). The pupal stage 4-6 weeks. Table VI-1-25 shows a comparison between the wood borers that affects Li Jiang spruce and the wood borers that attack western Sichuan spruce.

Table VI-1-24. Wood quality of western Sichuan spruce

Wood Type	Tree Type	Total Pieces	Wood Volume (m^3)	Pieces Investigated	Main Flaws
250 mm wide	Spruce	798	23.98	9	Dead knot, Live knot, Holes, Decayed knot, Discolored, Pitch knot, Cracks, Blunt edges
200 mm wide	Spruce	2078	47.38	11	Dead knot, Live knot, Holes, Blunt edges
Medium cubic board wood	Spruce	1920	90	15	Dead knot, Live knot, Insect holes
Large cubic board wood	Spruce	1442	104	10	Dead knot, Live knot, Blunt edges
Extra Large cubic board wood	Spruce	891	120	15	Dead knot, Live knot, Pitch pocket, Insect holes
Total				60	

Wood Type	Quality level of Wood (%)[1]			
	I	II	III	IV
250 mm wide	0	22	11.2	66.6
200 mm wide	0	18.1	54.6	27.3
Medium volume log	0	46.6	40	13.4
Large volume log	20	50	10	20
Extra Large volume log	53.5	26.6	13.4	6.5

[1]I=highest qualiy; IV=lowest quality.

Table VI-1-25. Comparison between two main wood borers of Li Jiang spruce

Name	Tree Type Affected	Body Length (mm)	Body Color	Antenna	Hatch and invade hole
Tetropium oreinum	Western Sichuan Spruce	9.2-14	Black or Black-brown	Scape shorter than third article	Dense
Tetropium castaneum	Li Jiang Spruce	10-16	Same As Above		Sparse

According to observations in the Changdu area, *Tetropium oreinum* occurs every year. Adults appear from May to July. From late May to early June, they lay eggs which invade the spruce wood.

The main pest that attacks western Sichuan spruce forests is spruce large bark beetle (*Dendroctonus micans* Kug.) This pest occurs commonly in western Sichuan spruce forests near the upper part of Lancang River. It lives on newly cut areas and affects healthy seed-trees. Turpentine starts to appear on the bark of infected trees. This is the characteristic effect of *Dendroctonus*. Adults are beetle-like, and the body is 5.2 – 7.9 mm long and is a shiny black color. The female adult tunnel is short and curved, with uncertain direction. One side of the female adult tunnel is chewed out for a common egg-laying room. Eggs are stored in stacks inside. After hatching, the larvae eat their way outside. With their growth, a big and irregular-shaped common tunnel is chewed out under the bark. This pest has a similar life style to *Dendroctonus micans* that affects the Armand pine of China. The difference in their appearances is described in Table VI-1-26.

Table VI-1-26. Comparison of appearances of two *Dendroctonus*

Pest Type	*Dendroctonus micans*	*Dendroctonus armandi*
Host	Spruce	Huashan Pine
Body Length	5.2 – 7.9 mm	4.4 – 6.5 mm
Prothorax, Back	Center slightly arches to the back, whole body almost straight-line or small-ripple shaped	Center protrudes to the back obtuse-angled, two sides slightly concave forward
Prothorax and Coleopterus Hair	Hair long and standing-up, hair on coleopterus slanted long	Hair length different, short hair on coleopterus slanted
Tunnel	Offspring-tunnels form common tunnel with female-tunnel, or family-tunnel	Female-tunnel is single vertical, Offspring-tunnels extend from the two sides, short

The bark beetle that attack western Sichuan spruce also include four-eyed bark beetle (*Polygrphus rudis* Eggers), Menshi (*Ips mansfili* Wachtl), Western Tibet bark beetle (*Ips nitidus* Eqqers), and spruce hairy bark beetle (*Dryocoetes hectographus* Reitter).

Table VI-I-27. Wood decay of cypress forest of Tibet

Forest Species Composition	Alti-tude (m)	Geography Location	An-nual Precip-itation	Forest Ecotype	Main Pathogen	No. of Trees Surveye d	Percentage of Wood Decay %
Big-fruit round cypress	3800	Changdu, Ritong	<400	Favorable, Dry-Cold Type	*Gymnospro rangium* sp.	240	5
Yunnan-Tibet square branch cypress	3200	Nielamu, Qu County	1453	Favorable, Wet-Cold Type	*Stereum* sp.	260	50

1.4 *Cypress forests diseases*

Cypress trees that occur in the forests of Tibet area include big-fruit round cypress (*Sabina tibetica* Kom.) and Yunnan-Tibet square branch cypress (*S. wallichiana* (Hook. F. et Thoms.) Kom.). The former is distributed in south and east Tibet, on the upper south slope of valleys with altitude 2800 – 4600 meters. The latter is distributed in south Tibet, often mixed with Tibet *Abies* forests. The two cypress forests are located in different geographical areas and have different ecotypes, so different harmful groups of diseases occur within each. Yunnan-Tibet square branch cypress forests have high humidity (dryness level 1.0 – 1.5). *Usnea*, a lichen, covers more than 80% of the tree, resulting in more leather fungus that likes the cool-humid weather. This fungus causes an increase in the cypress decay rate (Table VI-1-27). The infection rate can reach 50%. In seriously infected forests, 65% of the wood becomes firewood due to the decay (on average each hectare of firewood is 147 m^3). On the other hand, big-fruit round cypress grows on dry south slopes (forest dryness level under 1.0). The main disease is *Gymnosporangium japonicum* Syd. (Photo VI-1-19). The infection rate can reach 30%. The fungus causes cypress branches to form thick swollen mycocecidium. During the rainy season, a bright yellow gluey substance can be seen growing on the mycocecidium. This is the telia. *Gymnosproangium* sp. have Rosaceae as their alternate hosts. When expanding production of fruit-bearing trees in the apple family, attention should be paid to the prevention of the spreading of diseases by cypress forests.

1.5 *Hemlock forest diseases*

The hemlock forests of Tibet are mainly distributed on the south slope of the Himalayas,. The main tree species that form the hemlock forests is Yunnan hemlock (*Tsuga dumosa* (D. Don) Eichler). Often mixed in the forests is Yunnan red bean spruce (*Taxus yunnanensis* Cheng et L. K. Fu.). According to research on the hemlock forests of Small Jilong of Jilong County, Asangqiao of Yadong, and Zangmu of Lielamu, the ecological conditions of hemlock forests are helpful to the growth of fungi. These areas receive ample rainfall (annual rainfall above 1000 mm), warm temperature (hottest month average temperature 18 degrees Celsius), many clouds, and high relative humidity during the summer. These conditions are all suited for the infection by pathogens. Because the wood of hemlocks is decay-resistant, there are few diseases in the hemlock forests. Most hemlocks grow healthily and wholesomely. The forest age is over-mature, but usually the decay rate is only around 10 –20 % (Table VI-1-28).

Table VI-1-28. Decay Rates of Hemlock Forests of Tibet

Region	Forest Type	Altitude (m)	Forest Composition	Age (year)	Height (m)
Small Jilong, Jilong	Hemlock mixed with Purple spruce	2620	I: 10 hemlocks II: 10 purple spruce	150	34
Lixin, Lielamu	Hemlock	2380	10 hemlocks	210	35
Asangqiao, Yadong	Hemlock	2600	10 hemlocks	200	30

Region	Diameter (cm)	Density	Main Fungus	No. Investigated	Decay (%)
Small Jilong, Jilong	84	0.7	*Fomitopsis pinicola*	115	10
Lixin, Lielamu	140	0.6	*Phellinus pini* var. *abietis, F. pinicola*	206	20
Asangqiao, Yadong	44	0.7	Same as Above	200	10

The most important decay fungus of Hemlock white pocket rot is *Phellinus pini* var. *abietis*. Also important are hemlock Linzhi Murr, *Ganoderma tsugae*, fan-shape pore fungus (*Polyporus flabelliformis* Kl.), triangle-shaped polyporus fungus (*Polyporus triqueter* Fr.), rounded pore fungus (*Polyporus montanus* Quel.), and sulfur pore fungus (*Laetiporus sulphureus* (Bull. ex Fr.) Bond et Sing)

1.6 *High-mountain Oak Forest Diseases*

High-mountain oak is one of the most widely distributed tree types of Fagaceae in Tibet. Of this area, there are 20 commonly seen high-mountain oak types, and the main planted forest type is high-mountain oak (*Quercus semicarpifolia* Smith). High-mountain oak likes light and is resistant to drought. It is adapted to live in temperate-cool weather and fertile soil. This kind of weather and soil is best for the growth of many temperate-zone fungi. When the forest grows old, infection by decay fungi increases. The Tibet oak forests usually have lower decay rates than northeastern Mongolia oak forests. The oak forests of this area are usually distributed on mountains around 2500-3500 meters in altitude, with warm weather, ample rainfall, and rare frost. This results in a lower decay rate than other areas. Tibet area high-mountain oak forests not only can grow in the cold temperate zone, mixing with *Abies* forests, it also spreads downward into semitropical forest regions. Oak forests of different altitudes have different distributions of fungi. For instance, in Tibet high-mountain oak forests it is hard to see the cold-resistant annual poroid fungus *Polyporus dryophilus* Berk. And *Daedalea quercina* (L.) Fr. of the northeastern Mongolia oak forests. In most forests, *Phellinus igniarius* (L. ex Fr.) Quel. is the main fungus. There are some general distribution characteristics of basidiomycetes of Tibet oak forests. Oak forests at higher altitudes (2800-3500 meters) have mainly poroid fungi. Lower altitude oak forests (around 2500 meters) mainly have the leather fungus that likes warm and humid weather, and sometimes also poroid fungi of the subtropical zone, such as *Ganoderma oroflavum* (Lloyd) Teng. In 10 hectares of high-mountain oak (*Q.* semicarpifolia) forest, there usually are at least over 40 types of poroid fungi, far more than the leather fungus and the tooth fungus.

The oak forest is a type with the most valuable fungus resources. Fungi such as monkey head mushroom, linzhi, wood ear, and white wood ear fungus are all produced in different kinds of oak forests. The linzhi distributed in Tibet alpine oak forests have many varieties and high production. Some of the varieties are only found in Tibet. The alpine oak forests on the south slope of the Himalayas and near the big turn of the Yaluzangbu River is a region known for mass production of linzhi in China. Warm and humid air currents come from the south, annual average rainfall is 1000 mm, and relative humidity is above 85%. During the growth period of linzhi, the average temperature is 25-27 degrees Celsius. Research shows that there are many kinds of *Ganoderma* in the Himalayas. The *Ganoderma* type is distributed from the cold temperate zone to the subtropical zone. Most of the linzhi is distributed in Jilong, Milin, Linzhi, Bomi, and Nielamu. Of these, the Lixin village of Nielamu and the Qingduo region of Bomi are centers where the most types of population of Ganoderma lucidium are densely distributed. In many alpine pure oak forests at the big turn of the Yaluzangbu River, hundreds of linzhi populations are growing in a warm, humid, and half-shaded ecotype.

Table VI-1-29. *Ganoderma lucidum* Ecotype

Average Hectare per linzhi population	Average linzhi number per population	Forest Composition	Tree Height (m)	Density	Diameter (cm)	Forest Age (year)	Shrubs	Weeds
1	510	10 Alpine Oaks	21	0.3	43	200	Rose Cop[3] Clematis Cop[1] *Acanthopanax* Sol Lonicea Sol	Aster Cop[1] Wild *Fagopyrum* Sol. *Phlomis* Sol.

The types and amount of poroid fungi saprophytes of alpine oak forest are the most of all forests. Except for a few which can cause decay, many types are valuable resources, some edible, some medical, some types can also cause mild decay. Of these, the poroid fungi distributed in alpine oak forests of warm temperature zones include *G. lobatum* (Schw.) Atk., *G. applanatum* (Pers.) Pat., *Fomitopsis pinicola* (Sw. ex Fr.) Karst., *F. fomentarius* (L. ex Fr.) Kickx. (Photo VI-1-18), *Phellinus igniarius* (L. et Fr.) Quel, *Hirschioporus sector* (Ehrenb. et Fr.) Teng., *H. pergamenus* (Fr.) Bond. et Sing., *Polystictus meleagris* (Berk.) Cooke, *P. brumalis* (Pers.) Fr., *P. sulphureus* (Bull. ex Fr.) Bond. et Sing, *Lenzites tricolor* (Bull.) Fr., *Coriolus versicolor* (L. ex Fr.) Quel., *Trametes cinnabarinus* (Jacq.) Fr., *Hericium erinaceus* (Bull.) Pers., and *Schizophyllum commune* Fr.

Most Tibet orchards are planted after exploiting the oak forests. Consequently many diseases hidden in oak forests infect the orchards, such as Rosellinia disease and *Schizophyllum commune*.

According to research on the alpine areas of the middle and lower parts of the Yaluzangbu River in 1979, chestnut-yellow moth, *Trabala vishnou* Lefebure, occurs in Bomi and Yigong. It eats the leaves of the oak trees, tea trees, apple trees, and pear trees. On average, each tree has 50-60 larvae. Most densely infested was an apple tree with 246 larvae. Damage from this insect causes many fruit trees and tea trees to die.

The two sexes of the adult moth differ in color and size. Males are small. Wingspread of females is 53-72 mm, with a body color of yellow-green, orange, or green. Forewings are near triangle-shaped, with an inner horizontal nerve, extra horizontal nerve, sub-extra horizontal nerve, middle margin nerve dots, and middle discal dots plaga, which are all yellow-brown colored. In the middle of the underwing, there are two obvious yellow-brown colored horizontal stripes. The eggs are elliptical and grey-white colored. Mature larva body length can reach 86 mm, and has yellow hair all over the body. On two sides there are irregular-shaped black-brown spots. An obvious characteristic is that the first section cutex lamina of the midsection has a black-brown "][" shaped mark.

According to observations in the Bomi alpine oak forests, the life style of the moth is one generation per year, eggs that survive the winter, and larva start to affect the trees in May. At first they concentrate on the back of the leaves. When startled, they spin and hang down from the leaves, which allows them to spread around with the wind. In July, the appetite of the larvae increases drastically. After eating all alpine oak leaves, they move to nearby fruit trees and eat all the apple leaves. Towards the end of July, they often pupate in shrubs of the alpine oak forests. In the middle of September, adults appear, surviving winter after laying eggs.

Prevention Methods:

1. In the winter, collect the eggs by hand, or in spring, catch the larvae when they are concentrated together and are spinning webs.
2. After the beginning of September, use light to trap and kill adult moths.
3. At beginning of May, spray 2.5% trichlorfon to kill the larvae (Shi 1979. P. 423).
4. When the number of occurrences is high, use Tung tree oil plus BHC (Shi 1979. P. 46) to paint the tree, forming a toxic circle to prevent larvae from getting to the tree.

1.7 *Birch Broadleaf Forest Diseases*

Birch broadleaf forests are forests made up of trees that defoliate in the winter. Mainly distributed in Tibet are forests made of *Betula*, *Populus*, *Salix*, *Alnus* Mill., and *Hippophae*. Of these, *Betula* has the biggest distribution area. The following is a description of the main diseases and decay of southwest birch, white birch, and natural willow forests.

1.7.1 *Birch Decay Diseases*

The main birch of Tibet are white birch (*Betula platyphylla*) and its *variant B. platyphylla* var. *szechuanica*. There are also alpine birch (*B. delavayi*). Birch has the ability to resist cold and drought. It is often the frontier tree for conifer forests after digging or fire. Birch usually have low resistance to decay, except for the outer bark. The bast (phloem) and wood parts are both susceptible to decay. The wood starts decaying the summer after cutting. There is a common saying, "If the bark of birch is not peeled, after summer it melts into soil." Birch decay is also a serious problem of the birch resources of China. According to the comparison between the decay of birch of the northeast, northwest, and Tibet area of China, the decay rates of Tibet birch and the main birch forests of northwest China are roughly the same, and are usually less than that of northeast birch forests (Table VI-1-30).

The main fungus that causes birch decay is *Phellinus igniarius* (L. ex Fr.) Quel. It often forms asexual fruiting bodies under very cold conditions. There are also poroid fungi such as *G. applanatum* (Pers.) Pat., *F. fomentarius* (L. ex Fr.) Kickx., and *P. betulinus* (Bull. ex Fr.) Karst. Other than decay fungi, there is also birch leaf rust disease (*Melampsoridium betulinum* (Desm.) Kleb.) (Photo VI-1-20). Its occurrence rate is around 7%.

Table VI-1-30. Birch decay rates of a few regions of China

Distribution Area	Tree Types	Altitude (m)	Age (year)	No. Investigated	Decay Rate (%)		Main Fungus
					Phellinus igniarius	Wound	
Tibet, Jilong	*Betula utilis*	3600	60	220	4	31	*Phellinus igniarius, Piptoporus betulinus*
Tibet, Nielamu	*B. platyphylla* var. *szechuanica*	3200	80	80	10	33	Same As Above
Xinjiang, Tianshan	*B. tianschanica*	2900			10	17	Same As Above
Heilongjiang, Xiaoxinanling, Wutong River	*B. mandschurica*	500	99	350	4	78	Same As Above
Heilongjiang, Daxinanling, Geng River	*B. mandschurica*	300	70	472	10	56	Same As Above
Heilongjiang, Wanda Mountain, Rao River	*B. mandschurica*	210-450	73	205	11	39	Same As Above

1.7.2 *Populus davidiana* Decay and *Salix* Leaf Rust Diseases

There are many types of *Populus* in Tibet. Its vertical distribution is around 1500-4000 meters. Often seen under 3000 meters are Long Arrey aspen (*Populus pseudoglauca* Z. Wang et P. Y. Fu), marginal hair aspen (*P. ciliata* Wall. Ex Royle), and Yadong aspen (*P. yatungensis* (Wang et Fu) Z. Wang et S. L. Tung), often seen in mixed forests. The above *Populus* rarely have decay or poroid fungi. Above 3000 meters in humid semi-high mountain zones, there are clear-stream aspen (*P. rotundifolia* Griff var. *duclouxiana* (Dode) Gomb.), common aspen (*P. davidiana* Dode), and Miling aspen (*P. mainlingensis*). The main decay fungus is *Phellinus igniarius* (L. ex Fr.) Quel. f. tremulae. In semi-drought regions there grow Chair poplar (*P. wilsonii* Scheneid), Changdu poplar (*P. qamdoensis* Z. Wang et S. L. Tung), and Sichuan Tibet Poplar (*Populus szechuanica* var. *tibetica* Schneid.), which grow healthily and have high adaptability. When these trees are mixed, there are rarely diseases. The main pests of the young trees include *Populus* leaf beetle (*Chrysomela populi* L.) and *Populus davidiana* leaf beetle (*Chry. Tremulae* F.). When serious, they can eat up all the leaves.

Our research focused on Jilong aspen (*Populusciliata var. gyirongensis)*. It is distributed in Jilong, Yadong, and Nielamu. There are two main decay fungi of the *Populus* forests. One grows on plantation aspen trees (*Phellinus igniarius* (L. ex Fr.) Quel.). The other grows on *Populus davidiana* (*Phellinus igniarius* f. *tremulae*) (Photo VI-1-21). Because the *Populus davidiana* forests of Tibet are at high altitude, secondary forests are distributed in patches and are only minimally affected by frost and fire. They have a low decay rate (table VI-1-31).

Table VI-1-31. *Populus davidiana* forest stand wood decay

Distribution Area	Tree Types	Altitude (m)	Age (year)	No. Investigated	Decay Rate (%)	Main Fungus
Wanda Moutain, Heilongjiang	*Populus davidiana*	210-450	83	324	73	*Phellinus igniarius* f. *tremula*
Daxinanling, Heilongjiang	Same As Above	400-500	85	351	30	Same As Above
East slope, Xiaoxinanling, Heilongjiang	Same As Above	400	95	357	78	Same As Above
Yadong, Tibet	*P. ciliata* var. *gyirongensis*	2700	60	111	10	Same As Above

The decay characteristic of *Phellinus igniarius* that causes *Populus davidiana* decay is a black line separating the good wood from the decayed wood. Most of the decay causes the whole tree to become firewood.

Salix leaf rust disease is mainly caused by *Melampsora* rust fungi, and is mostly distributed on the north wing of the Himalayas and the south periphery of the plateau. Lhasa and Rikaze are representative areas (annual average temperature around 8 degrees Celsius, annual rainfall 450 mm). The weather is temperate-cool, dry, and the sun radiation is strong. *Melampsora* is often seen in high, cold, and dry areas, and can reach broadleaf trees and shrubs around 4000 meters. Some species are representative of the plateau, such as willow leaf rust (*Melampsora larici-capraearum* Kleb.). It affects the main forest-making tree types of the Lhasa, Jiangzi, and Rikaze areas generation after generation, such as Zuo Xuan willow and bamboo willow. The infection rate of willow rust can be as high as 85-95%.

Of the trees and shrubs, *Populus* and *Salix* are most easily infected by leaf rust. The uredial and telial stages are on *Populus* and *Salix*. Most of the alternate host spore stages are obligate parasites of *Pinus* and *Larix* (Table VI-1-32).

Salix leaf rust survives winter in its host as mycelium or in its asexual stage. In Tibet, most leaf rust species do not need an alternate host plant to occur. Their urediospore stage repeatedly infects *Populus* and *Salix* leaves and shoots, causing the leaves to drop.

Table VI-1-32. Rust disease of *Populus* and *Salix* of Tibet

Region	Fungus Name	Host	Alternate Host	No. Investigated	Infection Rate (%)
Rikaze	*Melmpsora laric-icapraearum*	*Salix oxycarpus*	*Larix Adans.*	679	90
Cangge of Lang County	*Melampsora larici-populina*	*Populus lasiocarpa*	Same As Above	51	40
Luozashen gere	*Melampsora tremulae*	*Populus tremulae*	Same As Above	100	20

1.8 *Mountain Subtropical Evergreen Broadleaf Forest Diseases*

This zone is distributed between altitude 1600-2000 meters, weather warm and humid, annual average temperature 17-20 degrees Celsius, and annual rainfall above 1000 mm. Commonly seen tree types include Cyclobalanopsis, *Quercus L.*, Lithocarpus, Castanopsis, *Cinnamomum Trew* of *Lauraceae*, and *Litsea Lam.* The fungi and high parasitic plants are also different from that of the temperate zone and exhibit characteristic of the tropical zone, such as the leather fungus and large *Discomycetes* have good growth. High parasitic plants are mainly *Loranthaceae* and *Balanophora*. Below are descriptions of each.

1.8.1 *Leather and Crustlike Fungi (Stereaceae and Thelephoraceae)*

Steraceae and *Thelephoraceae* are standing-wood decay fungi distributed in broadleaf forests mainly made up of *Lithocarpus*. *Stereum princeps* (Jungh.) Lev. (Photo VI-1-22) is in the *Thelephoraceae* of the higher *Basidiomycetes*. The fruit bodies are perennial, wood-textured, hard and brittle, large, and often layered. It causes sapwood of trees to decay. The decay rate around 20%. In semitropical mountain broadleaf forests, the heavy clouds and fog and the high humidity are favorable for the growth of leather and crustlike fungi which like humidity and warmth. It is often abundant in forests like these.

1.8.2 *Parasitic HigherPlants, including mistletoes*

There are many mistletoe species in Tibet, of which the subtropical mountain evergreen broadleaf forests have the most types. Often seen are *Viscum* (Photo VI-1-23) and *Loranthus* (Photo VI-1-24). In the broadleaf forest of Jiang Village, Jilong (altitude 1800-2000 meters), alpine oak is an important host plant, and other hosts include the branches or bark of species such as *Populus, Salix, Mallotus, Symplocos, Rhus, Cotoneaster, Ulmus, Lyonia, Elaeagnus, Viburnum, Cornus, Lesmodium,* and *Rhododendron*.

Leafy mistletoes (Photo VI-1-24) are evergreen shrub photosynthetic, plant-parasitic seed plants that belong to the families *Eremolepidaceae*, *Loranthaceae* and *Viscaceae*. The skin of its seeds has a gluey substance enabling the seeds to be distributed by birds. After birds eat the fruit, the seeds are expelled from their bodies and stick to the bark of the tree. Radicula grow out of the seeds and under the bark of the host plant, forming a cupule, and then forming the cupule root that connects to the vessel extracting water and nutrients from the tree. From the tree, a shrub-shaped parasitic plant grows out, infecting the tree and making it weak or wilting.

1.8.3 *Balanophora*

Balanophora, Cynomorium and *Rhopalocnemis* of *Balanophoraceae* are higher plants that are parasitic on tropical and subtropical evergreen broadleaf plant roots, reflecting the characteristic of the diseases of this kind of forest (Table VI-1-33). *Balanophoraceae* have no chlorophyll or roots. The stem grows out of the block-shaped roots formed with the host plant. The block-shaped roots grow on the roots of the host plant and depend on the water and nutrients of the host to live. In the hemlock forests of A-sang-qiao, Yadong, and Tibet, *Balanophora* is often seen on the roots of hemlocks. It has a harmful effect on the growth of hemlock trees.

Table VI-1-33. Tibet forest plant root Parasites - distribution of *Balanophora*

Name of *Balanophora* species	Host Plant	Altitude (m)	Distribution Regions
Balanophora dioica	Evergreen Broadleaf Forest	1800-2600	Chayu, Motuo, Chuona
Balanophora polyuandra	Same As Above	2400	Bomi
Balanophora laxiflora	Same As Above	2600	Chayu

2 METHODS TO CONTROL FOREST DISEASES

To effectively protect the existing forest resources, we list the following basic methods to control forest diseases. Most Tibet forests are virgin forests or sub-natural forests. Only a few are man-made forests. Because of the differences between natural geographical factors, such as the forest organisms and the local conditions, different forest populations are formed. All types of forests are composed of mainly perennial xylophyta. Compared to farmland and ranches, the growth cycle of forests is long and complicated. Different organisms live together year after year, communicating, adjusting, and connecting on an organic level. Each possesses certain space and living characteristics. Between each organism group and its environment, there is constant substance and energy exchange, forming a uniform forest ecosystem. A change in any part of the makeup will directly or indirectly affect the stability of the whole system, causing a change in the diseases, pests, and predators. Only by understanding and utilizing these natural processes can we implement appropriate and effective methods to protect the forest and control and prevent the occurrence and spread of diseases.

Forest disease is a big enemy of the development of forests. Some people pessimistically thought that forest diseases on a large scale were incurable. After the independence of China, extensive disease prevention work has acquired certain results and experiences that are important to the development and protection of the Tibet forests. Considering the natural and economical factors of Tibet and its forest characteristics, the main method for the prevention and cure of forest diseases should be prevention. According to the principle of reasonable management and utilization of the forests, and on the basis of scientific methods to manage the forests, an overall system that utilizes biological, physical, and chemical factors favorable to the healthy growth of the trees and keep the ecological balance of the forests should be developed and utilized. This is the only way to achieve the goal of protecting and improving forest development.

2.1 *The Correct Method to Organize Forest Cutting*

Reasonable forest cutting is not only a method to utilize the forest. It is also an important measure to improve the sanitation situation and adjust the makeup of the forest. The fact that Tibet forests are located in plateau valley regions and the upper part of big rivers, and have important functions such as reserving water resources, conserving the water and soil, and adjusting the weather need to be considered before harvests are implemented. In forests designated primarily for lumber production, the main cutting style should be selective cutting and progressive cutting. In forests that reserve water resources, serve as shelters, or are affected by diseases, the cutting style should be renovation and sanitation cutting. When cutting, the first trees to be cut should be disease-ridden trees, dying trees, decayed trees, wind-broken trees and over-mature trees. Cutting the good and leaving the bad is strictly forbidden. Forest regulations and minimum sanitation requirements should be followed. Reduce trunk height. Remove the cutting waste. Forbid storing or discarding logs on the cutting site. Log through summer in the forest. Following the rule that the cutting rate must be less than the growth rate, control the cutting amount and forbid over-cutting to keep the ecological balance of the forest. This will help to maintain the perpetual existence of the green mountain and the sustainable utilization of its resources.

2.2 *Suitable Forest Management Methods*

When regenerating or making forests on cutting sites or barren land, the principle of "good for the land and good for the tree" should be followed. Disease-preventing forest-making, and nurturing methods should be designed according to characteristics of the biology of the trees, conditions of the land, and occurrence patterns of diseases. Utilize the favorable factor that there are many tree types in the forests of this area to create mixed forests. Use reasonable mixing methods that help the predators of the pests and the prevention of spread of harmful fungi to improve the ability of the forest to resist diseases. Forbid destroying the forests, prevent the occurrence and spread of fire, and avoid developing pure man-made forests and economical forests that destroy forest ecological balance. Improve the management, tending, and protection of forest resources. Predict future occurrences of diseases and disease-ridden woods. The cause of the disease and its host and alternate host plants should be destroyed right away to keep a good forest sanitation environment.

2.3 *Division into Quarantine and Forest Protection Districts*

Speed up the improvement in basic forest protection to eliminate or decrease the possibility of the occurrence of dangerous diseases. Once introduced or developed, diseases need to be discovered at an early stage. Diseased areas should be marked out, and areas prone to infection should be marked as protection areas to improve quarantines and stop the spreading. Restrain the disease to its first stages to prevent it from causing real harm.

2.4 *Good Storage of Wood*

When categorizing the wood, prevent mixing healthy and diseased wood together. With above 50% non-economical wood, designate it as firewood. The wood storage area should be located in areas dry and ventilated. Wood stacks should have cushion wood underneath. Keep the cleanliness of the ranch by removing shrubs, weeds, wood dusts, and discarded wood. Store firewood, diseased wood, and other discarded wood in places with good ventilation and enough sunlight.

2.5 *Improved Utilization of Wood*

Utilize the bark and branches and the healthy and decayed parts of the tree to the maximum degree.

3 FOREST-PLANTATION AND ORNAMENTAL TREES DISEASES AND PREVENTION

The research on the diseases of forest-plantation and ornamental trees was conducted in Lhasa, Rikaze, Jiangzi, Zedang, Cangdu, Jiangda, and Luozha. The growth, diseases, and pest problems of the plants of suburb barren hills, blocks of forests, forest-plantations, gardens, temples, trees on sides of fields, and trees in nurseries of the cities of Lhasa and Rikaze were the focus of the research. There are about 70 forest-plantation and ornamental plants in Tibet, of which 43 are trees. Most of these are *Rosaceae*, *Leguminosae*, and *Salicaceae*, including wood-producing trees, oil-producing trees, trees for aesthetic values, and temperate zone fruit trees, the main diseases of which are described as follows.

3.1 *Tree Diseases and Prevention*

3.1.1 *Plateau Tree Maladjustment Diseases (Environmental Stress Disease)*

Lhasa and Rikaze are commonly known as the sun cities of the plateau. Here the sun radiation is strong, the difference between day and night temperatures is huge, weather is dry, and evaporation rate is usually 10 times the rainfall rate. Few trees are adapted to grow under such conditions. Especially in springtime, trees often have physiological drought and sunburn. This is most obvious on exotic trees. Damage caused directly or indirectly by environmental stress is termed plateau tree maladjustment disease. Such damage is most commonly seen on barren land or regions with few trees above 3500 meters. For instance, in regions west of Rikaze, whole trees often wilt and die.

Using the fruit trees and *Salicaceae* of Lhasa as an example, plateau tree maladjustment disease has the following main symptoms:

1. **Sunburn "half wilt"**. Because of the strong radiation of the sun, the side of the tree towards the sun is sunburned. The crown branches of this side wither year after year, first only a few branches, then spreading to half of the tree. This is often seen on apple trees.

2. **Tiger skin spots**. Because of insufficient water inside the trees, irregular-shaped spots start to form on the bark of the tree, and the tree seems to be dying.

3. **Trap strip block marks**. Because of the strong sun radiation, vertical concave strips 1-2 cm wide start to form on the bark (Photo VI-3-1).

3.1.2 *Whole Tree Water-loss Wither Disease(Spring Wilt Disease)*

Whole trees wither and die in the spring, usually in the middle or later part of May. Fruit trees with blooming flowers die in a few days. The flower petals are dry and crisp, and the flowers are still open instead of withered due to the high rate of evaporation. If wilt occurs during the seedling stage, the seedlings wilt and die while standing. After the above symptoms appear, canker diseases often occur in *Populus*, *Salicaceae*, and fruit trees.

The middle regions of Tibet experience strong winds during spring, with small amount of rainfall and low water quality. This is the cause of spring wilt disease, which in turn is the main cause of the plateau tree maladjustment disease. Using Lhasa as an example, the average rainfall of May is only 21.2 mm, but the monthly evaporation amount is 293.2 mm (Table VI-3-1). The relative humidity of the air is low. May is during the spring "forest-making" season, the time when young trees are growing roots. At the same time of the high evaporation, sunlight hours are long and sun radiation is strong. The difference between day and night temperatures is huge. The contrast of the cold and hot temperatures often causes the bark of the trees to be hurt. According to measurements of the bark temperature of the bamboo and willow trees of Lhasa, during the end of May, the bark temperature of the sunny side is higher than the shadowy side, causing harm to the sunny side or withering half of the tree.

Prevention Methods:

1. Improve planting technology, pay attention to irrigation and water storage to prevent drought.
2. Paint the bark white (for details see tree canker disease prevention method 2).

Table VI-3-1. Comparison of moisture in each region In May

Region	Altitude (m)	Rainfall (mm)	Evaporation (mm)	Sunlight Hrs.	Avg. Relative Humidity (%)
Lhasa	3680	21.2	293.2	69	41
Chayu	1590	125.2	153.3	31	70
Chengdu	500	81.0		13.9	78

3.1.3 *Tree Canker Disease*

Tree canker disease, caused by the imperfect fungus *Cytospora* (Photo VI-3-4), is expressed as necrosis and death of the bark of stems and branches. This fungus is the most commonly distributed fungus of the north temperate zone dry and semi-dry areas, often affecting trees with weak growth. This fungus has a close connection to trees introduced for forest plantations and tree improvement. There are usually two kinds of symptoms, dry bark and branch wilt. The former often occurs on main tree trunks. The latter occurs on side branches. In this area, poplar (*Populus)* and willow (*Salix)* often show the dry bark symptom, while the ornamental tree types show branch wilt. Canker disease mostly affects newly planted trees, and it is the main factor responsible for a low forest-plantation success rate.

According to observations, the distribution of *Cytospora* is closely related to ecological and geographical factors, which can be used as a sign to indirectly measure the success of tree improvement and planting of the trees. *Cytospora* is mostly distributed in Asian desert plant sub-regions, Eurasian grassland plant sub-regions, areas of Tibet plateau plant sub-region with good ventilation, and northeast and Huabei areas of the China-Japan sub-region. Using latitude and altitude to show the distribution, roughly divide the distribution of this fungus of China into the following areas (Table VI-3-2). North of latitude 36 degrees is an area of many diseases, while south of that is less disease areas or non-disease areas. However, there is also more disease in areas of low latitude and high altitude, including Tibet.

Table VI-3-2. Geographical distribution of *Cytospora*

Geographical Area	Non-disease Area	Little-disease Area	More-disease Area
North of latitude 36 degrees	-	-	North of Shanxi, Gan, Ning, Meng, Qing, Xin, Hei, Jin
South of latitude 36 degrees	Zhe, Gan, Min	Wan, Su, Lu, Yu	
Low latitude, high Altitude	-	-	West of Sichuan, Northwest of Dian, Tibet (middle, lower parts of Yaluzangbu River)

Table VI-3-3. Effects of different weathers on tree canker disease

Weather	More-disease Area	Little-disease Area	Non-disease Area
Absolute Low Temperature	Below –20 degrees Celsius	-15 degrees Celsius - -20 degrees Celsius	Above minus 5 degrees Celsius
Annual Rainfall (mm)	100-500	500-700	Above 700, below 100
Infected Areas	Rikaze, Lhasa, Zedang, Jiangzi, Changdu, Xining, Beijing, Urumqi, Yili, Taiyuan, Hohhot, Lanzhou, Yulin	Linzhi, Bomi, Yadong, Xuzhou, Jinan, Xi'an, Dalian	Chayu, Motuo, Nanjing, Chengdu, Nanchang, Xinjiangkuche

In an overview of the whole flora, dry bark caused by *Cytospora* is mainly distributed in the Northwest and Huabei regions of China which are north of 36 degrees (Table VI-3-3). In Tibet, it is mainly distributed in the north part of the Himalayas and regions near the Yaluzangbu River with warm, semi-dry weather, such as Rikaze, Lhasa, and Jiangzi. Although cold-resistant trees can be planted in these areas, there is much damage caused by the lack of water and warmth, low resistance of the tree to wounds, strong radiation of the sun, and very low temperatures. This greatly increases the infection rate of canker diseases such as the one caused by *Cytospora*. In the Lhasa and Rikaze regions cytospora cankers have been found on cyprus, papaw, willow, poplar, mulberry, Russianolive, Robinia, *Albizia*, apple, walnut, peach, plum, and crabapple trees. Cankers caused by *Tubercularia*, *Diplodia*, and *Shaeropsis* are also common (Table VI-3-4).

To illustrate the reasons for the severe canker problem, the situation for bamboo willow will be described. Bamboo willow is the main ornamental and forest plantation tree type in Tibet cities. The Cytospora canker incidence rate in bamboo willow is as high as 60-70% because of the low quality of the environment and poor tree care.

Cytospora canker spreads and infects mostly in April and May in Lhasa due to a combination of factors. First, *Cytospora* spores are easily and widely dispersed, either from nearby or from long distances. The spores are in and on trees and other plants in high quantities.

Second, conditions are favorable for infection and canker development. Bamboo willow (and many other tree species introduced for use as ornamentals and in plantations in Tibet) has low vigor as a newly planted tree, a situation making it highly susceptible to infection by *Cytospora*. Willows are mainly established by planting stem cuttings instead of seedlings. With the high spring evaporation rate, the bark moisture, and concommitly, the healing ability are low. Also, the cuttings are too thick (average of 7-8cm, up to 10cm), so after burying them in the soil, they hardly develop roots. Young trees that do grow are damaged by sunburn and frost in the spring in the Tibet plateau, allowing *Cytospora* to more readily infect them (manual inoculation experiments have shown the mechanical wounds are not the main pathway for the occurrence of *Cytospora* infection). Whether cankers occur on a large scale appears to depend largely on tree vitality (which may partially equate with resistance). A specific index of tree vitality is the amount of water stored in the bark. In the typical Tibet spring drought, the amount of water stored in the bark usually favors the occurrence of the disease. With low tree vitality, after a plant is infected, the ability of the tree to heal is reduced and the cankers can readily develop.

Table VI-3-4. Canker and wilt diseases of the Lasa and Rikaze regions

Investigation Location	Tree Type	Disease Name	Symptom Type	No. Investigated	Infection Rate (%)	Growth Condition
Lhasa	Arborvitae	*Cytospora cenisia*	Branch Wilt	81	20	Good
Lhasa	Walnut	*C. juglandis*	Branch Wilt	24	50	Good
		Diplodia juglandis	Branch Wilt	Same As Above	11	Good
Rikaze	Apple	*C. leucosperma*	Branch Wilt	426	90	Good
Lhasa	Tibet Papaw	*C. contoneastrii*	Branch Wilt	4	5	Good
Lhasa, Rikaze	Bamboo Willow	*C. chrysosperma*	Canker	374	80	Good
Lhasa	Silver-white Poplar	*C. chrysosperma*	Canker	45	70	Good
Lhasa	Mulberry	*C. atra*	Branch Wilt	3	40	Good
Lhasa, Rikaze	Peach	*C. leucosperma*	Branch Wilt	49	50	Good
		Tubercularia vulgaris	Canker	12	35	Good
Lhasa	Plum	*T. vulgaris*	Canker	11	5	Good
		C. leucosperma	Branch Wilt	11	60	Good
Lhasa	Russian-olive	*C. elaeogni*	Branch Wilt	8	5	Good
		T. vulgaris	Branch Wilt	8	14	Good
Lhasa	Robin	*T. vulgaris*	Branch Wilt	7	5	Good
		C. sp.	Branch Wilt	7	35	Good
Lhasa	Euonymus	*C.* sp.	Branch Wilt	5	40	Good
		Diplodia ramulicola	Canker	5	20	Good
Lhasa	*Albizia*	*T. vulgaris*	Canker	3	30	Good
Lhasa	Malus	*Valsa mali*	Canker	21	50	Good
Lhasa	Grape (Vitis)	*Sphaeropsis ampelos*	Branch Wilt	2	50	Weak

The perfect stage of *Cytospora* is *Valsa*. The sexual stage has not been found in the Tibet area. According to research in the little-disease areas and more-disease areas of Tibet, the fungus survives winter in its asexual spore stage. This is because the lowest normal temperature of the infection area is relatively high (-16.5 degrees Celsius), clearly higher than the temperature of the northern regions of China. It reflects the fact that the temperature increase of high altitude regions due to the sun radiation is good for the asexual spores to survive the harsh environment of winter.

Prevention methods:

To manage Cytospora canker, an integrative method should be based on the "good for the land, good for the tree" principle.

1. Good for the land, good for the tree. Select cold-resistant, drought-resistant, sunburn-resistant tree types that fit the ecological conditions of the region, such as plateau-native poplar and willow. Also Mongolian pine, Chinese pine, and alpine pine are cold and drought-resistant.

2. Improve tree-planting techniques and methods, such as caring for young forests to create good conditions for the growth of the trees. Pay special attention to improve cold-resistance and drought-resistance, give young trees reasonable irrigation, and prevent damage from spring drought, sunburn, and the cold of winter low temperature.

3. Chemical treatment prevention. In areas already infected, try chemical treatments.

(a). Paint the bark white. This is an effective method to prevent trees from sunburn, lower the difference in temperature, and conserve the water stored in the bark. It should be popularized in the canker prone areas of China. The preparation is 5 kilograms of quicklime, 0.5 kilograms of lime-sulfur mixture, 0.5 kilograms of salt, 2 ounces of animal oil, and 20 kilograms of water.

(b). Use "nail board" or knives to punch holes or cut stripes in infected bark to enable the soaking through of pesticides. After cutting out the infected spots, paint waste oil to kill the fungus. Then paint 50 units of "920" lanolin ointment to help healing. 10% potash can also be used (1.5 kilograms of 40% Asomate (Trade name from

Shi 1979 P. 25) and 1 kilogram of ping-ping Ja (Local folk's menu), each dissolved in hot water, mixed with 50 kilograms of water). Or use 70% thiophanate methyl (Shi 1979 P. 412), diluted 200 times or 400 times, plus 0.1 ping-ping Ja to spray the tree.

(3). When cankers occur on big or valuable trees, the infected spots can be scraped off. The bark of dead trees has to be completely scraped off so the tissues are shuttle-shaped after scraping. Then paint 50-100 times thiophanate methyl or carbendazol (Shi 1979 P. 68) or Zineb or Asomate etc.

3.1.4 *Willow Rust Disease*

Willow rust is mainly distributed in cold and semi-dry areas above 3500 meters of the Lhasa, Jiangzi, and Rikaze regions. It often occurs on Chang rui, Kangding, and big red willow. Seedlings and young trees are most seriously affected, often causing the side branches of young trees to wither and die, deforming and reducing the growth of the trees.

Table VI-3-5. Weather during Tibet willow rust infection time (April-May)

Region	Infection Area	Altitude (m)	Sunlight Hrs.	Sunlight (%)	Dry Degree	Avg. Temp.	Infection Rate (%)
Lhasa	More-infected Area	3680	244.6-289.2	63-69	0.04-0.16	8.3-12.5	54
Rikaze	Same As Above	3835	277.0-323.2	72-77	0.01-0.07	7.7-11.8	70
Jiangzi	Same As Above	4040	269.8-302.2	70-72	0.03-0.06	5.5-9.5	64.2
Chayu	Non-infected Area	1700	126.7-131.5	40-31	2.05-0.76	15.3-18.7	-

The willow rust fungus is specific to the plateau region. It is highly adapted to the semi-dry weather and ecological conditions of the plateau, and is drought-resistant, sun-loving, and sun radiation-resistant. From the middle of April to the middle of May, the urediospore stage of the willow rust fungus commonly occurs. Using Lhasa as an example, from April to May when the willow rust disease occurs, sunlight hours are at least two times that of non-infected areas (table VI-3-5). In the Chayu region, a few willow trees are manually planted and willow rust disease has not been found, showing that this disease mostly occurs under semi-arid ecological condition.

In the Lhasa region, from the beginning of May to mid-May, yellow flower-shaped leaf rust lesions occur on the side branches of willow trees in clusters (Photo VI-3-2). Uredia develop as orange pustules on both sides of infected leaves. In areas like Rikaze and Lhasa, willow rust does not form telia. According to observations made in Lhasa, it mainly survives winter in the form of mycelium or urediospores. Uredia form the next spring, which is the first source of infection. These characteristics of the infection cycle reflect the ecological characteristics of the rust disease of warm areas of semi-dry, rare-forest regions.

Prevention and cure method: Destroy the source of disease with the help of pesticides. Each year watch for the disease at the end of April. When yellow flower-shaped infected leaves form, pick them immediately and burn them to make sure spores are not released and spread. During the infection period, every 10-15 days, spray 1% Bourdeaux mixture (Shi 1979, p. 51) or sodium p-aminobenzene sulfonate (Shi 1979, p. 383) diluted by 200 times.

3.1.5 *Seabuckthorn (*Hippophae*) Forest Decay Disease*

The seabuckthorn forests of Tibet are mostly seen in valleys and riverbanks from 2800-4700 meters in altitude. This is the main forest-making tree type of the semi-arid regions of Tibet. In Longzi County, the seabuckthorn have already become a forest. Decay is very common in *Hippophae* forests (Table VI-3-6), mainly caused by seabuckthorn white rot (*Phellinus robustus* (Karst.) Bourd et Galz.) (Photo VI-3-3). The infection rate is high. For instance, over 80% of the trees in the *Hippophae* forest in Rikaze, Dongfeng, and Linka are decayed. Wood volume loss per tree reaches 70%. During the beginning stage of the decay, the wood changes color, then slowly changes to yellow-white decayed wood with obvious lines. Decayed wood often infects the sapwood, affecting the phloem, so the xylon stops growing. The long-term effect is the flattening of the trunk or healing of the canker tissue. If serious, decay can cause the tree to die.

Table VI-3-6. *Phellinus robustus* Decay Rate

Tree Type	Region	Altitude (m)	Avg. Tree Height (m)	Avg. Dia-meter (cm)	Avg. Age (year)	Infection Rate (%)	Avg. No. of Fruiting Bodies Per Tree
Hippo-phae rhamn-oides	Rikeza	3860	6	16	8	78	3
H. salici-folia	Bangxing & Jilong	2840	5	12	10	51	2
H. neuro-carpa	Leiwuqi	3400-4300	6	13	10	34	1

Prevention and cure methods:

1. Improve forest management methods. In areas like Shannan and Rikaze where the *Hippophae* are concentrated, tending and managing should be improved to prevent cattle from destroying the forest. Keep a good sanitation condition. Make sure enough air and sunlight go through

the forest. Pruning should be reasonable. Wounds from pruning should be painted with paint or oil to prevent infection by fungi.

2. Pluck fruiting bodies as they appear. This disease is perennial. Each year a new fungi body form on the old fruiting bodies and release basidiospores. This becomes the source of the seabuckthorn decay fungus, so it must be destroyed first.
3. Surgery. When conditions permit, the wounds of decayed big trees have to be scraped off, filled up, and painted with anticorrosive.

3.2 *Tree Insects and Their Control*

The main pests of the forest-plantation and ornamental trees are the following:

3.2.1 *Cerura manciana Coore*

This insect mainly affects poplar and willow tree leaves. It is distributed in regions where poplar and willow are ornamental tree types and is one of the more numerous pests of the poplar forest of Lhasa. According to Caigongtang Lhasa city plantation tree block research, an average of 49 cocoons form on each poplar tree per year.

The adult has a body length of about 24 mm, wing-spread of about 66 mm, and is gray-whitish. The base of the proala has a few black spots. The front has an elliptical-shaped circle. In the middle of the wings, there are a few rows of black wavy horizontal stripes. The outside part of the wing has 8 black dots. The underwing is gray-whitish, and the outside part has 7 black dots. The abdomen and back are gray-black. The egg is semi-circle shaped and auburn colored. The larva is 50 mm long. A pair of tail-branch out to the back of the body. The metathorax has an angular-shaped protuberance. The pupa is elliptical and brown-colored inside the hard-as-wood cocoon. The cocoon is gray-brown just like the color of the tree bark.

Prevention and cure methods:

1. In winter and spring before the adults appear, organize people to hit the cocoons with iron objects.
2. During the larvae stage, spray stomach insecticide or contact insecticide such as 50% Redlamone.
3. Use lights to attract and kill adults.

3.2.2 *Willow Chafer*

In the seed gardens of Lhasa, willow chafer occurs on a large scale in the end of May. On average each tree has 13 adult chafers. According to research, chafer occurred on a large scale in 1972. On each tree an average of 0.5 kilograms of larvae can be found. The adults mostly affect the young leaves at the top of the boji willow tree, eating all the leaves and causing the tree to wither.

There are two main chafers of the seed gardens of Lhasa. One is chafer (*Brahmina* sp.). Adults are 13 mm long. The sheath is yellow-greenish. Their prothorax, front of the back, and head are all black. Back of the back is yellow-greenish. The other type of chafer is 6.5-10 mm long, its prothorax and back are bright green, and the sheath is orange-brown.

Prevention and cure methods:

1. Before planting, amend the soil with 1.5-2 kilograms of 6% "666" mixed with 30 times fertilizer, spread evenly, then plough. 1.5-2 kg placed under each tree. 30 times fertilizer mean that 45-60 kg of fertilizer been used.
2. Spray insecticide during the adult stage; see *Leptomias acutus* Aslam prevention No. 2.
3. Manually catch adult chafers. Where electricity is available, use black light lamp to trick and kill chafers.
4. Grow castor in gardens and fields. Use the castor oil plant to paralyze and kill the chafer.

3.2.3 *Garden Pea Night Moth*

Garden pea night moth is a pest very adaptable to the plateau. In 1964 and 1975, it occurred on large scales in Rikaze. This moth eats many different things, other than affecting wheat, garden peas, potatoes, napa cabbage, turnip, clover, mallow, sunflower, broad beans, and all kinds of flowers. In seed gardens, it also affects the young leaves of willow trees, poplar trees, and fruit trees by eating all the young leaves and causing the young trees to die.

Garden pea night moth is distributed in Nanmulin, Jiangzi, and Lhasa. This moth survives winter in its pupa form. From the end of May to mid-June, it lays eggs. It likes to lay eggs on mallow in a neat order. Each egg-block has on average 55 eggs. July to August is the most harmful stage of the larvae. Usually during mid-July, young larvae spin and spread with wind. This stage lasts for 20 days to one month. Older larvae tend to fake death and have an increased appetite. Usually after August the effect clearly lessens.

Prevention and cure methods:

1. By the end of June, use lights to trick and kill adults that haven't laid eggs.
2. When the moth grows older, organize manual catching and killing by digging deep ditches in or around seed gardens to trap the moths and kill them.
3. During the occurrence stage, spray insecticide. Refer to Ceruridae prevention and cure method.

3.2.4 *Leptomias acutus and Leptomias semilircularis*

These insects used to be called big-leaved *Leptomias* collectively, including two *Leptomias* species. One is sharp-horned Himalayan weevil (*Leptomias acutus* Aslam) (Photo VI-3-5). The other is half-round Himalayan weevil (*L. semilircularis* Chao) (Photo VI-3-6). They affect nursery cuttings of willow, poplar, and young shoots of apple trees in seed gardens. According to research in seed gardens, sometimes more than 100 *L. semilircularis* can be found fallen off of 1 m^2 of *Salicaceae* shoots. This *Leptomias* is only found in the Himalayas.

Leptomias acutus is distributed in Chaya, Linzhi, Yadong, and Rikaze around 2100-3900 meters. *L. semilircularis* is distributed in Linzhi, Milin, Zhanang, Naidong, Qushui, Lhasa, and Renbu around 2900-3600 meters.

L. acutus has a long body, either elliptical or long-egg shaped. Males are more flat. The two sides of the coleopteroid body and the two sides of the femur have all red, golden, and green shiny scales. *L. semilircularis* is similar. The body is more flat. Males are 11-13 mm long and females are 11.0-14.5 mm long. They have obvious white marks on the sides of the prothorax and back. The scales behind the wing slope are less colorful but more densely forming a semi-circular-shaped mark.

Adults lay eggs in weeds near the seed gardens, larvae survive winter in the pupa stage, and adults fly into seed gardens, affecting the trees from May to July. Adults have the habit of faking-death. They usually appear in mass by nightfall.

Prevention and cure methods:

1. Take advantage of their habit of faking death to collect and destroy them at nightfall.
2. Use lights to trick and kill them.
3. Clean the weeds, and use poisoned bait to lure and kill them.

3.2.5 *Scale*

The scales of Tibet kill more than 30 tree and shrub species, especially the planted tree types poplar, willow, and fruit trees of the dry and semi-dry areas (Table VI-3-7).

The region of Lhasa has the most scales. The wax scale *Pulvinnaria targioni*-Tozzettithat affects apple, plum, apricot, and other flowers occurs commonly. 90% of the branches of the plum trees imported from India and 100% of the Hydrangea are attacked. The scale insect (*Rosanococcus* Kanda and wax scales (*Ceroplastes* Gray) attack and concentrate on the trunk and branches and suck the juices from the trees. When serious, they are like wool or wax densely distributed on the trees, causing the branches to wither and die.

The low temperature of the plateau is suited for the scales to spread its effects, and now spreading caused by humans is increasing daily. Seed gardens in Lahsa in recent years imported a massive number of scaled young trees and fruit trees without any quarantine before they were planted. Scales came to the Tibet Plateau with the seeds and packages. Under suitable conditions, very serious mortality can result. For instance, the citrus wax scales (*Pulvinaria citricola* Kawana) that came with fruit trees imported from India have massively reproduced and spread in Lhasa, affecting 6-7 tree types and flowers, some of which have been affected beyond cure (whole trees had died).

Prevention and cure methods:

1. Improve quarantine of the plants. The eggs of scales often spread with seedlings. When transporting young trees, check strictly, and if found, eliminate the infested trees immediately.
2. Treat with pesticide. Before budding in winter and spring, spray Lime sulphur, five percent solution. In summer, use 0.3-0.5 percent lime sulphur. Start spraying when young larvae first appears, and once every half month, thereafter for 2-3 consecutive times. Or use diluted dimethoate diluted 3000 times, 50% 1059 diluted 1500 times solution, fluoroacetamide diluted 1000-1500 times. These are all effective against young larvae.
3. Local biological control (using natural herbicides) can also be used, such as the root of langdu (*Euphorbia fisheriana*) plus alkali or boiled nitre. Spray to cure.

Table VI-3-7. Scale mortality and distribution on Tibet trees and shrubs

Name of Scale	Host	Seriousness	Altitude (m)	Region
Matsucoccus sinensis	Himalayan Pine	+	3300	Jilong
Crisicoccus pini	Himalayan Pine	+	3100	Yadong
Same As Above	Fir	+	3400	Tuodang
Pulvinaria vitis	Poplar, Willow, Rose, Birch, *Cotoneaster*,	+++	2600-3300	Tuodang, Jilong, Lhasa, Rikaze, Angren
Lepidosaphes ulmi	Poplar, Willow, Apple, Azalea, *Sorbus*	+	2500-3900	Yadong, Jilong, Rikaze, Lhasa, Zhangmu, Quxiang
Pulvinaria citricola	Tea, Camellia, Hydrangea, Mallow, Orange, Cherry, Plum	+++	2200-3680	Lhasa, Zhangmu, Jilong
Pseudococcus comstocki	*Cinnamomum*, Mulberry	+++	2100-2800	Zhangmu
Amnonostherium prionodes	Tea, Wild Jasmine	++	4500	Bashu
Drosicha corpulenta	Apple	+	3680	Lhasa
Chionaspis salicis	Willow	+++	3600-3900	Lhasa, Rikaze
Helmet Scale	*Amorpha*	+++	3680	Lhasa
(*******)	Cherry, Plum	+++	3680	Lhasa

3.2.6 *Poplar and Willow Aphis*

Aphis is mostly distributed on Poplar and Willow trees of forest-plantation regions. It usually only affects leaves and young branches, causing the leaves to curl and branches to wither, and the trees to defoliate early. It is often seen in Lhasa, Rikaze, Linzhi, and Bomi. The main types are listed in Table VI-3-8.

With dry weather, many generations can form in one year. The insects are often concentrated on the back of leaves, sucking the juices or forming branch galls, and causing branches and leaves to deform. There are alatus virginogenia female aphis and apterous virginogenia female aphis.

Prevention and cure methods:

1. In spring, patrol the seed gardens or planting grounds. If occasional infected leaves are found, pick and destroy.
2. In spring, dimethoate diluted 6000-8000 times can be used to spray once every week for 2-3 weeks.

Table VI-3-8. Poplar and willow *Aphis* mortality rates

Name	Host	Parts of Tree Affected	Infection Rate (%)	Altitude (m)	Region
Pterocomma bailangense	Green *Populus*	Small Branches	30	3800-4000	Rikaze
Pemphigus chomoensis	White *Populus*	Leaves, Insect Galls	40	2850	Yadong
Pemphigus tibetensis	Green *Populus*	Branches	30	2000-4020	Bailang
Callaphis nepalensis	Willow	Leaves	30	2800	Yadong
Cavariella thasana	Willow, Clover	Young leaves, Young stems	50	3400-4000	Rikaze, Dalong, Boli
Pterocomma tibetasalicis	White Willow	Young branches	50	3800	Rikaze
Elatobium chomoense	Willow	Young leaves and shoots	35	3800	Jilong

4 PEST PREVENTION AND CURE OF FRUIT TREES, WALNUT, AND TEA TREES

The special natural and geographical environment of Tibet creates suitable conditions for the development of many types of fruit trees. According to research, the resources of wild as well as planted fruit trees are very rich here. Almost all the fruit trees from the tropical zone to the cold-temperate zone can grow here. The highest garden is South Wood forest situated at 4100 meters above sea level. The quality, color, and sweetness of the apples grown on the plateau surpass those of the apples grown in inland China, giving them high economical values.

According to the natural conditions of the distribution of the fruit trees and the occurrence of diseases, the diseases of fruit trees of each ecological zone have the following characteristics.

Plateau semi-tropical zone: including Chayu, Chuona, Yadong, Nielamu, and Jilong. This is the north border of the tropical zone, altitude 1100-2500 meters, but the fruit trees planted are still mostly those of the temperate zone and subtropical zone. The main diseases and insects, such as apple leaf blight (*Phyllosticta pirina* Sacc.), pear venturia (*Venturia pyrina* Aderh.), grape white mold (*Plasmopara viticola* (B. et C.) Berl. et de Toni), and citrus scale, are very serious. Collectively speaking, because the planting history of fruit trees is short, diseases are still rare, but new diseases such as *Aleurtes* bacterial leaf spots have been found. Because of the relatively warm temperature of this region's winter, which is suited for pests to survive, fruit tree diseases seem to become more serious with time.

Plateau warm-temperate zone: This zone is the distribution area of the famous defoliation fruit trees of Tibet. Regions like Jiacha, Milin, Linzhi, Bomi, and Yigong have altitude 2500-3300 meters and are regions where fruit tree diseases regularly occur, such as apple leaf blight, apple powdery mildew, apple sclerotium*** root rot (*Sclerotium rolfsii* Sacc.), apple flower blight, apple canker, and peach leaf curl disease (*Taphrina defomans* (Berk.) Tul.), all of which often cause epidemics in this area. Pests include leaf-curling bug, chafer, and *Leptomias*. In gardens with a long history, there are also apple powdery mildew and peach leaf curl disease. This region should be a focus of prevention and cure of fruit tree diseases.

Plateau monsoon cool-temperate zone: including the middle parts of Yaluzangbu River, such as the wide valley areas of Lhasa River and Nianchu River, altitude 3300-4000 meters. Because of the sun radiation of the plateau, fruit trees are often burned by the sun.

4.1 *Fruit Tree Diseases, Prevention and Cure*

According to characteristics of symptoms of serious fruit tree diseases, the following key is listed:

1. Half of tree mortality — Apple Half Wither
1. Whole tree mortality
    2. Roots dead, fan-shaped mycelium inside of root bark. — White Sclerotium Root Rot
    2. Trunk mortality
        3. No disease symptoms on trunk, small shrubs parasites on trunk — Apple *Loranthus*
        3. Disease symptoms on trunk
            4. Disease symptoms are orange-colored fruit bodies — Apple Canker
            4. Disease symptoms are dark black fruit bodies
                5. Pustules present, bark concave in canker — Apple Dry Rot
                5. Pustules present, bark not concave in canker — Fruit Tree Canker
    2. Leaf mortality
        6. Disease symptoms on leaves
            7. Mostly dark colored symptoms on leaves — Apple Leaf Blight
            7. White disease symptoms on leaves
                8. White mycelium causing a dusty appearance, not shiny — Apple Powdery Mildew
                8. White spots causing a shiny appearance — Apple Discolored-leaves
        6. No disease symptoms on leaves
            9. Leaves dwarfed, no spots — Apple Small-leaf Disease
            9. Leaves turn yellow — Apple Yellow Disease
    2. Flower and fruit mortality
        10. Disease signs present
            11. Light brown, gray, or pink colored lesions, affects flower, fruit, and tip of leaves — Apple Flower Blight
            11. Black lesions, only affect fruit — Apple Anthracnose

10. No disease symptoms Apple Fruit-shrink Disease

4.1.1 *Apple White Sclerotium Root Rot (Sclerotium rolfsii Sacc.)*

The fruit trees of Zamu, Danka, Zalong, and Yigong of Bomi are mostly affected by this disease. The infection rate of the fruit orchards of Zamu Forest Management Station reaches 30%, and that of the fruit gardens orchards of Zalong reaches 36%. All apple varieties are attacked. No big differences appear between varieties, among most of the affected trees, the whole tree withers and dies.

According to observations made in fruit orchards of Bomi, Danka, and Zamu, symptoms of the initial stage are only small brown spots on roots. The trunk shows no symptoms. Mycelium spreads through the roots. When lesions have spread to encircle the main root, the skin of the side roots soften and decay. After some time, the leaves shrink and fruits stop growing. White fan-shaped mycelium appears inside the bark, which is the main characteristic of white silk disease. During later stages, the trees can fall with a push. Nearby trees are easily affected, and trees often wither and die in a cluster.

The cause of this disease is because the fruit trees are planted in high mountain oak forests, and white silk fungus (*Sclerotium rolfsii*) is common among the roots of these oak trees. It's very hard to destroy the fungus completely when establishing a fruit orchard, causing white silk disease to spread to the fruit trees.

Based on the occurrence pattern of the disease, the following suggestions are made for disease management. 1. Don't plant fruit orchards in high mountain oak forests. If necessary to plant in high mountain oak sites, make sure all the roots are removed and disinfect the soil before planting new trees. 2. Fruit trees already infected need to be checked regularly. Once infected trees are found treat immediately. Remove decayed roots as soon as possible. In winter, soak infected roots in sunlight and change the soil near infected roots. Paint 2.5% PMA (Shi. 1979 P. 334) diluted 300 times or PCP (Shi 1979 P. 312) diluted 250-300 times on infected trees, or use tung tree oil plus potassium permanganate to paint the roots. All of the above can be effective.

4.1.2 *Apple Flower Blight, Brown Rot*

Apple flower blight can infect the fruits, leaves, and branches of trees, but fruits are the most seriously damaged (Photo VI-4-1). It is collectively called flower blight. Other than apple trees, pear trees are also infected. The fruit orchards most damaged are those of Danka, Zamu, Yigong of Bomi, and Linzhi, Bayi New Village. The fruit garden of Danka loses 5000 kilograms of apples due to the flower blight every year. All varieties are infected, such as red jade, national light, green banana, and Zhuguang. Red top variety is most seriously damaged. The infection rate in Linzhi fruit orchards is 46.7%. Varieties that are resistant to the disease include red banana and sweet banana.

This pathogen belongs to the Ascomycotina, Helotiales order, Sclerotiniaceae family, *Sclerotinia* Fuck. Mainly, the conidiospore stage causes infections. Conidiophores are 3-4 μm long and grow in a cluster. Conidia are hyaline, branched or not, uniseriate, and get thicker after maturation. Large conidia are lemon-shaped, hyaline, single celled, and 12.0-16.4 X 8.4-13.0 μm. Sall conidia are globulose. The sclerotium is like rat pellets and black. Sometimes apothecia can be seen on diseased parts.

In Linzhi fruit orchards, the disease cycle is as follows. Leaf necrosis occurs during initial stages of leaf-opening (end of April to beginning of May). Red-brown small dots appear on leaves and the base of the petiole. After infection, the leaves wither and hang down. With high humidity, powder-shaped conidia form on the back of the leaves. The fungus spreads from the base of stalks to the base of flowers, causing the flowers and stalks to turn brown and rot, and the flowers to hang down. According to records, fruit rot is caused when the fungus infects through the petiole and reaches the gastrula through the pollen duct. Then through the ovary walls it reaches the surface. Disease symptoms start appearing on infected young fruits. A yellow gluey substance forms on the lesions, and the whole fruit quickly rots and ferments. After drying, the fruit appears shriveled and hard. When the fungus spreads from the stalk down to the rachis and then to the branch tips, branch rot develops in the form of cankers. Diseased areas sink in and dry. From leaf rot to fruit rot there is about 60 days (4/20 to 6/25) of concentrated infective period, especially before and after flowering. The flower blight fungus forms sclerotia on fallen fruits, infected leaves, infected fruits on the tree, and infected bark to survive winter. When the temperature is above 5 degrees Celsius and the water in the soil is between 30-40%, apothecia form. Ascospores are spread by wind, infecting leaves, flowers, and bark. Spores form on infected leaves and flowers and infect the petiole, causing fruit rot and flower rot. Because the fruit orchards are situated in damp plateau forest regions in the monsoon warm-temperate zone, the temperature and humidity factors are suited for the spreading of flower blight.

Prevention and cure method: According to the characteristics of the occurrence and effect pattern of flower blight in this area, between the end of April and the end of May. Spray 0.5 % lime sulphur once every 10 days for 4-5 times. Pay special attention to the prevention of infection before and after flowering. Prevention by fungicide treatment should be combined with destroying the inoculum sources. Before budding, prune out the branches that were cut down in winter. During the growth period when diseases are severe, pick and destroy all the infected branches, leaves, fruits, and flowers to reduce the spread of the effects.

Fruit gardens should be reasonably pruned to allow air and light to go through the tree crown. Pay attention to increasing the fertilization and cleaning out the weeds to increase the trees' ability to resist the disease. In newly planted gardens, different types of trees should be planted. Prevent planting the same infection-susceptible varieties.

4.1.3 *Apple Powdery Mildew*

Apple powdery mildew infects apples of the following regions most seriously: Zamu of Bomi, Danka, Qingduo, Linzhi, Milin, and Lakang of Luokong. In recent years it seems to be increasing. Infected varieties include red jade, national light, huanong No. 1, Zhuguang, wojing, and malus. Malus has the most mortality and is the first inoculum source for apple powdery mildew. Red banana, yellow banana, and green banana are less susceptible to infection.

Apple powdery mildew belongs to Ascomycotina, Pyrenomycetes (Muller et al., 1973), Erysiphales order, Erysiphaceae family, *Microsphaera* Lev. Genus. The sexual stage is rarely found. Spores of the asexual stage often distribute on the leaves in the form of white powdery mildew. Mycelium is hyaline, transparent and branched with membranes. Spores are hyaline. A single spore is elliptical-shaped. Appendages are usually dichotomously branched at the apex. The uniseriate size is16.4-26.4 X 14.4-19.2 μm.

Apple powdery mildew mainly infects inflorescences, young shoots of phyllotaxy, and early stages of the fruit. It also infects young branches.

The winter buds of infected branches are mostly diseased shoots, which are a little redder than healthy shoots. The distance between each segment of the infected branches is shorter. Leaves are long and narrow with edges curled up, and are thicker with a hard and brittle texture. If new leaves are infected, gray-whitish disease symptoms appear on both sides, causing the leaves to wrinkle and twist. During flowering in spring, the shoot is short and thick and cannot stretch out. Flowers are deformed. Petals shrink and are hard and brittle like wax, and can fall with one touch of the hand. Infected flowers cannot bear fruits, reducing the production. When fruits are infected, white powder appears on the bark, then disease symptoms show on the tip of the fruits, which turn into rust marks. The skin of the fruit becomes hard and rough, causing malnutrition of the fruits. With smaller sizes, there is lesser quality and less production.

In Linzhi and Bomi, apple powdery mildew occurs from late April to early June, with an infection period of about 50 days. The fungus hides in the scales of infected shoots in the form of dormant mycelium to survive winter. The next spring when leaves are just opening, white powder first appears on newly opened leaves and on the stamen and pistil. The spores on leaves, inflorescence, and new shoots are spread by wind to infect young shoots and leaves.

Prevention and cure methods: Experimental results show that reducing the inoculum is key to management of this disease. Especially in gardens located in virgin forest regions, first destroy all wild Chinese flowering crabapples. Then cut out infected branches and shoots with winter pruning, which can be repeated for a few years for seriously infected trees. Cut off at least 2/3 of the infected branch. Reducing the source of the fungus can reduce the disease. During spring when the disease occurs, cut out the infected shoots and branches in time, and combine continuous prevention and focused prevention. From opening of the leaves to the early stage of fruit maturation, spray asomate diluted 600 times once every 8 days for a total of 4 times. In areas with serious flower blight, 0.5 % lime sulphur should be sprayed once every 7 days for a total of 5 times, but during the peak period of 10 days, spray asomate instead of lime sulphur. In recent years, the following have also proven effective: 50% thiophanate diluted 800 times, 50% carbendazol diluted 600 times, 50% benomyl diluted 500 times, ferrisulphas 50% diluted 500 times, and ferrisulphas diluted 300 times. Also, the management of the gardens should be strengthened. Increase phosphorus and potassium in the fertilizer, improve forest soil, and improve ventilation and sunlight. Varieties of trees that are resistant to the disease should be planted in infected areas. Recently there have been reports on chemical treatments in winter, such as using pesticides to eliminate the dormant mycelium in infected shoots, but this has to be tested before being widely implemented.

4.1.4 *Apple Leaf-brown Spot Disease*

Leaf-brown spot disease is a spot disease of the leaves caused by a Deuteromycete. Each fruit orchard of Tibet has been infected to different degrees, of which the regions of Chayu and Bomi Zamu are most severely infected. When trees are infected, leaves often defoliate 1 to 1.5 months early.

According to research on the infection level of each variety of apples in different fruit orchards, Tanglizi, national light, yellow banana, and red jade are most infected, and red banana, huangkui, and bell fruit are most lightly infected (Table VI-4-1).

The main fungus that causes apple leaf-brown spots disease is *Phyllosticta pirina* Sacc. (Photo VI-4-3), which belongs to the Deuteromycetes, Coelomycetes (Sutton, 1973), Sphaeropsidales order, Sphaeropsidaceae family, and *Phyllosticta* Per. ex Desm. Genus. *Phyllosticta* symptoms are gray-white globulose, rounded brown spots, or irregular-shaped large brown spots. Conidiophore are 87-157 X 60-150 μm. Conidia are obrotund, hyaline, and 3.2-4.9 X 2.5-4.9 μm.

Phyllosticta pirina mainly infects leaves, but also branches, young shoots, petioles, and fruits. Often, different symptoms show due to differences in the development of leaves, the temperature, and the humidity. There are roughly two types of symptoms of the Tibet apple leaf-brown spots disease:

1. Brown spots: Most apples are this type. During the early stage, leaf spots are as small as a needle point. The light brown turns to gray-brown. On the spots, there are as many as 10 black pycnidia. Then the spots turn to purple-brown, especially on the edge, after which the spots turn lighter.
2. Big brown spots: Chayu fruit orchard has most of this type, big and irregular lesions. When serious, this type of disease development often causes leaves to wither.

The dormant fungus on leaves, branches, fruits, and cankered plants occurs in the form of mycelium or spores. The following year when the temperature and humidity are appropriate, it spreads with wind and rain. In Linzhi fruit orchards, early June is when the initial stage of the disease develops. Mid-July to late August is the peak time of development.

Prevention and Cure Methods:

To treat leaf-brown spot disease, first the disease source should be eliminated. In the fruit orchards, the fallen leaves should be cleaned in the fall and burned. The infected leaves and branches remaining on the trees should be pruned. In summer, the infected leaves should always be cleaned up. The management of water and fertilizer should be improved. Prune reasonably to increase the vitality of the trees. Choose disease-resistant tree types. Quarantine infected plants. In gardens already infected, choose how many times the pesticide is to be applied according to the type of the leaf-brown spot disease and the weather factor. The spray interval time and the type of pesticide to apply are based on the occurrence pattern of the disease in fruit orchards of the middle and lower parts of the Yaluzangbu River. Spray 10 days after the first day of the disease and once every 10 days thereafter to control the spread of the disease. Fungicides often used include Bordeaux mixture (1:2-4:200) (Shi 1979 P. 430), thiophanate methyl solution 50% diluted 500 times, 50% Tuzet diluted 800 times, and Feng Guan mei su, 50-100 units. Of these Bordeaux mixture has a long effective period and applies widely, having a good preventative effect. Each region should conduct experiments to predict the spread of the disease and understand the correct occurrence pattern to be able to receive anticipated effects.

Table VI-4-1. Brown-spot disease on the main Tibet apple varieties

Apple Variety	Yellow Banana	Red Banana	Green Banana	Na-tional Light	Huangku i	Wo jing	Zhu guang	Hong kui
Infection Level	M-S	L	L-M	S	L	M	M-S	S

Apple Variety	Pound Apple	Red jade	Xu	Jin zhu ma	White jade	Green sweet	Tangli zi	Bell fruit
Infection Level	L	M-S	M	S	M	M	S	R

Note: Resistant (R): no brown spots, Light (L): only a few brown spots; Medium (M): 1/3-1/2 of the leaf has brown spots; Serious (S): ½ to the whole leaf is covered with brown spots

4.1.5 *Peach Leaf-Curl (Taphrina deformans (Berk.) Tul.)*

This disease is found in Linzhi and Bomi fruit orchards. In the fruit orchard of Bomi Dange, 600 trees died due to the disease. In Linzhi fruit orchard, 700 trees died. It has not been found in Lhasa and Rikaze. According to observations of the Bomi Zamu fruit orchard, this disease mainly infects leaves, young shoots, and young branches. Infected leaves wrinkle and curl (Photo VI-4-2). With the opening of the leaves, the curling increases. Leaves are thick, at first gray-green, then turning peach-red, and then dark red. Infected leaves are covered with gray-white powder (sporangia and spores). This disease spreads fast. The leaves turn brown quickly and fall. The occurrence of this disease is closely related to ecological factors. From the beginning of May to the beginning of September, this disease occurs in the Bomi region. Humidity affects the occurrence of this disease. In different areas, the occurrence is very different. Usually it occurs most seriously under low temperature and humid weather. In Linzhi and Bomi, this disease has prevented peach trees from growing. These regions have stopped growing peach trees altogether.

Because of the dryness level of 1.5-5.0 in the plateau, the disease rarely occurs;. The dryness of Bomi is 1.0, and infection rate can reach 98% (Table VI-4-2).

Table VI-4-2. Infection rates of peach leaf-curl disease of different regions

Region	Fruit Tree Distribution Zone	Altitude (m)	Type	Dryness	Susceptibility Rate (%)
Lhasa	Plateau	3680	Gang white	1.5-5.0	0
Linzhi	Temperate Zone	3000	Same As Above	1.0-1.5	60
Bomi	Temperate Zone	2800	Same As Above	1.0	98

Prevention and cure methods:

1. In spring when peach flowers are just budding, carefully spray once lime sulphur 5% solution or 1:1:100 Bordeaux mixture to kill spores from winter and destroy the first infection source.
2. After flowering and opening of the leaves, spray 0.5% lime sulphur solution or Tuzet diluted 1000 times once every 10-15 days. Spraying a few times can control the occurrence of the disease.
3. In winter, cut off infected branches. In the first stage of the disease, check regularly, and once infected leaves are found, pick immediately and burn.

4.1.6 Sunburn

This is a type of physical disease. It usually occurs in regions above 3000 meters, and is seen mostly in Lhasa, Rikaze, Jiangzi, and Changdu. It often occurs on the inside of the fruit tree trunks and the trunks towards the side of the sun, causing the trunk, crown, and branches on the south side to dry. This is why it's also called "half dry." The affected parts often get dry canker disease. Dry canker disease includes dry canker disease and canker disease.

Dry canker disease is caused by strong sun radiation and the physical drying of spring. Fungi infect where the tree bark is burned, causing irregular brown spots on the surface of the trunk. Sometimes a dark brown liquid (may be gummosis) comes out of the diseased area. After the tree loses water, the spots on the infected area are concave, and many small black pustules form on the spots, which are the conididia. If the condition is serious, the lesions connect and circle the trunk or branch, causing the tree to wither and die.

4.1.7 Canker Disease

Caused by the canker fungus (*Cytospora spp.*), there are two kinds of symptoms. One is canker, occurring mostly on the sunny side of the main trunk and big branches. Infected tissue is soft and concave when pushed, and yellow-brown liquid flows out. The tissue is easy to rip and peel, and then sinks in to form big black-brown cankers. The other kind is branch wilt. This often occurs on small branches. Cankerss occur on the whole branch, causing the leaves to wilt.

The fungus infects by mycelium and conidia. Spores are spread by wind and rain, and infect through wounds. In Lhasa, the disease occurs most seriously during May to June.

Prevention and cure methods:

1. Strengthen management of planting to improve the resisting ability of the trees. Increase the usage of phosphorus and potassium fertilizer to increase the ability of the trees to resist cold and drought. Improve the soil to increase its water storing ability. During the dry season, remember to water the plants.
2. To decrease the effect of the plateau sun radiation, paint the tree trunk white. Ingredients of the paint: 5 kilograms of lime, 0.5 kilograms of salt, plus bluestone, mix into gluey liquid to paint the trunk.
3. Increase monitoring for the disease. Scrape and treat the diseased areas. Scraping should be thorough to eliminate pathogen's infection court. Paint the trunk with Asomate after scraping to disinfect.

4.2 *Fruit Tree Pests*

Fruit tree pests often seen in this area include leaf mites, leaf beetles, chafers, weevils, and scale insects.

4.2.1 *Leaf Mites*

According to research, in fruit orchards of this area with a long planting history, leaf mites have become more serious recently. They most commonly occurs in semi-drought regions such as Lhasa and Rikaze. Of the leaf mites that affect apples, Chinese flowering crabapples, and pears, hawthorn leaf mite is the most serious.

According to preliminary observations, only two generations of hawthorn leaf mite occur in Lhasa every year. In mid-April, massive amounts of nymphs and adults climb along the branches and trunks to the young shoots, leaves, and flower buds. The peak stage is the end of May. The first generation lays eggs in May, and the second generation in early July. The incubation period is long. From late July to early August, and even to early September, the eggs hatch.

Mites often live in colonies on the back of the leaves near the main vein and spin webs. After mid-October, some nymphs and adults concentrate on the leaves, bark, and soil cracks of the roots to survive winter. According to observations in Lhasa, the nymphs and adults that survive winter can directly affect the flowering and fruit bearing.

Prevention and cure methods: Before and after winter, concentrate and destroy the leaf mites.

1. Before winter at the beginning of October, spider mites prepare for winter by moving to their hiding places. Bind grass to the trunk and crotch of trees to lure hawthorn leaf mites to survive winter there. Burn them during winter, or after they concentrate, spray concentrated Dichlorvos (Shi 1979 P.134) on the trunk and crotch of the trees to kill the leaf mites.
2. Clean the fruit orchard. Collect the weeds and fallen leaves and burn them.
3. When spring comes and the soil is defrosting, mound up new soil to bury the leaf mites in the old soil.
4. After budding and before flowering, spray 0.6% lime sulphur solution or 45% Dimethoate diluted 2000 times to kill the leaf mites.
5. For additional pesticide control and cure methods, see poplar and willow aphis.

4.2.2 Rolling Leaf Beetle

According to preliminary understanding, leaf beetles mainly occur in Linzhi, Bomi, and Milin.

1. Leaf beetle: Forest armyworm causes the most serious damage, then yellow-mark leaf beetle, small leaf beetle, and brown leaf beetle.

During the early stage of apple leaf opening and flower bud appearance, rolling leaf beetle gets to the inside of the leaf cluster. They also get into flower buds to affect the stamen and pistil. The infestation rate is usually around 40% and the feeding period can last 40-50 days. Because of their big appetite, the multitude of parts that can be affected, and the long feeding period, they often cause a serious reduction in production if not treated. According to observations made in Linzhi fruit orchard, leaf beetles have an effect only during the early stage of the young fruit. During later stages, they may move to the forest to survive winter.

2. Cherry brown tortrix (*Pandemis ribeana* Hubner): Occurs in Lhasa and Linzhi. Mainly affects young shoots and mature leaves. More serious in fruit orchards with lots of weeds.

This moth mainly survives winter in weeds and fallen leaves as adults. The next year from the end of March to the beginning of April, adults appear and lay eggs on the two sides of the branch buds. Larvae feed on leaves from mid-April to early May. In Linzhi, there are 3-4 generations per year. They go into dormancy from mid to late October.

3. Apple leaf beetle: Occurs in fruit orchards of all regions in the warm, semi-humid zone. Usually has a minor effect.

The early stage of this moth mainly affects leaves, and sometimes inflorescences and young fruits too. In later stages, it ties affected leaves to the surface of the fruits, and mainly affects fruits. It usually first appears when the flower petals fall or the young fruits are just forming. Larvae spin to bind a few leaves together in the shape of flower buds. When serious, the flower buds wither and die. If the growing points are affected, the branches can't grow out. Old larvae form cocoons in the cracks of the tree bark to survive winter.

Prevention and cure method: Between the stage of leaf-opening and budding and the stage of young fruit-forming, spray DDT (Shi 1979 P. 122) solution plus a little BHC once every 10-12 days for 2-3 times for good prevention effects.

4.2.3 *Chafer*

Occurs in all regions. The types that have been found include big brown chafer, apple hair chafer, tea-colored chafer, and small chafer. Of these, tea-colored chafer and apple hair chafer are most serious.

1. Tea-colored chafer:

These chafers survive winter in soil as larvae. The next year in mid-May during the end stage of flowering, adults appear in massive amounts and damage fruit trees, poplars, willows, and elms. Damaged leaves are eaten into irregular shapes. Sometimes only the main veins are left. Adults often appear on sunny days, but are usually active during night. The adults are present for a month. In late June, they lay eggs in the soil.

2. Apple hair chafer:

According to observations in Lhasa, adults go through winter in the soil. The next year, they appear by the end of April. First they affect early flowering apricot, peach, and pear. In early May, they move to feed on flower buds, flowers, and leaves of apple. Adults habitually live in a colony, have strong appetites, and can eat up buds, flowers, and young leaves. By mid-May, adults lay eggs in soil. They are inactive at morning and night, and around noon when the temperature goes up, they fly around the trees in a cluster. If shocked, they fall on the ground and fake death. Other than feeding on fruit trees, they also affect poplar, willow, and elm trees.

Prevention and cure method:

For details, see prevention and cure methods for *Leptomias acutus* Aslam.

4.2.4 *Leptomias longicollis Chao*

This weevil mainly occurs in fruit orchards of Changdu and Chaya of Tibet. Newly planted fruit orchards are most seriously affected. At the beginning of June young shoots of apple leaves are completely eaten. When serious, there are hundreds of adults per tree, causing 20% of the trees to die. This is the main pest of newly planted fruit trees of semi-arid regions.

Males are 6.3-7.7 mm, long and 2.4-2.9 mm wide. Females are 7.4-9.0 mm long and 3.0-3.9 mm wide. The body is elliptical or long-egg shaped, black, with bronze scales, smooth surface, and they emit strong light. The abdomen and each coxae has a row of feather-shaped scales. Body hair is not obvious, only the hairs at the head, the front of the proala crustacea, and abdomen are obvious. The beak is flat. Antennas are short and thick.

They go through winter in the surface soil under weeds as adults or larvae. The next spring in April, they start to affect young shoots and branches. By mid-June, they eat all the apple leaves. Adults appear mostly around nightfall. They tend to fake death. If touched, they fall to the ground immediately, then shortly they are back on the trees again.

Other pests that damage fruit trees include apple branch wood borer and pear ruler moth, about 10 kinds. For the main methods of prevention, see Table VI-4-3.

Table VI-4-3. Regions of pest problems and main prevention methods

Pest Name	Distribution Region	Main Prevention Methods
Apple branch wood borer beetle	Changdu	Fill in "666" solution and block the hole
Ruler moth	Changdu	Use "666" or DDT solution
Carposina nipponcnsis waslsingham, *Carposina sasakii* matsumura	Yadong	Before larvae leave the soil, spray "666" diluted 2000 times during the egg stage
Apple aphis	Lhasa, Linzhi	Use "666" diluted 200 times
Scale	Lhsa, Linzhi	Squash, spray diesel diluted 100 times, or spray DDT
Cherry tree borer (Conopia hector Butler)	Chayu	Scrape off , use a knife to kill
****Green-yellow lappet moth	Yigong	Spray 50% DDV diluted 1000 times
Bean moth	Linzhi, Bomi	Spray DDT diluted 200 times
Peach gall aphis	Linzhi, Bomi	"666" diluted 200 times

4.3 *Tea Tree Diseases*

Commonly seen tea tree diseases of the Tibet area include the following: leaf brown spot disease, black mold disease, anthracnose, tea root disease, lichens, and moss. Also, scale that attack tea leaves. Detailed descriptions are as follows.

4.3.1 *Leaf mark disease*

Tea leaf white spots (*Phyllosticta theicola* Petch) is seen in tea lands in Nielamu, Yigong, and Dongjiu of Tibet, affecting young shoots and stems. Gray-white lesions are irregular -shaped and concave, 0.1-0.2 mm diameter, and with obvious brown edges.

Of the tea leaf white spots, other than tea white star disease, there are also tea anthracnose (*Colletotrichum camelliae* Mass.) and tea leaf brown spot (*Gloeosporium theae-sinensis* Miyake). For their prevention and cure, see apple leaf spot disease.

4.3.2 *Tea black mildews*

This disease is found in all the tea lands, Yigong region is most seriously infected. If continuously occuring for 4-5 years, it can cause a 1/3 decrease in the production of tea leaves. There are usually two sources for the disease; tea black mildew (*Neocapnodium theae* Hara) and black star mildew (*Zukalia nanloensis* Sawada). For their prevention and cure, see apple leaf spot disease.

4.3.3 *Tea Lichens and Mosses*

When planting tea trees in areas with high amounts of rainfall (above 1000 mm), especially near forest areas, the tea trees are often affected by lichens and mosses. In Yigong tea land, lichens are often seen on tea trees, causing the branches to die back, growth to slow down, and leaves become smaller and crisp, not suitable for making tea.

4.4 *Walnut Diseases*

The walnuts of Tibet have characteristics such as wide distribution, many varieties, high adaptability, and less diseases. Especially wild walnuts rarely have diseases or pest problems. But walnuts imported from other regions have already begun to be infected by diseases, some of which are very harmful to the development, such as walnut blight, walnut bacterial leaf spot, *Batocera horfieldi*, and *Lebeda nobilis*. Following are descriptions of each.

4.4.1 *Lappet Moth (Lebeda nobilis Walker)*

Lappet moth is also called pine large caterpillar. It attacks young Tibet walnut trees. This is a main type of lappet moth (Lasiocampidae). It often eats all the leaves. In the best scenario, it can cause a decrease of tree vitality, and in the worst scenario, it can cause trees to die. The walnut trees of Yigong imported from Xinjiang are commonly attacked by this pest, causing the trees to lose their vitality, then with the help of pests like woodborers, often cause large population of walnut trees to die (only seen distributed on Xinjiang walnut trees or in orchard in oak forests near Bomi).

The adult body color is varied, from yellow-brown to tea-brown to gray-white. Usually males are darker than females. Males have wingspans of 50-80 mm, and females 75-95 mm.

The life history is unclear. It overwinters as eggs. Young larvae live and feed together. July to mid-August is the peak period of their feeding. During the day, they usually hide on tree trunks quietly and do not eat. By nightfall or morning, they come out and feed.

Prevention and cure methods:

1. During winter, destroy egg blocks manually. In fall, collect the cocoons to kill the pupa.
2. From mid to late September, use lights to kill adults.
3. Before July, spray 6% "666" diluted 400 times to treat young larvae.

4.4.2 *Cloud spots Long-horned Beetle (Batocera horfieldi Hope)*

Cloud spots long-horned beetle is an important type of tree trunk pest which affects Tibet walnuts, especially imported ones. For instance, walnut trees imported from Xinjiang in the Yigong area are planted near high mountain oak trees, allowing the cloud spots long-horned beetle of the oak forests to move to the walnut trees. Infection rate can reach 100%, and 20% of the trees have already died.

Adults are 35-36 mm long and gray-black or black-brown. There is a vertical concave area in the middle of the head. The back of the prothorax has a pair of kidney-shaped white spots. The lateral spinae is big and thick. The sharp tip is slightly bent to the back. Eggs are milky white to bleached yellow. The prothorax of larvae has a Chinese written mountain word shaped brown spot. The pupa is 40-70 mm long, and milky white to light yellow.

One generation usually lasts 2-3 years. Adults appear from May to June, chew out circular or elliptical troughs on tree trunks or thick branches, and lay eggs in them. Newly hatched larvae eat their way into the phloem, and then the wood tissue. The holes are big, diameter 30 mm. Older larvae use the end of the passage for a pupa room.

Prevention and cure methods:

1. When the egg-laying sign is obvious, use a knife to scrape off or break egg blocks.
2. Wet cotton balls with 50% phosphamidom or 50% fenitrothion diluted 40 times and place into the holes to kill the larvae. Or use 5 units of clay with 50% DDT or one unit of 6% "666" plus water to mix into toxic mud, stick with wood particle into the holes to kill the larvae.
3. Get rid of dying trees.
4. Use iron hooks to kill the larvae if condition permits.

The occurrence and distribution of diseases of plateau fruit trees and economical forests are closely related to the geographical location and the ecosystem of the orchards. Most fruit trees are distributed in mountain temperate zone high mountain oak forests, and some are distributed in alpine shrubs-grassland. The fruit trees of the above areas have the following disease problems. First, when destroying forests and plowing land, many diseases and pests originally of the forests move to the fruit trees. Serious diseases occur on apple, pear, and walnut trees, such as chestnut yellow moth, cloud spot woodborer, apple powdery mildew, apple sclerotium root rot, silver-leaf disease, and apple trunk lichen. Second, because of the dry weather, low humidity, and not enough accumulated temperature, the orchards of Lhasa, Rikaze, and Changdu of the alpine shrub-grassland are usually less infected by diseases, but some pest specific to the Tibet plateau are serious, such as Himalayan weevil (*Leptomias)*, scales, and plateau ecological drought and plant physical imbalance. Third, because imported fruit trees from other regions were not quarantined, some dangerous diseases from the older fruit orchards are brought here, such as the smaller citrus cottony scale (*Pulvinaria citricola* Kuwana) that came from India in the 1950s, which has become wide-spread in Lhasa. There are also diseases from other parts of China such as flower virus and pear venturia leaf blight.

According to the occurrence of plateau fruit tree diseases, the following should be taken into consideration for prevention and cure:

1. Research on fruit tree and economical forest diseases show that there are many diseases and pests in the oak forests, and high mountain oak forests are often the source of diseases, so destroying forests and plowing land to become fruit gardens in forest regions should be strictly forbidden.
2. When importing cuttings, scions, or stock from other regions, the plants should be carefully quarantined.
3. Tibet fruit trees have characteristics such as strong resistance and high quality. Fruit tree types that are disease-resistant should be selected and bred.
4. To prevent sun radiation and other physical damage in fruit orchards in areas with few or no forests, the trunks should be treated or painted white to prevent sunburn.

5 REFERENCES

Shi, D.Z., Hu., X.X., & Chao, W.C. 1979. Dictionary of English-Chinese pesticides. Publisher House of Industrial Chemistry: Beijing.

Zhu, H.F., & Qin, D.D. 1991. English-Chinese Dictionary of Entomology (2nd ed.) Chinese Publisher House: Beijing.

Translated from: Forests of Tibet. (translated from Chinese) 1985. Chinese Academy Press: Beijing. Reviewed by Dr. James Walla

Fig VI-1-1. *Phellinus pini* (Thore ex Fr.) Ames

Fig VI-1-2. *Polystictus pergamenus* Fr.

Fig VI-1-3. *Poria cocos* (Schw.) Wolf.

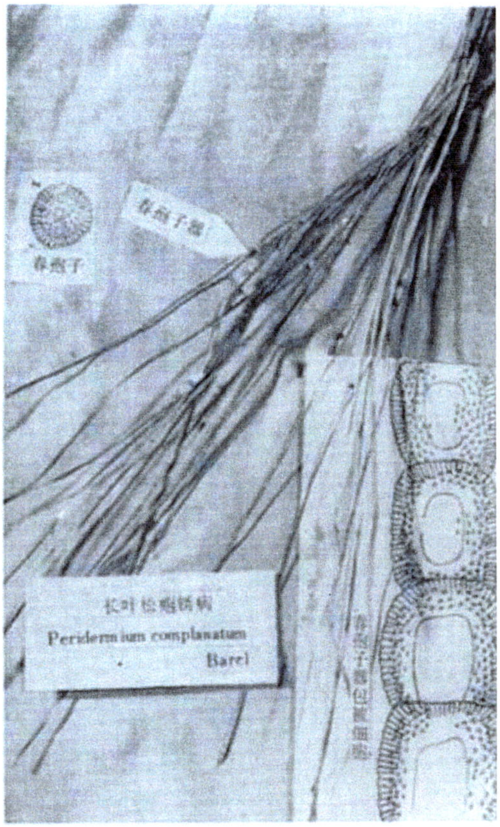

Fig VI-1-4. *Peridermium complanatum* Barel.

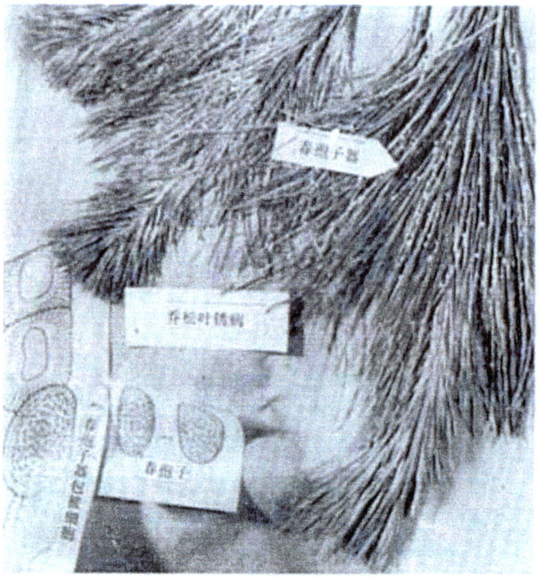

Fig VI-1-5. *Peridermium brevius*

Fig VI-1-6. *Arceuthobium oxycedri*

Fig VI-1-7. *Inonotus dryadeus*

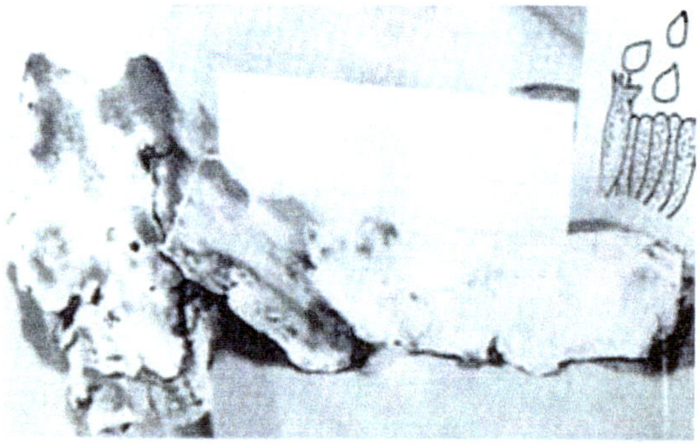

Fig VI-1-8. *Fomes annosus*

Fig VI-1-9. *Thekopsora areolata*

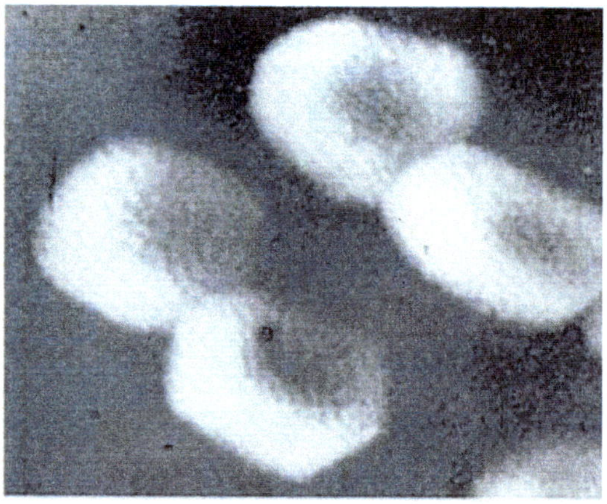

Fig VI-1-10. *Thekopsora areolata* 1000x

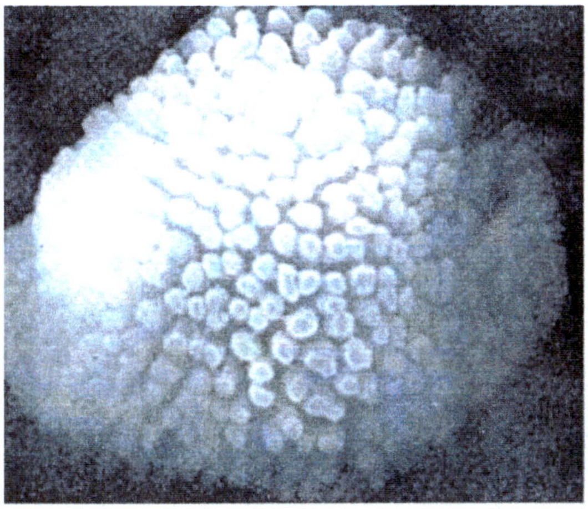

Fig VI-1-11. *Thekopsora areolata* 3000x

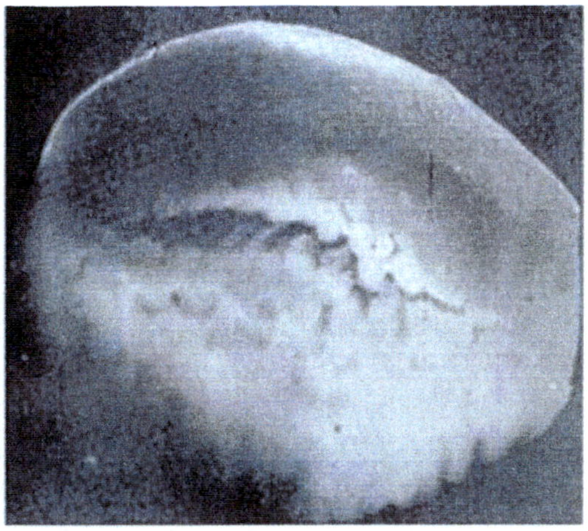

Fig VI-1-12. *Thekopsora* sp. 3000x

Fig VI-1-13. *Polyporus schweinizii*

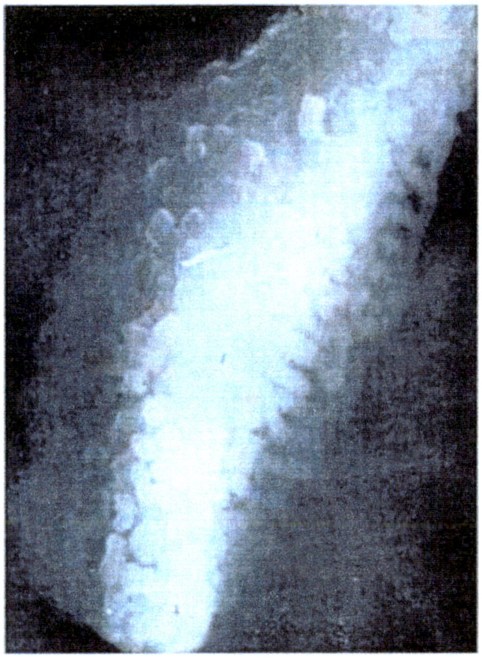

Fig VI-1-14. *Chrysomyxa picea*

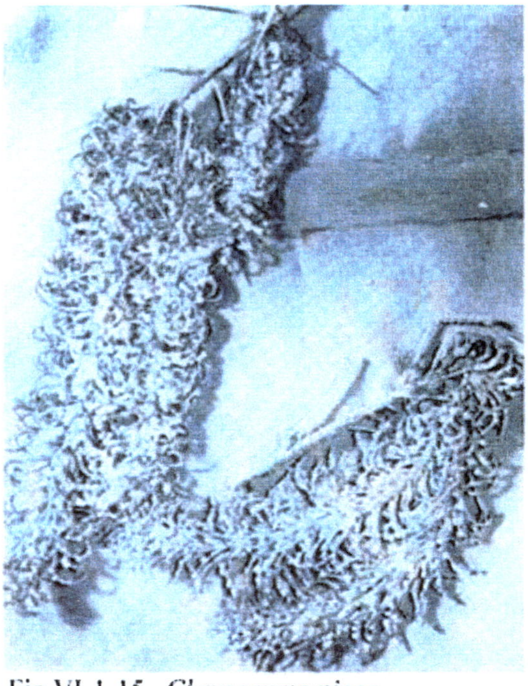

Fig VI-1-15. *Chrysomyxa picea*

Fig VI-1-16. *Phellinus pini* var. *abietis*

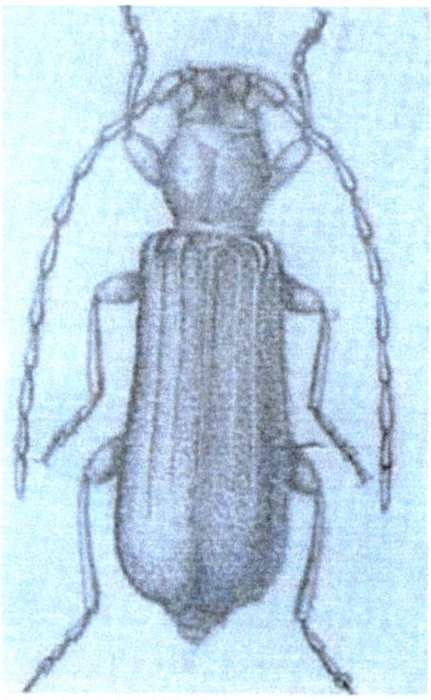

Fig VI-1-17. *Tertopium oreinum*

Fig VI-1-18. *Fomes fomentorius*

Fig VI-1-19. *Gymnosporangium japonicum*

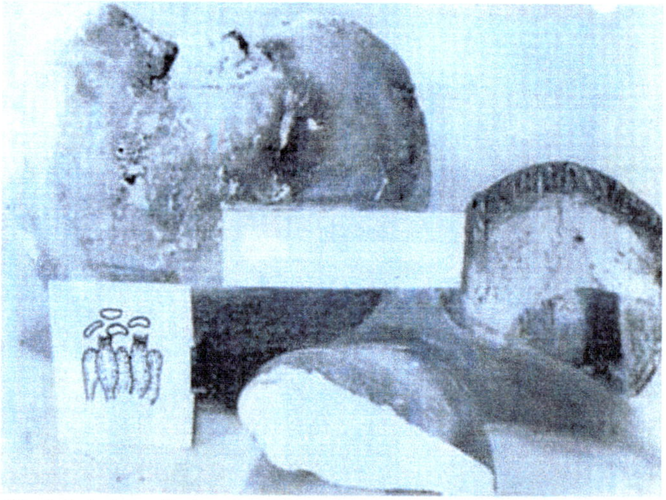

Fig VI-1-20. *Polyporus betulinus*

Fig VI-1-21. *Phellinus igniarius*

Fig VI-1-22. *Stereum princeps*

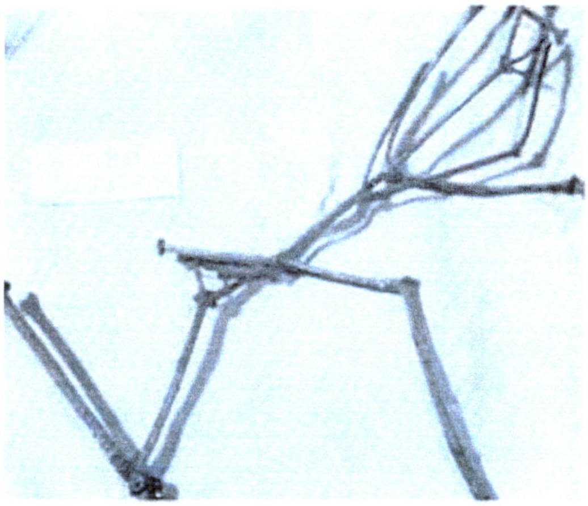

Fig VI-1-23. *Viscum* sp.

Fig VI-1-24. *Loranthus* sp.

Fig VI-3-1.

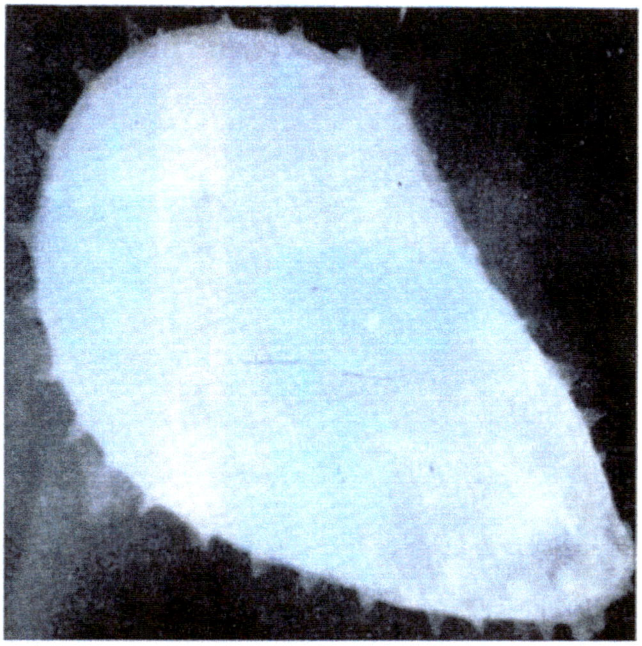

Fig VI-3-2. *Melampsora larici-caprarum* 6000x

Fig VI-3-3. *Phellinus robustus*

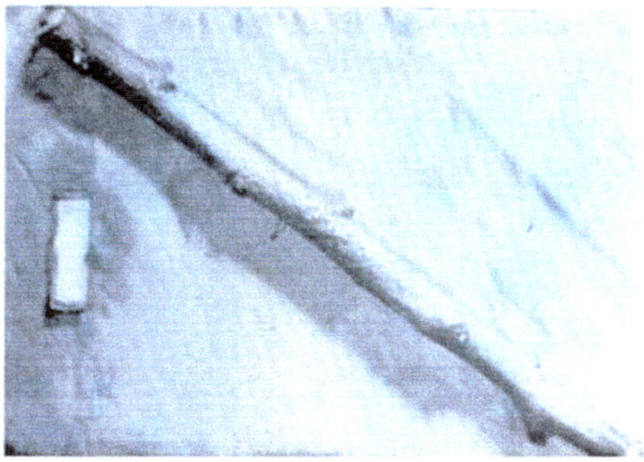

Fig VI-3-4. *Cytospra* sp.

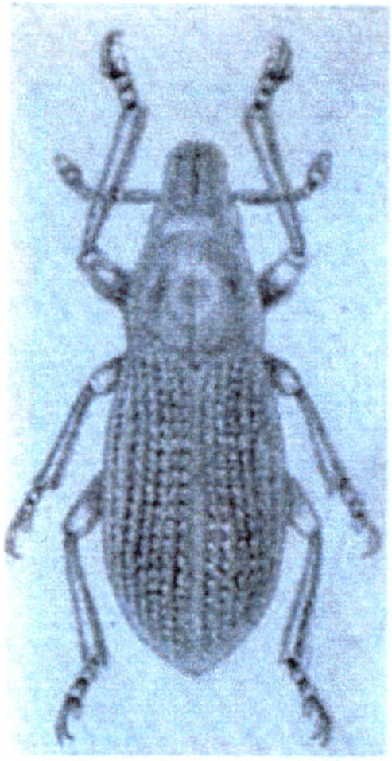

Fig VI-3-5.
Leptomias acutus

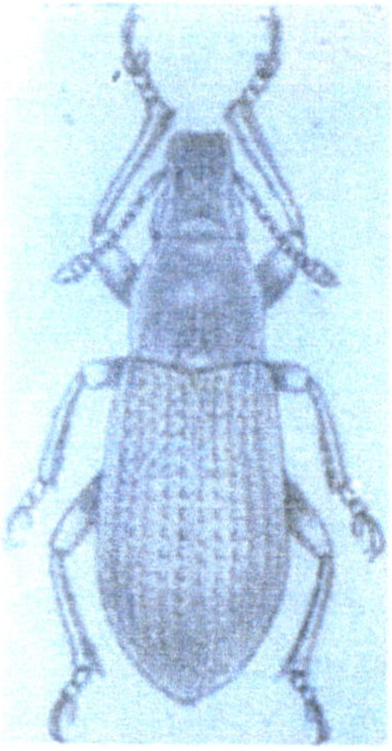

Fig VI-3-6.
Leptomias semicircularis

Fig VI-4-1. *Sclerotinia laxa*

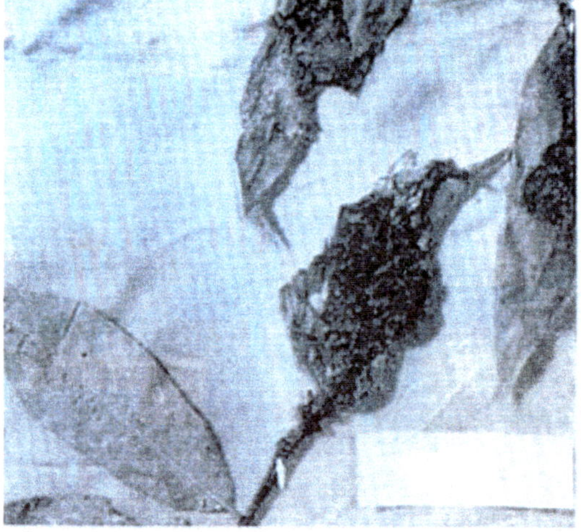

Fig VI-4-2. *Taphrina deformans*

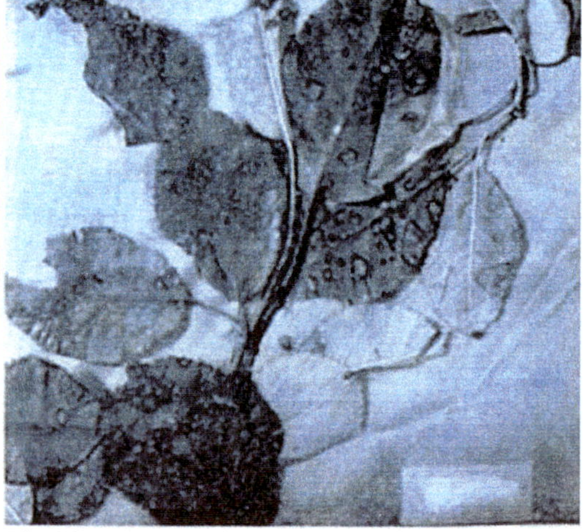

Fig VI-4-3. *Phyllosticia pirina*

Picea linzhiensis in Linzhi forest

Experimental Inoculation of Radiata Pine with Western Gall Rust

Perdermium harknessii, the causal agent of western gall rust, has long been recognized as a major disease of hard pine in North American (Parmeter and Newhook, 1967; Kiloch, 1972). It also poses a threat to susceptible pines growing elsewhere. The 2 year's epidemiological investigation at the Russell genetic experimental station, U.C. Berkeley's radiata pine plantation, large differences in susceptibility to western gall rust were found to exist between the gall populations of *Pinus radiata*. Studies of the genetic aspects radiata pine resistance to western gall rust require standardized methods for experimentally testing individual trees and of individual isolates for pathosystem. In the short periods the Gall formation is a significant stage for the expression of pathosystem under the controlled conditions upon which to base a virulence evaluation method. For this purpose the first step is to quantify the amount of western gall rust needed for standard inoculation techniques. In this test we successfully shortened the time for inoculation. Although successful artificial inoculation has been achieved on young trees, on greenhouse inoculation tests required at least many months for formation of standardized symptoms (Mcinecke 1929; Quick, 1966; Nelson, D.L., 1970; E. Allen, 1985). Field inoculation always takes at least one year to produce gall rust symptoms (Quick, C.R., 1966; Libby, 1985).

The ability to produce large numbers of infected root cuttings may permit standardized accurate scoring of the genetic interactions of Radiata pine and its rust under controlled conditions in short periods and analysis data.

METHODS

1) Root cuttings

Each of the 7 populations samples were used as test cuttings on inoculation. The seed foundation of *Pinus radiata* was from the following sources:

Source Population Sampled	Clone (Trees)	Clone (Trees)	Total
Ano Nuevo	10 (14)	3 (23)	37
Cambria	9 (11)	5 (93)	104
Cedro Island	6 (14)	2 (22)	36
Guadolupo Island	4 (10)	5 (15)	25
Monterey	8 (10)	4 (41)	51
NZ-AUS Selects	6 (26)	5 (42)	68
Inter-pop hybrids	11 (5)	7 (71)	76
Total	54 (90)	31 (307)	397

Using rooted cuttings of radiata pine will provide an efficient method for greenhouse inoculation and western gall rust development.

In May 1985 the cuttings were taken from "Gill Track" Libby's hedged ramets of Radiata pine population foundation. The freshly-cut ends of the cuttings were treated with indole-butyric acid (IBA) as a quick dip. The cuttings were placed in tubes containing a complex root medium of peat moss. Then cuttings were placed in a mixing room at approximately room temperature with water misted over the cuttings at frequent intervals (every 10 min.) during the daylight hours and day length was artificially extended to 16 hours with an incandescent lamp in order to insects *Rhgacionia* and after October (5 months) the cuttings were rooted and about 20-40-50 cm height then placed to lathhouse for shoot growth greenhouse work. Just at the time of these inoculations tests, the rooted cuttings were producing their flush of determinate growth the tender tissue shoot for inoculations tests because the primary growth when inoculated. It's more important physiological condition for produce the galls.

2) Preservation of Aeciospores

A study of the preservation of aeciospore viability in storage, a convenient method to supply inoculum for a long period. Galls were collected from specific radiata pine individual galls located in the Russell Tree Farm. We used Bernoulli's principle – designed for a spore collector. After collection, aeciospores were passed through an ethanol-cleaned 300 mesh sieve to remove peridial fragments and disperse spore masses. After screening, spores were placed in weighing paper, which were in turn placed in petri dishes with the lids of the petri dishes slightly ajar. Spores were air-dried for 24 hrs at room temperature and humidity. After drying, spores were pooled into small capsules and stored in the cold room.

The spore collection was made in mid-March to the end of April during the first Inoculation Fall 1985 and second Inoculation of early February. The viabilities were 90-95%.

3) Inoculation

As results, perhaps more consideration should be given to the quantity of inoculation. A large amount of inoculation is needed and the formation of gall and Witche's broom in WGR with special inoculation points techniques may be superior to express the patho-system levels.

1) The aeciospores were from mixture of various susceptible and resistant individual galls and stored at 0 ° C-1° C in cool room for up to 6 months.

2) Pines were inoculated with .5 mg of viable aeciospores (90% viability) per inoculate tree per test. After inoculation the procedure involved a misting chamber with distilled water. Cuttings were incubated continuously at 21° C room temperature. After 48 hrs replaced in greenhouse lath temperature 22 °C-23 °C for observation of development of gall. Four methods of inoculation were defined for find the appropriate inoculation skills.

4) Evaluation of symptoms

A. Gall formation determination

Two methods were used to determine the degree of western gall rust development in the inoculated root cutting: early infectional observation and gall formation determination.

The symptoms occurred as slightly brown discolored patches, when shoot beings to elongate at dincipient witches' occurred. Sometimes invisible stem swelling from the meristem of needle fascicles bred dormancy – gall start formation.

The first method consisted of description of gall formation. The symptoms indicate the degree of infection and the stage of external diagnosis. When the stem is swelling, the first importance is the diagnosis scores. As the cuttings elongated, tissue reaction was observed at the point of inoculation. Scores are classified in Table 1.

Table 1.

Stage	Symptoms	Score
0	No symptoms	0
I	Discolored spots on the lesions or resinous	1
II	Witches' broom, meristems of needle fascicles bred dormancy and shoots begin to elongate	2
III	Stem swelling	3
IV	Gall formation	4
V	Gall development and multiple galls	5

B. Early infectional observation

The second method is observation of early infection. After two weeks symptoms of infection occurred. The root cuttings were examined for any recognizable symptoms of infection. Plants suspected of being infected were studied microscopically by free hand section. The sections were stained with fluorescent brightener. As the tangential section's mycelium was stained and the brightener fluoresced, the cortex of the cell in which the reaction was observed at the points of inoculation. Scores and results are classified as follows:

Table 2.

Stage	Mycelium in the cortex	Grade
0	no hyphe	0
I	few hyphae	1
II	apparently mycelium brighteners less than 25% of total section	2
III	mass amount of mycelium 25-50% of total section	3
IV	full of all section	4

The examination included all the trees in the population. Fluorescent labeling was used on 18 trees by early infection of trees cut on the cross section. The fluorescent brightener was applied as a 12 % solution in 42 % aqueous cellosolve of the disodium salt of 4, 4'-bis-4-2-anilino-6-bis (1-hydroxy-mehtyl) amino-5-triazin-2-2-glamino-2-2-stilbenedisulfonic acid and obtained an optical brightening solution (Calcofluor white M2R polyscience ZNC—Warrinton, PA 18976) bezone mounting and finally a concentration of 0.025 ml of stock solution in 50 ul of distilled water was chosen as standard.

This caused hyphe to fluoresce brilliantly when examined under incident fluorescent illumination (Zeiss epi-fluorescence microscope equipped with HBO SOW mercury vapor lamp and filter to US excitation). The filter was set 48 77 01: exciter filter BP 365/12, Chromatic beam spliter F7 395, and barrier LP 397).

RESULTS AND DISCUSSION

1) The present study concerns the skills of successful inoculation of Western gall rust. Our major purpose was to develop an inoculation method able to express within a short period of time, a standardized symptom which double pursued in genetic structure of individual host-parasite interactions in the radiata pine gall rust pathosystem.

2) Four methods of inoculation were compared; the results are shown in Table 3.

Table 3. After Inoculated 82 Days the observed numbers of root cuttings inoculational skills

ITN-1

ski.ino	cuttings	ino.P.							Total Galls Produced	% successful inoculations
			0	1	2	3	4	5		
BDS	44	176	32	10	24	52	44	8	52	25 a
SPW+BDS	48	192	8	12	72	44	32	24	56	32 a
SSIA	22	88	20	20	24	16	8	0	8	12.5 b
SSIB	30	108	40	40	16	8	4	0	4	9.0 c

ITN-2

ski.ino	cuttings	ino.P.							Total Galls Produced	% successful inoculations
			0	1	2	3	4	5		
BDS	20	80	20	4	8	20	24	4	28	35 a
SPW+BDS	18	72	4	8	20	16	16	8	24	33.3 a
SSIA	14	56	20	16	8	8	4	0	4	7.1 b
SSIB	19	76	28	28	0	0	0	0	0	0 c

1 Values followed by the same letter are not significantly different at the 95% level.

As ITN-1and IN-2 from Table 2 indicate, after 82 days the inoculation methods BDS and SPW were more effective than SSIA and SSIB, but there was no significant differences between the BDS and SPW+BDS which caused an average of galls formed at 25% AND 32% (BDS), and 35% and 33.3% (SPW+BDS), respectively.
3) The first symptom, grade 1, which occurred was stem discoloration, spots or lesions and appearance of resin within two weeks of stimulation of needle fascicles meristems followed by witches broom after 30 days and finally swelling of the same stem area. The swelling of the stem occurred on trees within 30 days.

Table 4.

After Inoculation Date	After 6 series of inoculation the observed numbers of root cuttings Grade of Symptoms						Number dead or wilted	Number with grade 3-5	% successful inoculations	% dead cuttings
	0	I	II	III	IV	V				
7/1 week	362	0	0	0	0	0	0	0	0	0
14/2 weeks	357	1	4	0	0	0	0	0	0	0
30/1 month	290	3	21	47	1	0	2	48	13.3	.5
60/2 months	216	5	43	89	9	0	3	98	27.1	.7
90/3 months	181	8	18	49	48	58	34	155	42.8	12
180/6 months	121	5	9	13	7	60	147	168	46.4	40

4) The root cuttings were examined at six different dates: 7, 14, 30, 60, 90, and 180 days and the grades of symptoms were recorded for all seedlings present. At 7 and 14 days there were no significant symptoms, from 30 to 60 days the percentage of galls doubled and after 90 days visual symptoms occurred 63 percent more frequently than after 60 days, but only increased by 9 percent by 180 days.
After 90 days, the percentage of successful inoculations had reached a high level and continued up to 180 days. However, by 180 days the number of cuttings that died due to Western gall rust infection had increased dramatically. Data collection after 180 days will be difficult due to this high level of mortality.
5) Intensive study of the time after inoculation has led to a better understanding of the meaning of visual symptoms and their potential significance on host-rust interaction and gall formation in the days after 90 days under controlled conditions. Data obtained in this way are collected and the patterns of incidence of Monterey pine gall rust may then be monitored on other hard pine pathogen on a systematic basis.
6) Fluorescent labeling for observation on early infection of Monterey pine candle shoots was developed. The 540 free hand sections were made on inoculated points of the cuttings. The fluorescent brightener apparently was absorbed by the germinations and subsequent formation of the mycelium around the cortex cell's wall; the mycelium that carried the brightener fluoresced with a slightly blue color (Figure 1.2), which contrasted well with autofluorescence exhibited by the uninfected cross section (Figure 3).

REFERENCES

Allen, E. and Y. Hiratsuka. 1984. Artificial inoculation of young seedlings of lodgepole pine with *Endocronartium harknessi.*
Meinecke, E.P. 1929. Phytopathology 19: 327-342.

Nelson, D.L. 1970. The ecology and pathology of pine gall rust in California. Ph.D. Thesis. University of California, Berkeley, CA.

Parmeter, J.R., and F.J. New hook. 1967. New Zealand *pinus radiata* is susceptible to Western gall rust disease. New Zealand Journal of Forestry 12: 200-201.

Quick, C.R. 1966. Experimental inoculation of ponderosa pine with Western gall rust. Plant Disease Rep. 50: 550-552.

Zagory, D. and W.J. Libby. 1985. Maturation-Related Resistance of *Pinus radiata* to Western Gall Rust phytopathology 75: 1419-1421.

Kinloch. 1972.

Rust collection field trip

rust	stage	host	location	period
Peridermium harknessii	0	Pinus radiata	Russell Tree Farm, UCB	Feb.
	I	Monterey pine (P. radiata)	campus	Feb./25 –
	I	Monterey pine (P. radiata)	Russell Tree Farm	Mar./5 – Mar./20
			Ano Nuevo #100	Mar./5 – Mar./20
			Monterey # 500	Mar./5 – Mar./20
			Cambria # 200	Mar./5 – Mar./20
Cronartium sp.	II	California live oak (*Quercus agrifolia*)	Point Reyes	
Cronartium sp.	III, IV	California live oak (*Quercus agrifolia*)	Point Reyes	
Caeoma sp.	I	*Torrega* sp.	M.T. Tamalpais State Park	
Rust spp.		sugar pine oak		April/18 – 21
Melampsora sp.	III, IV	(*Salix* sp.)	S.F. China Beach	Mar./15 – 20
Croartium ribicola	0	sugar pine	Blodgett Forestry Station, UCB	Sept./15 – Oct./ 20
			Mi-Wuk-Village, Kings Mt. Nat. Forestry	
	I	sugar pine	Mi-Wuk-Village, Kings Mt. Nat. Forestry	Mar./15 – Jun./15
	II		Mi-Wuk-Village, Kings Mt. Nat. Forestry	Mar./2 - Aug./30
	III, IV		Blodgett Forestry Station, UCB	Sept./5 – Nov./10

Western Gall Rust Life Cycle in California

ABSTRACT:

All rusts are obligate parasites. Until recently attempts to culture them on non-living substrata were largely unsuccessful. Two types of life cycles have been proposed for western gall rust (short-cycle, endo-cycle) as alternative explanations of a non-host-alternating scenario. Another scenario proposed here is that western gall rust does alternate hosts. It is my view that gall rust of bishop pine (*Pinus muricata* D. Don) has two phases in its life cycle. The first phase is on coast live oak (*Quercus agrifolia*), as represented by telia collected in Point Reyes, California. The second phase is on bishop pine in the vicinity of infected coast live oak, as represented by numerous galls on which aecia develop. Another observation that supports a host-alternating scenario is my observation of pycnia on galls on Monterey pine (*P. radiata*) at the Russell Tree Farm, Western Gall Rust Experimental Station, University of California, Berkeley. These galls on the Monterey pine also have the aecial spore stage, and there are uredia on leaves of coast live oak in the surrounding area.

This life cycle would be representative of the pine-oak gall rust, *Cronartium quercuum*. It is proposed here that the pine-oak gall rust exists in California as an evolved ecotype for the Pacific West Coast in which diploidy and meiosis occur. The pycniospores and aeciospres evolved simultaneously with Monterey pine and the teliospores and urediospores simultaneously with coast live oak. The rust population with this life cycle must have evolved genetically with pine and oak in the California coast ecosystem, where oak appeared more recently than pine. This interaction typifies a genotype-environmental interaction. However, whether evolution is specifically linked between the hard pines and the California oaks still needs further investigation.

This paper describes some of what is currently known of the life cycle of gall rust in monterey and bishop pine. Although these observations indicate a possible alternative scenario for the life cycle of gall rust in this region, it is still not known how the life cycle of the gall rust functions.

1 INTRODUCTION

Western gall rust (WGR), a fungus that presently exists only in western and northern North America, but it poses a threat to susceptible pines growing elsewhere. At a single experimental site, large differences in susceptibility to WGR were found between the five native populations of *P. radiata*. The two island populations, from Guadalupe and Cedros Islands, were least susceptible. Of the three mainland populations, Ano Nuevo trees were substantially less susceptible than were trees from the Monterey and Cambria populations.

Researchers generally have a common knowledge about the rust disease that has been causing major Monterey pine forest diseases such as white pine blister rust and western gall rust (WGR). Rust diseases are caused by a group of fungi, all of which are obligatory parasitic. The fungi of the order *Uredinales* and of the class *Basidiomycetes* have evolved a complex life cycle not found in any other group of fungi. The obligatory parasitic complex life cycle usually involves two distantly related plants and five-spore stages. The scientific name in current usage is applied to the sexual stage in the *Uredinales* order of the class *Basidiomycotina*. Monterey pines' WGR disease is initially caused by galls (Fig. 2b, 2c), which result from basidiospore infections. Specifically, the gall rust infection of the Monterey pines is caused by Pycnia (#307-001, #307-005) and results in aeciospores. There is a concern that WGR is produced from the dikaryotic hyphae and are specialized for Monterey pine and live oak transfer. There is also a special concern that if the rust becomes established in other pine regions, the WGR may possibly be introduced to the extensive plantations of pines in the Southern Hemisphere and spread rapidly to cause serious damage. The host relationship and rust heteroeciuos species should give particular cause for concern (Old, 1981; Parmeter and Newhook, 1967). The pine-oak rust relationship is an example of a genotype-environment interaction. From the current pathological point of view, live oak is not necessary for the life cycle of the rust fungi. However, this study shows that telia, where present, plays a significant role in this life cycle, indicating a genetically related evolution.

2 MATERIALS AND METHODS

Details of tree propagation, location of the experimental site at the Russell Reservation and the arrangement of 5 replicate blocks in relation to a nearby source of inoculum of WGR are given elsewhere (Old, Libby and Russell 1985). The trees were planted in February 1982 at 2m square spacing.

The plantation was designed to maximize sensitivity of the between-population comparisons. Each plot was composed of ten randomly-arranged trees, as follows: one seedling from each of the five native populations; one seedling of an inter-population hybrid; one seedling from each of three different New Zealand-Australian select families: and one steckling (plantable rooted cutting) from one of the four clones. The four clones were systematically alternated among the 100 plots in the experiment, such that each clone sampled the environment of the entire plantation in a similar manner. Radiata pines at the University of California's Russell Reservation have been infected by WGR for over two decades, and there the disease has been at epidemic levels since about 1974.

3 LIFE CYCLE OF THE MONTERY PINE – OAK GALL RUST

This life cycle has a five-spore stage, which includes the Pycnia, Aecia, Telia, Uredinia, and Basidia stages. Uredinia, Telia and Basidia stages are produced in California Coast live oak (*Quercus agrifolia*). The basidiospores from the Basidia stage are dispersed through the air and infect Pine tissue. As a result of this infection, pycnia develop, serving as spermatia. This is followed by aecia growth. Aeciospores can infect the same Monterey pine, but, as this study shows, it also produces Uredinia in the California coast live oak. Thus, aecial hosts are those on which spermatia and aeciospores are produced. Telia hosts are those hosts in which uredinospores, teliospores, and basidiospores are produced.

3.1 *Pycnia*

Insects are attracted to the sweet pycnia's smell and taste and transfer them to other pycnia. In the spring of 1986, the pycnia appear on the 307-001 Monterey pine gall of the Russell experimental station. The pycnia as a fluid matrix are exuded in a sticky mass, in drops from minute ruptures in the gall (Fig. 3a). Within the pycnia, numerous tiny, sterigmated, one-celled, pear-shaped pycniospores develop (Fig. 3b, 3c).

These spores—each containing a single haploid nucleus Pycnia—are produced from haploid hyphae. The pycniospores serve as spermatia and are presumed to serve as male gametes, fertilizing the female flexuous hyphae around the periphery of the pycnium. Their function, however, is not known. The experimental data showed that the western gall rust might be able to produce aecia without having produced pycnia permanently). The spring following the appearance of pycnia (or the spring following infection), aecia appeared on the gall. Pycnia and aecia usually develop in alternate years as long as the gall remains alive.

3.2 *Aecia*

Blister aecia break through the bark surface where the pycnia previously appeared. The peridium wall of the aecium is fragile and soon breaks, releasing a mass of powdery yellow to orange colored aeciospores with verrucose walls (Fig. 4b, 4d). The aeciospores are produced from the dikaryotic hyphae and are specialized for Monterey pine transfer. They are similar to conidia but are unique in that they will form on live oak and will infect only Monterey pine.

The dikaryotic hyphae, produced upon germination, invade the Monterey pine, as did the haploid hyphae. Aeciospores resist drying and can germinate after being wind borne for long distances. The western gall rust lasts a long period of time and gradually disappears until fall in California. Aecia on globoid swelling galls of trunks and branches, form orange-yellow blisters that burst through the crevices of bark (Fig. 4a). The peridium is white and fall away immediately after bursting the obovoid aeciospores, which are 22-35x18-23nm. in size, hyaline or subhyaline, and rather coarsely verrucose. It has several layers of mycelial strands and a smooth area at the base or an abounded middle (Fig. 4c) with a wall of 2-3nm.

3.3 *Uredinia*

Uredinia are scattered, subgregarious, punctiform, and 130-300mn.in diameters. Uredinia are tiny yellow dome-shaped powder structures that open in the central pore to release yellow obovoid urediospores. They are 18-28x14-20nm in size, with a thick, echinulate wall hyaline of 2-3.5nm thick, and yellowish contents (Fig. 5a, 5b, 5c, 5d). They are produced within approximately two weeks and are collected from Ano Nuevo Live oak of pine-oak mixed forests (Fig. 1).

Urediospores re-infect the same host species from which they are produced. Thus, the uredinial stages of rust intensify during summer until late fall and from late February to March. The telia and uredinia always occur together, where telia follows uredinia production (Fig. 6a, 6b, 6c). This occurrence is seen in the evergreen California coast life oak in the mild temperatures of Point Reyes. Monterey pine-oak gall rust continues the life cycle.

3.4 *Telia*

From the same mycelium that gave rise to the uredinia, I found columns of teliopores that were produced in colder-than-normal winters in temperate California. In many cases, uredia can overwinter (remember, this is the evergreen oak). Telia occur around the urediospores. The telia are columnar, hair-like telium, which emerge from the hypophyllous lower surface of the Coast life oak (*Quercus agrifolia*) leaf (Fig. 6a). These telial filiforms, usually in columns of 3-5 a group, appear in the live oak leaves. They are dark brown or light brown (25m.), and large enough to be visible to the unaided eye (Fig 6d, 6e, 6f).

Telia are composed of chains of long obovoid teliospores joined side to side. They are light brown, have thick walls, and are 58-43x34-27 in size (Fig. 6c). In the teliospore stage, two compatible nuclei fuse to form a diploid nucleus. The teliospores, aggregated in columns germinate under fog mist and relative humidity in pine-oak mixed stands in coastal mountain areas where wet weather conditions are favorable. Teliospores swell somewhat to produce basidia that in turn give rise to basidiospores. As the basidium forms, the two nuclei in the mature teliospore fuse to form a single diploid nucleus. This nucleus then divides meiotically to give the haploid nuclei, which migrate into the basidiospores as they develop. Four basidiospores are produced from each teliospore.

3.5 *Basidia*

The basidiospores are delicate and short-lived. They are produced and released by wind to the Monterey pine host. Basidiospores are generally dispersed at night and early morning, and especially during nights that were preceded by rain during the day or evening. Spores appear when the relative humidity in the atmosphere approaches 100 percent. The basidia die from dessication or from the sun's radiation. The basidiospore is a haploid hyphae—produced upon germination of basidiospores—and invade between the cells of the new tissue. Nutrients are obtained from the living cells of the live oak without invading the cytoplasm. The metabolically active live oak cells allow a certain amount of nutrients to diffuse out to the specialized feeder hyphae called haustoria, which penetrate the cell wall but not the plasma membrane (These are four colorless, single-celled basidiospores with thin cell walls).

4 EVIDENCE OF VARIABILITY IN PATHOGENICITY OF WESTERN GALL RUST

Radiata pine (*Pinus radiata* D. Don) has become an important species for many purposes—such as commercial forests, parks, windbreaks, urban and amenity plantings, and Christmas trees—in many regions of the world. However, the native Radiata pine in California has a serious gall rust disease, caused by western gall rust. The fungus presently exists only in the western and northern parts of North America, but it poses a threat to susceptible pines growing elsewhere. It occurs on the radiata pine in three of its five native populations, and is commonly found on planted radiata pine in much of California. Damage is variable, but can be severe. WGR has not yet been reported outside of North America. Because of its wide host-range within the *Diploxylon* subgenus of *Pinus*, and because of its ability to spread from pine to pine without an intermediate stage on an alternate host, there is concern that WGR may spread rapidly and cause serious damage if it becomes established in other pine regions. Such an introduction is possible in the extensive plantations of pines in the Southern Hemisphere (Old, 1981; Parmeter and Newhook, 1967).

Radiata pines at the University of California's Russell Reservation have been infected by WGR for more than two decades, and there the disease has been at epidemic levels since about 1974. Repeated observations on its occurrence and spread have been made in radiata pine plantations installed at Russell Reservation for other purposes since 1969, and, since 1979, in a small clonal plantation designed specifically to test the relationship of WGR susceptibility to radiata pine's maturation state (Zagory and Libby, sub. Ms.).

By 1981, the accumulating observations at Russell Reservation and elsewhere increasingly suggested that there is substantial and perhaps usable variation in susceptibility to WGR among populations, among clones, and among maturation states of radiata pine (*Science* 1985). I joined a cooperative project that was set up between U.C. Berkeley, CSIRO Australia, and private forestry interests in California and New Zealand, to assess their susceptibility to WGR. To investigate the inheritance of WGR susceptibility, inter-population hybrids and heritabilty estimates within population samples were used.

In a common-garden experiment, it was found that large differences in susceptibility to WGR exist between the five native populations of P. radiata. The two island populations, from Guadeloupe and Cedars Islands, are the least susceptible. Of the three mainland populations, Ano Nuevo trees are substantially less susceptible than the Monterey and Cambria populations. New Zealand and Australian select trees are intermediate in susceptibility between the Ano Nuevo and the Monterey populations, from which the New Zealand and Australian land-races are derived. Significant components of variation between families within populations were most marked in the native Monterey population, among New Zealand—Australian selects, and among interpopulation hybrids. Heritability estimates and comparisons of full-sib families to half-sib or open-pollinated families indicate substantial levels of additive (narrow-sense) heritability about susceptibility to WGR, and provide no evidence for non-additive variance in this trait. The results suggest that breeding for resistance to the disease is likely to yield worthwhile changes in average susceptibility. The more virulent isolates were originally collected on the more susceptible hosts. From this evidence of western gall rust, it is clear that its sex states are hidden.

5 CLASSIFICATION HISTORY

In 1969, the endocyclic genus *Endocronartium* was established to include E. harknessii (J.P. Moore) Y. Hiratsuka (=*Peridermium harknessii* J.P. Moore) in North America and *E. Pini* (Persoon) Leveille emend Klebahn (*P. pini* (Persoon) Leveille emend Klebahn). In Europe, there are two autoecious pine stem rust, but the justifications for establishing this genus have been questioned. Cytological events in spores and germ tubes of *E. harknessii* (western gall rust) were reexamined. The number of nuclei and relative DNA contents in various stages of spore germination, number and nature of septa and branches, and mode of initial host penetration suggested that the two species' germlings function as metabasidia with nuclear fusion and meiosis, rather than as aeciospore germ tubes. The recognition of the endocyclic genus is thus concluded.

If we do not recognize these pine-to-pine forms as having an endocyclic life cycle, the use of *Peridermium* may be a reasonable solution and many researches are using this option. However, if the group of fungi is recognized to have an endocyclic life cycle, they cannot be called by the anamorphic generic name of *Peridermium*. Also, by definition, the use of the name *Peridermium* means that the teleomorph or telial state of the fungus is still unknown, as described in one of the tables in Vogler and Bruns (1998). However, these fungi obviously do not have a separate telial state on alternate hosts because the whole life cycle is completed on the *Pinus* host. The use of *Peridermium* does not indicate the endocyclic nature of the life cycle, and it may be confused with anamorphic taxa in which the life cycle is still unknown.

Class: *Basidiomycotina*
- Order: *Uredinales*
  - Family: *Melampsoraceae*
    - Genus: *Cronartium*
      - Species: *C. quercum*
        - Formae speciales: C. q. f. SP. radiatanae (primarily pathogenic on Monterey pine)

6 CONCLUSION AND THE SOUTHERN HEMISPHERE INSPECTION

The western gall rust fungus is a full life-cycle rust. It is the same pine-oak gall rust that causes the rust mane called the *Cronartium quercum* complex. From a pathologist's point of view, aeciospores will cycle and regenerate by themselves. Because Monterey Pine (*Pinus radiata*) is a diplophase pine, the pine-to-pine disease transfer is indeed an actual phenomenon in nature, and a product of human activity at plantation or residential areas. During the teliospore stage, uridia and telia coexist, which is an extraordinary phenomenon. Uridia exists over the winter, and its ability to survive over winter indicates that the California coastal weather is not hazardous to its developmental stage. Uridia is another rust that sometimes did not even need the telia over the coastal mild winter.

Therefore, from a genetic point of view, the pine-to-pine disease-transfer model is representative of the Mediterranean weather and vegetation type. The rust's ecotype is Mediterranean, and the foggy conditions in the forests provide humidity for the basidiospore to germinate. But substantially, for genetic reasons that produced the telia stage, the radial survey designed for pine and oak relationship is important evidence that can be found by surveying *Quercus* located radially from the infected pine. Specifically, when surveying, it was noted that uridia occurred in places that were closer to the gall rust. (Fig. 1).

6.1 *Oak uridia and telia*

From a genetic point of view, telia are produced from uridia. I have found that original live oak and Monterey pine occur in mixed pine-oak forests. In this situation, Aecia survive among the dead and dying pines (Fig. 7d) because as a tree gradually dies, aeciospores continue to produce.

I collected telia from the live oak area surrounding the dying tree and found numerous incidents of gall rust, which means that west gall rust does not have only pine-to-pine short life cycles. Monterey pines' western gall rust has the Pacific West Coast ecological characteristics. Aeciospores evolved simultaneously with Monterey pine while Teliospores evolved specifically with live oak. This life cycle must have evolved genetically with live oak, which appeared more recently than the pine. This interaction typifies a genotype-environment interaction.

7 REFERENCES

Bergdahl, D.R., & French, D.W. 1990. Basdiopore Infection Periods forCronartium Rusts on Jack Pine in Minnesota. Inf. Rep. NOR-X-317.

Blasdale, W.C. 1919. A preliminary list of the Uredinales of California.Univ. Calif. Publ. Bot. 7: 101-157.

Chen, M.M. & Cobb, Jr, F.W. 1985. White pine blister rust in young sugar pine plantations in the mid-elevation Sierra Nevada. Phytopathology 75(11): 1367.

Chen, M.M. 1963. Studies of pine-oak rust (*Cronartium quercuum*) on var. Scotch pine. Heilongjiang Institute of Forestry Design. 5 pp.

Chen, M.M. 1984. A new genus, Stilbechrysomyxa Chen Gen. Nov. Scientia Silvae Sinicae 20(3):268-270.

Chen, M.M. 1985. Analysis of the distribution of forest pathogens in the United States and China: two case studies. Phytopathology 75(11):1322.

Chen, M.M. 1988. Biogeography of Uredinales on Sino-Himalaya. Proceedings of the Symposium and Workshop of the Society of Chinese Bioscientists in America. p. 156 (abstract)

Chen, M.M. 1991. Pest-Risk Assessment of the Importation of larch from Siberia and Soviet Far East. USDA Forest Service Misc. Pub. No. 1495.

Chen, M.M. 1991. Scotch pine disease of the Soviet Far East. USDA Forest Service Misc. Pub. No. 1495.

Chen, M.M., et al. 1988. Development of greenhouse inoculation techniques for western gall rust on radiata pine with some early results. Phytopathology 78(12): 1603.

Chen, M.M., F.W. Cobb, Jr., W.J. Libby & D.R. Vogler. 1989. Evidence of variability in pathogenicity among isolates from an enzymatically monomorphic population of western gall rust. Proceedings of the 36th Annual Western International Forest Disease Work Conference. pp. 95-106.

Cheng, D.S., Lian, H.Y., Xue, Y., Pan, X.R., & Li, W.H. 1998. Genetic differentiation among three pine stem rust fungi from China as estimated by random amplified polymorphic DNA. Scientia-Silvae-Sinicae 34: 5, 53-60; 14 ref.

Christenson, J.A. 1969. Cytology of albino Peridermium harknessii. Mycologia 61: 398-401.

Epstein, L., & Buurlage, M.B. 1988. Nuclear division in germinating aeciospores and its taxonomic significance for the western gall rust fungus, Peridermium harknessii. Mycologia 80: 235-240.

Froelich, R.C. & Schmidtling, R.C. 1998. Survival of slash pine having fusiform rust disease varies with year of first stem infection and severity. Southern-Journal-of-Applied-Forestry 22: 2, 96-100; 15 ref.

Hedgock, G.G. 1913. Notes on some western Uredineae which atack forest trees. II. Phytopathology 3: 15-17.

Huber, D, White, T., Powell, G., Atwood, R., Lopes, U., Lopez-Upton, J., Lucero, V., Osorio, L., Sanhueza, R., Zhang, H., & Lu, P.X. 1999. Forty-first annual progress report, April 1999. Cooperative-Forest-Genetics-Research-Program No. 41, 46 pp.; 4 pp. of ref.

Kaneko, S., Kuhlman, E.G., & Powers, H.R. 1991. Morphological and physiological differnences in the Cronartium quercuum comlex. Pp. 69-75. In: rusts of pine: Proceedings of the IUFRO Working Party Conf., Sept. 1989, Banff, Alberta. Eds., Y. Hiratsuka, J.K. Samoil, P.V. Blenis, P.E. Crane, and B.L.

Laishley. Forestry Canada, Northern Forestry Centre, Info. Rep. NOR-X-317, Edmenton, Alberta. pp. 408.
Kinloch, B.B. Jr., & Meagher, C.H. 1990. Resistance to Fusiform Rust in Southern Pines: How is it Inherited? Inf. Rep. NOR-X-317.
Kojwang, H.O., & Van der Kamp, B.J. 1990. Infection of Lodgepole Pine Clones by Single-gall Isolates of Western Gall Rust Pathosystem. Inf. Rep. NOR-X-317.
Lee, Y.J., Lenhart, J.D., & Lee, Y.J. 1997. Estimating crown height for unthinned planted pines in East Texas., Southern-Journal-of-Applied-Forestry 21: 3, 130-133; 20 ref.
Liu, S.J., & Chen, M.M., et al. 1982. Studies on Cronartium. J. of Chinese Ministry of Forestry, Beijing (translated from Roger Peterson's English paper to Chinese), published by the Tottori Mycological Institute. The Japan Tokyo Research Center. Totter, Japan August 1973. pp. 203-223.
Meinecke, E.P. 1916b. Peridermium harknessii and Cronartium quercuum. Science 43: 75.
Meinecke, E.P. 1920. Facultative heteroecism in Peridermium cerebrum and Peridermium harknessii. Phtopathology 10: 279-297.
Meinecke, E.P. 1916a. Peridermium harknessii and Cronartium quercuum. Phytopathology 6: 225-240.
Old, K. M., Libby, W. J. and Russell, J. H. 1985 Patterns of Susceptibility to Western Gall Rust in Pinus radiata Proceedings of the Rusts of Hard Pines Working Party Conference S2.06-10 286
Old, K. M., Libby, W. J., Russell, J. H. and Eldridge, K. G. 1986 Genetic Variability in Susceptibility of Pinus radiata to Western Gall Rust Silvae Genetica 35: 145-146
Peterson, R.S. 1960. Western gall rust cankers in Lodgepole pine. J. For. 59: 194-196.
Peterson, R.S. 1960. Western gall rust on hard pines. USDA For. Serv., Washington, D.C. For. Pest Leaf. 50.
Peterson, R.S. 1967. The Peridermium species on pine stems. Bull. Torrey Bot. Club 94: 511-542.
Peterson, R.S. 1973. Studies of Cronartium (Uredinales). Rep. Tottori Mycol. Inst. 10: 203-223.
Powers, H.R. Jr., Kaneko, S., Kuhlman, E.G., La, Y.J., & Wang, Z.H. 1990. Susceptibility of Asian and American Oaks and Pines to Cronartium quercuum. Inf. Rep. NOR-X-317.
Pye, J.M., Wagner, J.E., Holmes, T.P., & Cubbage, F.W. 1997. Positive returns from investment in fusiform rust research. Southern-Research-Station,-USDA-Forest-Service No. SRS-4, 55 pp.; 42 ref.
Samuelson, L.J. 1998. Influence of intensive culture on leaf net photosynthesis and growth of sweetgum and loblolly pine seedlings. Forest-Science 44: 2, 308-316; 56 ref.
Schmidt, R.A. & Miller, T. 1999. Influence of inoculum concentration on production of and aecia on pine seedlings infected by basidiospores of Cronartium quercuum f.sp. fusiforme. Plant-Disease 83: 4, 367-370; 27 ref.
Stelzer, H.E., Doudrick, R.L., Kubisiak, T.L., & Nelson, C.D. 1999. Slash pine and Cronartium pedigrees for evaluation of complementary gene action in fusiform rust disease. Plant-Disease 83: 4, 385-389; 32 ref.
Tuskan, G.A., Walla, J.A., Lundquist, J.E., Wang, C.G., & McClean, P. 1990. The Use of Molecular Markers in Studying Ponderosa Pine-Peridermium harkessii Interactions. Inf. Rep. NOR-X-317.
Vogler, D.R. 1995. Uses of molecular techniques in rust systematics. pp. 9-15. In: Proc. 4th IUFRO Rusts of Pines Working Party Conf. Eds., S. Kaneko, K. Katsuya, M. Kakishima, and Y. Ono. Forestry and Forest Products Research Institute, and University of Tsukuba, Tsukuba, Japan. 197 pp.
Vogler, D.R., & Bruns, T.D. 1993. Use of molecular characters to identify holomorps: an example from the rust genus Cronartium spp. 237-245. In: The fungal holomorph: mitotic, meiotic and pleomorphic speciation in fungal systematics. Eds., D.R. Reynolds and J.W. Taylor. CAB International, Wallingford, U.K. 375 pp.
Vogler, D.R., B.B. Kinloch, F.W. Cobb, Jr., and T.L. Popenuck. 1991. Isozyme structure of Perdermium harknessii in the western United States. Canad. J. Bot. 69: 2434-2441.
Vogler, D.R., Cobb, F.W. Jr., Geils, B.W. and Nelson, D.L. 1996. Isozyme diversity among hard pine stem rust fungi in the western United States. Canad. J. Bot 74: 1058-1070.
Vogler, D.R., Cobb, F.W.Jr., & Epstein, L. 1989. Nuclei in mature aeciospores from two zymodemes of Peridermium harknessii in California (Abstr.). Phytopathology 79: 1164.

Vogler, D.R., Epstein, L. & Cobb, F.W. Jr. 1997. Nuclear behavior and evolution of two populations of the western gall rust fungus. Mycol. Res. 101: 791-797.

Vogler, D.R., Kinloch, B.B., Cobb, R.W., Libby, W.J., & Popenuck, T.L. 1988. Genetic architecture of western gall rust (*Peridermium harknessii*) in California pine forests. Phytopathology 78: 1555. (Abstr.)

Wagnener, W.W. 1958. Infection tests with two rusts of Jefrey pine. Pl. Dis. Rep. 42: 888-892.

Wagner, J.E., & Holmes. 1999. Estimating economic gains for landowners due to time-dependent changes in biotechnology. Forest-Science. 1999, 45: 2, 163-170; 24 ref.

Walla, J.A., Lundquist, J.E., Tuskan, G.A., & Freeman, T.P. 1990. Observations on the Spermogonial Stage of the Western Gall Rust Fungus. Inf. Rep. NOR-X-317.

Walla, J.A., Tuskan, G.A. & Lundquist, J.E. 1991. Comparison of white and orange-spored Peridermium harknessii in native North Dakota Pinus ponderosa. Mycologia 83: 251-256.

White, T.J., T. Bruns, S. Lee, and J. Taylor. 1990 Amplification and direct sequencing of fungal ribosomal RNA genes for phylogenetics. Pp. 315-322. In: PCR protocols: a guide to methods and applications. Eds., M.A. Innis, D.H. Gelfand, J.J. Sninsky, and T.J. White. Academic Press, Inc., San Diego. 482pp.

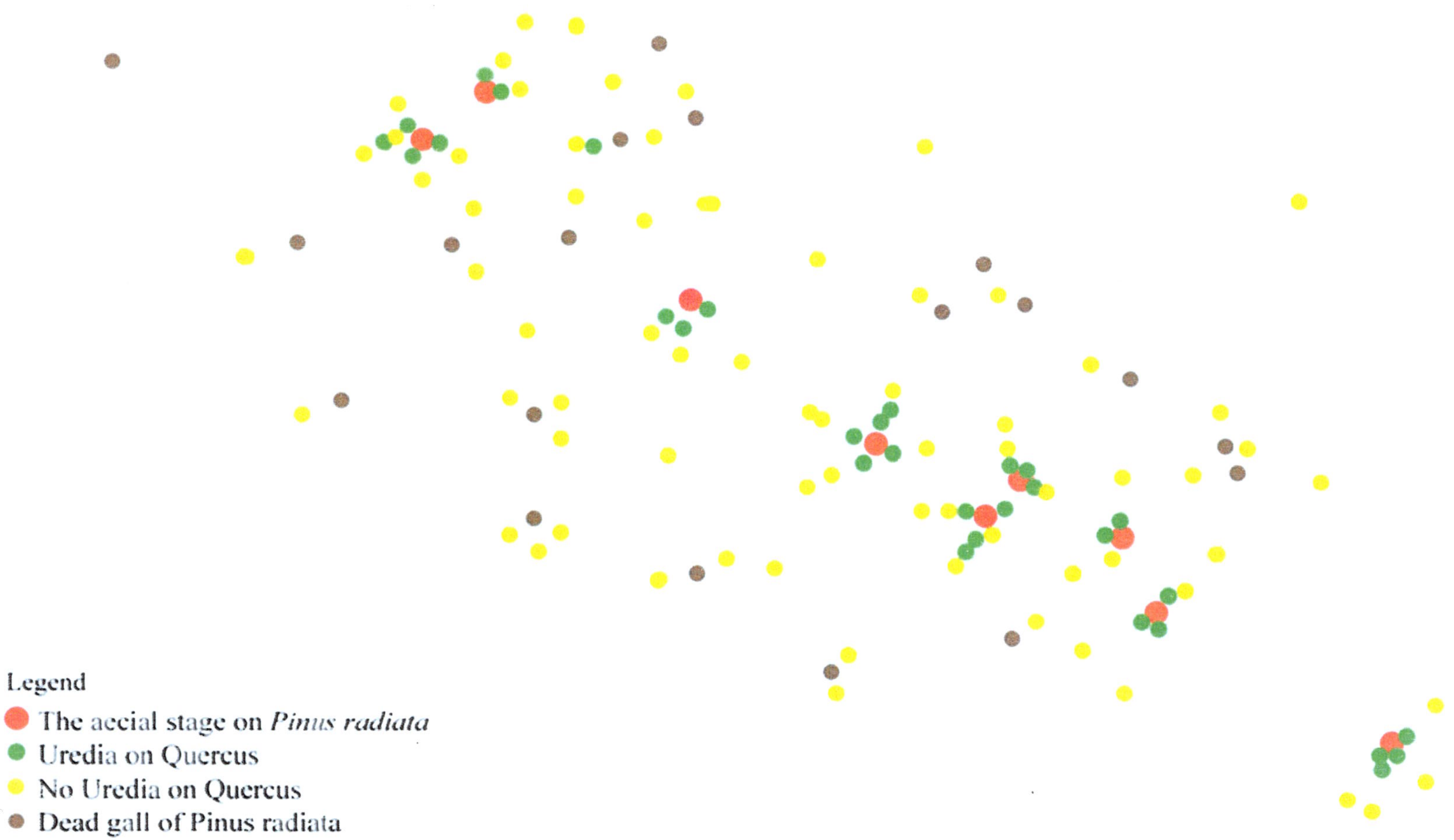

Fig 1. Map of Año Nuevo Coniferous forest (Data collected in 1985)

West gall rust on *Pinus radiata* research plantation at Russell Genetic Experiment Station, U.C. Berkeley.

Fig 2a. West gall rust aecia chain galls

Fig 2b. West gall rust causes breaking on trunk at gall sites. (20 year old trees)

Fig 2c. Peridium on globoid swelling gall.

Fig 3a. Pycnia on the 307-001 gall of *Pinus radiata* on Russell Genetic Experiment Station, U.C. Berkeleey.

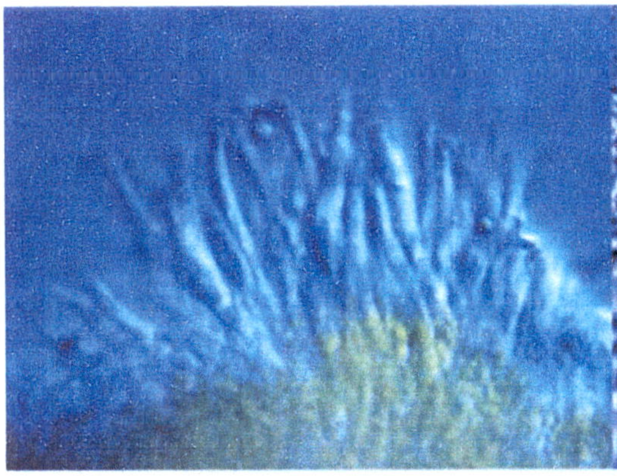

Fig 3b. Pycniospores sterigma.

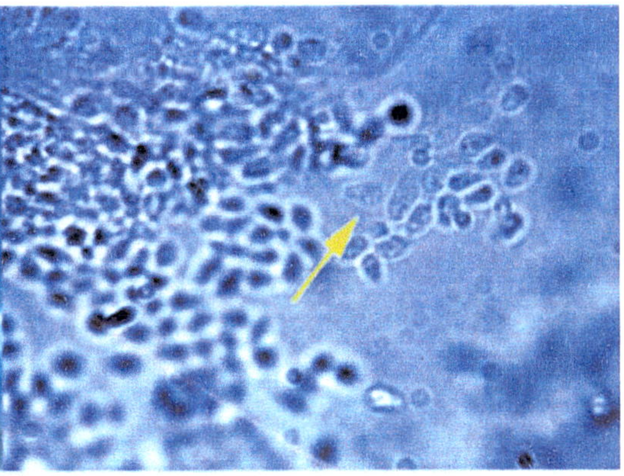

Fig 3c. sticky mass of pear-shaped pycniospores

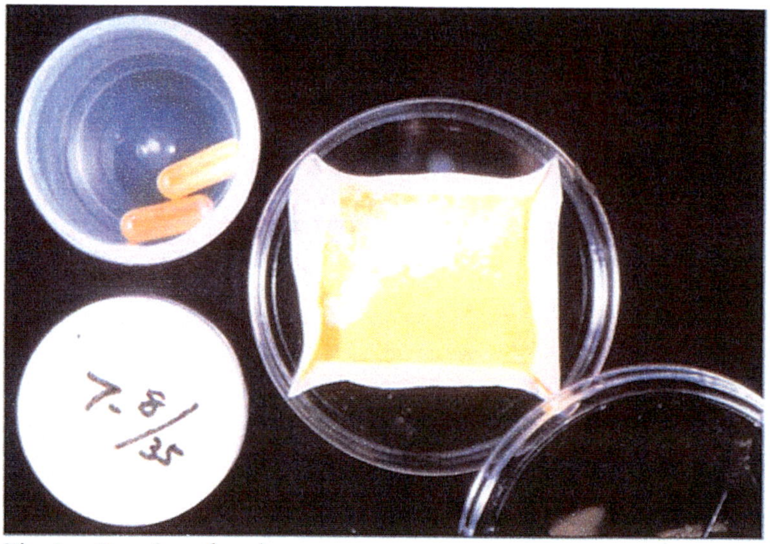
Fig 4a. Powder of aeciospores.

Healthy *Pinus radiata* at Monterey Bay, California

Fig 4b. Single spore structure under electron microscope.

Fig 4c. Verrucose walls of aeciospores.

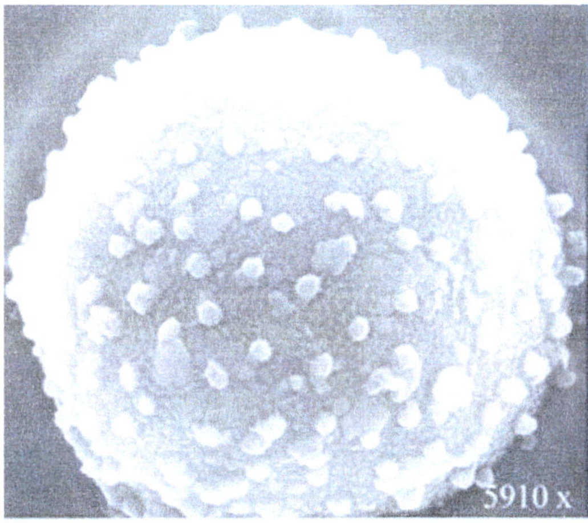

Fig 5a. Urediospore (Año Nuevo)

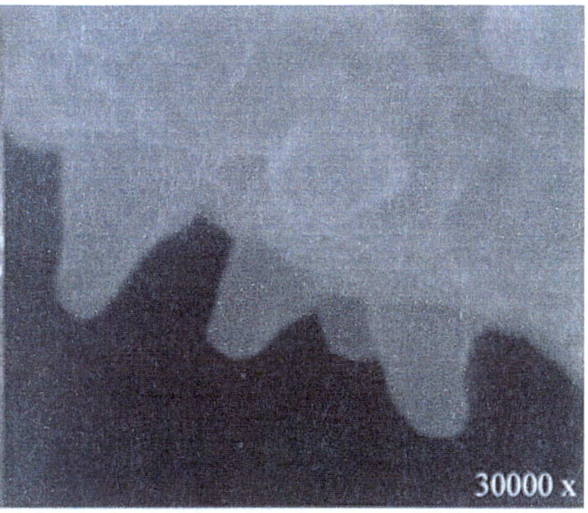

Fig 5b. Echinule on surface of urediospore

Fig 5c. Uredinia on live oak (*Quercus Agrifolia*) (Año Nuevo)

Fig 5d. Yellow dome-shaped powder structures of uredinia. (Año Nuevo)

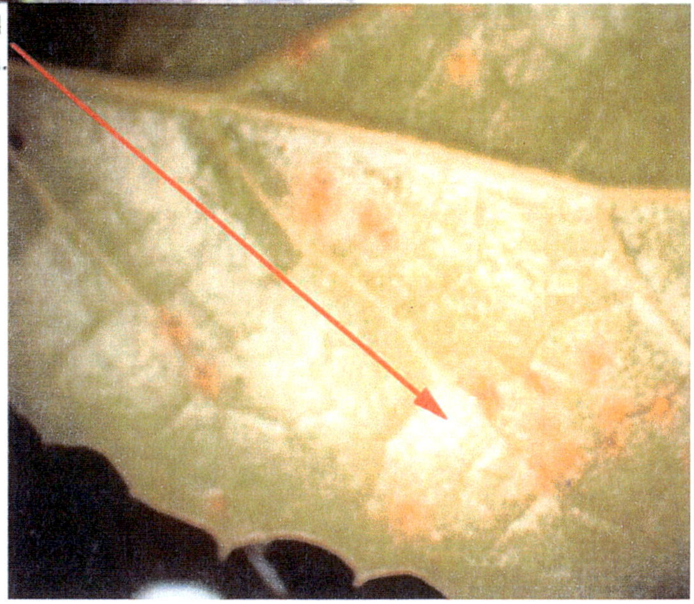

Fig 6a. Urediospores (Point Reyes)

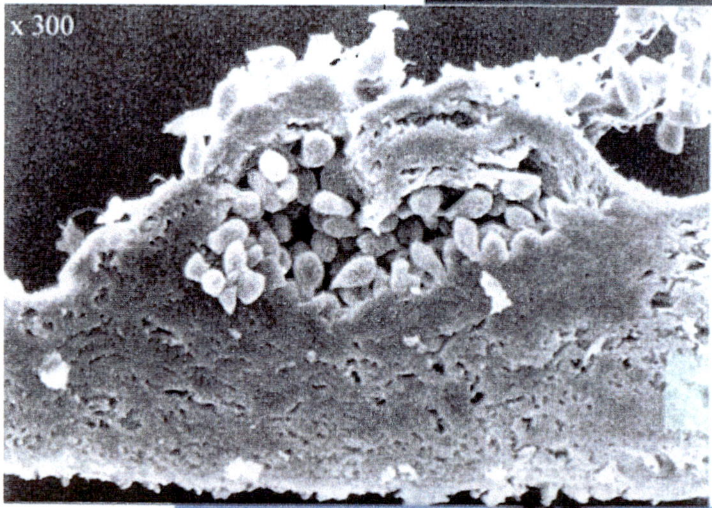

Fig 6b. Uredia slide structure. (Point Reyes).

Fig 6c. Yellow obovoid urediospores under microscope (Point Reyes).

Fig 7a. Coast live oak (*Quercus Agrifolia*) at Point Reyes, California.

Fig 7b. Telia and uredia.

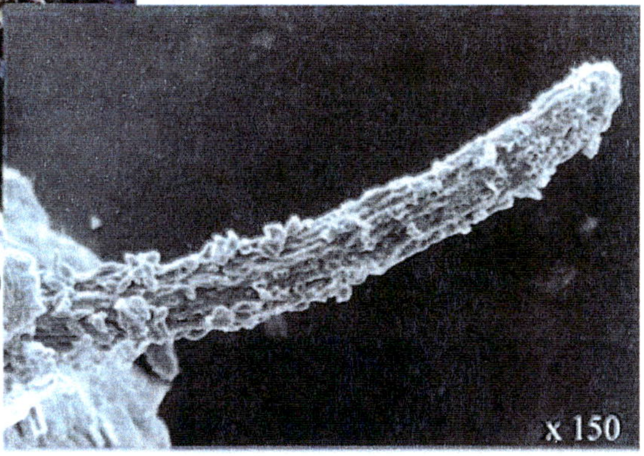

Fig 7c. Telia.

Fig 7d. Dying *Pinus muricata* caused by western gall rust in Point Reyes.

Fig 7e. *Quercus agrifolia*
(From The Jepson Manual of Higher Plants of California)

Fig 5f. *Quercus agrifolia*: a. leaf b. acorn
(From The Jepson Manual of Higher Plants of California)

Evidence of variability in pathogenicity among isolates from an isozymically monomorphic population of western gall rust

M.M. Chen[1], F.W. Cobb Jr.[2,3], W.J. Libby[4], and D.R. Vogler[5]

ABSTRACT: Inocula of western gall rust, collected from more than 100 different trees in a radiata pine common-garden experiment, were isozymically monomorphic at 13 or more loci and polymorphic at one. Inocula collected from 10 trees in this common-garden experiment proved to have significantly different levels of virulence in greenhouse tests. These differences cannot yet be clearly identified as to genetic and/or physiological causation. The host clones used in the greenhouse tests were drawn from the common-garden plantation and, with the possible exception of some Guadalupe Island, they generally maintained their relative levels of susceptibility in the field and in the greenhouse. Surprisingly, the more virulent isolates were originally collected on the more susceptible hosts, and vice-versa.

1 INTRODUCTION

In a preliminary sample of western gall rust [*Peridermium harknessii* J.P. Moore (previously called *Endocronartium harknessii,* see Epstein &Buurlage 1988)] from a genetic-architecture experiment of radiata pine (*Pinus radiata* D. Don), no genetic variation was found among spore collections from 60 different trees, based on 13 isozyme loci (pers. comm., Brian Racin, Spring 1986). Complete monomorphism was similarly round in a 1987 sample of 52 single galls from trees in the same experiment, using 14 well-defined and 5 less-well-defined isozyme loci (Vogler and others 1987).

The work reported in this paper is from a study to further develop greenhouse inoculation procedures (Nelson 1971; Chen and others, in press), to relate host susceptibility in greenhouse and field conditions, and to investigate in a preliminary manner whether variation in pathogenicity exists within the isozymically monomorphic rust population.

2 MATERIAL AND METHODS

2.1 *Host Plants*

In a 1000-tree genetic-architecture experiment at the University of California's Russell Reservation (Old & others 1985, 1986), western gall rust infection varied from a low level in the northwest part of the plantation to an intense level in the plantation's southeast corner. Thirty-one host clones were chosen

[1] Visiting Scholar, Dept. of Plant Pathology, University of California, Berkeley 94704 USA.
[2] Professor, Plant Pathology, University of California, Berkeley 94704 USA.
[3] Names of second and subsequent authors are shown in alphabetical order.
[4] Professor, Forestry and Genetics, University of California, Berkeley 94704 USA.
[5] Graduate Student, Dept. of Plant Pathology, University of California, Berkeley 94704 USA.

from this larger experiment for the study reported herein, and were meant to include a range of host susceptibilities. The seedling ortets of five of these clones were in host class "0" (no galls); four of these seedlings were in the heavily-infected southeast part of the experiment, and all five were free of infection after two years' exposure. Five ortets were in host class "5"; these were the most heavily-infected seedlings in the plantation (each had 45 or more galls after two years' exposure), and they mostly came from the central and southeastern parts of the plantation. Host classes "1-4" had five, four, six and six clones, respectively, and were defined by having 1-3, 4-9, 10-21, and 22-44 galls, respectively, on their ortets. Furthermore, the seedling ortets for classes "1" and "2" were mostly from the heavily-infected southeastern part of the plantation, and those for classes "3" and "4" were mostly from the moderately- or lightly-infected central and western parts of the plantation. In other words, classes "1" and "2" contained clones whose ortets were lightly and moderately infected in heavily-infected neighborhoods, while classes "3" and "4" contained clones whose ortets were heavily and very heavily infected in moderately- or lightly-infected neighborhoods.

Each of these seedlings had been cloned prior to being outplanted. Two rooted cuttings (ramets) of each clone were maintained as 15-cm-high hedges in clay pots in a rust-free environment. These were the cutting donors for the plants (ramets, or stecklings) used as hosts in the greenhouse inoculations reported below.

2.2 *Inocula*

Ten isolates of inocula were collected, each from a single gall, during mid-April and early May 1985. Five were collected from the most-lightly-infected class of trees (class"1", above, inocula "6-10") and the other five were from the two most-heavily-infected classes (four from class "4", inocula "21-24" and one from class "5", inoculum "26"). When choosing a gall for spore collection, care was taken to avoid areas with multiple galls, and galls with sectors or nodes, on the presumption that a single gall with uniform morphology was more likely to have resulted from infection by a single spore. The spores were collected when the gall cracked open and seams of orange-yellow aeciospores were visible (Figure 1). The peridium covering the aeciospores was punctured with hypodermic needle, and the spores beneath the surface were sucked into a sealed vial. Each isolate was then used, through inoculation and re-collection, to establish a "spore-line" for later experiments, and we refer to them as "spore-lines" in the remainder of this paper.

2.3 *Inoculation Procedures*

From one to twenty ramets of each of the 31 host clones (average 11.7 ramets per clone) were inoculated with the collected spores. In this experiment, each ramet received inocula from only one spore-line, from one collection. The spore-lines were assigned to clones in a systematic manner, arranged such that each class of host clones received approximately equal numbers of inoculations from each spore-line, but no one clone was inoculated with all ten lines (maximum five spore-lines per clone). Those four clones with only 1-5 ramets per cone had each ramet assigned to a different spore-line. The remaining 27 clones had more than one ramet assigned to a given spore-line, such that two spore-lines from the lightly-infected hosts (inocula "6-10") and two spore-lines from the heavily-infected hosts (inocula "21-26") were applied to approximately equal numbers of ramets in the clone. In 6 clones, 1-3 ramets were inoculated with a fifth spore-line, to achieve balance among spore-lines in total inoculations and within the six host classes with respect to the two inoculum classes. A few of the ramets died between inoculation and scoring, resulting in some imbalances at the time of scoring, but approximately equal numbers of inoculations within host classes between inoculum classes were achieved (Table 1).

Following preliminary trials using scarified and unscarified host tissue, and dry and aqueous-suspension applications of spores, the following procedure was found to be reliable and was used uniformly for the inoculations reported below. Host ramets were grown in a greenhouse until the root systems were well established and the main shoot was actively elongating behind a terminal meristem in

Table 1. Scored surviving ramets of *Pinus radiata* by spore-line inoculated and host class.

Host Class [2]	Spore-lines [1] 6	7	8	9	10	6-10-	21	22	23	24	26	21-26
0	5	5	3	2	2	17	5	4	6	2	3	17
1	8	5	6	5	3	27	6	4	7	5	6	28
2	2	3	3	2	1	11	1	3	3	3	0	10
3	6	7	7	10	12	42	7	5	9	11	9	41
4	10	8	9	8	11	46	11	8	9	7	11	46
5	7	10	7	6	10	40	12	6	5	8	6	37
Σ	38	38	35	33	39	183	42	30	36	36	35	179

[1] Spore-lines "6-10" from lightly-infected neighborhoods; and spore-lines "21-26" from heavily-infected hosts in lightly-or moderately-infected neighborhoods, or ("26") a heavily-infected neighborhood.

[2] Class of host clones: from uninfected ("0") to heavily-infected ("5") ortets in the plantation environment (4-6 clones per class).

free-growth phase. Dry spores were taken from the vial and brushed onto a region of approx. 1 cm length, 2 cm below the terminal meristem. Inoculations were performed on 6 September 1985 and on 6 February 1986. The newly-inoculated ramets were placed in a mist chamber at 21°C for 48 hours, and were then grown in the greenhouse for six months at ambient 23-24°C.

2.4 *Scoring*

Each ramet was classified as to gall symptoms three and six months after inoculation. Classes "0-2" were weakly infected or uninfected, as follows: Class "0" had no symptoms or injuries in the inoculated region; class "1" had a lesion or other suspicious symptom in the inoculated region but no visible swelling; and class "2" had suspicious short-shoot development in the inoculated region but no visible swelling. Classes "3-5" were infected, as follows: class "3" had one moderate swelling in the inoculated region; class "4" had more than one moderate swelling in the inoculated region; and class "5" had one or more large galls in the inoculated regions (Figure 2).

During the period between the 3-month and 6-month observations, there was a failure in plant care for a short period that resulted in the deaths of 147 of the 362 ramets scored at 3 months. Scores for the 215 ramets that survived and recovered healthy growth by 6 months were as follows. Of the 143 ramets scored in classes "0-2" at 3 months, 135 were again scored in classes "0-2" at 6 months; of the 72 ramets scored in classes "3-5" at 3 months, all 72 were again scored in classes "3-5" at 6 months. None of the 215 ramets was given a lower score at 6 months than at 3 months. Of the 16 ramets scored "3" at 3 months, 6 were scored "3" and 10 were scored "5" at 6 months; similarly, of the 25 ramets scored "4" at 3 months, 6 were scored "4" and 19 were scored "5" at 6 months. In sharp contrast, of the 143 ramets in the "0-2" classes at 3 months, only 1 was scored "5" at 6 months. Of the 128 "0" ramets at 3 months, 5 were still "1"s at 6 months, 1 was a "2", and 1 was the "5". Of the 8 "2" ramets at 3 months, 7 were still "2"s at 6 months and 1 was a "3".

These observations led us to conclude that classes "0-2" and "3-5" were useful groupings for "uninfected or weakly infected" and for "infected". Furthermore, few ramets scored "0-2" at 3 months were changed in status to "3-5" at 6 months, and the reverse (from "3-5" to "0-2") was not observed in 72 opportunities. We therefore opted to use the full data set for 362 ramets at 3 months (those in Table 1), rather than the slightly more accurate data set for only 215 ramets at 6 months.

Beginning 11 months after inoculation, the rust fungus in class "5" galls began to produce aeciospores, providing ongoing replicates of 9 of the 10 inoculum lines, to be used in further experiments.

2.5 *Statistical Analyses*

As explained above, this was a preliminary experiment performed before we had replicate collections of the spore-lines, and when we had only 1-20 available ramets of each of the 31 host clones. No host clone was inoculated with all 10 spore-lines; in fact most host clones were inoculated with either 4 or 5 spore-lines. Since we knew that these hosts had different susceptibilities in field conditions (Old & others 1985, 1986; Zagory & Libby 1985), we suspected that they would also differ in susceptibility in greenhouse conditions with artificial inoculations. Because of the small numbers of ramets per clone and the complication of different spore-lines being used on different clones, we did not attempt to characterize host-cline susceptibilities. Because there were similar numbers of ramets in each host-class inoculated with spore-lines "6-10" and "21-26" (Table 1), we did analyse infection by host classes, using the non-parametric Chi-square statistic.

Our main interest was in developing evidence whether these isozymically uniform spore-lines exhibited different pathogenicities. The null hypothesis is that there is no difference between spore-lines. To eliminate the expected effect of different susceptibilities among host clones, we made each host clone an experimental unit. Under the null hypothesis, the probability of infecting each ramet of a particular clone should be the same, no matter which spore-line contributed the spores. To evaluate this, we took the percentage of ramets infected by all spore-lines used on a host clone (typically two from spore-lines "6-10" and two from spore-lines "21-26") to generate a predicted number of infected ramets for each spore-line used on that host clone. We tested the null hypothesis using two non-parametric statistical methods. The first was a Sign Test. Each spore-line was used on 10-13 different clones that generated valid data. Under the null hypothesis, each spore-line should generate approximately equal numbers of plus and minus departures from the various host clones' predicted number of infections. The second was a Runs Test. In this, not only the sign but the size of the observed departure is taken into account. Data (departures from null-hypothesis expectations) from pairs of spore-lines are then combined and put in rank order, with the spore-line identity of each point also maintained. If the data for both spore-lines are drawn from the same set (the null hypothesis), the number of times adjacent ordered data entries are expected to be from the same or different spore-lines can be determined, with probabilities for various outcomes. If the data for the paired spore-lines are drawn from different sets, however, the frequencies of adjacent data-entries being from the same spore-line increase, while the frequencies of being from different spore-lines decrease (Dixon & Massey 1951).

3 RESULTS AND DISCUSSION

3.1 *Host Classes*

The upper part of Table 2 presents infection percentages of each of the 4-6 clones in each host class. Each clone's population of origin is coded to footnote 1 in the Table. Average infection percentage for the entire experiment was 43.1%.

The single-host-clone data are not reliable, being based on small numbers of inoculations per clone (1-20, avg. 11.7), and with different sets of four or five spore-lines used on each host clone. In spite of this unreliability, the variation is far from random. Of the 20 clones in host classes "0-3", only four exceed the average infection level of 43.1%. In sharp contrast, all six clones of host class "4" exceed 43.1% infection, as do three of the five host-class "5" clones.

TABLE 2. Clones in host classes "0-5" ranked by percentage of ramets in each clone in infection-classes "3-5".

	Host Class[1]					
	0	1	2	3	4	5
	L - 0%*	L - 0%*	Z-17%	TxM-8%	M-44%	T-7%
	T-12%	A-7%	Z-20%	Z-36%	L-46%	A-11%
	Z-33%	LxA-13%	A-29%	M-38%	LxC-65%	Z-47%
	L-33%	AxC-44%	LxT-67%	C-38%	C-71%	C-55%
	Z-38%	L-67%		M-42%	Z-76%	C-93%
				C-47%	C-87%	
Number Attempted Inoculations (Ramets)	34	55	21	83	92	77
(1) Percent of Hostclass Ramets in Infection-classes "3-5"	29%	27%	29%	35%	65%	47%
(2) Average of Clonal Percentages of "3-5" Ramets	23%	26%	33%	35%	65%	43%

[1] Host origin population: A- Año Nuevo, C– Cambria, L– Guadalupe Is., M– Monterey, T– Cedros Is., Z– Australia-New Zealand landrace, LxA – Guadalupe x Año Nuevo hybrid. Percentages based on (the ramets scroed in infection classes "3-5") ÷ (the number of surviving ramets inoculated) per clone.

*These two uninfected clones were not used in further data analyses.

(1) (Total number of ramets scoring "3-5") ÷ (number of surviving ramets in the host class)

(2) (Sum of the "3-5" ramets percentage by clone) ÷ (number of clones in the host class)

In the field experiment where the ortets of these clones were selected, the two most susceptible native population-samples were from Cambria and Monterey (Old & others 1985, 1986). The six Cambria-origin clones in Table 2 average 65.2% infection, with only one of them being below 43.1%. The three Monterey-origin clones average 41.3% infection. The Año Nuevo population-sample was much less susceptible than those of Cambria and Monterey in the field, and the three Año Nuevo-origin clones average only 15.7% infection. The Australia-New Zealand landrace was intermediate in susceptibility between the Monterey and Año Nuevo populations, its populations of origin, in the field experiment. The seven clones of this landrace in Table 2 average 38.1% infection. The Cedros and Guadalupe Island population-samples were the least susceptible in the field experiment. The two Cedros-origin clones average 9.5% infection, but the five Guadalupe-origin clones average 29.2% infection. Furthermore, a hybrid Guadalupe x Cedros clone was more of the two host-class "0-3" clones with greater than 50% infection, a Guadalupe clone being the other. While these clones were not random draws from the different population-samples, it does appear that their rank-order of susceptibility is almost identical to that of their random population-samples of ortets in the field. Thus there is little evidence of interaction of maturation-state (the ramets of the host clones were more mature in this experiment than were their field ortets, see Zagory & Libby 1985), growing environment (field vs. greenhouse) or inoculation method (natural vs. artificial) with host susceptibility. The possible exception to this is that some clones of the Guadalupe population may be relatively more susceptible to artificial inoculation and/or in the greenhouse environment.

The lower half of the Table 2 presents the percentages of ramets in infection classes "3-5" by hose class. These numbers are more reliable than those for individual clones, both because they are based on larger numbers of ramets (34-92 per class), and because similar proportions of the ten spore-lines were applied to the clones of each host class (see Table 1). Statistical analysis indicated that the differences in infection among the six host-classes (line [1], Table 2) are highly significant (X^2=18.1, 5 d.f., p<.01). Host classes "0-3" are all below average in infection percent, while host classes "4" and "5" are both above-average. Unweighted averages of clonal infection percentages provide a similar result (line [2], Table 2).

The ortets of the five clones in infection class "5" all were growing in blocks in the most heavily-infected portion of the field experiment. The ortets of the six clones in infection class "4", by contrast, were growing in lightly-infected or moderately-infected parts of the field experiment. Table 2 provides some evidence that the criteria used to choose clones for host class "4" more consistently identified susceptible hosts than did picking heavily-infected ortets in heavily-infected parts of the plantation (host class "5").

It may be noted that the clones in host classes "0-2" provided fewer ramets per clone (6.8, 11.0 and 5.2 respectively) than did the clones in host classes "3-5" (13.8, 15.3 and 15.4 respectively). This is due to a combination of numbers of cuttings per donor hedge, rooting percentages of the cuttings and survival of the newly-rooted cuttings. Statistical analysis indicated that the differences in ramets per clone among the six host-classes are highly significant (X^2=39.6 5 d.f., p<.001). This relationship between susceptibility and rate of clonal expansion has been noted in one other experiment (WJL—unpublished data), although not consistently. The opposite result (faster clonal expansion of the more resistant clones) has not been observed. If the relationship is real, it may be related to differential maturation rates of the different clones (see Fig. 1, Burdon and Bannister 1973), and the relationship of maturation state to vegetative proliferation (Libby, Fanger-Vexler & Russell 1985), to rooting effectiveness (Libby & Conkle 1966), and to susceptibility to western gall rust (Zagory & Libby 1985).

3.2 *Spore-lines*

Figure 3 shows the cumulative above-expected-numbers of ramets per clone in infection classes "3-5" (positive bars), and below-expected-numbers of ramets per clone in infection classes "3-5" (negative bars). Not that none of the ten spore-lines produced infection rates that were 100% above-expected or 100% below-expected on all host clones.

Figure 4 shows the net departures of the ten-spore-lines from expected infection rates. From this, it appears that the spore-lines may be ranked and clustered as follows: spore-lines "6", "8", "21" and "9" in a group of below-average infectivity; spore-lines "7", "10", "23" and "24" in a group of about average infectivity; and spore-lines "22" and "26" in a group of above-average infectivity.

Table 3 presents spore-lines in the above rank order, and a Sign Test for each spore-line. Four of the ten spore-lines had statistically unlikely departures from average infectivity, as tested by a Sign Test. The Sign Test supports the groupings based on net departures, above, with spore-lines "6", "8", "21" and "9" all having more than twice as many minus as plus departures, spore-lines "7", "10", "23" and "24" all having about equal numbers of plus and minus departures, and spore-lines "22" and "26" both having 5 times as many plus as minus departures.

Runs Tests on these data (see Materials & Methods) also tended to support these groupings. Spore-lines "6" and "8" were both highly significantly different from spore-lines "22" and "26" (p<.01).

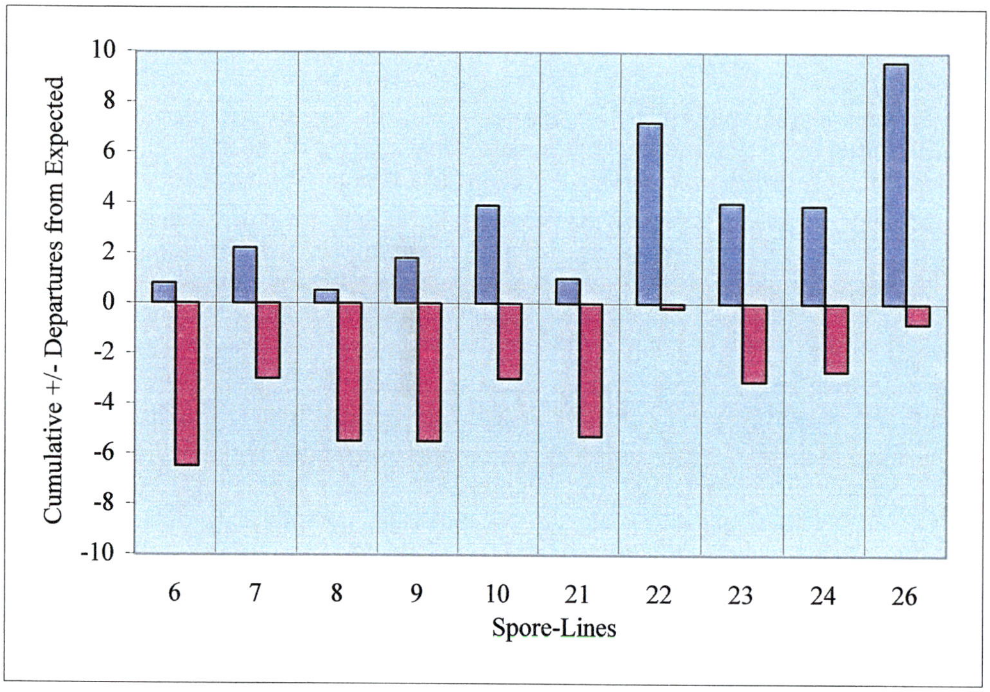

Figure 3. Departures from expected infection rates of the ten spore-lines. The positive bar indicates that an infection rate above the host-clone average was observed for one or more host clones, with the height of the bar summing large and small positive departures from expected. The negative bar similarly indicates summed departures of infection rates below host-clone averages.

Spore-line "21" differed from "26" ($p<.01$) and "22" ($p<.05$). Interestingly, spore-line "9" did not significantly differ from "26" or "22", but spore-line "7" significantly differed from "26" ($p<.05$). No other pairs were significantly different.

At the time that we collected these spore-lines, we thought it possible that the spore cloud was of similar composition (although of different density) over the entire plantation. Thus, in any given area, the more-heavily-infected trees were permissive hosts for many kinds of spores, while the lightly-infected trees were permissive hosts for only a small subset of those. We therefore thought that only the more virulent spore-lines from the more susceptible hosts would have been less virulent. The data do not support this idea (see Figure 4). The two most infective spore-lines ("22" and "26") are from heavily-infected trees in host classes "4" and "5", while the two least infective spore-lines ("6" and "8"), are from host class "1". Spore-lines "22" and "26" were both collected from host seedlings of the Cambria population (the most susceptible), while spore-line "6" came from a Cedros host and spore-line "8" came from a Guadalupe host, the two most resistant populations in the field experiment.

4 GENERAL DISCUSSION AND CONCLUSION

This preliminary experiment has served to expand our supply of the inoculum lines, through multiple infections and re-collections. We have also begun to characterize the 31 included host clones as cutting production from their donor hedges was increasing. We have also tested our inoculation procedure, and have gained experience and confidence in early scoring of infection success.

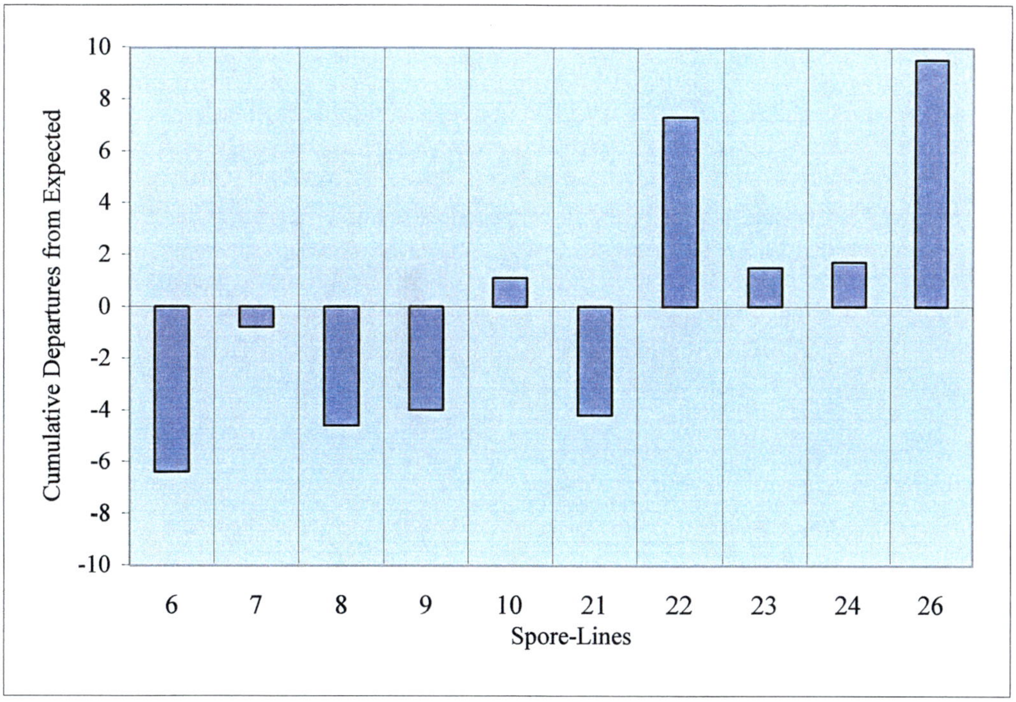

Figure 4. Combined departures from expected infection rates of the ten spore-lines. Figure 4 combines the positive and negative bars of Figure 3 into a net departure bar. Spore-lines "6-10" are from lightly-infected hosts and "21-26" are from heavily-infected hosts.

TABLE 3. Rank-order of spore-lines by net departures from expected infection rates[1].

	Spore-lines[1]									
Departures[2]	6	8	21	9	7	10	23	24	22	26
+	2	1	3	3	5	6	5	6	10	10
-	11	11	10	7	8	5	7	7	2	2
Probability[2]	<.05*	<.01**	<.10	>.25	>.25	>.25	>.25	>.25	<.05*	<.05*

[1] Based on Figure 4.
[2] For each clone inoculated with the spore-line, either more ramets than expected were infected (one +) or less were (one -). For example, 13 clones were inoculated with spore-line "6". Line "6" spores caused greater than clonal-average infection rates in 2 clones, but less than clonal-average infection rates in the other 11 clones, a result that is statistically unlikely (<.05 and significant -*) according to the Sign Test, given the null hypothesis that all spore-lines were equally infective.

The data produced during these activities indicate that the relative susceptibility of naturally-infected seedling ortets in the field was generally maintained by their artificially-inoculated steckling ramets in the greenhouse. Numbers of available ramets per clone did not permit analyses of specific host-clone/spore-line interactions, but contrasting host clones and spore-lines were identified to make such planned experiments more effective.

The analyses used produced statistically convincing evidence that all ten spore-lines did no behave identically on the host clones. Rather, we found two strongly-infective spore-lines and two weakly-infective spore-lines, and all four of these were of origins opposite to those anticipated. We emphasize here that, while we have demonstrated performance differences among the spore-lines we have not demonstrated genetic differences as the basis of such performance. It might be that there were physiological differences among the spore-lines, as a result of differences in storage environment, or differences in condition at the time of collection, or even differences somehow imposed by their relatively resistant and susceptible ortet hosts. The re-collection of these lines from a variety of hosts, and their testing in larger numbers on a set of increasingly-known host clones, should serve to provide evidence on the presence of genetic differences in virulence among them.

5 ACKNOWLEDGEMENTS

This work was supported in part by PSW/UCB Coop Agreement-PSW-870004CA; manuscript preparation was supported in part by USDA Competitive Grant 86-FSTY-9-0191. We thank Weiching Wang and Kun Jin for statistical help and advice, Tina Popenuck for help in collecting the spores, and Ang-he Shang and Evelyn Chung for help in raising the plants.

6 REFERENCES

Burdon, R.D., and M. H. Bannister. 1973. Provenances of *Pinus radiata*: their early performance and silvicultural potential. N. Z. J. For. 18: 217-232.

Dixon, W.J., and F.J. Massey, 1951. Introduction to Statistical Analysis. Chap. 17.3. McGraw Hill. NY. 370 pp.

Epstein, L, & Buurlage, M.B. 1988. Nuclear division in germinating aeciospores and its taxonomic significance for the western gall rust fungus, *Peridermium harknessii.* Mycologia 80: 235-240.

Nelson, D.L. 1971. The Ecology and Pathology of Pine Gall Rust in California. PhD dissertation. Univ. Calif. Berkeley. p. 160

Libby, W.J., & M.T. Conkle. 1966. Effects of auxin treatment, tree age, tree vigor and cold storage on rooting young Monterey pine. Forest Science 12: 484-505.

Libby, W.J., Fanger-Vexler, L.J. & Russell, J.H. 1985. A population study of Monterey pines evaluated as Christmas trees. Final Report to the Calif. Christmas Tree Growers Assn. p. 19.

Old, K.M., Libby, W.J. & Russel, J.H. 1985. Patterns of susceptibility to western gall rust in *Pinus radiata.* Pages 285-305 *in* Proc. Of the Rusts of Hard Pines Working Party Conference. *IUFRO* S2.06-10. U of Georgia, Athens. *Eds.* J.H. Barrows-Broaddus and H.R. Powers. p. 331.

Old, K.M., Libby, W.J., Russell, W.J., & Eldridge, K.G. 1986. Genetic variability in susceptibility of *Pinus radiata* to western gall rust. Silvae Genetica 35:145-149.

Vogler, D.R., Kinloch, B.B., Cobb Jr., F.W. & Popenuck, T.L. 1987. Isozyme patterns in *Endocronartium harknessii* from several pine hosts and stands in California. (Abstr.). Phytopathology 77: 1242.

Zagory, D. & Libby, W.J. 1985. Maturation-related resistance of radiata pine to western gall rust. Phytopathology 75:1443-1447.

Taken from:
Proceeding from the thirty-sixth Annual Western International Forest Disease Work Conference. 1988. Park City, Utah.

Fig 1. Gall on yearling branch resulting from infection approximately 11 months earlier, shortly after collection of spores from the aeciospore-filled fissures.

Fig 2. Ramets showing various symptoms 3 months after inoculation:

2a. Short-shoot development, but little or no swelling - class "2".

2b. Moderate swelling in the inoculated region - class "3".

2c. Several swellings in the inoculated region - class "4". Note also the developing short-shoots.

2d. One and two large swellings in the inoculated region - class "5". Note continuing long-shoot development from short-shoots on the galls.

MEDICINAL FUNGI IN CHINA, M. M. Chen, Department of Plant Pathology, University of California, Berkeley, CA 94720

The use of fungi in Chinese traditional medicine has a history dating back several thousand years (Shen Nung, 2600 B.C.; Hua To, 206 B.C.). According to more recent investigations, there are approximately 117 species of fungi in China that are used medicinally. Many of these are known throughout the world as wood decay fungi or as edible species. Few, however, are used medicinally in western countries. Fungi are used as stimulants, sedatives, refrigerants, purgatives, emollients, diuretics, and tonics. Several species are produced commercially on large scales, including: (as anti-cancer agents) *P. versicola*, *Ganoderma applanatum*, *Pyropolyporus fomentarius*, *Schizophyllum commune*; (as tonics) *Cordyceps sinensis*, *Tremella fuciformis*, *Hericium erinaceus*, *Armillaria mellea*, *G. lucidum*, *Lentinus lepideus*; (as a diuretic) *Sclerospora graminicola*. Fungi are prepared for use in a variety of forms, including soups, teas, pills, plasters, and medicinal liquors. (Taken from First SCBA International Symposium and Workshop Poster Abstract Form)

THE STUDY OF RUST FLORA ON SINO-HIMALAYAN FORESTS, M. M. Chen, Department of Plant Pathology, University of California, Berkeley, Berkeley, CA 94720

In the period from 1975-1980 the author made pathological and fungal investigations in the forest in a region located at an altitude of 1,800-6,000 m. in the Sino-Himalayan forests where a complete and comprehensive spectrum of vertically distributed forest was seen. Here is provided a unique natural laboratory in which there exist very good conditions for the study of the ecological aspect of rust. This preliminary study of the rust flora contains 20 genera and is called the Sino-Himalayan rust flora. The evolution of rust is also discussed.

This rust flora is classified into three types according to conditions of water and temperature: 1) the plateau type, 2) the temperate zone type, 3) the sub-tropical zone type. Each type may be characterized by species belonging to it. The analysis that has been made shows that the region investigated has the following features: It is a place of origin for a new rust type, the rust flora in this region has affinities to those in other regions, and thus it is possible for this rust flora to constitute a potential source of forest rust disease. (Taken from First SCBA International Symposium and Workshop Poster Abstract Form)

FOREST PATHOLOGY RESEARCH & MANAGEMENT IN CHINA. M. M. Chen, Department of Plant Pathology, University of California, Berkeley, Berkeley, CA 94720.

China and North America have similar flora and, since the recent establishment of plantations in China, have similar forestry practices as well. Therefore, many pathogens are of interest to pathologists on both continents. Research in silviculture, management and protection has been an important concern of forestry institutions in China, and has focused on detection, distribution, impact assessment and management of major pests causing conifer, poplar, and paulownia diseases, rusts, nematode problems, etc. Recently, we proposed research on the biology, evolution, biogeography and management of these pests in order to establish a network of scientific communication permitting development and evaluation of integrated pest management strategies and their transferability between continents. (Taken from American Phytopathology Society 1988 Annual Meeting Abstract Form)

COMPARISON OF THE DISTRIBUTION OF PATHOGENS OF WISCONSIN AND HEILONGJIANG, CHINA. M. M. Chen, Department of Plant Pathology, University of Wisconsin-Madison, Madison, WI 53706.

The forested regions of Wisconsin and Heilongjiang are at similar latitudes (42°-46° N), but different longitudes (87°-92° and 127°-132°, respectively). Similar genera of forest trees and their pathogens are found in both locations. Wisconsin's pine species include *Pinus storbus*, *P. resinosa*, and *P. banksiana*, and their major pathogens include *Cronartium ribicola*, *C. quercuum*, *Phellinuspini*, and *Armillaria mellea*. These same four pathogens are important in Heilongjiang, although in this hemisphere the hosts are *P. koraiensis* and *P. sylvestris* var. *mongolica*. (Taken from a lecture for the department of Plant Pathology, Universtiy of Wisconsin-Madison)

WHITE PINE BLISTER RUST IN YOUNG SUGAR PINE PLANTATIONS IN THE MID-ELEVATION SIERRA NEVADA. M. M. Chen, Department of Plant Pathology, University of California, Berkeley, Berkeley, CA 94720
Sugar pine seedlings were included in mixed-species plantations on Blodgett Research Forest (av. elevation 1330 m) beginning in 1976. In 1983, the disease was detected on planted saplings, and a study was initiated in 1984. A total of 1277 saplings in 11 plantations and 476 scattered naturally-occurring saplings was examined; 16 percent of the planted saplings (1-53% of seedling per individual plantation) and 12 percent of the natural saplings were infected. The years of origination of the infected internodes were determined: 60% of the infections were on 1981 tissue; 15% on 1982; 10% on 1980; 6% on 1979; and less than 1% on tissue of other years. Local weather data indicated that 1981 (with 11 days) and 1982 (with 6 days) were the only years since 1977 with more than one day of conditions favorable to infection. There was a strong correlation between infection and distance to Ribes. Both mature pycnia and aecia were detected during Fall, 1984. (Taken from APS Annual Meeting Abstract Form 1985)

ANALYSIS OF THE DISTRIBUTION OF FOREST PATHOGENS IN THE UNITED STATES AND CHINA: TWO CASE STUDIES, M. M. Chen, Department of Plant Pathology, University of California, Berkeley, Berkeley, CA 94720
There are 740 major forest tree species in China and 1260 known forest pathogens, 93% of which are fungi. Mudanjian, in NE China, and Wisconsin are at similar latitudes and share genera of forest trees and many pathogen species. Both regions are horizontal mountains. Many forest pathogens are also common to the Sierra Nevada and the eastern margin of the Tibetan plateau. They are similar in that both form large rain shadows, but the climate of the Sierra Nevada is influenced by the Pacific Ocean and Southeast China's climate is continental and strongly influenced by the high elevation of Qinghai-Xizhang (Tibetan) plateau. In spite of the great climate differences between the two regions, they share several coniferous genera and hardwood pathogens: dwarf mistletoes, rusts, decays and root diseases. The geographical distribution of fungi has been analyzed. A scientific analysis of the biogeography of pathogens is crucial to understand where and how pathogens may become a problems and for the application of quarantines. (Taken from APS Annual Meeting Abstract Form 1985)

A GREENHOUSE INOCULATION TECHNIQUE FOR TESTING HOST-PATHOGEN SPECIFICITY IN RADIATA PINE--WESTERN GALL RUST PATHOSYSTEM. M. M. Chen, Department of Plant Pathology, University of California, Berkeley, Berkeley, CA 94720
Field inoculations with *Peridermium harknessii* have proven to be difficult, and nearly a year is required for reliable development of symptoms. Three greenhouse inoculation methods were tested: dry spores applied with brush, no wounding; dry spores applied to tissue after a light sandpapering; and injection of aqueous spore suspensions. Protocols with dry spores only, developed for single-isolate inoculations of host clones in the greenhouse, provide greater reliability of inoculum identity, and reliable symptoms could be scored in less than six months. The first symptoms appeared within 30 days, and these increased in number through 90 days, with infection number becoming stable by 120 days. Aeciospores were produced on some of these infections within 11 months after inoculation. (Taken from APS 1987 Annual Meeting Abstract Form)

BIOGEOGRAPHY OF UREDINALES ON SINO-HIMALAYA. M. M. Chen, Department of Plant Pathology, University of California, Berkeley, Berkeley, CA 94720

The Sino-Himalayan plateau (Xizhang), 27°-31° north latitude and 5°-98° west longitude, is the largest and highest topographic plateau on earth. The summit of the highest peak is 8842 m and elevations commonly range from 1100 to 6000 m. Located in the southwest of China, Xizhang is sometimes called "the roof of the world" and it is one of the few remaining places on the earth that has not yet been fully explored by mankind. Therefore, it has long attracted attention in international scientific, rust floristic studies of unexplored areas of the world. The plateau is a critical region for solving problems in geoscience and evolutionary biology.

Biogeography of rust and the forests' flora are classified into three zones according to conditions of water and temperature: 1) the plateau zone; 2) the temperate zone; and 3) the sub-tropical zone. A distinct group of rust species is characteristically found in each zone. Endemic rusts are common and the area is a place of origin of new rust species. The biogeography of rust in East Asia has affinities to that in North America. (Taken from Second SCBA International Symposium and Workshop Poster Abstract Form)

MORPHOLOGY OF ISOLATES OF *CERATOCYSTIS ULMI* FROM WISCONSIN. M. M. Chen and E.B. Smalley, University of California, Department of Plant Pathology, Berkeley, CA 94720.

Ceratocystis ulmi isolated from branches and bark beetles of recently killed elms in southern Wisconsin (175 samples, 13 locations) were examined and the colony characteristics of the non-aggressive and aggressive strains were described. *Sporothix* conidiophores from vegetative hyphae often extended at right angles. Conidiophores are slender, tapered and produce conidia sympodially on short denticles along the conidiogenous portions of hyphae. The conidia taper to a truncate base and accumulate in minute, mucilaginous droplets. *Graphium* produced erect synnemata composed of tightly packed, pigmented hyphae elements comprising the stalk of synnemata, diverge at the apex, forming separate condiogenous branches. Differences in the fungi carried by bark beetles in Wisconsin and China (Xinjiang) are discussed. (Taken from Mycological Society of America, Application for a place on the 1988 MSA Program at the University of California, Davis – August 14-18, 1988)

OCCLUSION OF SPAWOOD BY FUNGI CARRIED BY *DENDROCTONUS* SPECIES

M.M. Chen, Chinese Academy of Forestry, Beijing, J.R. Parmeter, Department of Plant Pathology, University of California, Berkeley, and D.L. Wood, Deparment of Entomology, University of California, Berkeley.

Bark beetles (*Scolytidae*) generally carry fungi that colonize host sapwood and contribute to tree killing by interfering with water flow through the stem. Vascular occlusion was easily observed by standing infected stem sections in dye solution and determining pattern of straining. Studies in California showed that when only outer sapwood rings were colonized and occluded, water continued to flow in deeper uncolonized rings. Only fungal penetration of the entire sapwood depth prevented flow through deeper rings. The ability to penetrate sapwood varied within and among fungal species and among trees. Two isolates of *Leptographium terebrantis* (from *Dendroctonus valens*) penetrated the entire sapwood depth of small (10-24 cm dbh) inoculated ponderosa pines (*Pinus ponderosa*) 33% and 54% of the time; whereas *Ceratocystis minor* isolates (from *D. brevicomis*) did so 12% and 13% of the time and *C. ips* (from *D. valens*) only 7% and *%. Study of the effects on water conduction associated with numbers of infections and depths of sapwood penetration may help to explain why some trees survive beetle attacks while others are killed. (Taken from Forest Pathology Symposium in China, 1988)

OBSERVATIONS ON THE DEVELOPMENT AND MORPHOLOGY OF *MELAMPSORA ROSTRUPII* WAGNER ON *POPULUS TOMENTOSA* L.

G.P. Ge, Y Jing, M.M. Zhan, Q.C. Shi, Northwest Agricultural College.

Melampsora rostrupii Wagner severely infected the seedlings and young trees of *Populus tomentosa* L. which caused a sharp decreasing in the growth-rate both in length and in width. Generally, the mycelia of the pathogen overwintered in the winter buds, or in the cambium region of the current twigs occasionally. Although the teliospores were existed, yet since no alternate host had been found, that they were of little importance in the occurrence of the disease.

Uredospores were produced abundantly on young leaves, leafscars, or infected stems, beginning from the end of March. Such initial infection would soon become an outbreak of the disease from May, until the end of September.

Uredospores germinated quickly on the back of leaves after inoculated at 19.8-23.6° C in 1-2 days at a relative humidity of 100 per cent. However, none or very few spores germinated on the upper surface of the leaf. The development of the epidemic required a monthly mean temperature 18.2-26.8° C, and it would be of little influence when a relative humidity was below 4.1%. Thus, temperature was the main environmental factor which influenced the development of this epidemic.

Uredinia hypophyllous, scattered or somewhat confluent, roundish, small 1mm in width; naked when young, somewhat pulverlent, orange-yellow, light-yellow on the opposite sides of the pustules, and amphigenous on infected leaves.

Uredospores globose, sub-globose or oval, 15-22.5x17.5-27.5 u, wall colorless, 3 u thick, even verrucose with fine papillae; paraphyses capitate or clavate, smooth, 50-75 u long, head 15-23 u broad, wall 3-6 u thick.

Telia hypophyllous, sometimes amphigenous, scattered, around the old uredinia, irregular roundish, small ½-1 mm width, subepidermal, slightly elevated, grayish-brown; teliospous prismatic, 1-celled, flat or round at the head, broader in top, 7.5-10x37.5-50 u, light brown, smooth, wall 1-2 u thick.

Based on these characters, this fungus is considered to be *Melampsora rostrupii* Wagner, and it is believed to be the first collection in this country.

(Taken from Scientia Silvae Vol.9, No.3)

美国柏克莱大学学术演讲："美国主要森林病害"(向到访贵州等16省森林病虫专家)在报告中作者介绍在美国研究的几种国际流行病虫害及防治 ：1)由抗病育种已防治了的荷兰榆病，2)南半球重点国际检疫病害–美国辐射松西方锈瘤，3)美国传入亚洲之松材线虫病，4)区系型白松疱锈病，5)加州太浩湖成片枯死松蓝，黑变病，6)迅速传布的流脂病等,其中重点对中国杨树天牛及刺吸式害虫以及防治进行了讨论和交流(包括造林树种选择,IPM应用新型内吸剂，信息素运用等)．并专题收集杨天牛生防资料,其中包括北美天牛危害乔灌木专著，小蠹虫，信息素，及桉树天牛防治方法等著作．

学术报告还讨论森林保护与法律，分子生物学，以及电脑模拟与病虫害防治和生态系 统管理森林等之间的关系问题．

主要森林病害概述：

根病，树干腐朽病，白松疱锈病，西方锈瘤，槲寄生为主要病害，造成木材价值降低，此外有相当大的部分枯萎多为非生物性自然因子引起的灾害，常是树木枯萎的直接因素，然后腐生,半腐生物相继作用．美东部山区常见树种如白蜡，山毛榉，桦，来木，榆和冷杉，槭，橡，云杉等．

东部广泛分布的橡树枯萎现象是由许多综合因子引起的，大气候灾害，树木生理以及病虫等，近年来木炭疽病较为严重，但至今基本的地理分布和致病因子尚未弄清楚．美西地区近五年来发生在太浩湖森林大量枯萎，对湖水涵养影响颇大.调查表明小蠹虫入侵并带入真菌病原，造成大量菌丝堵塞树木输导组织，导致树木的衰退，而枯萎的树又急剧增加小蠹虫群数量，致使虫害及树木生理衰弱之间恶性循环，追究其原，乃因旅游地区管理不善，引起诸多社会问题，因太浩湖森林介于加州和内华达州之间，赌博,滑雪地盘造成纠纷，引起联邦及国会重视，今年七月克林顿总统及高尔副总统均亲赴现场强调保护环境的重 要及采取措施．

此外旧金山蒙特利湾的辐射松流脂病以及针叶树根病在加州，埃达荷州，蒙他那州，奥立冈州和华盛顿州以及南部山区随着林木年龄增加损失．

一 ，根部病害：

美西北约 8%商品材被根病危害,该类树易受小蠹虫害，总计北方大约三百万英亩为根病，美国林务局认为这是美西在森林管理上之最重要的问题．

二， 树干腐朽病

几乎美国针阔叶林均有此病，造成商品材产量和质量上损失，尤其在旅游地区还会造成腐朽树打伤死人等事件发生，诉诸法律事件造成国有或私人公园管理上巨额赔偿。

三， 白松庖锈病

由于皆伐后林地 Ribes 增加而大发生．糖松疱锈病产原地称着：太平洋地中海型

该病以加州及俄勒冈州糖松为主要病害，幼林令约 40~60%受害枯死，糖松纯林造林地 10 年左右树木死亡高达 80%，该病还蔓延至爱达荷州及华盛顿州；已繁殖出之抗锈原种已在一些地区应用作造林树种，但也有报道有不久失去抗病性现象发生，但中北部及东部地区白松庖锈病近年来并不严重。

在国家冰川公园地区，北美大灰熊啃食白皮松树皮现象严重。

四 槲寄生

它为寄生性植物，是逐渐吸食树木营养，造成生长缓慢、蓄积量减少，危害程度和分布广度是十分惊人的，据统计西部森林受害最重，针叶树有 2 千 2 百万英亩受害，其中危害最厉害的有七种槲寄生，受害树种包括扭叶松、花旗松、西部落叶松，真冷杉，西部铁杉和两种西黄松，清除此病美国人说最简单方法是把树全株砍掉。

五 梭形锈病

受害树种主要为火炬松和湿地松，大概感病率为 30%，其中 10%为树干长条状溃疡斑，影响材质，火炬松和湿地松每年大约因此病损失 89%。

六 近期发现来木炭疽病蔓延迅速，1984 年仅在马里兰州，87 年发展到九个州（麻州到乔治亚州），到 1991 年此病已发展到 16 个州（新泽西州到乔治亚州），树木 100%枯死，特别是在南方高海拔处严重，此树美学鉴赏价值高，所以认为此珍贵树木应当加以特别保护。

七 西方锈瘤

美国人形容"辐射松比竹子长得快"，但辐射松受此病危害最严重，发源地在旧金山蒙特利湾，因此病日益严重，现移种至南半球，尤以新西兰造林成功，长得快，尚无病，美国有 1/3 木材是从新西兰进口辐射松，但新西兰十分重视西方锈瘤检疫问题，目前尚未发生此病，近期据称中国投资 20 亿在新西兰种植辐射松，每年可盈利 15%，中国必需注意防止此病传入中国，虽辐射松为典型地中海型树种，但中国有的地方也可种此树，但应了解此病基本情况，每年新西兰大公司主管病虫害检疫专家来美，调查该病发展，采取阻止病害入侵的种种方法。

八 松流脂病

为辐射松新发生之病害，原仅在蒙特利湾地区，现已发展到 17 县，辐射松树林密度大的 3 个县发生严重，现已传至有花旗松的国有林，现该病害正处在严密监视中，由加州林务局直接领导管理辐射松森林病虫害，拨款加强研究、教育和管理，该病系由某种自然因素（现不甚清楚）引起树势衰弱，继而发生大量星坑小蠹虫危害树木，并携带镰刀菌而引起树木流脂及枯萎。

美国主要病虫害名单：

Major Forest Insects of the U.S.

Common Name	**Latin Name**	**Chinese Name**
Gypsy Moth	**Lymantria dispar L.**	
Southern Pine Beetle	**Dendroctonus frontalis Zimmerman**	
Spruce Budworm	**Chorstoneura fumiferana**	
Mountain pine Beetle	**Dendroctonus ponderosae Hopkins**	
Western Spruce Budworm	**Choristoneura occidentalis Freeman**	

加利福尼亚州森林保护工作是如何组织领导的？

加州自 1951 年起成立了加州森林病虫害顾问委员会，并作出文件及出版报告。

1， 病虫害调查：每年调查整理编辑有关当年森林病害、虫害及动物损害资料报告；

2， 组织损失评估：集中主要病虫害问题进行技术经济损失评估；

3， 提出病虫害防治措施及实施方法（私有林地占加州总林地 1/2）；

4，　总结病虫害防治政策、法律和研究工作，在提交联邦的建议上批注并找到该病虫害本州之专家，参加领导及技术指导。

顾问委员会为加州森林健康保护主要领导组织，顾问委员会每年开年会一次，凡加州私有林主、公众林业工作者、造林者以及对昆虫、病害、生物杂草有兴趣者，均可参加此会。顾问委员会成员参加每年西部森林顾问委员会和自然保护学会会议。

森林病理学家

谌谟美

1997 年 12 月 10 日

二十一世纪初加州旧金山海岸山发生几种橡树枯萎，简称 SOD 取得连联邦巨款，成为西部森林病害重点项目，指 *Phytophthora ramorum* 为主要病原，重点放在分子生物学和遥感模拟．不少学者指出：这是生态病害，是温室效应等大气物理直接因素，不拟夸大病菌致病力，应综合研究生态系全局，并使用 IPM 等经济有效措施效防止，合理经营恢复．　(2003，2，柏克莱大学植物标本馆)

Fungi of the Alaskan Inland Ecosystem

PREFACE

During the American Phytopathological Society's annual meeting in San Diego in 1988, I was invited by plant pathologist Professor Jennifer Huang McBeath of the University of Alaska to cooperate with her on the Alaska Inland Economy Fungi Research Project (Alaska map). After I accepted this invitation, I started thinking how I could use my past knowledge and experience of taiga forests to understand the taiga forest ecosystem in Alaska and what role fungi play in it (Fairbanks, Alaska).

After I graduated in plant pathology in the early 1950s, I worked for many years with forest ecologists and learned from them much about forests and forestry techniques, but I noticed that ecologists do not often think of fungi as a part of the forest ecosystem when conducting experiments or making observations. Similarly, mycologists do not pay adequate attention to ecosystem studies when they conduct systematic or evolutionary studies of fungi. So this project was a good opportunity for me to study the two fields together and to understand the ecology of Alaskan forest fungi in comparison with other Taiga forest fungi in order to provide suggestions for future forest management.

The study of fungi has increased exponentially in the past 100 years, but fungi are still being ignored or neglected in many fields of study. For example, more than 90 percent of fungi species have never been screened for useful species or compounds. However, some fungi play very important roles in the ecosystem. Many fungi are associated with other organisms in their evolutionary history. I developed an idea of basing my research plan on the tree-rust relationship within the taiga forest ecosystem. In the vast biomass of forests, fungi are a vital part of the food web as decomposers and as pathogens. Many types of fungi are excellent scavengers in the taiga forest, breaking down dead materials into simple compounds that become available to evolute in a healthy forest ecosystem. Also, the fungi are valuable because of their mutual relationships with trees. Fungi associate with tree roots and make nutrients available to the tree roots; this is important because taiga forest trees grow in nutrient-poor soil in this cool temperature zone. The microbiological activities of fungi in the root zones appear to be significant in the maintenance of a healthy taiga forest.

Many botanists and foresters have studied the trees and shrubs of Alaska. During the fieldwork survey, we learned much from the previous work of Dr. Leslie A. Viereck, a very knowledgeable expert on the Alaska forest. There are important works in this field that proved useful in our survey. They include a mycological study of the Alaskan Arctic reprinted from Annual Report of the Institute for Fermentation (Osaka No. 3, 1967) by Yosio Kobayasi, Naohide Hiratsuka, Richard P. Korf, and colleagues, which was useful in providing some mushroom species references. Also, one small book was recently written on Alaska's mushrooms, *A Practical Guide* written by Harriette Parker. *A Checklist of Alaskan Fungi* by Cash and *Identification of Destructive Alaska Forest Insects* were also very useful, but the latter publication dealt mainly with the more damaging forest insects of southeastern Alaska. Recently, more complete catalogues of Alaskan fungi have focused on the southern gulf forests of Alaska and non-inland areas. We, on the other hand, needed to focus on the inland of Alaska, therefore this expedition was focused on the characteristics of northern Alaska and inland forests near Fairbanks. It was the first time fungi had been categorized into edible and medical fungi and their ecosystem described by the resource value of the fungi. It was also the first time the Alaskan inland fungi species and their populations had been studied with regard to their evolutionary relationship to flora. The conservation and management of fungi species according to the ecosystem were summarizde and the major diseases generalized. The study also emphasized the importance of mycorrhizae in a healthy forest and the need to take this into account in the intelligent design of cultivation species.

From the 1950s to 1970s I had joined several scientific research teams on expeditions into virgin forests of Northeast and Southwest China. Conditions were arduous as we camped in tents with primitive equipment far from civilization, but I had greatly enjoyed being in the wilderness with the daily opportunity to collect rare fungi and fruiting body specimens. I was very pleased to be a part of a research project that promised to offer similar opportunities in the taiga forests of Alaska.

1 METHODS OF RESEARCH AND INVESTIGATION

1.1 *Plan for the Collection of Information*

Besides collecting information from the University of Alaska library, I obtained much information from the Biological library and the University and Jepson Herbaria at the University of California, Berkeley. In designing the research plan I collaborated with Professor Dick Parmeter and Professor Fields Cobb of the plant pathology department of UC Berkeley, specialists in fungi, and Professor Jennifer Huang McBeath of the University of Alaska. The research period was set for two summers (1989, 1990), during the growth period of fungi, especially mushrooms, from the beginning of July to the end of August. Near Fairbanks the mushrooms appeared and collections were made from the beginning of August to around the twentieth of August. The preliminary analysis of the collection of specimens was done in the same year; the specimens were categorized and listed in a catalog of fungi. The specimen symbol was AKMC (Alaska McBeath, Chen).

1.2 *Survey, Forms*

The survey was based on the sample spot areas. We utilized our self-assessment of the ecological factors and a traditional standard survey form to record the forest ecological factors and their characteristics. Survey information recorded included the type of forest ecosystem, the geographical location of the area, the composition of the forest tree species, the height, (m), the DVH (cm), the age of the trees, the latitude, longitude, altitude (m), the slope, the aspect and location of the slope, the identification of the forest diseases species, their common names and scientific names, the characteristics of the sample spot areas, the specimen number, the photograph number, the slide number, the collector, the researcher, the date, the month, the year of the specimen, and the signature of the collector. Pictures were also included to illustrate the location of the sample area. Furthermore, detailed forms were designed for the edible and medicinal fungi samples to record their ecosystem and the sample area number for future analysis of the behavior of these fungi based on ecological factors. The forms included field recording, the collection numbers of the mushrooms, and the year AKMC 89 or AKMC 90. They also recorded the scientific and common names of the mushrooms and detailed description of the characteristics of the mature fruiting body such as size, color, texture, and even odor and taste. Other characteristics which were recorded included cap, gills, stalk, veil, annulus, volva mycelium spore, spore prints, and also the sample area number, slide number, photograph number, sampling date, and the signatures of the specimen's collector and researcher.

1.3 *Utilization of the US Long Term Ecosystem Observation Information:*

The US Long Term Ecological Station Survey (LTESS) sites are distributed all over North America. One of the boreal forest ecosystem station research offices and labs is located in the United States Agricultural Department situated in Fairbanks, Alaska, a neighbourhood of the University of Alaska. The LTESS collects research data from their settled spots around inner Alaska. The forest spot's name, FP #A, the weather, and soil observation information were first recorded into a computer in 1960. From 1989 to 1990 I worked in the main ecotype of forest spots such as the FP3A (Cotton Wood + White Spruce), FP1A (Bank Young Willow Plantation), FP4A, FP2A (Closed Shrub + Young Cotton Wood), FP5A (Black Spruce), FB3B (Young White Spruce) and other local systems to research the categories and the ecological characteristics of fungi and to utilize the long term ecological information (Fig. Main ecotype of forest spots on Tanana River, P.237).

Under the organization of Professor Huang, we conducted field exploration, collection and arrangement of fungi samples for two summers. We explored the Alaska Fairbank country and the Tanana Riverbank for forest pathogens and visited 21 sample spots (average 0.5-1 hectares each), and collected 305 fungi specimens.

Of the above data collections, using the forests and vegetation ecosystems as units, we collected information on the population of fungi. Every sample spot area and specimen was recorded in detail in a total of 629 pages of notes including the evaluation of every species of fungus ecotype and its location and role in the ecosystem. In 1989 and 1990 winter, I submitted annual survey reports (including edible medicinal mushrooms and forest disease pathogen's name list) of the Alaskan Inland Ecosystem fungi to Jennifer McBeath Huang /University of Alaska, Fairbanks.

1.4 *Boundaries of the survey*

Because of the vastness of Alaska, besides the sample areas, we also utilized roads, households, and fields for the survey and examination of gardens and nurseries for tree and shrub diseases and even for mushrooms. Distribution: From as near as around the households to as far as the Mount McKinley National Park several hundred miles away, we observed the species of trees and fungi to determine the ultimate northern boreal Distribution:

1.5 *Specimens: collection and preservation*

After many years of vast field survey I have acquired a general knowledge of the various groups of fungi to provide a background for further discussion and study. To preserve fungi in their essential characteristics, most specimens retained for long-term study were immediately and properly pressed, dried, and mounted on special papers or bags (if leaf or smaller twig) or on microscope slides. Most pores fungi spores print of most fungus specimens were carefully collected. These near permanent herbarium specimens are stored in a box or bag in a herbarium. An extensive collection of fungi specimens, useful for both teaching and public reference, was considered an important contribution to the development of an Alaskan herbarium. The preservation of taiga fungi specimens is vital for future taxonomic, evolutionary, and genetic research. Copies of specimens were also given to University and Jepson Herbaria at the University of California in Berkeley.

1.6 *Fungi Specimen Identification and It's Correction*

Whether researching forest ecosystems or systematic evolution, it is important to take into accoun the evolutionary of species. Usually, the fungi were identified according to species or varieties using publications on the flora of local fungi and the collected specimens to produce a list of hosts of fungi. Fungi were further divided into forest pathogenic, edible, medicinal categories. Then, following traditional methods, every specimen was processed in well-equipped labs. We also contacted professionals who were familiar with the authority of the taxa at Bend Oregon. Professor Robert L. Gilbertson identified specimens of *Hymenomycetes*, *Basidiomycotina,* conifer rusts were identified by Dr. Roger S. Peterson, Professor Jennifer Huang McBeath, and Dr. Michelle Tiseidl participated in the identification of most of the mushrooms species and special *Cortinarius species.* We asked for advice of inland forest and the northern salix classification, this team experts (including Dr. Leslie A. Viereck and scientists from Canada and Russion) regarding the identification of willow rust of Melampsoraceae hosts and other hosts. (Fig. Fairbanks forest inland fungi team, P. 237)

1.7 *Alaskan Inland Fungi Knowledge Exchange and Spread*

Because the state of Alaska is situated at high latitude, research on the characteristics and geographical distribution of fungi is especially valuable. We used all kinds of knowledge exchange and spread methods, such as academic reports, the exchange and donation of samples, interviews on specific topics, and expedition of the geographic areas by famous fungi specialists.

1) My seminar at the University of California, Berkeley, Plant Pathology graduate department, on September, 1991, was entitled "Alaska Fungi in Taiga Forest."

2) The second seminar was "The Forest tree and fungi on Biogeography comparison of US/China and Alaska/Siberia) at The Society of Women Geographers, Berkeley, 1992.
3) The US Department of Agriculture invited me to give a speech on the USSR timber importation from Siberia and Far East, The Pest Risk Assessment 09/1991 at Oregon. The work done in Alaska was a highly valuable reference.
4) During the second year of this research, Professor Dick Parmeter and his wife Anita Parmeter drove from Berkeley to the city of Fairbanks in Inland Alaska and met with Jennifer and me. They produced 6 videotapes of the geography and wildlife from Berkeley to Alaska, and on the 13-15th of August went with us to the White Birch Forest for the collection and identification of fungi and mushrooms. We also exchanged our views and analysis of the Alaska Inland Fungi.
5) In 1994 Professor Wendy and John Helms went to Alaska to attend the SAF Conference. Wendy collected 3 specimens of mushrooms out of her immense interest in the Alaskan fungi.
6). In 1990 the researchers of the Alaskan Inland Fungi donated some of the samples to the Fungi specimens the University Jepson Herbaria of the University of California at Berkeley. It was accepted and processed by the administrator of fungi specimens at the time. This set of specimens in University and Jepson Herbaria/UC Berkeley is very useful to the students and professors for research and to the public to refer to the purpose of the studies of Alaskan flora. We showed Fungi of the Alaskan Inland Ecosystem on the 2002 CAL open house day.
7). Collecting information among the people and using their households for research and collecting specimens. During the two years, I visited 50 households, recorded verbal accounts of the history of Alaska and accumulated information on the pathogenic fungi on household plants and building wood decay, the relationship of the fungi to the surrounding ecosystem and vegetation, as well as experiments on the kinds of fungi on plants. In these verbal accounts I learned about the social history of Alaska before and after it was sold to the US, its forest fires, and the evolution of its forests.

2 REGION OF INVESTIGATION

2.1 *Geography*

Alaska is 2100 kilometers long and 3500 kilometers wide (Alaska map). Because the state spans different ecosystems, its landscape, weather, geology, geophysics and vegetation vary greatly. The total area of Alaska is 14.6 trillion hectares including 48 million hectares of forested lands. Its weather varies from warm temperate zone to freezing, from "the rain forest of the Boreal " to arid wilderness; the inland temperature can reach 27.5 centigrade thermomter (83 degrees Fahrenheit). The amount of rainfall per year ranges from less than 250 mm to more than 3800 mm. It is the wealthiest ecological and geographical region near the North Pole. The vegetation that fungi parasitise on is distributed very unevenly: the vegetation of the coastal forests (also called the "non-inland" forest) varies from the towering fast growing forests and fungi of the south-eastern coast, extending through the low, slow-growing boreal forests of the interior, to the treeless tundra of the north and west.
This fungi research program is centered at the geographical area of this inland on the region of Fairbanks, north latitude between 60 and 66, east latitude between 146 and 152. All the fungi ecological sample areas are located along the Tanana River and the Yukon Valleys, routing survey (traveling by car, we sometimes stopped on the two sides of road surveys, and made collections) reaching from the Kanai Peninsula in the north to the Brooks Range in the south. The annual average temperature of Fairbanks is -7 degrees Celsius to -1 degree Celsius. In the winter the temperature usually reaches -40 degrees Celsius. The annual average rainfall is 150 mm to 300 mm. The highest average temperature occurs in July, at 16 degrees Celsius, and the lowest average temperature is in January at -23 to -29 degrees Celsius. The absolute highest temperature is 30 degrees Celsius, and the absolute lowest -36 degrees Celsius.

Because fungi are not green organisms, the distribution of the fungus populations at high latitude near the North Pole clearly represents the geophysics, chemistry, exposure to sun, and other characteristics of the region. When summer comes (around June 20th), day length is extremely long. After the extreme cold of the winter, starting in May, mycelial organisms in the northern forests, especially fungi, wake up and start growing under the prolonged sunlight. The mycelium matures during June and July, and mushroom gestation occurs until August rain cause mushroom buds to form fruiting bodies underground. Mushrooms mature from underground and fruiting bodies appear on tree trunks. For fungi near the North Pole, where amount of sunlight time is shortered, the amount of precipitation (snow and rain) and the cool fresh temperature are very suitable for the basidiospores, such as those of boreal forest mushrooms. Because of this, they appear only during August. It is especially hard to find mushrooms in September. This is a short collection time for mushrooms, which is geographically characteristic of Alaskan Inland mushrooms.

2.2 *Forests Ecological types*

The Fairbanks inland forest systems have been classified with the purpose of management for fungal biodiversity. Currently, national vegetation classifications in the United States and Canada are under way in an effort to merge classifications into a uniform format, set of terms, and of communities. In the United States, the collaborative efforts between state and federal agencies and The Nature Conservancy may promote better communication between management specialists throughout the country. The vegetation types in interior Alaska form a mosaic of patterns that are related in part to past fire history, slope and aspect, and the presence or absence of permafrost. Most forest stands are mixtures of two or more tree species but are usually classified by the dominant species.

Forests were classified into two flora and six ecotypes:

1). Fairbanks Mountain foot hills flora:

- A. *Picea glauca/ Rosa aciculoides*
- B. *Populus tremuloides / Viburnum / Linnaea borealis*
- C. *Betula papyrifera / Viburnum edule*

2). Tanana River Bank Flora:

- D. Cottonwood-Balsam popula (*Populus balsamifera*)
- E. Black Spruce-*Picea mariana / Betula papyrifera / Ledum groenlandicum*
- F. Willow-*Salix spp.*

3 THE FUNGI OF THE FORESTS ECOSYSTEMS (FLORA)

3.1 *White Spruce*

Here, We often see wood decay fungi, rust diseases, and very diverysity species of mushrooms. Some of them are valuable food and medicinal resources. Wood decay fungi which mainly belong to *Basidiomycetes* are often seen on forest trees of over-mature condition. They are called higher fungi; the fruiting bodies are often within human view.

We generally see genera of *Fomitopsis, Fomes, Trichaptum, Diplomitoporus, Pycnoporellus, Oligoporus, Crytoporus, Gloeophyllum, Antrodia,* etc. Also spruce tree rust *Chrysomyxa ledicola.*

We see *Sarcodon imbricatus;* Cortinarious groups to three: cinnamon brown cort, rusty brown cort; common *Clitocybe, Spathularia, Gomphidius, Suillus, Amanita, Agaric* genera mushrooms. Slippery Jack (*Suillus luteus*) is mainly of White Spruce's flora mycorrhizal species *Agaricus, Gomphidius, Lacctarius.*

3.2 *Quaking aspen (Populus tremuloides)*

After a conifer fire, aspen forest proides a pioneer forest which constitutes a very rich environment for fungi. The wood decay mainly consists of this species: *Phellinus tremula.* Other genera are: *Ganoderma, Pholiota, Peniophora, Fomes, Trametes, Cerrena, Trichaptum, Skeletocutis, Bjerkandera, Coriolopsis, Crepidotus,* etc.

Aspen leaf rust belongs to Melapsora rust. There are also several Cortinarius species, and others such as: *Hydnellum, Lycoperdo, Amanita, Agaric, Russula, Leccinum, Leccinum, Leccinum, Clavicorona, Clitocybe, Gomphus, Gomphus, Pluteus, Pluteus, Pholiota, Panus, Russula, Clavariadelphus, Sarcodon, Amanita, Amanita, Armillariella, Russula, Laccaria, Gomphus, Agaricus,* etc. Many species are edible and may be valuable medicinal resources.

3.3 *Paper Birch (Betula paperifera)*

Paper birch is the common invading tree after fire on east- and west-facing slopes and occasionally on north slopes and flat areas. This species grows either in pure stands or more often mixed with White Spruce, aspen, or Black Spruce. Average diameter of 20-22 cm and height of 18-24 m. is more common in the interior birch stands dominated by paper birch that occupy about 2 million hectares.

Main wood decay species on live birch is Birch trunk pocket rot (*Phellium igniarius).* Other common genera are: *Piptoporus, Lentinus, Fomes, Trametes, Fomes, Ganoderm.,*

Fallen birch trees easily rot over the summer. Wood structure rots but the white bark is preserved for several years. The birch's main leaf rust is *Melampsora betulina.*

There is also powdery mildew on the birch. Mycchorizal fungus is *Leccinum aurantiacum.* The most common poisonous mushroom seen is *Amanita muscaria.* The most delicious mushroom is *Boletus edulis.* Other mushroom genera are: *Lyophyllium, Russula, Agaricus, Agaricus, Clitocybe, Fuscoboletinus.*

3.4 *Balsam Poplar (Populus balsamifera* L.*)*

Common decay genera are: *Polyporus, Spongipellis, Trametes, Hopalopilus, Ciborinia, Ganoderma, Fomes, Bjerkandera, Trametes, Fomitopsis, Gloeoporus, Cerrena Spongipellis.*

There are several species of *Cortinarius* mushrooms. One of *Cortinarius brunneus* group mushrooms grows in circles on the forest ground, and is a very important species that promotes cottonwood/elder forest succession procession. Other mushrooms such as *Suillus luteus* and *Cortinarius brunneus* are mainly mycorrhiza of birch and White Spruce flora.

There are many other genera such as *Pleurotus, Lactarius, Suillus, Hygrophorus, Clitocybe, Tricholoma, Hericium, Russula, Clavariadelphus, Lycoperdon, Melanomphalia, Agaricus* and puffballs.

3.5 *Black Spruce* (*Picea mariana*)

The Taiga Black Spruce forest fungi flora forms a very unique ecological habitat in the inland of Alaska. On north-facing slopes and poorly drained lowlands, forest succession leads to open Black Spruce and bogs, usually underlain by permafrost. The Black Spruce are slow growing and seldom exceed 20 cm in diameter, usually being much smaller; a tree 5 cm in diameter is often 100 years in age. The Black Spruce comes in abundanty after fire because its persistent cones open after a fire and spread abundant seeds over the burned areas. A thick moss mat, often of sphagnum mosses, sedges, grasses, and heath or ericaceous shrubs usually make up the subordinate vegetation of the open Black Spruce. In the wet bottomland is the slow-growing tamarack. As with the Black Spruce, it is of little commercial value, seldom reaching a diameter of more than 15 cm.

Because Black Spruce usually grows on permafrost ground, there are fewer mushrooms in this ecosystem than in other flora. It is common to see the genera: *Gloeophyllum, Trichaptum,*

Coniophora, Lenzitus, Tyromyces. Main rust species are *Chrysomyxa ledicola* and *Chrysomyxa arctostaphyli.*

Leccinum alaska is representative of the Alaska taiga forest fungi flora.
Suillus luteus mainly mycorrhiza same as White Spruce 1034
Leccinum alaska same as 0268, mainly mycorrhiza; *Suillus grevillei* (Klotzch) Singer This species grow in *Lenzites sepiaria* after fire*; Suillus grevillei* (Pk.) Singer var. *clintonianus*; other mushrooms such as *Clitocybe, Lycoperdon, Amillaria...*
Obviously the species decreased at this flora.

3.6 *Willows* (*Salix spp.*)

3.6.1 *Salix Diseases List*
Rhytisma salieis, Uncinula salicis, Melampsora sp., Uncinula salieis, Rhytisma sulicis, Melampsora sp. anthracnose, Rhytisma salicinum, Erysiphe polygoni, and Cytospora;

3.6.2 *Mushrooms*: *Cortinarius*

Cortinarius, Clitocybe odora (Fr.) Kummer; *Hericium ramosum, Galerina navcina* or *G. oiner.* Lots of this species and others; *Clitocybe, Hebeloma.*

4 WHITE SPRUCE-PICEA GLAUCA/ ROSA ACICULOIDES/FUNGI

The White Spruce, along with the paper birch, comprises 32% of the total Alaskan forests, about 42.4 million hectares. The weather varies distinctively, the earth surface freezes in the winter season, being frozen to a depth of about 140 mm, and snow-covered between mid-October and end of April. During June, under 24-hour sunlight exposure, the plants grow rapidly. The inland forest type characteristics have fire hazard as an index. It is distributed in mosaic blocks related to the slope, the direction of the slope, and the existence of the frozen layer at the burned site. This kind of forest is often seen on the south slope near riverbanks; it has rich soil and no frozen layers. Its plants include roses, alder, and willows. Spruce can usually reach 100 to 200 years old, with diameter of 25 – 60 cm. There are about average 58 cubic meters per hectare. In the inland area White Spruce (*Picea glauca*) dominated, also some Paper Birch (*Betula papyrifera*) and Balsam poplar (*Populobals amifera*). Furthermore, there are many other shrubs such as: Red fruit bearberry (*Arctostaphylos rubra*), Crowberry (*Empetrum nigrum*), Narrow-leaf labrador-tea (*Ledum decumbens*), American red currant (*Rihes triste*), Prickly rose (*Rosa acicularis*), Feltleaf willow (*Salix arbusculoides*), Bebb willow (*Salix bebbiana*), Buffalo berry (*Shepherdia canadensis*), Mountain cranberry (*Vaccininum vitisidaea*), Bog blueberry (*Vaccinium uliginosum*), High bush cranberry (*Viburnum edule*), and White Spruce / Rose Forests (*Picea glauca / Rosa acicularis*).

In the main type of the Alaskan inland virgin forest ecotype, mainly spruces, the most important tree of the spruce-birch interior forest is *Picea glauca* (Moench) Voss, English name: White Spruce. White Spruce in our Parks Loop South plots were usually 30-33 meters in height, average diameter 20-55 cm, at the most reaching 76 cm; needles are short-stalked, spreading on all sides of twig but missing on top near ends, 4-angled sharp-pointed, stiff, bluish green, needle-like leaves growing around the sides; the twigs slender, hairless, smooth, cones near stalkless, handing down, round-sided ellipse-shaped cones, on the crown tip round-shaped small branches hangs down scattered. The bark is gray and smooth and the sample wood is white. The wood is near white, and sapwood is hard to recognize. Just compare weight, unevenness and mull of the texture, and the cross-section annual rings are easy to recognize.

The Alaska Forest Disease Survey plot named Parks Loop South, forest type was old growth White Spruce, the Forest ecological type was White Spruce–Seather Moss, located at Bonanza Creek

Long Term Ecological Research Area, Latitute 64 46 N, Longitude 148 19 W Elevation 1150 ft. 26-32% slope and 180 Aspect upper position on slope about one ha size; The tree composition were 91% White Spruce (Height 30m, DBH 27m. 206 yrs.) 8% Black Spruce (height 13-20 cm DBH 20 cm 140 yrs.)(Fig. 1-1, 1-2).

White Spruce wood decay mostly belongs to Aphyllophorales order (Polyporales, TALBOT et al., 1973), Polyporaceae family, includes nine species belonging to these genera *Fomitopsis, Trichaptum, Diplomitoporus, Pycnoporellus, Oligoporus, Crytoporus, Gloeophyllum, Antrodia.*

Ecological habitat:

White Spruce -*Picea glauca*/*Viburum edule* Forest ecological type, the tree 140 years average height 22m DBH 20 cm located Tanana River, Fairbanks, 64 45'N latitute and 148" 00 W longitude 120-470 m Elevation on bank, related one kind mycorrhizal mushroom is *Suillus leteus*, White Spruce mixed with Cotton wood this specimen under a cotton wood tree.

4.1 *Fomitopsis pinicola* (Swartz: Fr.) Karst. Krit. Finl. Basidsv., p.306. 1889. Boletus pinicola Swartz: Fr., Syst. Mycol. 1:372, 1821.
Bank of Tanana River, M. M. Chen, August 21, 1990.
Type of rot. Brown cubical rot of living and dead White Spruce. Wood decay fungus, anti-cancer.

Basidiocarps darkening to grayish brown or blackening, often with a red, resinous margin, not chalky or bitter, sclerids absent in context. Basidiospores cylindric-ellipsoid, hyaline, smooth, IKI-,6-9 x 3.5-4.5 mcm.

Distribution: Throughout the North American coniferous forest regions but absent or rare in the southern pine region. (15, 23, 32, 36)

4.2 *Fomitopsis rosea* (Alb. et Schw.:Fr.) Karst.
Bank of Tanana River, M. M. Chen, August 21, 1990.
Brown cubical rot of dead White Spruce. Wood decay fungus, anti-cancer, can be used for medicinal purposes. Pore surface, context and tubes rose pink to pinkish brown. Basidiospores cylindric, straight, cystidioles absent. Boreal or Alaskan high elevation species, on White Spruce. Basidiospores narrowly cylindric, slightly curved, hyaline, smooth 6-8 x 1.5-2.5 μm. (15, 23, 32, 36)

4.3 *Trichaptum abietium* (Dicks.: Fr.) Ryv.
Norw. J. Bot. 19:237, 1972. – *Boletus abietinus* Dicks., Plant Crypt. Brit., fasc. 3:21, 1793. – *Polyporus abietinus* Dicks.: Fr., Syst. Mycol. 1:370, 1821.
M. M. Chen, 1989.
Basidiocarps Hymnophore poroid, sometimes with lacerate dissepiments, Pores angular, 1-3 per mm; cap, if present hirsute to coarsely strigose. Pore surface purplish, pale brown to buff; temperate-boreal species. Cap tomentose to veluntinate; basidiocarps up to 3mm thick; pores 3-6 per mm; widespread species. Basidiocarps resupinate, effused-reflexed to broadly sessile, rarely more than 1 cm wide, on coniferous wood, very rarely on hardwoods. Distribution: Throughout coniferous forest regions of North American and global in the North Temperate Zone. (15, 23, 32, 36)

4.4 *Diplomitoporus crustulinus* (Bres.) Dom.
Tanana River Bank, M. M. Chen & Mcbeath, August 3, 1989.
Acta Soc. Bot. Pol. 39:192, 1970. – *Poria crustulina* Bres., Mycologia 17:75, 1925. – *Poria chromatica* Overh. Pa. Acad. Sci. Proc. 13:123, Stud. Bot. Cech. 3:3, 1940 (based on the type of Overholts' invalidly named species).
Pore surface straw-colored when dry, often cracked in polygons, skeletal hyphae not gelatinized in KOH, without an amyloid reaction. Distribution: Widespread in North America in conifer forests regions and circumpolar in the conifer-zone. (15, 23, 32, 36)

4.5 *Pycnoporellus fulgens* (Fr.) Donk.

Tanana River Bank, M. M. Chen & Mcbeath, August 3, 1989.
Persoonia 6:216, 1971. – *Hydnum fulgens* Fr., Ofvers. Kung. Vet. Akad. Forh. 9:130, 1852. – *Polyporus fibrillosus* Karst., Syd. Finl. Polyp., p. 30, 1859.
Basidiocarps sessile, pores 2-3 per mm; basidiospores ellipsoid, 5-6 μm long.

Distribution: Transcontinental in the northern US and Canada from Newfoundland to Alaska. Apparently absent in the southern states even in high mountains of Arizona and New Mexico. (15, 23, 32, 36)

4.6 *Oligoporus tephroleuca*
Wood decay fungus. Basidiocarp pileate. Basidiocarps sessile to effused-reflexed, conspicuous on surface of substratum, exceeding 1 cm in width. Cap and pore surface white, buff to unevenly brown; spores not amyloid. Context rarely more than 1 cm thick, firm, more or less homogeneous. Basidiocarp white, buff to yellowish, not changing color when bruised or dried. Basidiocarps sessile to effused-reflexed, applanate; upper surface white, yellowish, buff to pale unevenly brown. Cystidia absent in the hymenium. Spores allantoid, 4-6 x 1-1.5 μm. Pores 3-7 per mm; margin flat and even. Cap cream to mouse grey, mostly strigose.

Distribution: Widely distributed in forest ecosystems of Eastern and Western North America. (15, 23, 32, 36)

4.7 *Crytoporus volvatus* (Pk.) Shear.
Bull. Torrey Bot. Club 29:450, 1902. – Polyporus volvatus Pk., N.Y. State Mus. Ann. Rept. 27:98, 1877.
Wood decay fungus, anti-cancer, can be used for medicinal purposes. This is a newly recorded species from Fairbanks, Alaska. Basidiocarps annual, sessile, ungulate, solitary or in large numbers, up to 4 x 5 x 4 cm, upper surface cream-colored to yellowish or tan, azonate, glabrous, often coated with a clear, lacquer-like layer, smooth or rugose, margin concolorous, continuous with a volva-like structure which completely encloses the pore surface except for a small hole at the base, pore surface pale to dark chocolate brown, the pores circular, 4-5 per mm, with thick, entrie dissepiments, context ivory-white, azonate, soft-corky, up to 2 cm thick, tube layer pinkish buff, trama continuous with context, tubes up to 6 mm long, taste slightly bitter.
Hyphal system trimitic, contextual generative hyphae thin-walled, with clamps at all septa, with occasional branching, mostly 3-7 μm in diameter but with inflated portions at branches up to 15 μm in diameter, contextual skeletal hyphae thick-walled, hyaline, nonseptate, with occasional branching 2.5-8 μm in diameter, binding hyphae thick-walled, noseptate, much branched, 1.5-2.5 μm in diameter, tramal hyphae simlar. Cystidioles fusoid, not projecting, thin-walled, 20-28 x 5-7 μm, with a basal clamp. Sexuality: Heterothallic and tetrapolar.

Substrata: Recently killed conifers, commonly fruiting a year after trees are killed by fire, bark beetles or other factors and then displaced by other saprophytic wood-rotting fungi. Species of *Pinus* are the most common substrata through out range of *C. volvatus* and *Abies* and *Psudotusga* also commonly colonized in the West.

Distribution: Widely distributed in coniferous forest regions of North America with the exception of the southern pine forests of the southeastern US. Also known from the East Asia.
Remarks: Harrington and Shaw found that *C. volvatus* released wind dissepiments spores through the perforation in the volva in quantities comparable to other polypores. They interpreted the volva as an adaptation to retain moisture and high relative humidity for sporulation during dry summer periods. Young, developing basidiocarps have context tissue composed almost entirely of generative hyphae with conspicuous clamps. Context tissues of mature basidiocarps are composed almost entirely of skeletal and binding hyphae and clamped generative hyphae are difficult to find. (15, 23, 32, 36)

4.8 *Gloeophyllum saepiarium* (Walf.: Fr.) Karst.
Finl. Hattsv. 2:80, 1879. – *Daedalea sepiaria* Fr., Syst. Mycol. 1:333, 1821.

Wood decay fungus, anti-cancer, can be used for medicinal purposes. Hymenophore lamellate to distinctly deadaleoid, occasionally mixed with poroid parts, hyphal system trimitic with rare binding hyphae. Lamellae and/or deadaleoid pores mostly 3-4 per mm, upper surface of basidiocarp soft and smooth, evenly colored in some wide sulcate zones. Hymenophore with more or less straight lamellae and with few to many deadaleoid to elongated pores, upper surface yellowish rusty brown, lumber to black at the base. Basidiocarps usually distinctly lamellae with relatively few pores, upper surface hirsute to scrupose, cystidia abundant in the hymenium, boreal species.

Distribution: Widespread in North America and seemingly present wherever there are coniferous forests. Circumglobal thorough USSR, Japan and China to Europe. (15, 23, 32, 36)

4.9 *Antrodia arida* (Fr.:Fr.) Karst. Var. *suffocata* (Perk.) Ginns *Antrodia* Karst. Medd. Soc. Fauna Fl. Fenn. 5:40, 1880.
Causes a brown rot, mostly in coniferous wood. Basidiocarps annual to perennial, resupinate to effused-reflexed, more rarely sessile and shelf-like, mostly light-colored and tough to hard, hyphal system dimitic, generative hyphae with clamps, skeletal hyphae present, hyaline or slightly tinted in a few species, usually non-amyloid, variably amyolid in a few species, cystidia not present, small ventricose and pointed cystidioles often present among the basidia, spores cylindrical to oblong ellipsoid, hyaline, thin-walled, smooth and non-amyloid. Large cosmopolitan genus. Under the White Spruce forest has weeds and shrubs oblige fungi alone with the ecosystem; the microbial effectors relationship with forest species should be addressed in future forest management and development. (15, 23, 32, 36)

4.10 *Chrysomyxa ledicola* lagh. (Spruce needle rust) (Fig. 1-3)
Spruce needle rust is an important disease of the spruce distributed near the North Pole. There are 13 known diseases of the spruce needles in the world. The one in North Alaska is on White Spruce (*Picea glauca*); its mortality levels vary. In virgin spruce adult trees, the susceptibility rate is only 5%. It can also cause witches' broom (often seen on Black Spruce). When serious it can cause the crown to brown wilt. When spruce is used for the plantation nursery seedlings or for Christmas trees, if this disease occurs, the whole tree will lose its usefulness.

The susceptible tree species is White Spruce. This disease has an alternate host called Labrador-tea (*Ledum groenlandicum* Oeder). On the Labrador-tea leaves, this disease produces uredia and telia period. This disease is often seen in White Spruce and Black Spruce forests near Fairbanks because Labrador-tea is also part of the forest ecosystem here. Y. Hiratsuka of Canada describes pycina, which grows on the surface of the Labrador-tea leaves, aeciaspore size 22–34 x 27-46 μm.

Method of spreading: usually uredia spores mature on the White Spruce leaves in June, and with the wind breeze inoculated on the Labrador-tea leaves. Urediaspores and teliaspores can both grow on the upside or underside of the leaves, The plot XII specimen # AKMC 0075 Uredia specimen was collected on *Ledum groenlandicum* at White Spruce forest at early August, 1990 and 0036,0037 VI Ledum leaf rust. This collection shows good evidence the spruce needle rust was Picea rust and Ledum rust co-evolution in Alaska spruce rust flora. Often uses telia over winter. Next year basidia appear and produce basidiospores. The basidia cell wall is very thin, sometimes can spread by wind or rain into the mistems of the White Spruce leaves.

4.11 Prickly rose (*Rosa acicularis*) leaf rust (*Phragmidium sp.*) telia and teliospores White Spruce ecotype forest (*Picea glauca/ Rosa aciculoides*/Fungi) July 24, 1990.

4.12 *Puccinia porphyrogenita* (Cornus rust)
Pyenia unknown, probably not formed. Aecia and uredia wanting. Telia hypophyllous, in groups, chocolate-brown; teliospores oblong or clavate-oblong, 16-23 by 45-64 μm acute or acuminate above, narrowed below, slightly constricted at septum; wall chestnut-brown, 1.5-3 μm thick in sides, 7-20

μm above, smooth; pedicel yellowish-brown, about as long as spore. The # 0130 (90) collection Uredia at August 21.
On Cornaceae: *Cornus Canadensis*. Range: Newfoundland to southern Alaska, southward to northern New York and northern Oregon; also in Japan.
**Puccinia volkartiana.*
#0010(90) rust spores collected from Epilobium auqustifolium back of leaf at July.
Pycnia unknown, probably not formed. Aecia and uredia wanting. Telia and teliospores as in Puccinia extensicola, with the spores usually narrower, 10-17 μm wide.
On Onagraceae: from Range Northern Manitoba to southern Alberta and northern Washington; also in northern Europe.

4.13 *Salix alaxensis* leaf rust (*Melamsora sp.*). Birch rust (*Melmsoridium sp.*). *Salix bebbiana (Uncinula salicis)* 0040 (90).

4.14 *Cortinarius sp.*
Tanana River, Fairbanks, Alaska, 64 45'N latitude and 148" 00 W longitude and 120-470 m elevation on river bank, plot XVIIIB, M. M. Chen, 8/11/1990, specimen # AKMC 1040, 1042 (with spores print).
Cap 2.5 cm. Cinnamon brown when moist viscid and glabrous, stipe 2.5 x 0.6 cm. Gills cinnamon brown and sinuate. Spores prints yellowish brown spores 5.6-11.2 x 4.2-8.4 μm elliptical. Ecological habitat: White Spruce with birch / *Viburnum edule* forest ecological type, the white spruce trees average 140 years old and height 22 m, diameter 20 cm. This species grows together with another kind of mycorrhizal mushroom *Suillus leteus*.

4.15 *Cortinarius* subg. Phlegmacium sect. Multiformes.
Geophysics Institute & Arctic Health Bldg, plot XIIIB, Fairbanks, Alaska, M. M. Chen, Richard Parmeter, J. H. McBeath, 8/13/1990, specimen # AKMC 1057.
Cap 2.5-6.5 cm cinnamon brown when moist viscid and glabrous when dry, stipe 2.5-6.5 cm some times upward a cortinate hairy-fibrillose zone on stalk which stained rusty brown by spores, gills cinnamon brown and sinuate to adnated, spores 8.4-15.4 x 5.6-8.4 μm cinnamon brown elliptical. Ecological habitat is White Spruce (*Picea glauca/Viburum edule*) mixed aspen - spruce forest.

4.16 *Cortinarius brunneus* Fr.
Tanana River, Fairbanks, 64 45'N latitute and 148" 00 W longitude 120-470 m elevation on bank, M. M. Chen, 8/22/1989, specimen # AKMC 0203.
Chinese call it "Brown Cort. Mushroom". *Cortinarius* is the largest genus of gilled mushrooms, with an estimated 1000 species – many of them still unclassified. It's an ectomycorrhizae. Most terrestrial plant species have in their roots a symbiotic association with soil fungi called mycorrhizae. Ectomycorrhizal fungi are known to enhance the uptake of water and plant nutrients, increase resistance to root pathogens, and promote plant growth. David Arora mentioned in *Cortinarius evernius* included this species. "A widespread nondescript conifer-lover (white spruce-lover in inland Alaska) with an equal to club-shaped, dull brown stalk whose apex sometimes has a slight violet or vinaceous tinge". (Arora, *Mushrooms Dymystified*) Cap 8-8.5 cm gray coffee-brown, stipe black brownish 7.5 cm x 1.2 cm in stipe remained patches of cinnamon brownish collapsed cortinae and stained cinnamon-brown by spores, gills adnated, gills coffee brown; spore prints cinnamon-brown, spores yellowish brown elliptical with warty or rough.

Ecological habitat: White Spruce-*Picea glauca/Viburum edule* Forest ecological type, the trees 140 years, height 22 m DBH 20 cm. A related kind of mycorrhizal mushroom is *Suillus leteus*, white spruce mixed with cottonwood. This specimen grows under cotton wood trees.

4.17 *Cortinarius* sp.
FP4A, M. M. Chen, 8/7/1990, specimen # AKMC 1091.

Medium size, cap 4.2-4.7 cm, cinnamon-brown, viscid when moist, cap appears dry fine corrugated with crackle when dry, stipe 6.2-7.6 cm upward tapered 0.5-0.8 cm gills adnexed coffee redish brown downward portion wider, at low portion around mass mycelium with bulb and mosses; spores 8.4-11.2 x 5.6-7.0 μm cinnamon brown ellipsoid, often occur on White Spruce forests ground.

4.18 *Cortinarius* sp.
FP4A, M. M. Chen, 8/21/1990, specimen # AKMC 1075 (with spore print).
FP2A, M. M. Chen, 8/21/1990, specimen # AKMC 1080 (with spore print).
Cap 3.5-4.0 cm surface dark brown viscid when wet, corrugated when dry then wrinkled, brown when wet became expanded, stipe 4.5-6.0 cm spongy stuffed then hollow cortine hairy-fibrillose zone on stalk which stained rusty brown by spores, gills adnated or attached, dark cinnamon-brown remained collapsed cortinae at gills; spores 8.4-11.2 x 5.6-8.4 μm cinnamon brown, elliptical and spore print brownish.

4.19 *Sarcodon imbricatus* (L. ex Fr.) Karsten syn. *Hydnum imbricatum* L. ex Fr. (Fig. 1-4)
Plot XIIIB, M. M. Chen and Richard Parmeter, 8/13/1990, specimen # AKMC 1055.
Chinese call it "Wild Deer mushroom". Ectomycorrhizae, edible but poor, can be used for medicinal purposes to reduce chlesterol in the blood. It's easy to be collected and processed, and contains polysaccharose. (*The Macrofungi in China*, Mao, Xiaolan, 2001, p. 155). Cap 5-20 cm across, flattened-convex then depressed; dark reddish to purplish brown against paler pinkish flesh; velvety then cracked into coarse overlapping scales. Spines on undersurface 1-10 mm long; white to purplish brown. Stem 50-80 x 20-50 mm, tapered or swollen at base; whitish to purplish brown. Flesh firm; white. Odor not farinaceous, not distinctive. Taste soon bitter. Spores ellipsoid, tuberculate, 7-8 x 5-5.5 μm. Deposit brown. Habitat coniferous woods. Often found around Fairbanks, Alaska, suburban northern and eastern North America.

4.20 *Clitocybe odora* (Fr.) Kummer (Fig.2-1a and 2-1b)
FP4A, M. M. Chen, 8/21/1990, specimen # AKMC 1065, 1078.
Edible, but some records say not edible. Cap 3-9 cm across, convex at first with a low, broad umbo, later expanding and becoming irregular and wavy at the margin; dingy green to bluish green; grayish, bluish, or nearly white; finely matted with silky hairs or sometimes with a hoary bloom. Gills slightly decurrent, close or crowded, broad; whitish tinged with cap color. Stipe 30-70 x 5-15 mm, solid becoming hollow, sometimes curved and enlarged toward base; whitish tinged with cap color, base spongy and covered in fine whitish down. Flesh thin, firm; whitish to pale tan. Odor strongly of aniseed. Taste strongly aniseed. Spores ellipsoid, smooth, nonamyloid, 6-7.5 x 3-4 μm. Deposit whitish pink. Habitat singly, scattered, or in groups on leaf litter under hardwoods, especially oak. Found widely distributed in Fairbanks, Alaska.

4.21 *Spathularia flavida*
FP5A, M. M. Chen, 8/21/1990, specimen # AKMC 1069.
Ascocarps gregarious, fleshy, 3-8 cm high; ascigerous portion yellow, compressed-obovate, decurrent on opposite sides of the stem, 1-2 cm broad, often undulate and radiately rugose; stem subconcolorous or slightly paler, subcylindric, 3-5 mm thick; asci clavate, 90-120 x 10-13 μm; spores fasciculate, hyaline, clavate-filiform, multiseptate, 35-48 x 2.5-3 μm; paraphyses filiform, 2 μm thick.

4.22 *Gomphidius glutinosus*
Plot VIII, M. M. Chen, 8/7/1990, specimen # AKMC 1033, 0256.
Ectomycorrhizae, edible. Cap fleshy, convex, 4-6 cm broad, viscid, glabrous, brownish ochraceous; margin narrowly involute; flesh yellowish; stipe solid, whitish or pallid, thickened at the base, 5-7 cm long, 6-10 mm thick; gills close, adnexed, olivaceous, becoming rusty brown; spores broadly ovoid to subglobose, rough, 6.5-8 x 5.5-6.5 μm.
Comments by S. C. Teng: *Cortinarius glutinosus*.

4.23 *Suillus luteus* (Fr.) S. F. Gray. "Slippery Jack" (Fig. 2-2)
FP4A, FP3B, M. M. Chen, 8/22/1990, specimen # AKMC 1034, 0247.
Ectomycorrhizae, anti-cancer. Edible - good after peeling and removing slime. Cap 5-17 cm across, rounded becoming convex and flatter in age; chestnut to sepia; smooth, slimy and covered with brown gluten, shiny on drying. Tubes adnate to subdecurrent; lemon yellow to straw-colored. Pores round; yellow becoming brown-dotted. Stipe 30-90 x 10-30 mm, solid, equal or slightly tapering at base; pale straw-colored with darker pinky-brown. Veil finely webbed, shiny white; leaving a purplish sleeve-like ring around the stipe. Flesh white or pale yellow, often slightly pinky toward the stipe. Odor not distinctive. Taste not distinctive. Spores subfusiform to elongate ellipsoid, 7-10 x 2.5-3.5 μm. Deposit dull cinnamon. Habitat scattered to gregarious under white spruce, Alaska. Common. Widespread in Fairbanks, Alaska.

4.24 *Amanita muscaria* var. *muscaria* (L. ex Fr.) Pers. Fly Agaric (Fig.2-3)
Plot XIIIA, M. M. Chen, 8/13/1990, specimen # AKMC 1054.
Poisonous ectomycorrhizae. Cap 5-25 cm across, convex to flatter, sometimes slightly wavy or depressed with a lined margin; blood red to orange-red, becoming lighter toward the margin; smooth, a bit sticky when moist, dotted with flaky patches of whitish volval remnants sometimes almost in concentric rings. Gills free to adnexed, crowded, broad; whitish. Stipe 50-180 x 3-30 mm, sometimes enlarging toward rounded basal bulb; spores 9.4-13 x 6-8 μm, broadly elliptical, smooth, colorless, spore print white. Often seen in middle to late August. (3, 27, 36)

4.25 *Clavariadelphus truncatus* (Quel.) Donk "Flat bat coral mushroom"
FP4A, plot V, M. M. Chen, 8/7/1990, specimen # AKMC 1036,1017.
FP4A, M. M. Chen, 8/21/1989, specimen # AKMC 0125.
Edible - good. Fruit body 5-15 cm high, 3-8 cm wide at the top, club-shaped, often broad and flattened at sterile top, narrowing down to a bulbous base; yellowish ochre to dark apricot orange; wrinkled. Stipe indistinct; white-hairy at base. Flesh firm to spongy; whitish to ochre, darker on bruising. Odor mild. Taste sweet. Spores ellipsoid, smooth, 9-12 x 5-8 μm. Deposit pale ochre. Habitat scattered or in groups or clumps on the ground in coniferous woods. Widely distributed throughout inland Alaska. (Mao, 2001).

4.26 *Agaricus auqustus* Fr. (Fig.2-4). "Big Violet Agaricus"
FP4A, M. M. Chen, 8/22/1989, specimen # AKMC 0223 (with spore print).
Edible. Subcespitose; cap 3-15 cm, broad, convex to plane, dry, fibrillose-scaly, purplish brown; flesh white, thick; stipe 8-11 cm long, 2-2.5 cm thick, stuffed, cylindric, with bulbous base, white and glabrous above the annulus, fibrillose-scaly below, but soon glabrescent; annulus double, superior to submedian; gills free, close, finally dark purplish brown; spores ellipsoid to subovoid, 7.5-10 x 5-6 μm; sterile cells present on edge of gills. Prominent hilar. (36)

4.27 *Lactarius deliciosus* var. *deliciosus* (Fr.) S.F. Gray., Nat. Arr. Brit. Pls. 1:624. 1821; - *Agaricus deliciosus* Fr., Syst. Mycol. 1:67. 1821. (Fig.3-1)
FP3A, M. M. Chen, 8/21/1990, specimen # AKMC 1064.
Edible ectomycorrhizae. Cap 5 – 14 cm across, broadly convex with a depressed disc and a distinctly inrolled margin, becoming funnel-shaped with a wavy margin in age; pale flesh or rosy buff tinged greenish in places, with numerous purple-brick or salmon-colored blotches arranged in narrow concentric bands; sticky when moist. Gills adnate-decurrent, crowded; pale salmon, bruising pistachio green. Stipe 30-70 x 10-25 mm, stuffed then hollow, pinched off at base; buff or red-orange to salmon, sometimes with darker, spotlike depressions, becoming green in places; brittle with mycelium on base, pitted, with a distinct bloom. Flesh rigid at first then fragile; cream, yellowy then carrot-colored. Later orange, fading to orange-yellow then gray-green. Odor slightly fruity. Taste mild or slightly bitter. Spores ellipsoid, amyloid, 7-9 x 6-7 μm; ornamented with minute warts and ridges

forming a partial reticulum. Deposit cream. Habitat scattered to gregarious under white spruce forest ground.

4.28 *Amanita ceciliae* (Berk. & Br.) Bas "Strangle Scale Amanita"
FP4A, M. M. Chen, 8/3/1989, specimen # AKMC 0105 (with spore print).
Disagreement on edibility, medicinal value: Anti-eczema. (*The Macrofungi in China*, Mao, 2001) David Arora comments in *Amanita constricta*: "*A. inaurata* (=*A. strangulata, A. ceciliae*) is a closely related species "complex" whose volva is also "strangled", usually forming a belt of grayish tissue around the stalk base. It has a gray to grayish-brown to blackish cap decorated with gray to charcoal-gray warts (which may wear off in age) and it has round spores (10-14 µm). This ectomycorrhizae is fairly common in white spruce forests ground.

5 QUAKING ASPEN (POPULUS TREMULOIDES MICHX.) / VIBURNUM / LINNAEA / BOREALIS

Aspen is one of the pioneer trees which rapidly because established to grow after a fire in conifer forests in the North Temperate Zone (Fig.3-3). It is mostly distributed on south slopes, well-drained benches, and creek bottoms throughout interior Alaska, to about 914 m in altitude. It often grows in dense pure stands, because it easil propagaties from root suckers from roots, especially following forest fires. Quaking aspens are usually 60 – 80 years old, besides being pure aspen forests on higher altitude on the hill and dry land, on lower slope often mixed with White Spruce, on riverbanks and lower grounds often subrogate Black Spruce forests in permafrost soil. These kinds of aspen forests are mainly distributed in the center of Alaska, area around 960,000 hectares. The composition of forest species contain trees such as Quaking Aspen (*Populus tremuloides*)(Fig.3-4), White Spruce (*Picea glauca*), and Black Spruce (*Picea mariana*). Monogolian aspen plantation has experienced more than 10 years its composition will replace to spruce forest.

Common shrubs include: Bearberry (*Arctostaphylos uva-ursi*), Prickly rose (*Rosa acicularis*), Bebb willow (*Salix bebbiana*), Scouler willow (*Salix scouleriana*), Bullaberry (*Shepherdia canadensis*), Mountain cranberry (*Vaccinium vitis-idaca*), English name: American aspen, trembling aspen, popple, squawstone, Latin name: *Populus tremuloides* var. *aurea* (Tidestr.) Daniel

Usually grows to 6 – 12 meters, can reach 24 meters at the tallest. Diameter 7.5 – 30 cm, can reach 46 cm at the most. The bark is white, green, and gray in color, smooth, and has black stripes. It has wide white and light gray; the texture of the wood is thin, light, soft, and brittle.

This kind of tree is distributed most widely in Northern America, from Alaska to Newfoundland, Canada, to New Jersey, Virginia, and Missouri of northeastern US, to the south and west mountain regions until Transpecos in Texas, and to California and Mexico. (*The Ties that Bind Fungi in Ecosystem* (James M. Tiappe and Daniel L. Luoma, Oregon State University, Corvallis, Oregon). Introduction: P. 17 (Q. K. 604.2 C64 F86 1992 BIOS))

Latitude 64° 65' and longitude 148° 19' elevation 850 ft. slope 5%, aspect 180 Position: mild. Plot name Mile 3.2 Aspen, forest type old aspen and forest ecological type Aspen-buffalo berry located at Bonanza Creek LTER tree composition Aspen 50%, White Spruce 45%, birch 5%.

	Birch	White Spruce
Height (m)	22-25	28-32
DBH (cm)	11	28
Years old	100-120	100-120

5.1 *Phellinus tremulae* (Bond.) Bond. & Boriss.(Fig. 3-2)
in Bond., Polyporaceae Eur., USSR and Caucasia, p. 358, 1953. – *Fomes igniarius* f. tremulae Bond., Fungi in the Bryansk Forest. p. 22, 1912.

On angiosperms. Basidiocarps pileate. Basidiocarp perennial, rusty brown to black, tomentose to glabrous, often cracked or sulcately zoned, spores ellipsoid to globose, common to rare species. Spores broadly ellipsoid to globose, hyaline, to colored. Spores hyaline to pale golden brown. Spores 3-4 µm in diameter, cap brown, tomentose with a black zone below the tomentum, tropical species. Hymenial setae present. Setal hyphae or tramal setae absent. Basidiocarp effused-reflexed to sessile, often with strongly sloping cap, gray to brown, smooth to glabrous, on *Populus* and *Prunus*. On *Populus*, fruit body from old wounds in the trunk or along the lower side of branches, often radially cracked.

Distribution: Probably wherever aspen grows in North America. Remarks: *Phellinus tremulae* has not been segregated form the *Phellinus igniarius* complex by some American authors. Niemela gives a thorough account of the basidiocarps and cultural morphology of *P. tremulae*. Besides the macroscopic difference in basidiocarps, *P. tremulae* is microscopically different from other members of the *P. igniarius* complex included here because of the parallel arrangement of its tramal skeletal hyphae. Culturally it differs in the slow growth rate and sweet wintergreen odor. (7, 15, 23, 32, 36)

5.2 *Ganoderma applanatum* (Pers. ex Wallr.) Pat.
Soc. Mycol. France Bull. 5:67. 1889. – Boletus applanatus Pers., Obs. Myc. 2:2. 1799.- Polyporus applanatus (Pers.) Wallr., Flora Crypt. Germ. 4:591. 1833.
Wood decay fungus, can be used for medicinal purposes, anti-cancer. Context a definite thick layer, often developed between tube layers. Pilear crust thick; basidiocarps perennial. Cap surface not yellow to reddish and laccate, with a thick crust not composed of a palisade of closely packed clavate end cells. Distribution: Cosmopolitan species apparently throughout the forest regions of North America.

It is a medical fungus and its fruiting bodies are collected in the summer and the fall. After they are collected they are dried. If the *Ganoderma applanatum* is collected from the Chinese honey locust and prune trees, it can be used as an anti-cancer medication. They can be cultivated and deep-layer fermented to produce more Ganoderma applanatum. (Chen, Collection of Mushroom Prescriptions, 2000, p.389) These fruiting bodies can be collected from the southern slopes of the aspen forests. (15, 23, 32, 36)

5.3 *Pholiota squarrosa* (Muell. ex. Fr.) Quel.,
Champ. Jura Vosg. 1:126. 1872; Rolland, Atlas Champ. 61. pl. 62. 1910; Rea, Brit. Basid. 117. 1922.
Edible, anti-cancer, but poisonous. Cepitose; Cap 3-6 cm Broad, convex, viscid, cinnamon-buff, covered with erect pointed tawny scales, denser at center; flesh white, rather thick; stipe 4.5-7 cm. Long, 5-8 mm. thick, equal, solid, subconcolorous with the cap, lower two-thirds covered by coarse tawny scales, whitish and smooth above the evanescent annulus; gills close, adnate, pallid, becoming tawny olive; cystidia scattered, clavate, 20-35x9-13µm; spores ellipsoid, smooth, 4-6 x 2.5-4 µm. Though they cause trunk decay, these fruiting bodies are edible and also medicinal. (Fungi of China, S.C. Teng, Mycotaxon, LTD. Ithaca, New York)

Experimental tests show this species is anti-cancer. According to tests, 70% of small tumors found in laboratory rats were cured by this species. (*The Macrofungi in China*, Mao, 2000) (15, 23, 32, 36)

5.4 *Fomes fomentarius* (L.:Fr.) Kickx. "The tinder fungus"
Flore Crypt. Flandres 2:237, 1867.- Polyporus formentarius L.:Fr., Syst. Mycol. 1:374, 1821.
Wood decay fungus, medicinal and anti-cancer. Temperate, boreal species, ungulate, grayish on the cap, spores 15-20 µm long.

Distribution: Widely distributed on all broad leaf forests. They are also on dead trunks and fallen logs. It is a medicinal fungus. During June and July it is collected and used for medicinal purposes, such as anti-stomach and ovarian cancers. (Chen, *Collection of Mushroom Prescriptions*, 2000, p.340) (15, 23, 32, 36)

5.5 *Trametes ochracea* (Pers.) Gilbn. & Ryv. comb. nov.
Basionym: *Boletus ochraceus* Pers. Ann Bot. (Usteri) 11:29, 1794. – *Polyporus zonatus* Nees.:Fr., Syst. Mycol. 1:368, 1821.
Basidiocarps rigid to hard; cap usually in white to brown shades; pores 3-4 mm, spores 6-8.5 x 2-2.5 µm. Cap adpressed velutinate and dull to subshiny or soon becoming glabrous except for margin; context homogeneous although a cuticle may develop form the base with age. Pores regular, 3-8 per mm, round to angular, more or less entire.
Distribution: Widely distributed in boreal forest regions of North America, south in the Rocky Mountain forests to Arizona. (15, 23, 32, 36)

5.6 *Cerrena unicolor* (Bull.:Fr.) Murr.
J. Mycol. 9:91, 1903.- *Daedalea unicolor* Bull.:Fr., Syst. Mycol. 1:336, 1821.
This is a new record for Alaska's *Populus(Fig. new record of Cerrena unicolor P.193)*. Basidiocarps annual, sessile, effused-reflexed or rarely resupinate, pilei often in imbricate clusters, dimidiate, up to 10 mcm wide, upper surface pale brownish to gray, hirsute to almost glabrous, often green due to algae, sucate, pore surface ivory to pale buff on young specimens, becoming darker with age, the pores daedaleoid, variable, 3-4 per mm, in parts larger, dissepiments at first thick and tomentose, becoming thin and splitting, context duplex, up to 3 mm thick, corky, ale brown lower layer separated from soft, spongy, darker upper layer by a thin dark zone, tube layer continuous with lower context, up to 1 cm thick.

Widely distributed in North America. Widespread in Asia and Europe. (15, 23, 32, 36)

5.7 *Trichaptum subchartaceum* (Murr.) Ryv.
Norw. J. Bot. 19:237, 1972.- *Coriolus subchartaceus* Murr., North Am. Flora 9:24, 1907.
Cap hirsute-strigose; basidiocarps up to 1 cm thick; pores 2-4 per mm, on *Populus*; boreal species. Pore surface purplish, pale brown to buff; temperate-boreal species. Pores angular, 1-3 mm; cap, if present, hirsute to coarsely strigose. Hymenophore poroid, sometimes with lacerate dissepiments.

A boreal fungus, in northern parts of eastern North America, and throughout the range of aspen in Western mountains from Arizona to Alaska. (15, 23, 32, 36)

5.8 *Skeletocutis nivea* (Jungh.) Keller.
Persoonia 10: 353, 1979.- *Polyporus niveus* Jungh., Berh. Batav. Genootsch. 17: 48, 1839.- *Tyromyces semipileatus* (Pk.) Murr., N. Am. Flora 9:35, 1907.- *Polyporus semipileatus* Pk., N.Y. State Mus. Ann. Rept. 34:43, 1881.
Type of rot-White rot of dead hardwoods. Basidiocarps annual, effused-reflexed or often resupinate, rately sessile, pilei solitary or imbricate, dimidiate to elongate, sometimes laterally fused, up to 3 cm wide; uper surface white to cream colored, azonate, finely tometnose to galbrous, pore surface white to cream, the pores circular to angular, 3-10 per mm, with thin, entire dissepiments; context white, zaonate, up to mm thick; tube layer white to pale buff, distinct from context, easily sectioned, up to 2 mm thick.

Hyphal system trimatic; contexual generative hyphae thin-walled, nodose-septate, with occasion al branching, 2-3.5 µm in diameter; contextual skeletal hyphae thick walled, aseptate, with rare branching, 3-4 µm in diameter; contexual bindingn hyphae developing form lateral branches on generative hyphae, thick-walled, much branched, non septate, 1.5-2 µm, compactly arranged and difficult to separate.

Cystidia absent; fusoid cystidioles rare and inconspicuous; hyphal pegs present, usually abundant. The small pores, glancing pore surface, and narrow spores are distinctive characters of *S. nivea*. Basidia clavate, 4-sterigmate, 11-17 x 3.5-5 µm, with a basal clamp. Basidospores allantoid, hyaline, smooth, negative in Melzer's raegent 3-5 x 0.5 –1 µm.

Substrata: Dead wood of numerous genera of hardwoods, rarely on conifers. Distributed in eastern and western forest regions of North America from the southern states in Canada. Circumglobal in temperate forest regions. (15, 23, 32, 36)

5.9 *Bjerkandera adusta*

Wood decay fungus, medicinal and anti-cancer. White rot of hardwood logs and slash occasionally on conifers, positive in gum guaiac solution. Basidiocarps annual, sessile, effused-reflexed, or occasionally rsupinate under logs, often in imbricate clusters, tough, reflexed up to 3 cm, uper surface of cap cream to buff, tomentose or strigose t oglabrous with age, azonate or faintly zonate, pore suface gray to black, the pores angular, regular, 6-7 per mm, dissempiments thin, entire, context pale buff, azonate with distinct thin upper layer of tomentum up to 6 mm thick, tube layer smoky gray, distinct from context, up to 1 mm thick. The smoky gray to black color of the pore surface and the cram colored cap are distinctive field characters. Bjerkander fumosa is similar and differs in having thicker and wider basidiocarps and broader spores. Bjerkandera adusta is particularly common in aspen and is always a conspicuous fungus in older aspen stands.

Hyphal system monomitic, contextual hyphae thin to moderately thick-walled, with abundant clamps, 3-5 µm in diameter, with occasional branching, tramal hyphae similar but densely compacted and agglutinated.

Cystidia or other sterile hymenial element slacking, dark brownish vascular hyphae sometimes present in subhymenium and in hymeial layer. Basidia clavate to napiform, 4-sterigmate, 22-25 x 5-6 µm, with a basal clamp. Basidiospores short-cylindric, hyaline, smooth, negative in Melzer's reagent, 5.6 x 2.5-3.5 µm.

Distribution: Many genera of hardwoods, rarely on conifers. Everywhere in Alaska. (15, 23, 32, 36)

5.10 *Coriolopsis gallica*

Causes a white rot. Basidiocarps annual, pileate, broadly sessile, up to 10 cm wide, 7 cm broad and 1 cm thick, semicircular or elongated, often several imbricate pilei from a common, effused resupinate part, corky to tough, cap densely hirsute to hispid, first brownish, but soon dirty gray, zonate or azonate, more hispid at the base than at the margin, the hirsute tometusm is clearly differentiated towards the brown context, pore surface brown or gray, pores angular, thin-walled, 1-3 mm in diameter, in larger and older specimens often radially elongated and deeply split, tubes up to 15 mm long, whitish to gray on the inner walls, trama brown, context mostly thin, more rarely up to 10 mm thick, rusty to umber brown, first black in KOH, then fading back to almost the original color.

Hyphal system trimitic, generative hyphae thin-walled, hyaline and clamped, 2.4.5 µm in diameter, binding hyphae tortuous, thick-walled to almost solid, light golden brown, 2.5-4.5 µm wide, skeletal hyphae thick-walled to solid, golden brown in trama and context, hyaline in the tomentum, 2.5-6 µm in diameter. When typically developed this is an easy species to recognize because of its quite thick basidiocarps with a hispid to villose, often grayish cap, large pores, and a brown pore-surface and context. Trameters trogii may macroscopically be somewhat simlar. Coriolopsis rigida is normally thinner, has smaller pores and a more tomentose to velutinate cap, Overholts treated this species as Trametes hispida Bagl.

Cystidia or other sterile hymenial elements absent. Basida clavate, 20-40 x 5.5-8 µm. Basidiospores cylinddrical, hyaline, thin-walled, smooth, IKI-, 10-16 x 3-5 µm, sometimes varying considerably even with in the same basidiocarp. On dead hardwoods, most common on Salix and Populus species, very rarely on conifers. Distribution from Southern Canada and throughout Alaska except for the southeastern part. (15, 23, 32, 36)

5.11 *Crepidotus mollis* (Jelly Crep; Flabby Crepidotus)

Edible wood decay fungus. Cap 1-5, 8 cm broad, fan-or kidney-shaped to nearly round in outline, convex to plane; surface gelatinous in wet weather beneath a dense to rather sparse coasting of fulvous to rusty-ochre to brown fibrils (hairs) or small fibrillose scales; in age often smooth or with very few fibrils and varying in color form tawny to pale ochre to brown, or fading to whitish.

Flesh soft, thin, pallid, soon flaccid. Gills close, whitish becoming brown or dull cinnamon; radiating from base of cap. Stalk absent or rudimentary. Spore print dull brown to yellowish-brown; spores 7-11 x 4.5-6.5 μm, elliptical, smooth.

Usually in groups or overlapping tiers on the bark of dead hardwoods (or sometimes conifers); it is very widely distributed and common. (15, 23, 32, 36)

5.12 *Cortinarius* sp.

Plot V, M. M. Chen and J. H. McBeath, 8/7/1989, specimen # AKMC 0136 (with spore print).

Cap 4.5 cm cinnamon brown viscid when wet then dry hygrophanous, stipe 6.4 x 7.0 cm clay-color to rusty brown, bulb at base; gills adnated, confluent with cap cortinae at upward remained and stained cinnamon-brown by spores, gills 6.5 x 0.6 cm brown to cinnamon; spores 11-12 x 7.0-8.0 μm and elliptical. Found in aspen forest ground on a hill.

5.13 *Cortinarius* sp.

Menley hot spring, plot VII, Fairbanks, Alaska, M. M. Chen, 7/31/1990, specimen # AKMC 1023.

Cap 3.5-6.1 cm. Small to middle size when moist viscid and cinnamon brown with smooth shine and corrugated, velvety; context brown, stem pallid 6.5-7.0 x 0.5-0.8 cm remained and stained cinnamon-brown by spores, gills adnated, some with a fine cobweb cortinae at upward between the stipe and cap edge, tasted not bitter, spores 12-13.2 x 8.4-9.6 μm rusty brown smoothly and piriform. Found in foothill pure aspen forest, fairly common in aspen forests, often also seen in birch forests.

5.14 *Hydnellum aurantiacum* (Fr.) Karsten (Fig. 4-1) "Golden Hydnellum"

J. H. McBeath home yard, hardwood trunk base, M. M. Chen, 7/15/1990, specimen # AKMC 1010.

Ectomycorrhizae, not edible. Fruit body often fused together. Cap 3-15 cm across, flattened-depressed; orange-brown to rusty cinnamon; tomentose-velvety, often with coarse lumps and protrusions at center when mature. Spines on undersurface white then brownish with white tips. Stipe 30-60 x 10-20 mm; orange to dark brown. Flesh distinctly zoned; orange to cinnamon. Odor fragrant, persistent. Taste not distinctive. Spores strongly tuberculate, 5.5-7.5 x 5-6 μm. Deposit buff. Habitat under conifers, often in masses. Found throughout Fairbanks, Alaska.

5.15 *Lycoperdon perlatum* Pers. syn. *Lycoperdon gemmatum* Batsch (Fig.4-2)

FP3B, M. M. Chen, 8/22/1989, specimen # AKMC 0205, 1002.

Edible when flesh is completely white; excellent. Medicinal values: blood coagulent and tumor inhibition. (*Collection of Mushroom Prescriptions*, Chen, 1999). Fruit body 2.5–6 cm across, 2-9 cm high, subglobose with a distinct rudimentary stipe; white at first, becoming yellowish brown; outer layer of short pyramidal warts, especially dense at the head, rubbing off to leave an indistinct meshlike pattern beneath, which opens by a pore. Spore mass white, then olive-brown at maturity. Sterile base spongy, occupying the stipe. Spores globose, minutely warted, olive-brownish, 3.5-4.5 x 3.5-4.5 μm. Habitat singly, scattered, or in clusters in waste areas and open woods and along wood edges. Found widely distributed in Fairbanks, Alaska. (36)

5.16 *Russula emetica* (Schaeff.:Fr.) Pers. ex S. F. Gray (Fig.4-3)

Plot VII, M. M. Chen, 7/31/1990, specimen # AKMC 1012.

Poisonous but anti-cancer ectomycorrhizae. If eaten, it causes strong stomach pain, diarrhoea, vomiting, and nausea, when serious can cause facial muscle tic, heart or blood circulation failure, leading to death. (*The Macrofungi in China*, Mao, 2001).

Scattered; cap 5-9 cm broad, fleshy, fragile, convex, becoming plane, and in age depressed, viscid, glabrous, light jasper-red to jasper-red; margin sulcate-striate, light coral-pink; flesh thin, white, reddish under the cuticle; stem cylindric, 4-7.5 x 1-2.2 cm, spongy-stuffed, white or tinged flesh-pink; gills equal, white, subdistant, adnexed, interspaces venose; spores hyaline, subglobose, echinulate, 8-10 x 7-9 μm; cystidia lanceolate, 60-110 x 10-15 μm, projecting 8-18 μm beyond the hymenium; taste very acrid.

5.17 *Leccinum insigne* Smith, Thiers, & Watling (Fig.4-4a and 4-4b) "Aspen Scaber Stalk"
Plot IV, M. M. Chen, 7/24/1990, specimen # AKMC 1004, 1005,1007.
Edible ectomycorrhizae. Cap edible, but not stem. Cap 4-16 cm across, round to broadly convex becoming flatter; color dull rust-orange; dry, smooth or minutely hairy and sometimes scaly, then pitted in age. Tubes adnate or decurrent; whitish then olive-gray bruising pinkish brown. Pores whitish bruising yellow to olive-brown. Stem 60-150 x 10-25 mm, solid, swollen at the base; whitish when young, but covered with numerous small projecting scabers which turn from reddish brown to blackish in age, base often bruises blue; tough, fibrous. Flesh thick, soft; white turning violet-gray or dingy brown when cut. Spores spindle-shaped to ellipsoid, smooth, 11-18 x 4-6 µm. Deposit brownish or yellowish-brown. Habitat scattered or in groups under aspen or birch in woods and along wood edges. Widely distributed in Aspen forest ground.

5.18 *Leccinum aurantiacum* (Fr.) S. F. Gray (Red-cap Bolete)
Plot XIII, Lorson, M. M. Chen, 8/10/1990, specimen # AKMC 1039.
Edible ectomycorrhizae. (Comment by R. P. Korf: This species was not treated in the Chinese version, Chung-kuo Ti Chen-chun, 1963. Teng added it to the revised key, but the portion of this manuscript containing the description of the fungus and any references as to where it was collected or reported from China has been lost.) Cap 5-15 – 10-20 cm across, convex; bright orange-brown to reddish orange, often with pallid areas where covered by leaves; dry, woolly-fibrillose, margin with sterile band of tissue up to 1cm deep hanging in irregular segments. Tubes 1-2 cm long; olive-buff then darker brownish. Pores small, pallid when young, then pallid olivaceous. Stipe 80-160 x 20-30 mm, narrowly clavate, solid, fibrious, very firm; entirely pallid at first with finely woolly-scabrous surface, soon discoloring brown, finally blackish, especially in basal half, apex remaining pallid. Flesh white then rapidly staining grayish vinaceous, then fuscous, and in base of stipe often bright blue mixed with reddish brown. Odor pleasant. Taste pleasant. Spores subfusoid, 13-16(18) x 3.8-4.5(5) µm. Deposit deep yellow-brown. Habitat: under aspen and pine. Mixed birch/spruce collection. Rather common in China, distributed from the north to Tibet.

5.19 *Clavicorona pyxidata* (Pers.:Fr.) Doty "Crown Coral Mushroom"
Plot VI, M. M. Chen, 7/31/1990, specimen # AKMC 1035, 0118.
Edible. Gregarious, 6-7 cm high, whitish or pale yellowish; stem slender, enlarging upward and dividing simultaneously like an umbel into several branches with cup-shaped apex; from the margin of the fruiting body spring another whorl of branches, thus branching repeatedly and finally ending in small cups with little points on the rims, after drying then pliant; spores hyaline, smooth, ellipsoid, 3.5-4.5 x 2.5 µm; cystidia present, hyaline, fusoid, 5-7 µm thick, projecting up to 30 µm beyond the hymenium. Found on aspen dead wood. (36)

5.20 *Clitocybe odora* (Fr.) Kummer
FP4A, M. M. Chen, 8/21/1990, specimen # AKMC 1065, 1078.
Edible, but some records say not edible. Cap 3-9 cm across, convex at first with a low, broad umbo, later expanding and becoming irregular and wavy at the margin; dingy green to bluish green; grayish, bluish, or nearly white; finely matted with silky hairs or sometimes with a hoary bloom. Gills slightly decurrent, close or crowded, broad; whitish tinged with cap color. Stipe 30-70 x 5-15 mm, solid becoming hollow, sometimes curved and enlarged toward base; whitish tinged with cap color, base spongy and covered in fine whitish down. Flesh thin, firm; whitish to pale tan. Odor strongly of aniseed. Taste strongly aniseed. Spores ellipsoid, smooth, nonamyloid, 6-7.5 x 3-4 µm. Deposit whitish pink. Habitat single, scattered, or in groups on leaf litter under hardwoods, especially oak. Found widely distributed in Fairbanks, Alaska.

5.21 *Gomphus clavatus* (Fr.) S. F. Gray (Fig.5-1) "Earth Top"
Aspen Forest, Fairbanks, Alaska, M. M. Chen, 8/9/1989, specimen # AKMC 0128.

Edible ectomycorrhizae. Fruit body 2-10 cm wide, up to 15 cm high, compressed and partially fused, the cap flat with a sunken center and wavy margin; violet becoming yellowish buff; smooth, moist then dry felty becoming scaly on the disc. Fertile undersurface shallow, wrinkled, sometimes with folds or pits; violet when young, becoming duller and more brownish in age. Stipe 10-50 x 10-20 mm, very short, often curved, sometimes fused with adjacent stems; buff to pale lilac; smooth to minutely hairy. Flesh solid; whitish to pale pink. Odor none or faintly earthy. Taste mushroomy. Spores ellipsoid to narrowly ovoid, warty, 10.3-15.5 x 4.3-7 μm. Deposit ochre to dark olive-buff. Habitat growing singly or in overlapping cluster or arcs or circles of up to 40 fruiting bodies, under aspens. CHANTERELLES: These mushrooms have primitive foldlike or even absent gills.

5.22 *Pluteus petasatus* (Fr.) Gillet
Aspen Forest, Plot VI, M. M. Chen, J. H. McBeath, 8/9/1989, specimen # AKMC 0135.
Wood decay fungus, edible. Cap 5-18 cm across, convex to obtuse; white to ivory with dark brownish fibrils or scales at center. Gills free, crowded, rather obtuse, blunt at outer margin of cap; white for a long time before turning pink. Stipe 40-100 x 7-15 mm, slightly broader at base; white to streaked below with darker fibrils. Flesh white. Odor pleasant. Taste pleasant. Spores ovoid, smooth, 6-7.5 x 4.5-5 μm. Habitat on rotten logs, stumps, or litter pile on aspen forest ground.

5.23 *Pholiota squarrosoides* (Pk.) Sacc. “Scaly Pholiota”
Plot IV, M. M. Chen, 8/7/1989, specimen # AKMC 0080 (with spore print).
Wood decay fungus, edible, but not recommended. Cap 2.5-10 cm across, obtusely convex with an umbo becoming flat; whitish becoming cinnamon with downcurved tawny scales scattered near the margin, clustered over the disc; sticky beneath the dry scales, veil remnants often hanging form the margin. Gills adnate becoming sharply adnexed, close to crowded, moderately broad; whitish, changing slowly to rust-brown as spores mature. Spores 8-6 μm long; cystidia clavate; stipe 50-150 x 5-15 mm, stuffed or solid; whitish at the top, pale buff below the ring, with coarse down-curving ochre-tawny scales, sometimes staining rusty brown near the base; silky above the pale, fibrous. Grows on aspen living tree trunk. (36, 27)

5.24 *Panus rudis* Fr., Epicr. 398. 1838; Farl. & Burt, Icon. Farl. 46, pl. 42, 1929; Teng, Sinensia 7:542, 1936. (Fig.5-2).
Plot I, M. M. Chen, 7/27/1989, specimen # AKMC 1003.
Wood decay fungus, edible and anti-cancer. Cespitose; cap 2-9 cm broad, top funnel-shaped little depressed, thin, ascending; infundibuliform, eccentrically or sublaterally stipitate, vinaceous fawn when young and fresh, becoming buckthorn-brown or ochraceous tawny to cinnamon-brown, tough, coriaceous, strigose-hirsute; context light brown; stem off-center, solid, short, 0.5-2 cm long, 2-10 mm thick, clothed and colored like the cap; gills pinkish white when fresh, subconcolorous with the cap when dry, narrow, crowded, decurrent; hyphae simple; cystidia hyaline, cylindric-clavate, 45-50 x 11-14 μm, projecting 30-35 μm beyong the hymenium; spores ellipsoid, 4.5-6 x 2-3 μm, smooth, hyaline. (36)

5.25 *Russula xerampelina* (Schaeff.) Fr.
Aspen Forest, Fairbanks, Alaska, M. M. Chen, J. H. McBeath, 8/9/1989, specimen # AKMC 0086, 0133, 0129.
Edible ectomycorrhizae that tastes like crab or shrimp, also anti-cancer. Cap 5-8 cm broad, Pompeian-red, darker at disk, fading in part to yellowish or deep olive-buff, firm, convex, then plane-depressed, subglabrous, dry, subviscid when moist; margin even; flesh white, becoming yellow; stem equal, glabrous, spongy-stuffed, 4-5.5 x 1.5-1.8 cm, flesh pink; gills whitish, then yellowish, close, adnate, often forked, interspaces venose; spores pale yellowish, globose, echinulate, 9-10 μm diameter; cystidia lanceolate, 80-100 x 8-12 μm. One of the least appreciated of edible fungi, perhaps due to the mediocrity of its brethren. The young, nutty caps are superb stuffed with grated cheese, chives,

walnuts, and parsley and then boiled. Fishy odor at maturity, which is accentuated by cooking or drying. May be a composite speices. (3, 36)

5.26 *Clavariadelphus truncatus* (Quel.) Donk "Flat bat coral mushroom"
FP4A, plot V, M. M. Chen, 8/7/1990, specimen # AKMC 1036,1017.
FP4A, M. M. Chen, 8/21/1989, specimen # AKMC 0125.
Edible - good. Fruit body 5-15 cm high, 3-8 cm wide at the top, club-shaped, often broad and flattened at sterile top, narrowing down to a bulbous base; yellowish ochre to dark apricot orange; wrinkled. Stipe indistinct; white-hairy at base. Flesh firm to spongy; whitish to ochre, darker on bruising. Odor mild. Taste sweet. Spores ellipsoid, smooth, 9-12 x 5-8 μm. Deposit pale ochre. Habitat scattered or in groups or clumps on the ground in coniferous woods. Widely distributed throughout inland Alaska. (Mao, 2001).

5.27 *Sarcodon imbricatus* (L. ex Fr.) Karsten syn. *Hydnum imbricatum* L. ex Fr.
Plot XIIIB, M. M. Chen and Richard Parmeter, 8/13/1990, specimen # AKMC 1055.
Chinese call it "Wild Deer mushroom". Ectomycorrhizae, edible but poor, can be used for medicinal purposes to reduce chlesterol in the blood. It's easy to be collected and processed, and contains polysaccharose. (*The Macrofungi in China*, Mao, Xiaolan, 2001, p. 155). Cap 5-20 cm across, flattened-convex then depressed; dark reddish to purplish brown against paler pinkish flesh; velvety then cracked into coarse overlapping scales. Spines on undersurface 1-10 mm long; white to purplish brown. Stem 50-80 x 20-50 mm, tapered or swollen at base; whitish to purplish brown. Flesh firm; white. Odor not farinaceous, not distinctive. Taste soon bitter. Spores ellipsoid, tuberculate, 7-8 x 5-5.5 μm. Deposit brown. Habitat coniferous woods. Often found around Fairbanks, Alaska, suburban northern and eastern North America.

5.28 *Amanita muscaria var. muscaria* (L. ex Fr.) Pers. Fly Agaric. (Fig.5-3)
Plot XIIIA, M. M. Chen, 8/13/1990, specimen # AKMC 1054.
Poisonous ectomycorrhizae. Cap 5-25 cm across, convex to flatter, sometimes slightly wavy or depressed with a lined margin; blood red to orange-red, becoming lighter toward the margin; smooth, a bit sticky when moist, dotted with flaky patches of whitish volval remnants sometimes almost in concentric rings. Gills free to adnexed, crowded, broad; whitish. Stipe 50-180 x 3-30 mm, sometimes enlarging toward rounded basal bulb; spores 9.4-13 x 6-8 μm, broadly elliptical, smooth, colorless, spore print white. Often seen in middle to late August. (3, 27, 36)

5.29 *Armillariella mellea* (Vahl ex Fr.) Karsten. "Honey Fungus"
FP3A, M. M. Chen, 8/22/1989, specimen # AKMC 0224.
Wood decay fungus and ectomycorrhizae, edible, timid, medicinal and anti-cancer. Cap 3-15 cm across, very variable, convex then flattened and centrally depressed or wavy; ochre, tawny to dark brown, often with an olivaceous tinge; covered in dark hairy scales, especially at the center. Gills attached or slightly descending stalk, nearly distant, narrow; whitish, then yellowish becoming pinky-brown and often spotted darker with age. Stipe 60-150 x 5-15 mm, stuffed to hollow; whitish becoming reddish brown. Veil partial veil leaving a thick whitish to yellow cottony ring on upper stipe. Flesh white. Odor strong. Taste astringent. Spores ellipsoid, smooth, nonamyloid, 8-9 x 5-6 μm. Deposit whitish. Habitat in small or large clusters at the bases of trees or near stumps. Commonly found widely distributed throughout Fairbanks, Alaska. Edible but must be cooked. Some cases of severe stomach upset have been reported after eating this mushroom. Eat in small quantities the first time you try it.

Comment: This fungus spreads by long black cords called rhizomorphs, resembling bootlaces, which can be found beneath the bark of infected trees, on roots, or in the soil, where they can travel long distances to infect other trees. This is one of the most dangerous parasites on living trees, causing an intensive white rot and ultimately death. There is no cure, and the fungus is responsible for large losses of timber each year.

5.30 *Russula rosacea* (Bull.) Fr., Epicr. 351, 1838; Bres., Icon. Myc. 9:405, 1929; Teng, Sinensia 7:550, 1936. *Agaricus rosaceus* Bull., Herb. Fr. pl. 509. 1790; Pers., Syn. Fung. 439, 1801. (Fig.5-4) University of Alaska campus, M. M. Chen, 7/31/1990, Specimen # AKMC 1001, 1012.
Edible ectomycorrhizae. "Bitter Russula" or "Rose Russula". Cap 4-9 cm, broad, convex, then expanded, finally depressed, subviscid, about jasper-red when fresh, drying Pompeian-red; margin even; flesh white; stipe subequal, solid, white, even, rosy red in places, 4-7 cm long, 1-2 cm thick; gills white, close, equal, narrowly adnexed; spores hyaline, white in mass, echinulate, subglobose, 7-8 x 6-7 µm in diameter; cystidia abundant, fusoid, 50-85 x 12-20 µm; taste bitter and acrid. Spore print milk yellow.

Often grows in clusters or scattered in Fairbanks birch / spruce mixed forests ground.

5.31 *Laccaria laccata* (Scop. ex Fr.) Berk. et Br., Ann. Mag. Nat. Hist. V. 12:371. 1883; - *Agaricus laccatus* Scop. : Fr., Syst. Myc. 1:106. 1821. (Fig.5-5) "Lackluster Laccaria"
Aspen Forest, Plot III, Fairbanks, Alaska, M. M. Chen, 8/3/1989, specimen # AKMC 0088.
Edible and anti-cancer ectomycorrhizae. Cap 1-6 cm across, convex then flattened, often becoming finely wavy at the margin and centrally depressed; tawny to brick red drying paler to ochre-yellow; margin striate when moist, surface often finely scurfy. Gills adnate to short decurrent, distant, broad; pinkish, dusted white with spores when mature. Stipe 20-80 x 3-10 mm, often compressed or twisted; same color as cap; tough and fibrous. Flesh thin; reddish brown. Odor not distinctive. Taste slight. Spores ornamented globose to broadly ellipsoid, spiny, 7-10 x 6-8 µm. Deposit white. Habitat scattered or in groups in sandy or pool soil in wasteland and under conifers. Found widely distributed in Fairbanks, Alaska.

6 BIRCH-BETULA PAPYRIFER /VIBURNUM EDULE

This kind of paper birch forest is often distributed on the east or west slopes of the hill after a fire. It's also seen sometimes on the north slopes and level ground. Usually the forest is pure birch, but it sometimes includes White Spruce or Black Spruce trees. The shrubs are similar to that of the aspen forests, but Labrador–tea and mountain–cranberry are often seen. Average tree height is 18 – 24 meters, and average diameter 20 – 22 cm. This kind of forest is distributed in the inland area, about 2 million hectares, most often seen in Susitna River Valley of North Fairbanks.

Common wooden species include: Paper Birch (*Betula papyrifera*) (Fig. 6-1), White Spruce (*Picea glauca*), Black Spruce (*Picea mariana*). Common shrubs include: Narrow-leaf labrador-tea (*Ledum decumhens*), Labrador-tea (*Ledum groenlandicum*), American red current (*Ribes triste*), Prickly rose (*Rosa acicularis*), Bebb willow (*Salix bebbiana*), Scouler willow (*Salix scouleriana*), Banclay willow (*Salix barclayi*), Mountain-cranberry (*Vaccininum vitis-idaca*), High bush cranberry (*Viburnum edule*), Dwarf blueberry (*Vaccininum caespitosum*).

Around University campus at birch forest often seen the mushroom cicle of fly Agaric (*Amanita muscaria* var. muscaria (L. ex Fr.) Pers.)

In Paper birch forests, the most common species of wood decay is cosmopolitan, some species are distributed all throughout the Northern hemisphere. In Taiga forests, for instance, *Phellium igniarius* occurs on living trees, *Piptoporus betulinus, Lentinus betulina, Fomes fomentarius* and *Trametes versicolor* etc., on fallen trees. These are all saprophytic species that are the decomposers of the wooden forests. Many of the fruiting bodies formed may be excellent sources of medicine, and can be cultivated. Other fruiting bodies are the food of Alaskan moose in the winter. Some others are powdery mildew on the leaves and leaf rust on paper birch trees. Because the Alaskan paper birch forests are still young, these diseases are not very severe.

6.1 *Phellium igniarius* (L.:Fr.) Quel.

Ench. Fung., p.172, 1886.- *Polyporus igniarius* L.:Fr., Syst. Mycol. 1:375, 1821.- *Fomes igniarius* (L.:Fr.) Kickx, Fl. Crypt. Flandres 2:237, 1867.
Type of rot - uniform white rot of the heartwood of living hardwoods. Fruiting bodies often seen on old living trees, this is related to the age of the tree, the older it is, the more fruiting bodies formed.

Basidiocarps perennial, sessile or rarly effused-reflexed, ungulate or sometimes applanate, up tp 11x 20 x 8 cm; upper surface gray or blackish, glabrous, sulcate, becoming deeply rimose, incrusted; margin concolorous and glabrous or yellowish brown (Saccardo's umber or Snuff Brown); pore surface pale cinnamon brown to dark purplish brown, the pores circular, 5-6 per mm, with thick, entire dissepiments; context dark redish brown, zonate, woody, up to 2 cm thick; core absent or present next to substratum, with white tissue intermixed; tube layers concolorous with context, the tubes white-stuffed, in distinct layers, each up to 4 mm thick.

Contexutal hyphae of two types, some brown in KOH, thick-walled, distinct, with rare branching, aseptate, 2-5 µm in diameter, some hyaline, thin-2awlled with occasional simple septa, very indistinct; tramal hyphae similar, 2-3 µm in diameter.

Setae ventricose to subulate, abundant to rare, 14-17 x 4-6 µm; core setae present in some specimen, irregularly lobed and branched; thick-walled, up to 15 µm in diameter. The *Phellinus igniarius* complex has been another difficult taxonomic problem, such as Aspen *Phellinus tremulae*, which has already been classified as Aspen white pocket rot. Birch white pocket rot is different from that of Aspen.

Basidia broadly clavate, 4-sterigmate, 9-10 x 6-7 µm, simple-septate at the base. Basidiopores broadly ovoid to subglobose, hyaline, smooth, thick-walled, negative in Melzer's reagent, acyanophilous, 5-6.5 x 4.5-6 µm.

Living hardwoods of many genera, continuing decay in dead trees. Found transcontinentally from Newfoundland to Alaska but absent or rare in the southern tier of states, apparently a boreal species. Distribution records in Arizona and New Mexico given by Overholts are apparently based on *P. tremulae*, which he lumped in with *P. igniarius. (*15, 6, 36 Gilbertson, Bondarzew, Teng) (7, 15, 23, 32, 36)

6.2 *Piptoporus betulinus* Pk. Bull. Torrey Bot. Club. 26:69, 1899. (Syn. *Polyporus betulinus*) (Polyporaceae)
Wood decay fungus, edible and anti-cancer. Type of rot. Brown cubicla rot of the sapwood of dead birches. Occurs only on birches (black, gray, paper, European white, and yellow), usually on dead trees and rarely on living ones. Its distinctive pale brown to nearly white annual basidiocarps with smooth, rounded upper surfaces up to 25 cm wide are unlike those of any other fungus on birch. It occurs throughout the range of birches in North America and Eurasia, causing a red-brown cubical rot. (Sinclare, 1987) In winter the snow in paper birch forests can reach a meter high, fruiting bodies can provide food for moose in the cold weather. In Siberia this fungus is called "Moose's bread" fungus.

Basidiocarps dimidiate to substipitate; stipe short, stout, glabrous, often resinous, whitish to brown, up to 6cm long and 5 cm thick; pilei often pendent, usually dimidate orreniform, solitary, up to 15 x 25 x 6 cm; upper surface whitish to mouse-colored or brownish usually with a pellicle that breaks up to give a pitted or scaly appearance, glabrous azonate; margin concolorous usually extending down below the pore surface; pore surface white at first, becoming pale brownish with age, the pores circular to annular, 3-5 per mm, with thick, entire dissepiments that split and clump together with age to give a hydnaceous appearance; context white, tough when fresh, during soft-corky, azonate , up to 5cm thick; tube layer easily separated from the context when fresh, up to 1cm thick.

Hyphal system ditrimic; contextual generative hyphae thin-walled, hyaline, with clamps, reately branch, 2.5-4 µm in diameter; contextual skeletal hyphae hyaline in KOH, thick-walled, nonseptate, some much branched, others with reate or occasional branching, 2.5-5 µm in diameter; tramal hyphae similar. Cystidia or other steile hymeia lelmestn absent. Basidsi clavate, 5.-stetrignmate, with a basal clamp; 10-12 x 5-6 µm. Basidiospores cylindric, slightly allantoid, hyaline, smooth, IKI-, 5-6 x 1.5-1.7 µm. This species is scarely comparable to any other considered here. The restriction to birch, the smooth, usually pelliculose upper surface, the easily separated tube layer, and the margin extending below the pore surface are reliable field characters.

Heterotlalic and bipoloar. Restricted to Betula. In North America particularly common on *B. papyrifera* (paper birch) and *B. alleghaniensis* (yellow birch). A true boreal fungus, *P. beulinus* is found throughout the range of paper birch in Fairbanks, Alaska. (7, 15, 6, 32, 36)

6.3 *Lenzites betulina* (Fr.) Fr.

Epicr. P.405, 1838.- *Daedalea betulina* Fr., Syst.- Mycol. 1:333, 1821.

Type of rot. White rot of dead hardwoods. Basidiocarps annual, single to a few together, pileate, dimidiate to semicircular or broadly attached with a pearly reupinate, effused part, 1-5 x 2.8 x 3-2 cm, margin even to lobed or incised, corky and coriaceous, upper surface tomentose to hispid in concerntric, partly sulcate zones, first white, later grayish to coral, old specimens often with a greenish tint because of algae in the tomentum, hymnophore lenzitoid with thin radial lamellae, toward the margin new lamellae arise by dichotomous forking of old ones, but also individually between older ones, when young and along the margin straight, 10-15 per cm measured tangentially, about 100-200 µm thick, in older parts and when dry, mostly undulating or flexuous, thus, the distance between individual lamellae may vary considerably, first white, later cream to ochraceous, lamellae up to 12 mm deep at the base, context thin, 1-2 mm thick, fiborus and white, distinctly lighter than the lamellae.

Hyphal system trimitc, generative hyphae hyaline and with clamps, in the subhymenium 2-4 µm wide and thin-walled, in the context rather scattered, somewhat thick-wall to thin-walled and up to 5 µm wide, skeletal hyphae solid to thick-walled, 3-7 µm wide, totally dominating in the tomentum, almost solid, at least in old specimens, in the trama with a more distinct lumen, binding hyphae very common in both the context and trama, hyaline, thick-walled to solid, tortous and much branched, up to 10 µm wide (in KOH) in the context with thin and whip-like branches, in the trama with stouter branches and below the subhymium with straight, thick-walled, swordlike branches, more or less parallel, partly pointing into the lower part of the hymenium, but in fertile specimens never above the dense palisade of basidia.

Cysitida none, but in collapsed hymenia the sword-like branches of the binding hyphae may easily be mistaken for thick-walled cystidia unless a careful examination is undertaken. Basidia clavate and with 4-sterigmate, 15-20 µm long.

Basidiospores cylindrical, often slightly bent, hyaline, thin-walled and IKI negative, 5-6 x 2-3 µm. Sexuality: Heterothallic and tetrapolar. Substrata: On hardwoods, preferably on *Betula sp.,* but recorded form any genera, occasionally also on conifers such as Picea, Pinus, Thuja, Tsuga.

Remarks: The species is easy to recognize because of the hirsute to tomentose-zoned cap and the lamellate hymenophore. (7, 32, 15, 6, 36)

6.4 *Fomes fomentarius* (L.:Fr.) Kickx.

Flore Crypt. Flandres 2:237, 1867.- *Polyporus fomentarius* L.:Fr., Syst. Mycol. 1:374, 1821

Wood decay fungus, medicinal and anti-cancer. Type of rot. White mottled heart rot of living hardwoods. Continues fruiting on dead and fallen trees. Fruiting body contains polysaccharide, such as 5, 6-Dimethoxyphthalide and 6-Carbomethoxyphthalide etc., can be used for medical purposes. In Chinese medicine it's said to be a cure for cancer. (Chen, Collection of Mushroom Prescriptions, 339-340, 2001) Basidiocarps perennial sessile, ungulate, up to 15 cm wide, tough, woody upper surface of cap quickly developing a hard glabrous crust, older part gray, zonate and shallowly sulcate, marginal part light brown, also zonate, minutely tomentose, pore surface concave, pale brown, the pores 4,5 per mmm with thick tomentose dissepiments, context yellowish-brown, tough-fibrous, azonate, up to 1 cm thick, granular core of varying size developing at upper part of the context next to the substrate, mottled with a mixture of pale and darker areas, tube layers indistinctly stratified, comprising most of the interior tissue of the basdiocarp, light brown and becoming stuffed with white mycelium, context tissue usually a relatively thin layer between the surface curst and the old tube layers.

Hyphal system trimitic, contextual generative hyphae thin-walled, nodose-septate, 2.4 μm in diameter, inconspicuous, contextual skeletal hyphae thick-walled, pale yellowish brown in KOH, aspetate, 3-8 mcrometer indiam, ocntexutal binding hyphae thick-walled, much-branched, aseptate, 1.5-3 mcrometer in diam, granular core a mixture of binding hyphae, narrow skeletal hyphae, and irregularly shaped, thick-walled sclerids that are brown in KOH, tramal hyphae similar, except for sclerids.

Cystidioles thin-walled, fusoid, 24-37 x 3.5-7.5 μm, with a basal clamp, also cystidia-like elements near dissepiment edges, thaese up to 120 μm long and 3.5 μm in diameter, projecting to 55 μm, some lightly incrusted. Basidia with swollen base, 4-sterigmate, 23-25 x 7-9 μm with a basal clamp. Basidispores cylinfdric, hyaline, smooth, negative in Melzer's reagent, 12-18 (-20) x 4.7 μm.

Substrata: Living and dead trees in several genera of hardwoods, particularly common on *Betula* and *Alnus*. Distribution: Transcontinental in Canada and the northern US, in the eastern mountains south of Kentucky and N. Carolina.

Remarks: *F. fomentarius* is a cosmopolian species in the boreal regins of the world. It is quite similar to *F. fasciatus*, a southern and stubropical species in America, but has larger spores. The distinctive sclerids in the granular core of both species have not been reported in earlier literature (15, 7, 32, 6, 36).

6.5 *Trametes versicolor* (L. ex Fr.) Pilat.

Atl. Champ. Eur. 3:261, 1936.- *Boletus versicolor* L., Sp. Plant., p.1176, 1753.- *Polyporus versicolor* L.:Fr., Syst. Mycol. 1:368, 1821.- *Coriolus versicolor* (L.:Fr.) Quel., Ench. Fung., p.175, 1886.

Type of rot-White rot of dead hardwoods. In the US it's called "turkey's tail". Fruiting bodies have 14 different types. Basidoicopars annual, sessile or effused-reflexed, dimidiate, often in large imbricate clusters; upper surface hirsute to tomentose, highly variable in color, with sharply contrasted concentric zones of various shades of brown, buff, reddish-brown or bluish colors; pore surface cream-colored to cinereous, the pores angular to circular, 4-5 per mm, dissepiments thick; context cream-colored tough-fibrous, with a think black layer below the surface tomentum, up to 5 mm thick; tube layer concolorous and continuous with the context, up to 3 mm thick. Hyphal system trimitic; contextual generative hyphae thin-walled, with clamps, 2.4-3 μm in diameter; contextual skeletal hyphae thick-walled, nonseptate, 4-6 μm in diameter; contextual binding hyphae thick-walled, nonseptate, much branched, 2-4 μm in diameter; tramal hyphae similar.

Cystidia or other sterile hymenial elements lacking. Basidia clavate, 4-steigmate, 15-20 x 4-5 μm, with a basal clamp. Basidiospores cylindric, slightly curved, hyaline, smooth, IKI, 5-6 x 1.5-2 mm. Positive in gum guaiac solution. Sexuality-Heterolthallic and tetrapolar.

Substrata: Dead wood of numerous gerna of hardwoods, occasionally on conifers.

Distribution: Throughout forest regions of North America. Circumglobal species.

Pharmacology: Pharmacological activities that may be due to the protein-bound polysaccharide PSK include the inhibition of sarcoma 180 (Hirase et al, 1976a; Ueno et al, 1978;Yan, 1985); improvement in the functioning of blood vessels (Ito and Kidaka, 1980a); support of hepatic function (Ito and Hidaka, 1980b); restoration of serum lysozyme content and normalization of spleen index in irradiated mice (Cai et al, 1987); immune function enhancement (Iwaguchi, 1985), and the possible prevention of liver cancer (Wang, 1989). Against lethal cytomegalovirus infection, the action of PSK appears to be through NK cell activation (Ebihara and Minamishima, 1984). Also, nitrogen-containing polysaccharides extracted from *T. versicolor* mycelia increase antibacterial potency and prolong antibacterial effects of antibiotics and can increase antibiotic sensitivity in antibiotic-resistant bacteria (Kureha Chemical Industry Co., 1978).

Medical effects: Tumor-resistant, strengthens the immune system, protects the liver system, proven to protect the kidney. Contained in 14 Chinese prescriptions and many non-prescription drugs: CVP, PSP, etc. (Chen, Collection of Mushroom Prescriptions, 339-340, 2001) (15, 7, 32, 6, 36)

6.6 *Fomes pinicola* (Swartz:Fr.) Karst.

Type of rot. Brown cubical rot of living and dead conifers and hardwoods.
Krit. Finl. Basidsv., p. 306. 1889. *Boletus pinicola* Swartz, Svenska Vetensk.-akad. Handl. 1910, p.88. 1810. *Polyporus pinicola* Swartz:Fr., Syst. Mycol. 1:372, 1821.
Basidiocarps perennial, usally sessile, rarely effused-reflexed or resupinate, woody, applanate to ungulate, up to 38x20x15cm upper surface at first with a sticky reddish brown resinous layer this often persisting over the younger marginal areas, becoming glabrous, sometimes laccate, grayish to brown or blackish, smooth to sulcate, pore surface cream colored, the pores ciruclar, 5-6 per mm, with thick, entire disspeiments, context cream colored to buff, corky to woody, azonate to zonate, up to 12 cm thick, tube layers indistinctly stratified, concolorous with the context, sometimes separated by a thin layer of context, up to 6 cm thick. Hyphal system trimitic, contextual generative hyphae thin-walled, with clamps, 2-5 μm in diameter, contextual skeletal hyphae thick-walled, hyaline, thick-walled, nonseptate, much branched, 1-5-4 μm in diameter, tramal hyphae similar. Cysitida hyphoid, often thick-walled at the base, thin-walled toward the apex, attenuated at the tip or not, up to 150 mm long, 3-10 mm in diameter, projecting up to 90 mm. Basidia short-clavate, 4-sterigmate, 17-24 x 7-8.5 mcm with a basal clamp.
Basidiospores cylindric-ellipsoid, hyaline, smooth, IKI-, 6-9 x 3.5-4.5 mm. Conspicuous white mycelial felts develop in shrinkage cracks of the decayed wood, negative in gum guaiac solution. Sexuality: Heterothallic and bipolar.

Substrata: Common in dead conifers and occasionally causing heart rot of living conifers, a major heart rot fungus in black cherry, also occasionally on aspen and birch.
Distribution: Found throughout the coniferous forest regions of North America but absent or rare in the southern pine region.

Remarks: *F. pinicola* is one of the most widely distribute polypores in coniferous forest regions of the Northern Hemisphere. It is a major factor in the production of brown rot residues that are a stable soil component in coniferous forest ecosystems. (15, 7, 32, 6, 36)

6.7 *Ganoderma applanatum* (Pers.) Pat.
Wood decay fungus, medicinal and anti-cancer. Type of rot. White mottled root and butt rot of living aspen, also found on dead standing or fallen hardwoods. Basidiocarps perennial, sessile, woody to corky, applanate, really ungulate, pilius surface crustose, grayish to black or brown, usually covered with a layer of chocolate-brown spores, appearing dusty, irregular to tuberculate, margin thin, white, context purplish brown, corky, pore surface white on fresh specimens, quickly bruising brown when handled, becoming dull buff with age, pores circular, 4-5 per μm, tube layers concolorous with context, separated by a layer of context tissue, each layer up to 13.5 mm thick.
Hyphal system trimitic, contextual generative hyphae inconspicuous, thin-walled, with clamps, 2-5 μm in diameter, skeletal hyphae thick-walled, brown aspetate, 3-6.5 μm in diameter, occasionally branched, the extremities tapering to acute apices, binding hyphae few.

Cystidia and other sterile hymenial elements lacking. Basidia clavate, 4.-sterigmate, some tapering to a narrow base, 20-45 x 8-10 morometer, arising from intercalary positions on the subhymenial hyphae, with a basal clamp.
Basidiospores ovoid, truncate at the distal end, with two walls, connected by inter-wall pillars, brown negative in Melzer's reagent. (8) 9-12 x 6.5-8.....[enclosure]
Landis and Eans examine dwindthrown aspen following a severe storm in Colorado. They found 86% of the windthrows exhibited signs of G. applanatum at the base.
Substrata: Dead standing trees, stumps, and living trees of numerous genra of hardwoods, particularly common as a cause of root and butt rot of aspen (*Populus tremuloides*). Also common on conifers in the Pacific Northwest and on conifers elsewhere. Distribution . Apparently throughout the forest regions of the US and Canada. Cosmopolitan species.
Remarks: *G. applanatum* is commonly known as "artist's fungus" or "artist's conk" because drawings can be made on the fresh pores surface with a sharp instrument. The bruised tissue undergoes an immediate oxidation and turns brown permanently, azonate or with a few broad, weakly sulcate

zones, the inner ones usually smoother than *Armillariella mellea* (Vahl ex Fr.) Karsten often (15, 7, 32, 6, 36)

6.8 *Melmpsoridium betulinum* (Desm.) Kleb.
Leaf rust (9)

6.9 *Microsphaeria penicillata*
Powdery mildew

6.10 *Cortinarius sp.*
University of Alaska campus, plot XIIIA, M. M. Chen, 8/10/1990, specimen # AKMC 1020.
Cap fleshy, 10 cm light brown to tawny; stipe slender 9 x 1.1 cm stuffed then hollow with grey to light dark silk-fibrillose and rusty spots and scattered; gills browny somewhat separated; spores elliptical with apicule 8.4-12.6 x 5.6-7.0 μm and spores print dark-brown. (3, 27, 36)

6.11 *Cortinarius sp.*
Capmad to University of Alaska campus roadside, M. M. Chen, 8/10/1990, specimen # AKMC 1030 (With spores print).
Cap 5 x 4 cm brightly yellowish brown when moist viscid and yellowish glabrous; stipe long 4.5 cm thick 1.5 cm hollow, silky-fibrillose; gills brown to cinnamon, equal, close gills brown and sinuate; spores average 5.6-12.6 x 5.6-8.4 μm warted, similar it is a common species in White Spruce-birch forests. (3, 27, 36)

6.12 *Cortinarius* subg. *Myxacium cf. mucosus.*
Geophysics Institute and Arctic Health Bldg. Fairbanks, Alaska, M. M. Chen and Richard Parmemter, 8/13/1990, specimen # AKMC 1050.
Cap small to middle size 2.5-4.5 x 3.0-6.5 cm when moist viscid and cinnamon brown with smooth shine and corrugated, velvety, stipe, 0.8-1.0 cm light brown with some what swollen at base cortinae forming superficial hairy-fibrillose on upward stalk stained cinnamon-brown by spores, gills dark rusty- brown when young cover with silky grayish cobweb by veil or cortina; spore 8.4-16.8 x 5.6-8.4 μm very light yellow and ellipsoid; often seen in birch forest.

6.13 *Cortinarius sp.*
Geophysics Institute and Arctic Health Bldg. Fairbanks, Alaska, M. M. Chen, Richard Parmeter, 8/13/1990, specimen # AKMC 1052.
Cap 1.5-2.0 cm viscid when wet, corrugated when dry, pale violet, lilac-white, bell-shaped umbonate and brownish central umbilicate, stipe 4.0 cm half upward, color same as cap; gills universal veil forming thin, soft, brown silky sheath cover gills and stained by brown spores; spores first violaceous, soon to rusty brown color, obovate or ellipsoid 5.6-11.2 x 5.6-7.0 μm on campus birch forest ground. (3, 27, 36)

6.14 *Leccinum aurantiacum* (Fr.) S. F. Gray (3, 27, 36)
Plot XIII, Lorson, M. M. Chen, 8/10/1990, speciemen # AKMC 1039.
Edible ectomycorrhizae. For detail see Aspen 5.18.

6.15 *Lyophyllum decastes* (Fr.) Sing., Agaricales 165. 1949; - *Agaricus decastes* Fr., Obs. Myc. 2:105. 1818; - *Lyophyllum aggregatum* (Schaeff. ex Secr.) Kuhn., Konr. & Maubl., Agaricales 1:366. 1948.
Fried Chicken Mushroom (Fig.6-3)
University of Alaska campus, plot XIIIA, M. M. Chen, Richard Parmeter, J. H. McBeath, 8/15/1990, specimen # AKMC 1058, 1063.
Edible – delicious. Top quality wild mushroom, also has medicinal value. Property: tumor-inhibition (*Medicinal Mushrooms*, Arora, Kamasuka). Cap 6-12 cm across convex then expanded becoming

wavy; gray-brown to yellowish brown with silky or silvery streaks; smooth, soapy. Gills adnate to slightly decurrent or sinuate, moderately broad, unequal, close to crowded, attenuating toward the margin; white to grayish. Stipe 3-10 x 1-2.5 cm, thick, solid, glabrous, white, pruinose at apex, the outer rind subcartilaginous; spores globose, smooth, hyaline, 5-6.5 μm in diameter, with alarge central gutta; odor mild, taste mild. Flesh firm, white. Habitat in clusters on the ground in waste places, grassy areas, wood edges, and paths. Common and sometimes abundant. Found widely distributed in Fairbanks, Alaska. Comment: This is a much sought after edible mushroom but should be tried with caution as there have been some reports of gastric upset. (27, 3, 36)

6.16 *Boletus edulis* Bull. Ex Fr. (Fig.6-4)
University of Alaska campus, plot XIIIA, M. M. Chen and Richard Parmeter, 1990, specimen # AKMC 1053.
Nicknamed King Bolete and Porcini. Ectomycorrhizae, very delicious edible mushroom, also medicinal and anti-cancer. Many varieties. Property: tumor inhibition. Traditionally called by the Chinese "delicious beef liver mushroom" (Its color is like beef liver). Mushroom meat tender and crisp. Odor: pleasant. Mycorrhizae on oak (*Quarcus*). Contains PA1 and PA2, mannose. Contained in 9 Chinese prescriptions. (*Collection of Mushroom Prescriptions*, Chen, 1999). Cap 8-25 cm, broad, convex, dry, glabrous, yellowish brown; flesh thick, white, unchanging, taste pleasant; stipe subequal or enlarged below, 6-12 cm long, 2-3 cm thick, paler than the cap, delicately reticulate in the upper portion; tubes yellow, deeply depressed around the stipe, mouths small, 2-3 per mm; spores yellow, smooth, fusiform-elliptic, 12-17 x 4.5-5.5 μm. (27, 36)

6.17 *Amanita muscaria* var. muscaria (L. ex Fr.) Pers.
Plot XIIIA, M. M. Chen, 8/13/1990, specimen # AKMC 1054.
In July and August, this species can be seen everywhere in Fairbanks, Alaska.
(See 4.24)

6.18 *Lactarius sp.*
M. M. Chen, 8/13/1990, specimen # AKMC 1049.
Ectomycorrhizae. Cap 3.3-4.0 cm, when moist viscid and glabrous, first red-brownish then become brown egg shell color, margin at first in carved; stipe 2.5-2.8 cm long and 0.8-1.3 cm thick forming hairy-fibrillose on stalk and upward remain cinnamon-brown cortinae; gills light yellowish-brown adnated to sinuate, fairly close or rather well spaced, between margin and gills covered the thickly and light yellowish brown cortinae; spores obovate hyalin 8.0-8.5 μm with warts. (3, 27, 36)

6.19 *Russula sp.*
Plot XIIIA, M. M. Chen, 8/10/1990, specimen # AKMC 1037.
Ectomycorrhizae. Seen in birch forests. Cap 10 cm, chestnut-brownish on center yellowish at margin; stipe brawny 7.5 cm long, 3.7 cm thick, smoothly no cortina and lacking basal bulb, stipe hollow in center, upward and downward equal, egg shell color; gills not attached or free, close, spores unknown. (3, 27, 36)

6.20 *Russula rosacea* (Bull.) Fr., Epicr. 351, 1838; Bres., Icon. Myc. 9:405, 1929; Teng, Sinensia 7:550, 1936. *Agaricus rosaceus* Bull., Herb. Fr. pl. 509. 1790; Pers., Syn. Fung. 439, 1801.
University of Alaska campus, M. M. Chen, 7/31/1990, Specimen # AKMC 1001, 1012.
Edible. "Bitter russula" or "rose russula". Cap 4-9 cm, broad, convex, then expanded, finally depressed, subviscid, about jasper-red when fresh, drying Pompeian-red; margin even; flesh white; stipe subequal, solid, white, even, rosy red in places, 4-7 cm long, 1 -2 cm thick; gills white, close, equal, narrowly adnexed; spores hyaline, white in mass, echinulate, subglobose, 7-8 x 6-7 μm in diameter; cystidia abundant, fusoid, 50-85 x 12-20 μm; taste bitter and acrid. Spore print milk yellow. Often grow in clusters or scattered in Fairbanks birch / spruce mixed forests ground.

6.21 *Hebeloma sp.*
University of Alaska, Geophysical Institute and Arctic Health Bldg., M. M. Chen, 8/13/1990, specimen # AKMC 1048.
Cap 9-10 cm, broad, coin-shaped at first then becoming convex, then umbonate to plane, surface smooth markedly hygrophanous center, brownish marginal egg shell to violet-tinged; stipe 7 cm long, 1.3 cm thick, tined color smoothly, free or adnate some what bulb or swollen with base; gills not attached or free olive-brown, fairly well-shaped every gill recurvate when dry; spores 7.0-11.2 x 4.2-5.6 μm; elliptical no warts, often seen in birch forests. (Videotaped by Richard Parmeter) (3, 27, 36)

7 COTTONWOOD-BALSAM POPULA (POPULUS BALSAMIFERA)

Closed spruce/Cottonwood, the Alaska, balsam poplar type, (Fig.7-1) is an importance hardwood forest in the interior which reaches its greatest size and abundance on the flood-plains of the meandering glacial rivers. It invades sandbars and grows rapidly to heights of 24-30 m. and diameters of 60 cm before being replaced by White Spruce. Commercial stands occupy 840,000 hectares, primarily along the Tanana River and others rivers, the wood plants of this type include White Spruce.
Picea glauca/ Viburnum edule/Equisetum arvense/Hylocomium splendens (FP3A), Black cottonwood (*Populus trichocarpa*), White Spruce (*Picea glauca*), American greeen alder (*Alnus tenuifolia*), Stika alder (*Alnus sinuata*), Littletree willow (*Salix arbusculoides*), Feltleaf willow (*Salix alaxensis*), High bushcranberry (*Viburnum edule*)

Roots in flooded areas die of anoxia (oxygen deficiency). Damage occurs not only to plants on obviously wet sites but also to those in planting holes along the city streets and in landscapes where soil drainage is impeded by high clay content. Most trees and shrubs cannot grow for long in waterlogged soil, and some perish if flooded for only a few days during the growing season. Mycorrhizal fungi, which associate with plant roots symbiotically are also adversely affected, further suppressing plan uptake of mineral nutrients, especially phosphorus. Internal water deficit in some plants increases until they die, but many kinds of plants regain the normal degree of hydration while their stomata remain closed during flooding. Stomata of some tolerant plants reopen as the plant adapts to flooding.Tolerant: black and sandbar willows and cottonwood Intermediate tolerant to flooding. (37, 6, 32)

Cottonwood wood decay Fungi:
7.1 *Polyporus elegans* Alnus Bull.: Fr.
Epicr. Syst. Mycol. P.440, 1838.
Type of rot. White rot of dead hardwoods, poisitve in gum guaiac solution. Basidiocarps annual, centrally to laterally stipitate, usually solitary; cap cirucular to reniform or flabelliform, up to 6 cm wide and 1 cm thick, usually 2-3 mm thick, upper surface tan to chestnut brown, azonate, glabrous; margin concolrous; stipe up to 7 cm long and 5 mm thick, black at the base, upper part concoorous with cap, glabrous; pore surface pale buff, the pores circular to angular, 5-7 per mm; dissepiments entire; context pale buff, ocrky, zonate, up to 7 mm thick; tube layer concolrous with context or slightly darker, up to 2 mm thick. No radial lines across enterie cap but may be netate edge.

Hyphal system dimitic; contextulay genreateive hyphae thin-walled, with clamps, 2.5-5 mcm in diam; contextual binding hyphae thick-walled nonseptate, much branched, 2.5-5 mcm in diam; tramal hyphae similar.
Cystidia or other sterile hymenial elements lacking. Basidia broadly clavate, 4-sterigmate, 15-20 x 6-7 mcm, with a basal clamp.
Cultural characteristics: Unknown.
Sexuality: Unknown.
Substrata: Dead hardwoods occasionally on conifers.
Distribution: Widely distributed throughout forest regions of North America.

Remarks: Although they occasionally grow up to 9 cm in diameter, basidiocarps of *P. elegans* usually range from 1-4 cm wide, and are usually smaller than those of *P. meanopus* and *P. badius*. In addition, *P. melanopus* differs in its terrestrial habit and a black, velvety stipe, *P. badius* in the dark brown or balckish-brown upper surface and lack of clamps, and P. varius in the radially striate cap. (15, 32, 23, 36)

7.2 *Spongipellis delectans*

Type of rot. White mottle rot of living or dead hardwoods. Basidiocaprs sessile or slightly effused, pilei solitary or imbricate, dimidiate, applanate to ungulate, up to 7 x 15 x 4.5 cm; upper surface white, discoloring to pale brownish and often streaked with light reddish-brown after drying, zonate tomentose or short-hispid to glabrous; margin concolrous; ore suface white when fresh, becoming pale buff to ochraceous, the pores large, circular to angular, sometimes daedaleoid, 1-2 mm; with thin dissepiments that soon become lacerate; context e white to ochraceous, very faintly zonate, corky below, soft, spoingy aove up to 2 cm thick; tube layer concolorous and continuous with te context, up to 1 cm thick.

Hyphal system monomitic; contextual hyphae hyaline in KOH, thin to thick-walled, occasionally branched, with clmaps, 4-7 mcm in diam, tramal hyphae similar.

Cystidia or other sterile hymenial elements absent.

Basidia clavate, 4-sterigmate, 20-30 x 7-9 mm, with a basal clamp.

Basidiospores broadly ellipsoid to subglobose, hyaline, smooth, IKI- , 7-9 x 5-7 mcm.

Sexuality: Heterothallic and tetrapolar.

Substrata: Dead standing or fallen hardwoods in several genera, also causing a heart rot in living trees, particularly species of *Populus*.

Distribution: Throughout eastern hardwood forests from Canada to the southern oak-pine forest; also in the Pacific Northwest and Alaska. Also in Europe.

Remarks: Basidiocarps of *S. delectans* resemble those of *S. unicolor*, which differ in having narrower spores, and usually, larger pores. Trametes cervina also has macroscopically similar basdidiocarps but has cylindrical spores and dimitc hyhal system. (15, 23, 32, 36)

7.3 *Trametes pubescen* (Schum.:Fr.) Pilat

Atl. Champ. Europ. 3:268, 1939.- *Polyporus pubescens* Schum.:Fr., Syst. Mycol. 1:367, 1821.

Type of rot. With rot of dead hardwoods.

Basidiocaprs annual, sessile or effused-reflexed, up to 6 cm wide, dimidiate, often in imbricate cluster, thin, coriaceous; upper surface tomentose to finely pubescent or almost glabrous, cream color to warm buff, azonate or fiantly zonate; pore surface cream color to light ochraceous buff or sometimes becoming cinereous with age, the pores angular 3-5 mm in diameter, dissepiments becoming thin; context white to cream, tough-fibrous up to 5 mm thick; tube layer cream colored to pale buff, up to 4 mm thick.

Hypahl system trimitic; contextual genreative hyphae thin-walled rarely branche, with clamps 2-3 mcm in diam; contextual skeltal hyphae thick-walled with occasional branching, nonseptate, 5-10 mcm in diam; contextual binding hyphae thick-walled, non-septate much branched, 1.5-3 mcm in diam; tramal hyaphe similar.

Cystidia or other sterile hymenial elements lacking; hyphals pegs usually present.

Basidia clavate, 4-sterigmate, 14-18 x 4.5 –6 mcm, with a basal clamp.

Basidiospores cylindric, slightly curved, hyaoline, smmmoth, IKI-, 5-7 x 1.5-2 mcm.

Sexuality: Heterothallic and tetrapolar.

Substrata: Dead wood of numerous hardwood genera, rarely on conifers.

Distribution: Transcontinental species from the Artic to the southern states.

Remarks: The uniformly cream to buff, azonate and tomentose cap distinguishes basidiocarps of T. pubescents from those of other species in the so-called Coriolus group. The microscopic characteristics of all these are virtually identical. (15, 23, 32, 36)

7.4 *Trametes versicolor* (L. ex Fr.) Pilat 0254,0057

Type of rot-White rot of dead hardwoods. Positive in gum guaiac solution.
Basidoicopars annual, sessile or effused-reflexed, dimidiate, often in large imbricate clusters; upper surface hirsute to tomentose, highly variable in color, with sharply contrasted concentric zones of various shades of brown, buff, reddish-brown or bluish colors; pore surface cream-colored to cinereous, the pores angular to circular, 4-5 per mm, dissepiments thick; context cream-colored tough-fibrous, with a thick black layer below the surface tomentum, up to 5 mm thick; tube layer concolorous and continous with the context, up to 3 mm thick.
Hyphal system trimitic; contextual genrative hyphae thin-walled, with clamps, 2.4-3 µm in diameter; contexutal skeletal hyphae thick-walled, nonseptate, 4-6 µm in diameter; contextual binding hyphae thick–walled, nonseptate, much branched, 2-4 µm in diameter; tramal hyphae similar.
Cystidia or other sterile hymenial elemnts lacking.
Basidia clavate, 4-steigmate, 15-20 x 4-5 µm, with a basal clamp.
Basidiospores cylindric, slightly cureved, hyaline, smooth, IKI, 5-6 x 1.5-2 mcm.
Cultural characteristics: See Davidson etal.
Sexuality: Heterolthallic and tetrapolar.
Substrata: Dead wood of numerous gerna of hardwoods, occasionally on conifers.
Distribution: Throughout forest regions of North America. Circumglobal species.
Remarks: This is probably the most common wood rotting fungus on dead hardwoods throughout North America. Its basidiocarps are extremely variable in the color and other characters of the upper surface. *T. pubescens* basidiocarps are usually lighter in color, are thicker, and lack the blackish layer separating the upper tomentum. (15, 23, 32, 36)

7.5 *Trametes ochracea* (Pers.) Gilbn. & Ryv. Comb. Nov. Basionym: *Boletus ochraceus* Pers. Ann. Bot. (Usteri) 11: 29, 1794.- *Polyporus zonatus* Nees.:Fr., Syst. Mycol. 1:368, 1821.
Type of rot. With rot of dead hardwoods.
Basidiocaprs annual or reviving, sessile or effused-reglecsed, dimidiate to elongated, tough-forbrous; upper surface finely tomentase to almost glabrous, vinaceous-buff to avellaneous with zonaes of reddish brown (ferruginous) or pale buff with faint darker zones; pore surface cream clored to cinereous, the pores circular, 3-4 per mm, with thick dissepiments; context cream colored, tough-fibrous, azonate, up to 5mm thick; tube layer concolorous and continuous with the context, up to 4 mm thick.
Hyphal system trimitic; contextual generative hyphae thin-walled, with clamps, 2-3.5 mcm in diam; contextual skeletal hyphae thick-walled, nonseptate, 4-8 mcm in diam; binding hyphae thick-walled, non-septate, much branched, 2.5-5 mcm in diam; tramla hyphae similar.
Cystidia or toerh sterile hymenial elements lacking.
Basidia clavate, 4-sterigmate, 15-20 x 4.5 mcm, with a basal clamp.
Basidiospores cylindric, slightly curved, smooth, IKI-, 6-8 x 2.25 mcm.
Cultural charactersiscs…
Sexuallity: Heterothallic and tetrapolar.
Substrata: Dead hardwoods of numerous genera of hardwoods, rarely on conifers.
Distribution: Widely distributed in boreal forest regions of North America, south in the Rocky Mountain forests to Arizona. Also in Europe.
Remarks: Basidiocarps of T. ochracea are usually much paler in color than those of *T. versicolor*, less strongly zonate, and they lack the black layer seen in the uper context of the latter species. *T. pubescens* basidiocarps may also be similar but have an azonate or very faintly zonate upper surface. (15, 23, 32, 36)

7.6 *Hopalopilus nidulans* (Fr.) Karst.
Rev. Mycol. 3:18, 1881.- *Polyporus nidulans* Fr. Syst. Mycol. 1:362, 1821.- *Polyporus rutilans* Fr. Syst. Mycol. 1:368, 1821.- *Polyporus pallido-cervinus* Schw. Trans. Am. Phil. Soc. Ser. II. 4:156, 1832.
Type of rot. Causes a white rot.

Basidiocarps annual, pileate, broadly sessile to effused reflexed, mostly convex, often almost triangular in section, up to 10 cm wide and long, but usually smaller, up to 4 cm thick at the base, soft and watery when fresh, light and somewhat brittle when dry, all parts of the fruitbody light violet to purplish with KOH, upper surface cinnamon to ochraceous, first finely tomentose to scrupose with small adpressed tufts of hyphae, soon completely smooth, azonate or with a few broad, weakly sulcate zones, the inner ones usally smoother than the distal ones, margin acute and entire, pore surface ochraceous to cinnamon brown, usually with a distinct sterile edge towards the substrate and margin, pres thin-walled and angular, 2-4 per mm, the pore surface often with a few larger cracks in larger fuitbodies, tubes up to 10 mm deep, ochraceous or white due to cottony sterile hyphae, context light cinnamon, mostly distinctly lighter in color towards the cap, soft and fibrous and quite brittle, up to 4 cm deep at the base.
Hyphal system monomitic, generative hyphae with clamps, in the context large, up to 10 mcm wide and with conspicuous clamps, distinctly thick-walled and richly branched, mostly smooth, but also covered partly with amorphous substances mixed with poygonal, light pinkish to browish crystals, in the trama and hymenium more straight and narrow, up to 6 mcm wide.

Cystidia none fusoid cystidioles present, 18-22 x 4-5 mcm with a basal clamp.
Basidia clavate, 4-sterigmate 18-22 x 5-6.5 mcm with basal clamp. Basidiospores ellipsoid to cylidrical, hyaline, thin-walled smooth and nonamyloid, 3.5-5 x 2-2.5 (3) mcm.
Cultural characteristics: Unknown. Sexuality: Unknown.
Substrata: Dead hardwoods in numerous genera, very rarely reported from coniferous wood in eastern North America, but not uncommon in Pinus and Pseudotsuga in Arizona.
Distribution: Common in Eastern US and in the southwest, but rare elsewhere in Western North America. Circumglobal in the Temperate Zone.
Remarks: The species is usually easy to recognize by its cinnamon sappy fruitbody with a vivid violet reaction with KOH. (15, 23, 32, 36)

7.7 *Ciborinia folicola* 0020

7.8 *Ganoderma applanatum* (pers.) Pat.0253, 292 (b)
Type of rot. White mottled root and butt rot of living aspen, also found on dead standing or fallen hardwoods. Basidiocarps perennial, sessile, woody to corky, applanate, realy ungulate, cap surface crustose, grayish to black or brown, usually covered with a layer of chocolate-brown spores, appearing dusty, irregular to tuberculate, margin thin, white, context purplish brown, corky, pore suface white on fresh specimens, quickly bruising brown when handled, becoming dull buff with age, pores circular, 4-5 per m, tube layers concolorous with context, separated by a layer of context tissue, each layer up to 13.5 mm thick.
Hyphal system trimitic, contextual generative hyphae inconspicuous, thin-walled, with clamps, 2-5 μm in diam., skeletal hyphae thick-walled, brown aspetate, 3-6.5 μm in diam., occasionally branched, the extremities tapering to acute apices, binding hyphae few.

Cystidia and other sterile hymenial elements lacking. Basidia clavate, 4.-sterigmate, some tapering to a narrow base, 20-45 x 8-10 μm, arising from from intercalary positions on the subhymenial hyphae, with a basal clamp.
Basidiospores ovoid, truncate at the distal end, with two walls, connected by inter-wall pillars, brown negative in Melzer's reagent. (8)9-12 x 6.5-8 μm.
Landis and Eans examined windthrown aspen following a severe storm in Colorado. They found 86% of the windthrows exhibited signs of G. applanatum at the base.
Sexuality: Heterothallic and teterapolar.
Substrata: Dead standing trees, stumps, and living trees of numerous genera of hardwoods, particularly common as a cause of root and butt rot of aspen (*Populus tremuloides*). Also common on conifers in the Pacific Northwest and readily on conifers elsewhere.
Distribution: Throughout the forest regions of the US and Canada. Cosmopolitan species.

Remarks: *G. applanatum* is commonly known as the "artist's fungus" because drawings can be made on the fresh porous surface with a sharp instrument. The bruised tissue undergoes an immediate oxidation and turns brown permanently. (15, 23, 32, 36)

7.9 *Fomes fomentarius* 0046

White mottled heart rot of living hardwoods, continues fruiting on dead and fallen trees.

Basidiocarps perennial sessile, ungulate, up to 15 cm wide, tough, woody upper surface of cap quickly developing a hard glabrous crust, older part gray, zonate and shallowly sulcate, marginal part light brown, also zonate, minutely tomentose, pore surface concave, pale brown, the pores 4,5 per mm with thick tomentose dissepiments, context yellowish-brown, tough-fibrous, azonate, up to 1 cm thick, granular core of varying size developing at upper part of the context next to the substrate, mottled with a mixture of pale and darker areas, tube layers indistinctly stratified, comprising most of the interior tissue of the basidiocarp, light brown and becoming stuffed with white mycelium, context tissue usually a relatively thin layer between the surface crust and the old tube layers.

Hyphal system trimitic, contextual generative hyphae thin-walled, nodose-septate, 2.4 µm in diam, inconspicuous, contextual skeletal hyaphae thick-walled, pale yellowish brown in KOH, aseptate, 3-8 µm in diam, contextual binding hyphae thick-walled, much-branched, aseptate, 1.5-3 µm in diam., granular core a mixture of binding hyphae, narrow skeletal hyphae, and irregularly shaped, thick-walled sclerids that are brown in KOH, tramal hyphae similar, except for sclerids.

Cystidioles thin-walled, fusoid, 24-37 x 3.5-7.5 µm, with a basal clamp, also cystidia-like elements near dissepiment edges, these up to 120 µm long and 3.5 µm in diam, projecting to 55 µm, some lightly incrusted.

Basidia with swollen base, 4-sterigmate, 23-25 x 7-9 µm with a basal clamp.

Basidiospores cylindric, hyaline, smooth, negative in Melzer's reagent, 122-18(-20) x 4.7 µm. Living and dead trees in several genera of hardwoods, particularly common on Betula and Alnus. Distributed transcontinentally in Canada and the Northern US, in the eastern mountains south to Kentucky and N. Carolina. (15, 23, 32, 36)

7.10 *Bjerkandera adusta* (Willd.:Fr.) Karst.

Medd. Soc. Fauna Fl. Fenn. 5:38, 1897.- *Polyporus adustus* Willd.:Fr., Syst. Mycol. 1:363, 1821.

Type of rot. White rot of hardwod logs and slash. Occasionally on conifers, positive in gum guaiac solution.

Basidiocarps annual, sessile, effused-reflexed, or occasionally resupinate under logs, often in imbricate clusters, tough, reflexed up to 3 cm, upper surface of cap cream to buff, tomentose or strigose to glabrous with age, azonate or faintly zonate, pore suface gray to black, the pores angular, regular, 6-7 per mm, dissepiments thin, entire, context pale buff, azonate with distinct thin upper layer of tomentum up to 6 mm thick, tube layer smoky gray, distinct from context, up to 1 mm thick.

Hyphal system monomitic, contextual hyphae thin to moderately thick-walled, with abundant clamps, 3-5 µm in diameter with occasional branching, tramal hyphae similar but densely compacted and agglutinated.

Cystidia or other sterile hymenial elements lacking, dark brownish vascular hyphae sometimes present in subhymenium and in hymenial layer.

Basidia calvate to napiform, 4-sterigmate, 22-25 x 5-6 µm, with a basal clamp.

Basidiospores short-cylindric, hyaline, smooth, negative in Melzer's reagent, 5.6 x 2.5-3.5 µm.

Substrata: Many genera of hardwoods, rarely on conifers.

Distribution: Circumglobal in the Northern Hemisphere.

Remarks: The smoky gray to black color of the pore surface and the cream colored cap are distinctive field characters. Bjerkander fumosa is similar and differs in having thicker and wider basidiocarps and broader spores. Bjerkandera adusta is particularly common in aspen and is always a conspicuous fungus in older aspen stands. (15, 23, 32, 36)

7.11 *Trametes zonatella* 0147

7.12 *Fomitopsis pinicola* (Swartz:Fr.)P. Karst.

Type of rot. Brown cubical rot of living and dead conifers and hardwoods. Basidiocarps perennial, usally sessile, rarely effused-reflexed or resupinate, woody, applanate to ungulate, up to 38x20x15cm upper surface at first with a sticky reddish brown resinous layer this often persisting over the younger marginal areas, becoming glabrous, sometimes laccate, grayish to brown or blackish, smooth to sulcate, pore surface cream colored, the pores ciruclar, 5-6 per mm, with thick, entire dissepiments, context cream colored to buff, corky to woody, azonate to zonate, up to 12 cm thick, tube layers indistinctly stratified, concolorous with the context, sometimes separated by a thin layer of context, up to 6 cm thick.

Hyphal system trimitic, contextual generative hyphae thin-walled, with clamps, 2-5 µm in diam, contextual skeletal hyphae thick-walled, hyaline, thick-walled, nonseptate, much branched,, 1-5-4 µm in diameter, tramal hyphae similar.

Cysitida hyphoid, often thick-walled at the base, thin-walled toward the apex, attenuated at the tip or not, up to 150 mcm long, 3-10 mcm in diam, projecting up to 90 mcm. Basidia short-clavate, 4-sterigmate, 17-24 x 7-8.5 mcm with a basal clamp.

Basidiospores cylindric-ellipsoid, hyaline, smooth, IKI-6-9 x 3.5-4.5 mcm.

Conspicuous white mycelial felts develop in shrinkage cracks of the decayed wood, negative in gum guaiac solution.

Sexuality: Heterothallic and bipolar.

Substrata: Common in dead conifers and occasionally causing heart rot of living conifers, a major heart rot fungus in black cherry, also occasionally on aspen and birch.

Distribution: Throughout the coniferous forest regions of North America but absent or rare in the southern pine region.

Remarks: *F. pinicola* is one of the most conspicuous and widely distributed polypores in coniferous forest regions of the Northern Hemisphere. It is a major factor in the production of brown rot residues that are a stable soil component in coniferous forest ecosystems. (15, 23, 32, 36)

7.13 *Gloeoporus dischrous* (Fr.) Bres.

Ann. Mycol. 14:230, 1916. – Polyporus dischrous Fr., Syst. Mycol. 1 :364, 1821.

Type of rot. – Causes a white rot.

Basidiocarps annual, resupinate to pineate, often effused-reflexed, mostly imbricate with several small shelf-like, narrow and elongated pilei rarely above 4 cm wide, 10 cm long and 5 mm thick at the base, upper surface white to cream, first finely tomentose, later more scrupose to smooth or hispid with tufts of hyphae according to weathering and active periods of growth, with concentric zones in different shades, margin sharp and undulating, pore surface, pores round to angular, 4-6 per mm, often not more than a reticulate pattern, up to 1 mm deep, hymenium developed over the dissepiments as in Merulius, concolorous with the pore surface, tube layer gelatinous and tubes rubbery when fresh, resinous to horny when dry and old, above the tubes a thin and distinct zone of the same color and consistency as the tubes, context pure white, up to 4 mm thick, cottony to loose, distinctly thicker than the tubes.

Hyphal system monomitic; generative hyphae in the context distinct and thick-walled with large clamps, up to 6 µm wide, moderately branched, the tubes and the resinous zone above the tubes strongly agglutinated, thin-walled and mostly collapsed in dry specimens, up to 3.5 µm in diameter.

Cystidia and other sterile hymenial elements absent. Basidia clavate, 4-sterigmate, 14-20 x 3-4 µm, with a basal clamp.

Basidiospores allantoid to cylindrical, hyaline, thin-walled, smooth, negative in Melzer's reagent, 3.5-5.5 x 0.7-1.5 µm.

Cultural characteristics: See Davidson et al. 1938, Nobles 1948, 1965, Bakshi et al. 1969.

Sexuality: Heterothallic and tetrapolar, reported by Westhuizen (1971), David (1972) and Ginns (1976).

Substrata: Reported from dead wood of numerous genera of hardwoods, occasionally on conifers and dead polypores. In North Europe commonly found on Betula sp. Attacked by Inootus obliquus, but this type of connection has not been reported from North America (1, 3, 8, 20, 25, 26, 27, 153, 165, 190, 191, 194, 195).

Distribution: Very widespread throughout North America. Cosmopolitan species, but reports from the tropics should be treated with caution as it may be confused macroscopically with dark specimens of G. thelephoroides, which differs in having simple septate hyphae.

Remarks: Usually this species is easy to recognize because of the deep reddish pore surface and the white and cottony context and cap. It is separated from both *G. taxicola* and *G. thelephoroides* by its clamps on the generative hyphae. In fresh condition the gelatinous to rubbery pore layer is rather characteristic and may be peeled off the fungus with a fingernail. (15, 23, 32, 36)

7.14 *Cerrena unicolor* (Bull.: Fr.) Murr.

J. Mycol. 9:91, 1903.- *Daedalea unicolor* Bull.:Fr., Syst. Mycol. 1:336, 1821.

Type of rot. White rot of dead hardwoods. Basidiocarps annual, sessile, effused-reflexed or rarely resupinate, pilei often in imbricate clusters, dimidiate, up to 10 mcm wide, upper surface pale brownish to gray, hirsute to almost glabrous, often green due to algae, sucate, pore surface ivory to pale buff on young specimens, becoming darker with age, the pores daedaleoid, variable, 3-4 per mm, in parts larger, dissepiments at first thick and tomentose, becoming thin and splitting, context duplex, up to 3 mm thick, corky, pale brown lower layer separated from soft, spongy, darker upper layer by a thin dark zone, tube layer continuous with lower context, up to 1 cm thick.

Hyphal system trimitic, generative contextual hyphae thin-walled nodose-septate, 2-4 mcm in diam, binding hyphae thick-walled, aseptate, much branched, 2-4 mm in diameter; tramal hyphae similar.

Cystidia hyphoid, thin-walled 40-60 x 4-5 mm with a basal clamp.

Basidia clavate, 4-sterigmate, 20-25 x 5-6 mm, with a basal clamp.

Basidiospores cylindric-ellipsoid, hyaline, smooth, negative in Melzer's reagent, 5-7 x 2.5-4 mcm.

Sexuality: Heterothallic and bipolar.

Substrata: Dead wood of many genera of hardwoods, rarely on conifers.

Distribution: Widely distributed in North America. Widespread in Asia and Europe.

Remarks: Cerrena unicolor was identified as the fungal symbiont of the wood wasp *Tremex columba* on *Fagus grandifola* in eastern Canada by Stillwell. This is the only polyporc symbiont of a wood wasp. *C.unicolor* is easy to recognize because of the hirsute cap, the black line in the context and the labyrinthine hymenophore. (15, 23, 32, 36)

7.15 *Spongipellis delectans* (Pk.) Murr.

North Am. Flora 9:38, 1907. - Polyporus delectans Pk., Bull. Torrey Bot. Club 11:26, 1884.

Type of rot. - White mottled rot of living or dead hardwoods.

Basidiocarps sessile or slightly effused, pilei solitary or imbricate, dimidiate, applanate to ungulate, up to 7 x 15 x 4.5 cm; upper surface white, discoloring to pale brownish and often streaked with light reddish-brown after drying, azonate, tomentose or short-hispid to glabrous; margin concolorous; pore surface white when fresh, becoming pale buff to ochraceous, the pores large, circular to angular, sometimes daedaleoid, 1-2 per mm; with thin dissepiments that soon become lacerate; context white to ochraceous, very faintly zonate, corky below, soft, spongy above, up to 2 cm thick; tube layer concolorous and continuous with the context, up to 1 cm thick.

Hyphal system monomitic; contextual hyphae hyaline in KOH, thin-to thick-walled, occasionally branched, with clamps, 4-7 µm in diam, tramal hyphae similar.

Cystidia or other sterile hymenial elements absent.

Basidia clavate, 4-sterigmate, 20-30 X 7-9 µm, with a basal clamp.

Basidiospores broadly ellipsoid to subglobose, hyaline, smooth, IKI-, 7-9 X 5-7 µm.

Cultural characteristics: See David 1969; Staplers 1978.

Sexuality: Heterothallic and tetrapolar (David 1969).

Substrata: Dead standing or fallen hardwoods in several genera, also causing a heart rot in living trees, particular species of Populus.
Distribution: Throughout eastern hardwood forests from Canada to the southern oak-pine forest; also in the Pacific Northwest and Alaska. Also in Europe.

Remarks: Basidiocarps of S. delectans resemble those of S. unicolor, which differ in having narrower spores, and usually, larger pores. Trametes cervina also has macroscopically similar basidiocarps but has cylindrical spores and a dimitic hyphal system. (15, 23, 32, 36)

Canker
7.16 *Cytospora chrysosperma* (Pers.)Fr.
Canker of cottonwood is subject to *Cytospora chrysosperma*, the conidial states of *Valsa sordida*. This fungus is an opportunistic pathogen, quick to attack plants that have been predisposed by heat, drought, winter damage, or infection by other pathogens. It also colonizes dying or dead bark as saprophytes. The fungus causes these symptoms: rapidly spreading necrosis of weakened bark or, on stems of intermediate susceptibility, localized annual cankers or slowly expanding perennial cankers. It occurs around the Northern Hemisphere, may also cause losses in nursery seedbeds, storage, newly established forest plantations, and landscape or shelterbelt plantings. Cankers on branches often arise at the bases of dead twigs. New infections in smooth bark appear as brownish sunken patches that may girdle the stem. Blackstem disease of cottonwood seedlings and cuttings, attributed primarily to *V. sordida*, causes severe losses in storage and sometimes in nursery beds. Most infections are assumed to be initiated by conidia because conidia are much more abundant than ascospores and are liberated throughout the year except perhaps during midwinter. Buds, nodes, lenticels, and wounds of all kinds are sites of infection.

Tree breeders have searched for poplar species and clones resistant to *V. sordida*, and some workers have reported heritable resistance. To date, however, heritable resistance has not been distinguished from heritable avoidance of predisposing stress at test sites. The best approach to control of this disease remains the planting of stock known to be well adapted to the planting site.

Rust disease
7.17 *Melampsora medusae* 0131,0135
The most widespread and important of the Melampsora rusts in Alaska inland and North America is *M. medusae* (syn. *M. albertensis*), which occurs throughout nearly the entire range of poplars around the world. It causes the leaves of highly susceptible clones to shrivel and drop prematurely, reducing growth. Growth loss due to rust, often masked by the intrinsically rapid growth of poplars, can be detected by comparing growth of infected trees to that of non-infected ones protected by fungicides. In one test involving natural infection, the average annual growth loss (wood volume) of five clones was 31-42%, and the volume loss in highly susceptible clones ranged up to 57%. The spermagonial and aecial states of *M. medusae* develop on cottonwood. The cycle and features of *M. medusae* typify host-alternating Melampsora species. In spring, basidiospores from telia in dead cottonwood leaves on the ground infect young current-year needles of coniferous hosts.

Resistance of cottonwood to *M. medusae* is under strong genetic control. Therefore, selection and breeding programs now emphasize rust resistance as well as rapid growth. In general, resistance is most strongly expressed in warm weather, for example at 25 degrees Celsius. Many races of *M. medusae* exist that vary in virulence patterns on cottonwood clones.

Powdery mildew
7.18 *Uncinula salicis*
The powdery mildew fungi are a family of ibobligahely parasitic Ascomyestina, the Erysiphaceae, the growth is mainly on the surface of cottonwood leaves and on other tender aerial plant parts and spores are produced on wefts or mats of white or light-colored mycelium. This superficial growth often appears powdery. Most of these fungi penetrate and parasitize only the epidermal cells of their hosts, producing within the living host cell an absorbing structure (haustorium) that diverts water and food

materials to the fungus. Powdery mildews are unimportant in young cottonwood forests with large trees.

The characteristic fruiting structure of powdery mildew fungi is the cleistothecium, a tiny sphere usually 0.1-0.2 mm in diameter that forms on or in the mycelial mat on the plant surface and at maturity liberates ascospores from microscopic sacs (asci) that develop within the sphere. Cleistothecia are at first colorless, then yellow, brown, and finally black in most species. They have microscopic appendages of various types that diagnosticians use together with internal features for identification. In a few species, cleistothecia are so distinctive as to be identifiable with just a hand lens. Those of Uncinula on cottonwood are easily visible to the unaided eye.

7.19 *Cortinarius brunneus* group. (Fig.7-2)
Tanana River bank-Bonanza creek LTER forest station, FP3A, Fairbanks, Alaska, M. M. Chen, 8/22/1989, specimen # AKMC 0248.
Mycorrhizae species. Cap 7-8 cm broad convex then becoming plane and cinnamon brown corrugated with crackle, stipe 6.0-7.0 long 1.8-2.0 thick gray and composed of fine fibers with swollen base, more or less equal inner hollow cortinae at upward remained and stained cinnamon-brown by spores, gills dark rusty brown sinute, spores rusty-brown 8.4-11.2 x 5.6-7.0 μm elliptical; this mushroom is found growing close to an elder tree and forming a mushroom circle. It's a very important mycorrhiza species at spruce - cottonwood / alder (*Alnus tenuifolia, Alnus sinuata*) forests.

7.20 *Cortinarius* sp. (subgenus *Leprocybe*)
Tanana river-Bonanza creek LTER station, Fairbanks, Alaska, FP3B, M. M. Chen, 8/22/1989, specimen # AKMC 0245, 0260.
Mycorrhizae species. Cap 3.8 cm, light cinnamon brown and cap appears dry & smooth fine corrugated when dry with crackle, stipe 6.8 x 3.0 cm base with bulb and forms light cinnamon brown a zone of collapsed veil hairs on the upward of stalk and stained cinnamon-brown by spores, gills bright cinnamon brown, margin narrowing involuted, and the edge toward inner rolling; spores rusty brown, 7.0-11.2 x 5.6-8.4 μm & elliptical to subglobose & spores print cinnamon-brown.

7.21 *Cortinarius sp.*
Bank of Tanana River Bonanza Creek LTER station forest, FP3B, Fiarbanks, Alaska, M. M. Chen, 8/22/1989, specimen # AKMC 0270.
Mycorrhizae species. Cap 2.8-4.0 cm viscid when wet and small or middle size, gray brown and broad or conical with wrinkles especially toward margin, stipe 6.0 x 2 cm base with bulb forms and mosses, free to not attached, gills date brown & the veil or cortinae silky layering and covering in gills (or rupturing); spores cinnamon brown 8.4-11.2 x 5.6-8.4 μm elliptical and thick wall with warts. Can be found in cottonwood/white spruce forests.

7.22 *Cortinarius* subgenus *Phlegmacium.*
Bank of Tanana River-Bonanza creek LTER station forest, FP3A, Fairbanks, Alaska, M. M. Chen, 8/7/1990, specimen # AKMC 1025.
Mycorrhizae species. Cap 4.0-2.5 cm gray-brown with many wrinkles especially toward at margin viscid, stipe 6-7 cm base with bulb or swollen forms cortinae collapsed hairs or spores remained and stained cinnamon-brown, gills dark-red-brown or purplish date and adnated and brownish silky cortinae covering in gills, spores 7.0-11.2 x 5.6-7.0 μm rusty brown and elliptical. Habitat: on spruce/cottonwood mixed forests ground.

7.23 *Cortinarius* sp.
FP3A, M. M. Chen, 8/7/1990, specimen # AKMC 1059.
Main edible mycorrhizae species. Cap 4-7 cm in diameter, gray-brown with many wrinkles especially toward at margin, viscid when wet, context eaten by insects, stipe 8 cm in diameter, cover gray with light hair, spore print brownish ellipsoid 8.4-14.0 x 5.6-8.4 μm, main mycorrhizal at young white

spruce mixed with large cottonwood forests ground. Important species for the improvement of white spruce regeneration from replacement by cottonwood forest.

7.24 *Pleurotus ostreatus* (Jacq. Ex Fr.) Kummer Oyster Mushroom
FP3A, M. M. Chen, 8/22/1990, specimen # AKMC 0249, 0269.
This is oyster mushroom, very common delicious mushroom source. Cap 6-14 cm across, shell-shaped, convex at first, then flattening; rather variable in color, light brown. Gills slightly decurrent on the short stem, crowded, narrow; white to pale cream-ochre. Stem 20-30 x 10-20 mm, eccentric to lateral, or absent; white with a woolly base. Flesh white. Odor pleasant. Taste pleasant. Spores subcylindric, 9-10 x 4-4.5 μm. Deposit lilac. Habitat in large clusters on stumps, logs, and trunks of deciduous trees. Found throughout inland of Alaska. Can be cultivated in low-grade forest and agriculture substrate, only takes 2 or 3 weeks to be ready for harvest. (See *Fungi Treasure*, M. M. Chen and Jo-Sing Yang)

7.25 *Gomphidius glutinosus* (See 4.22)
FP3B, M. M. Chen, 8/22/1989, specimen # AKMC 0256, 0263.

7.26 *Cortinarius castaneus* Fr.
FP3B, M. M. Chen, 8/22/1989, specimen # AKMC 0267.
Cap: 2-5 cm across; medium to large sized very hygrophanus, dark chestnut brown when wet, edge white; silky. Gills adnexed; cinnamon brown. Stem 20-40 x 4-8 mm, equal; slightly tapering at base; dirty whitish. Flesh whitish brown where really wet. Odor pleasant. Taste slight. Spores ovoid to pipshaped, rough, 8-10 x 4.5-6 μm, quotient 1.7. Deposit rusty brown. Habitat: in coniferous or mixed woods. Found in cottonwood forests ground in Tanana River bank.

7.27 *Suillus grevillei* (Klotzch) Singer Larch Bolete
Plot XIV, M. M. Chen, J. H. McBeath, Richard Parmeter, 8/15/1990, species # AKMC 1060.
Edible – good after peeling and removing slime. Cap 5-15 cm across, convex to nearly flat; yellow to chrome or bright rusty red; smooth, slimy, and sticky with pale lemon gluten and removable pellicle. Tubes adnate to subdecurrent; bright yellowish becoming greenish and bruising pinky-brown. Pores small, angular; lemon yellow bruising rust. Stem 40-100 x 15-30 mm, solid, pale yellow above the ring, flushed cinnamon below, bruising brownish. Veil white to yellowish; leaving cottony ring on stem. Flesh pale yellow in cap, darker lemon chrome in stem, bruises pinkish. Odor none or faintly metallic. Taste mild or very slightly bitter. Spores subfusiform-ellipsoid, 8-10 x 2.8-3.5 μm. Deposit olive-brown to dull cinnamon. Habitat gregarious or growing in dense tufts under larch (tamarack). Common. Found in northern and eastern North America. Comment: The specimens in the photograph were found in the West, where you usually find the red color form; in the East the yellow color form is generally found. Good source of deer food. (27, 36)

7.28 *Lactarius deliciosus* var. *deliciosus* (Fr.) S.F. Gray., Nat. Arr. Brit. Pls. 1:624. 1821; - *Agaricus deliciosus* Fr., Syst. Mycol. 1:67. 1821. FP3A, M. M. Chen, 8/21/1990, specimen # AKMC 1064.
Edible – delicious. Very popular mushroom. In China it's collected in recipes, its tender mushroom and chamellia together can be made into "mushroom oil", not only a great dish, it's also a top ingredient. Cap 5–14 cm across, broadly convex with a depressed disc and a distinctly inrolled margin, becoming funnel-shaped with a wavy margin in age; pale flesh or rosy buff tinged greenish in places, with numerous purple-brick or salmon-colored blotches arranged in narrow concentric bands; sticky when moist. Gills adnate-decurrent, crowded; pale salmon, bruising pistachio green. Stem 30-70 x 10-25 mm, stuffed then hollow, pinched off at base; buff or red-orange to salmon, sometimes with darker, spotlike depressions, becoming green in places; brittle with mycelium on base, pitted, with a distinct bloom. Flesh rigid at first then fragile; cream, yellowy then corrot-colored. Latex orange fading to orange-yellow then gray-green. Odor slightly fruity. Taste mild or slightly bitter.

Spores ellipsoid, amyloid, 7-9 x 6-7 μm; ornamented with minute warts and ridges forming a partial reticulum. Deposit cream. Habitat scattered to gregarious under spruce.

7.29 *Hygrophorus tephralercus*
FP3A, M. M. Chen, 8/21/1990, specimen # AKMC 1070.

7.30 *Clitocybe aurantiace* (Wulf.) Stud (Fig.7-3)
(= *Hygrophoropsis aurantiaca*)
Cottonwood with young spruce forest ground, FP3A, M. M. Chen, 8/21/1990, specimen # AKMC 1085.
The Europeans think it's poisonous and can cause hallucinations. In Japan it's listed as edible. (*Poisonous Mushrooms*, 1975, Chinese Science Press; Mao, 2001) Chinese use *Clitocybe aurantiace*. Cap fleshy; margin incurved when young; stipe central, spongy-fleshy to fibrous, confluent with the cap; gills decurrent or broadly adnate; spores hyaline, elliptic or globose; growing on ground, rarely on wood.

Comment by David Arora: "The false chanterelle was originally placed in *Cantharellus*, and is listed in many mushroom books as *Clitocybe aurantiaca*, but the forked gills, frequently off-center stalk, and dextrinoid spores connote a closer kinship to *Paxillus*." (3, 36)

7.31 *Clitocybe odora* (Fr.) Kummer
FP3A, M. M. Chen, specimen # AKMC 1026, 1065, 1089.
Edible. Cap 3-9 cm across, convex at first with a low, broad umbo, later expanding and becoming irregular and wavy at the margin; dingy green to bluish green; grayish, bluish, or nearly white; finely matted with silky hairs or sometimes with a hoary bloom. Gills slightly decurrent, close or crowded, broad; whitish tinged with cap color. Stipe 30-70 x 5-15 mm, solid becoming hollow, sometimes curved and enlarged toward base; whitish tinged with cap color, base spongy and covered in fine whitish down. Flesh thin, firm; whitish to pale tan. Odor strongly of aniseed. Taste strongly aniseed. Spores ellipsoid, smooth, nonamyloid, 6-7.5 x 3-4 μm. Deposit whitish pink. Habitat singly, scattered, or in groups on leaf litter under hardwoods, especially oak. Found widely distributed in Fairbanks, Alaska. Comment Said to be edible, but I would avoid eating it. (3, 27, 36)

7.32 *Tricholoma vaccinum* (Pers. Ex Fr.) Kummer
FP3A, M. M. Chen, 8/21/1990, specimen # AKMC 1084.
Some records say it's not edible – possibly poisonous. But some record it to be not only edible, but also has tumor inhibition property. Cap 2-7 cm across, obtusely conic expanding to convex then flatter and slightly umbonate, with an incurved margin hanging with hairy flaps of veil remnants, disappearing later as margin expands; flesh-brown or rusty tan and darker toward the center; dry, minutely hairy, then breaking into woolly scales. Gills adnate becoming notched, close, broad; white at first, later pallid flesh-color. Stipe 30-80 x 6-22 mm, silky-hairy, sometimes with small brown scale. Flesh often hollow in the stipe; pallid to rosy. Odor mealy. Taste bitter. Spores ovoid to ellipsoid, smooth, nonamyloid, 6.2-7.6 x 4.3-5.2 μm. Deposit white. Habitat scattered or in dense tufts under conifers, especially pine and spruce. Common and often abundant. Found widely distributed in Fairbanks, Alaska. (27, 36)

7.33 *Hericium coralloides* (Scop.) Gray, *A Natural Arrangement of British Plants* (London) 1: 652 (1821)
(= *Hericium ramosum* (Bull.: Merat) Lebellier) (Fig.7-4)
Mycologia 27: 367. 1935; Coker & Beers, Stip. Hydnums East. U. S. 14. pl. 8, 55, f. 1-3. 1951.
Tanana River bank, FP2A, M. M. Chen, 8/7/1990, specimen # AKMC 1016.
Edible. Delicious "Coral monkey head" mushroom, a super Chinese cuisine dish containing abundant protein, can also be used for medicinal purposes, anti-cancer. Similar to *Hericium erinaceus*. This species is already on the Chinese market. Basidiocarps fleshy, white when fresh, brownish when dry,

consisting of a rooting base with several stout main branches from which short and more slender branchlets arise, the latter densely covered with pendent clusters of spines; spines terete, acute, 5-15 mm long; spores hyaline, smooth, subglobose, 5-6 x 4.5-5.5 µm, 1-guttulate. On cottonwood trunk peripheral of white spruce forest willow ground. (27, 36)
(See M. M. Chen and Jo-Sing Yang, *Fungi Treasure* (2), pg. 17-21)

7.34 *Russula nigricans* Fr.
FP3B, M. M. Chen, 8/22/1990, specimen # AKMC 0261.
Edibility suspect – not advisable; other members are poisonous. Cap 10-20 cm across, broadly convex, soon funnel-shaped; dirty white soon staining brown and finally black as if burnt, bruising reddish. Gills adnate, thick, widely spaced; pallid to straw, eventually black. Stipe 30-80 x 10-40 mm; white then soon stained like cap. Flesh white, becoming blood red then gray to black. Odor fruity. Taste slowly hot. Spores broadly ellipsoid, 7-8 x 6-7 µm; warts under 0.5 µm high, with partial reticulum. Deposit white. Habitat under deciduous trees. Common. Found in Tanana River bank cottonwood forests mixed with young spruce trees.

7.35 *Lactarius deliciosus* var. *deliciosus* (Fr.0 S.F. Gray) "Orange Milkly Lactarius"
FP3A, plot VIII, M. M. Chen, 8/21/1989, specimen # AKMC, 0250.
FP3A, plot VIII, M. M. Chen, 8/7/1990, specimen # AKMC 1028, 1064.
Edible ectomycorrhizae. Cap 5–14 cm across, broadly convex with a depressed disc and a distinctly inrolled margin, becoming funnel-shaped with a wavy margin in age; pale flesh or rosy buff tinged greenish in places, with numerous purple-brick or salmon-colored blotches arranged in narrow concentric bands; sticky when moist. Gills adnate-decurrent, crowded; pale salmon, bruising pistachio green. Stipe 30-70 x 10-25 mm, stuffed then hollow, pinched off at base; buff or red-orange to salmon, sometimes with darker, spotlike depressions, becoming green in places; brittle with mycelium on base, pitted, with a distinct bloom. Flesh rigid at first then fragile; cream, yellowy then corrot-colored. Latex orange fading to orange-yellow then gray-green. Odor is slightly fruity. Taste mild or slightly bitter. Spores ellipsoid, amyloid, 7-9 x 6-7 µm; ornamented with minute warts and ridges forming a partial reticulum. Deposit cream. Habitat scattered to gregarious under conifers, especially pine. Common. Found widely distributed throughout Fairbanks, Alaska. Comment in another variety of this species, Lactarius deliciosus var. *deterrimus* Groger, the flesh stains wine-red.

7.36 *Clavariadelphus truncalus* (Quel.) See 4.25
FP4A, plot V, M. M. Chen, 8/7/1990, specimen # AKMC 1036,1017.
FP4A, M. M. Chen, 8/21/1989, specimen # AKMC 0125.

7.37 *Lycoperdon perlatum* Pers. syn. *Lycoperdon gemmatum* Batsch
FP3B, M. M. Chen, 8/22/1989, specimen # AKMC 0205.
Edible when flesh is completely white; excellent. Medicinal values: blood coagulent and tumor inhibition. (*Collection of Mushroom Prescriptions*, Chen, 1999).

Fruit body 2.5–6 cm across, 2-9 cm high, subglobose with a distinct rudimentary stipe; white at first, becoming yellowish brown; outer layer of short pyramidal warts, especially dense at the head, rubbing off to leave an indistinct meshlike pattern beneath, which opens by a pore. Spore mass white, then olive-brown at maturity. Sterile base spongy, occupying the stipe. Spores globose, minutely warted, olive-brownish, 3.5-4.5 x 3.5-4.5 µm. Habitat singly, scattered, or in clusters in waste areas and open woods and along wood edges. Found widely distributed in Fair banks, Alaska. (36)

8 BLACK SPRUCE-PICEA MARIANA / BETULA PAPYRIFERA / LEDUM GROENLANDICUM / PLEUROZIUM SCHREHERI

Low growing, open forests primarily of Black Spruce are often interspersed with tamarack, paper birch, and willows, locally interspersed with treeless bog. Often seen on overcast slopes and level low grounds, or near the riverbanks of Tanana River, they are distributed on the permafrost soil frozen layer. (Fig.8-1) The Black Spruce is slow growing, it seldom exceeds 20 cm in diameter and is usually much smaller; a tree 5 cm in diameter is often 100 years in age. The Black Spruce comes in abundantly after fire because its persistent cones open after a fire and spread countless seeds over the burned areas. These seeds land in soil, germinate, and then the natural regeneration flourishes. A thick moss mat, often of sphagnum mosses, sedges, grasses, and heath or ericaceous shrubs usually make up the subordinate vegetation of the open Black Spruce forests. In the wet bottomland is the slow-growing tamarack. As with the Black Spruce, the tamarack is of little commercial value, seldom reaching a diameter of more than 15 cm.

Common wooden tree species: Black Spruce (*Picea mariana*), Tamarak (*Larix laricina*) Paper birch (*Betula papyrifera*), White Spruce (*Picea glauca*) Common shrubs: Red fruit bearberry (*Arctostaphylos rubra*), Crowberry (*Empetrum nigrum*), Labrador-tea (*Ledum groenlandicum*), Prickly rose (*Rosa acicularis*), Littletree willow (*Salix arbusculoides*), Bebb willow (*Salix bebbiana*), Grayleaf willow (*Salix glauca*), Blueberry willow (*Salix myrtillifolia*), Diamonleaf willow (*Salix planifolia spp.* Pulchra), Scouler willow (*Salix scouleriana*), Bog blueberry (*Vaccinium uliginosum*) Mountain-cranberry (*Vaccinium vitis-idaca)*

Alaska forest Disease Survey plot name FP5A Forest ecological type Black Spruce forest, located at Tanana River Bananaza Creek experimental Station; tree composition was 100% of Black Spruce height 14 m. DBH 12cm. Latitude 64° 45’ N, Longitude 148° 00’ W elevation 120-470 near bank.(37, 38)

Black spurce decay Fungi:

8.1 *Gloeophyllum saepiarium*

Tanana River Bananaza Creek experimental Station, M. M. Chen and McBeath, 1989 and 1990, specimen # 769 or 775.

Wood decay fungus, medicinal and anti-cancer. Type of rot. Causes a brown rot.

Basidiocarps annual to perennial, pileate, broadly sessile, dimidiate or rosette shaped, often imbricate in clusters from a common base, or fused laterally to compound fruitbodies, up to 7 cm wide, 12 cm long and 6-8 mm thick at the base of the cap, tough and flexible, margin sharp and slightly wavy, upper surface first finely grayish to black, when young and along the margin finely tomentose, in age the hyphae agglutinate and the surface becomes tufted, hirsute to hispid or scrupose with coarse protuberances, finally more or less smooth in zones mixed with narrow, more persistent hispid bands, narrowly to broadly zoned reflecting different stages of growth and thus, the zones form the margin to base are in old specimens often differently colored, hymenophore lamellate with anastomosing, dense lamellae, 15-20 per cm behind the margin, more rarely mixed with poroid areas with rounded to irregular, sinuous, radially elongated pores, about 1-2 per mm, edges of lamellae light golden brown in active growth, later umber brown, side surface of lamellae ochraceous to pale brown, usually distinctly lighter than the context and trama, up to 7 mm deep, context dark brown, denser next to the tubes than towards the upper surface, up to 5 mm thick, black in KOH.

Hyphal system trimitic, generative hyphae thin to thick-walled and with clamps, 2.5-4 µm in diam, skeletal hyphae dominate in the fruitbody, especially in the upper context and trama, golden brown, straight, thick-walled, up to 6 µm in diameter, binding hyphae few, tortuous and with relatively, up to 6 µm in diameter, binding hyphae few, tortuous wand with relatively short branches, only seen in older parts of the context, light golden brown, up to 4.5 µm in diam at the base.

Cystidia abundant in the hymenium, subulate to obtuse, thin to thick-walled, by age, some extremely elongated, not or only slightly projecting. 26-95 x 3-7 µm usually smooth, more rarely with a small crown of crystals.

Basidia narrowly clavate, 18-30 x 4.5-7 µm some elongate 110 µm, with clamp.

Basidiospores cylindrical, 9-13 x 3-5 µm, hyaline, smooth, IKI-.

Sexuality: Heterothallic and bipolar.
Substrata: Usually on dead coniferous wood, such as Abies, Cupressus, Juniperus, Larix, Picea, Pinus, Psuedotsuga, Taxodium and Tsuga, more rarely noted on angiosperms like Alnus, Betula, Cratageus, Populus, Prunus and Salix rather common on aspen. Often found on processed boards.

Distribution: Widespread in North America and seemingly present wherever there are coniferous forests. Circumglobal thorough USSR, Japan and China to Europe.
Remarks: Normally this is an easy species to recognize because of the yellowish to rusty color and the lamellate hymenophore. G. Abietinum has much coarser and more distinct wavy lamellae and normally a more irregular hymenophore with scattered lamellae grading into poroid parts, or some zones lamellate while others are poreid. Furthermore, basidiocarps of this species are softer and smoother on the suface, evenly brown when young, paling with age. *G. sepiarium* is a major factor in the decay of dead coniferous wood and formation of brown rot residues. (15, 23, 32, 36)

8.2 *Trichaptum abietium* Bob
Tanana River Bananaza Creek experimental Station, M. M. Chen, 1990.
Type of rot. White pocket rot of dead sapwood of conifers. Basidiocarps annual, usually effused-reflexed, sometimes sessile or resupinate, pilei solitary or imbricate, often laterally fused, up to 1.5 x 8 x 0.2 cm; upper surface gray, hirsute, azoate, smooth; margin concolorous pore surface bright purplish, fading to ochraceous, rough, the pores anular, 4-6 mm, with thick, entire dissepiments that become thick and deeply lacerate with age, context usually less than 1 cm thick, duplex, the upper layer whitish, floccose, soft, the lower layer white, firm, tough-fibrous; tube layer concolorous and continous with the lower layer of the context, up to 1.5 mm thick.

Hyphal system dimitic; contextual skeletal hyphae thick-walled hyaline, with rare branching, nonseptate, 2.5-5 μm in diameter; contextual generative hyphae thin-walled, hyaline, rarely branched, with clamps; 2-4 μm im diameter; tramal hyphae simlar.
Cystidia abundant, usually capitately incrusted, imbedded or projecting to 15 μm, 4-7 μm in diameter, arising from imbedded tramal skeletal hyphae that curve out into the hymenium; hyphal pegs also present.

Basidia clavate, 4-stigmate, 12.5-1.4 x 5-6 μm, with a basal clamp. Basidiospores cylindric, slightly curved, hyaline, smooth, IKI-, 6-7.5 x 2.5-3 μm. The pockets are hollow and the wood becomes fragile and lacy in the late stages of decay. Cultural characteristics:
Sexuality: Heterothallic and tetrapolar.
Distribution: Substrate is dead sapwood of conifers. Overholts (1953) reports occasional occurrence on hardwoods. Distributed throughout coniferous forest regions of Fairbanks, Alaska and cirumglobal in the North Temprate Zone.

Remarks: Overholts' concept of *T. abietinum* included the taxon in cluded here as *T. fusco-violaceum* and as *P. abietinus var. abietis*, the taxon included here as *T. laricinum*. We accept the opinion of Macre (1967), who carried out infertility tests and other comparative studies and concluded they are best considered as three distinct species. Magasi also reported the porid, irpiciform, and lamellate forms were intersterile populations. *T. abietinum* is very similar to the common *T. bifrome* found on hardwood slash. Basdiocarps of the latter species tend to be much wider than those of T. abietinum, are usually attenuated to a narrow base, and have a tomentose or glabrous upper surface. *T. subchartaceum* differs in having much thicker basidiocarps with a persisitently poroid lower surface and is restricted to Polpulus. The restriction to sapwood is remarkbly consistent, and *T. abietinum* basidiocarps often form a complete ring over the entire sapwood surface on the ends of large logs. (15, 23, 32, 36)

8.3 *Tyromyces chioneus* (Fr.) Karsten Bob. 785
Rev. Mycol. 3: 17, 1881. – *Polyporus chioneus* Fr., Syst. Mycol. 1:359, 1821. – Polyporus albellus Peck., N. Y. State Mus. Ann. Rept. 30:45, 1876.
Type of Rot – White rot in dead hardwoods.

Basidiocarp annual, pileate, applanate to slightly convex, tubes usually not decurrent, broadly attached to semicircular and dimidiate, more rarely spatulate, single or a few specimens together, up to 8 cm broad and 10 cm wide, 0.5-2 cm thick, soft and fleshy when fresh, drying rather hard and brittle, taste mild and with a slight aromatic scent when fresh; upper surface azonate, first whitish and finely tomentose, soon becoming glabrous as the hyphae agglatinate, then finely scrupose and warted, cream, light yellowish, or pale grayish to straw-colored as the agglutination proceeds there develops a smooth pellicle which on drying becomes radially to irregularly wrinkled, in old specimens rather distinct in section, in age somewhat discolored and then dirty yellowish to pale sordid gray; pore surface white to pale cream, slightly shiny, drying somewhat darker, pores angular to round and mostly thin-walled, 3-4(-5) per mm; tubes concolorous with pore surface, up to 8 mm deep; context white and dense in dry condition, usually distinctly thicker than the tubes, up to 1.5-2 cm thick at the base.

Hyphal system dimitic; generative hyphae with clamps, in the context intricately branched and twisted and difficult to separate in long sections, side branches partly as tube-like hyphae, often separated by a septum, but also with repeated branchings, these hyphae are very characteristic and diagnostic for the species, they are randomly oriented, occasionally mixed with more unbranched, long hyphae, both types with rather numerous clamps, thin- to thick-walled, 3-8 μm wide, in parts collapsed, in the trama more or less parallel and more straight, mostly 2-4 μm wide; skeletal hyphae straight, thick-walled, 2-4.5 μm, present only in the trama.

Cystidia absent, fusoid cystidoles present, 9-13 x 4-5 μm.

Basidia clavate, 4-sterigmate, with a basal clamp, 10-15 x 4-5 μm.

Basidiospores cylindrical to slightly bent, hyaline, smooth, thin-walled and IKI-, mostly 4-5 x 1.5-2 μm.

Cultural characteristics: – See Nobles (1948); Stalpers (1978); David (1980).

Sexuality: Tetrapolar (David 1980).

Substrata: On dead wood of numerous hardwood genera, especially common on *Betula*.

Distribution: Common in Eastern and Northern North America. Known also from Europe.

Remarks: The slightly applanate, short-lived basidiocarp, frequently with a thin yellowish pellicle, provides the basis for a field determination. Microscopically the branched generative hyphae of the context and the slightly wider spores will separate it from the often confusingly similar *O. tephroleucus*. (15, 23, 32, 36)

Rust disease

8.4 Black Spruce witches' broom (Fig.8-2)

Black Spruce FP5A plot, East Town, M. M. Chen and J. H. McBeath, 1990.

Also called large-spored spruce-Labrador tea rust (Y. Hiratsuka, 1987), this refers to White Spruce needle rust, *Ledum groenlandicum* leaf rust. The most widely noticed disease in this group is broom rust, or low witches' broom, caused by *C. arctostaphyli*. This occurs across Canada and the northern United States on Black Spruces. In the mountainous West it occurs from Alaska into Mexico. Brooms commonly grow to lengths of meters. Spruces with numerous large brooms often have dead or broken tops, grow slowly, and die prematurely. Aecia develop in mid to late summer; the needles are shed in autumn, leaving the broom devoid of foliage in winter. Witches' broom causes Black Spruce branch to break down, and wood decay by *Trichaptum abietium*.

8.5 *Pucciniastrum vaccinii* (Rab) Joerst host *Vaccinium vliginosum*

Fairbank east town pure spruce forests, M. M. Chen, J. H. McBeath and R. Parmeter, 1990.

Called bog blueberry leaf rust. Spores bright yellow color.

8.6 *Hedysarum* rust (*Uromyces sp.*)

It's weed leaf but very common seen. Uredia spores collected under pure spruce forests, Fairbank, Tanana River bank, M. M. Chen, 1990.

8.7 Willow leaf tar disease. (*Salix lasiandra*)
M. M. Chen, 1990.
Willow leaf tar disease produces on *Salix lasiandra*.
see 9.1

8.8 *Melmpsoridium betulinum* (see Birch leaf rust)
M. M. Chen, 1990.
Beluta papyrifera is host.

8.9 *Uncinula salicis* (Powdery mildew)
see 9.2
Chena lake, M. M. Chen, 1990
Salix bebbiana is host.

8.10 *Microspharea sp.* Shorter birch rust
Chena lake, M. M. Chen, 1990.

8.11 *Cortinarius* sp.
Tanana River bank, M. M. Chen, 8/15/1990, specimen #AKMC 1061.
Cap 3.5-4.0 cm surface dark brown viscid when wet, corrugated when dry then wrinkled, stipe 3.5 cm long and 1.2 cm thick inner hollow silky and shining with dark brown fibrils, gills brown-coffee color margin narrowly involute, cortinae hairy-fibrillose zone on stalk which stained rusty brown by spores, main mycorrhizae species on Black Spruce forests. Spores 5.6-11.2 x 4.2-5.6 μm.

8.12 *Leccinum alaska* (Fig.8-3)
Tanana River bank, pure black spruce forest (height 6, dbh 6, age 100), plot XIV, M. M. Chen, J. H. McBeath, Richard Parmeter, 8/15/1990, specimen # AKMC 1062.
Main mycorrhiza seen in Taiga forests, hundreds can often form together. Fruit bodies short, on permafrost soil; thick moss mat, about 10-15 cm thick. This is a cold-tolerant species. Most plants in Taiga forests on earth have a symbiotic association in their roots with a permafrost soil mushroom Taiga species known as Taiga boreal forest mycorrhizae. Scientists believe that mycorrhizae are beneficial to the growth and health of Black Spruce. In Siberia the main species is *Larix*, here it becomes shrubbery tamarack, because there are less mycorrhizae.

This species is an ectomycorrhizal fungus. Fungi are known to enhance the uptake of water and plant nutrients, increase resistance to root pathogens, and promote plant growth. Initial evidence for the role of ectomycorrhizal fungi in wood decay fungi disease suppression was provided by a number of field observations that showed seedlings or trees of both angiosperms and gymnosperms to be more resistant to root pathogens than their nonmycorrhizal counterparts.

8.13 *Suillus grevillei* (Klotzch) Singer
Plot XIV, M. M. Chen, J. H. McBeath, Richard Parmeter, 8/15/1990, species # AKMC 1060.
See 7.27

8.14 *Armillariella* sp.
FP5A, M. M. Chen, 8/21/1990, specimen # AKMC 1082.
Edible. Cap 8 x 8.5 cm – 4.2 x 3.5 cm Cap fleshy, firm; stipe fleshy, continuous with the cap, central or eccentric; annulus present, persistent, membranous or subarachnoid; gills adnexed, adnate, common separated or decurrent; spores smooth, hyaline, white in mass. Stipe off-center.

8.15 *Lycoperdon* sp.
Bonanza West LTER station, plots IX, XIX, M. M. Chen, 8/16/1989, speciemen # AKMC 0199.

Many species of this puffball genus have medicinal values such as blood coagulent: *L. asperum, L. pyriforme, L. atropurpureum, L. umbrinum*...etc. (*Collection of Mushroom Prescriptions*, Chen, 1999).

Basidiocarps globose, ovoid or pyriform, with or without a sterile base; exoperidium usually composed of fugacious spines, warts or granules; endoperidium membranaceous, dehiscing by an apical aperture; capillitium composed of long simple or branched threads connected with the peridium. (36)

9 WILLOWS (SALIX SPP.)

Willows are well represented in the Fairbanks area; (Fig.9-1) Shrubby willows are widely distributed almost throughout Fairbanks, Alaska as in other far northern lands. In habit they vary from prostrate or creeping dwarf shrubs to erect bushes 0.6-2 m. willows are the largest genus of trees here. Some species seem to intergrade or hybridize, often making identification difficult. The species we know of are as follows: *Salix glauca* L., *S. myrtillifolia* Anderss., *S. bebbiana* Sarg., *S. lasiandra* Benth, *S. interior* Rowlee, *S. brachycarpa* Nutt., *S. pseudomonticola, S. alaxensis* (Anderss.) Cov., *S. novae-angliae* Anderss. Ten more species of willow are very common. The great variation in willows and their tendency to hybridize make it difficult to identify leaf rust.

The global warming phenomenum occurs in Tanana Riverbank, which is the up river's deglaciation reaction to the forest bank open area spruce forest replacement by various willows. The willows' ecological habitat: they are the undergrowth of the open spruce-birch forest of interior Fairbanks, and form thickets on sandbars and other porous soils along streams. Although young trees are not suitable for lumber because of their small size, shrubby willows provide important summer and winter food for many game animals, especially moose and ptarmigan. (37)

9.1 *Melampsora paradoxa* Diet. & Holw.

Near Easter (10 miles from Fairbanks), M. M. Chen and J. H. McBeath, 7/25/1989.

Willows are hosts of several formae speciales of *Melampsora* that cannot yet be differentiated in their uredinial or telial states. All North American willow rusts are currently assigned to the group species *M. epitea*. This includes the former *M. paradoxa* with aecia on tamarack (*Larix laricina* (Du Roi) K. Koeh.). Aecia caemoids lacking peridia; aeciospores with a smooth spot, walls of aeciospores not thickened, bilateral on opposite sides, mean spore size 17.6 x 24.3 μm, also found spermogonia on current year's needles. The uredinial state can be found on these three species of willow: *Salix alaxensis* (Anderss.) Cov. (1989,1990), *Salix novae-angliae* Anderss (1989, 1990), *Salix glauca* L. (1990). Uredinia and telia chiefly hypophyllous, uredinospores are 18.4–21.9 x 18.1–27 μm. (32, 18, 20, 37, 38)

9.2 *Uncinula salicis* (DC. ex Merat) Wint. Powdery mildew (Fig.9-2)

Tanana River bank willow shrubs, M. M. Chen and J. H. McBeath, 8/18/1989.

Powdery mildews are characterized by the presence of grayish white, powdery fungal growths on the surface of leaves. In late summer, they produce small brown to black spherical fruiting bodies (cleistothecia). Perithecia form on the back of leaves, average 129.6 μm. Appendages avg. 148.5 μm, ascospores avg. 24 x 14 μm (Fig.) Host plants include three willow species and a poplar: *Salix glauca* L. P95 (1990), *Salix myrtillifolia* Anderss. P99 (1989, 1990), *Salix bebbiana* Sarg. P. 116(1990), *Populus deltoids* (1990). (32, 17, 36)

9.3 *Rhytisma punctatum* Willow leaf tar (Fig.9-3)

Bananaza creek experimental forest long term ecological research station, M. M. Chen and J. H. McBeath, 8/9/1989, specimen # AKMC 0166.

Tar spots are among the most showy and least damaging foliar diseases. Caused by species of *Rhytisma* (*Rhytismatales, Rhytismataceae*), they occur on several kinds of willow plants: *Salix lasiandra* Benth P126 (1989, 1990), *Salix interior* Rowlee P126 (1990), *Salix brachycarpa* Nutt. P97

(1990), *Salix glauca* L. (1990), and *Salix pseudomonticola*(1989). It's distributed throughout the world, including Eastern Asia and Europe. Spots on willow arise in late spring. At first infected tissues turn slight green or yellowish green, then during mid to late summer, black stomata develop on the upper leaf surface. (32, 17, 16)

9.4 *Cytospora chrysosperma* (Pers.) Fr., Syst, Myc. 2:542. 1823; Sacc., Syll. Fung. 3:260. 1884; Grove, Brit Stem- & Leaf-fungi 1: 272. 1935. Hardwood dieback and canker
Tanana River bank, M. M. Chen, 7/17/1989.
Willows are most commonly affected, but the disease occurs occasionally on hardwood: elders, aspen, and cottonwood. It is generally considered to be an opportunist that causes damage to plants stressed by environment, injuries, or other diseases. It is common also as a saprobe on dead twigs. Although the fungus is a normal inhabitant of the bark of willow, it may become parasitic if the host is weakened. Under forest conditions it is of no consequence except in stands on an unfavorable site or injured by drought, frost, or fire. Tree vigor is the most important factor in resistance. In inland Alaska, poplars and willows are valuable shade and ornamental trees, especially in Alaska open semiarid regions. Heavy losses occur among trees that have been weakened by neglect or by frost.

The disease appears as lesions or cankers on the trunks and large limbs of affected trees. Cankers attack the sapwood of tree trunk barks, mainly because the low temperature of Alaska weakens the trunks. In spring, the fruiting bodies of the conidial state of *Cytospora* occur. Small branches and twigs are usually killed without a definite canker being formed. Cankers are developed by gradual killing of the diseased bark in more or less circular areas. On smooth-barked shoots young infections can be recognized by the presence of brownish shrunken patches. The diseased area may be fairly regular or quite irregular in shape, gradually enlarging until the stem is girdled. Water sprouts often develop profusely just below a canker on a large stem; most of them are soon killed. When old trunks and large branches with rough bark are attacked, the typical cankered appearance is seldom developed, and it is consequently difficult to recognize the diseased condition until spore tendrils appear. The diseased inner bark blackens and emits a disagreeable odor. The cambium is killed, the sapwood is also infected, becoming watery and reddish brown in color, and the heartwood is sometimes discolored. Severely attacked trees from 1 to 2 cm dbh die in 2 or 3 years. On willow the disease functions more as a dieback, killing trees branch by branch.

Stromata scattered, conic-depressed, about 1.5 mm broad, 1 mm high, covered by and slightly elevating the epidermis, then erumpent by a grayish to blackish disc; loculi several, often irregular, with a common central ostiole; conidia allantoid, 4-7.5 x 1-1.5 μm, oozing out in long yellowish to reddish tendrils; conidiophores filiform, 10-15 x 1 μm.

In cold areas, perfect stage = *Valsa sordida.*

Keeping the trees growing vigorously is the best method to prevent this disease. When planting, the right soil and trees, healthy seeds and vigorous seedlings should be selected; pay special attention to protecting the root system. After planting, proper management and nursery are required. If it's a young plantation, regular drainage and irrigation, prevention of pests, regular pruning, no dead trunks, and protection of wounds are all important. If necessary, low-toxin pesticides can be used. (7, 32, 15, 23, 36)

9.5 Cortinarius subg. *Sericeocybe caninus* group. (Fig. 9-4)
Tanana river - Bonanza Creek experimental station, (LTER) Fairbanks, Alaska FP3A, M. M. Chen, 8/22/1989, specimen # AKMC 0245.
Cap 4.2-7.0 cm light cinnamon brown and cap appears dry (possibly subgenus *Leprocybe*) & smooth, fine corrugated with crackle when dry, stipe 9-4x 0.8-1.0 cm base with bulb or swollen remain cinnamon cortinae, cortinae forming superficial hairy-fibrillose zone on stalk. Stained cinnamon-brown by spores, gills bright cinnamon brown, margin narrowing involuted, spores 7.0-11.2 x 5.6-8.4 μm, elliptical to subglobose, rusty brown with clamps, no apical germ pore or plage warty spores, print cinnamon-brown.

Ecological habitat: this species grows between two elder trees, forms a large mushroom circle of over 230 fruiting bodies, it is a very important species.

9.6 *Cortinarius spp.*
Tanana river - Bonanza Creek experimental station, (LTER) PB4A, M. M. Chen, 8/21/1990, specimen # AKMC 1075, 1080 (brown spore prints)
Cap 5.5cm, cinnamon-brown, when wet become convex then expanded, context fleshy, stipe 6.2 cm long and 0.8 cm thick; base with bulb or swollen parts, gills brown close adnated or notched becoming brown and finally rusty-brown in age, cortinae is appearing up stalk parts fulling spores brown color; spores 7.0-13.0 x 5.6-8.4 μm.

9.7 *Clitocybe odora* (Fr.) Kummer.
FD3A, M. M. Chen, 8/7/1990, specimen # AKMC 1078, 1079.
FP2A, M. M. Chen, 8/21/1990, specimen # AKMC 1081.
Edible. (Said to be edible, but I would avoid eating it.) Cap thin milk white, 7-5 x 5-3.4 cm across, convex at first with a low, broad umbo, later expanding and becoming irregular and wavy at the margin; dingy green to bluish green; grayish, bluish, or nearly white; finely matted with silky hairs or sometimes with a hoary bloom. Gills close or slightly decurrent, close or crowded, broad; whitish tinged with cap color. Stipe off-center, 3-6 cm long, solid becoming hollow, sometimes curved and enlarged toward base; whitish tinged with cap color, base spongy and covered in fine whitish down. Flesh thin, firm; whitish to pale tan. Odor strongly of aniseed. Taste strongly aniseed. Spores ellipsoid, smooth, nonamyloid, 6-7.5 x 3-4 μm. Deposit whitish pink. Habitat singly, scattered, or in groups on leaf litter under hardwoods, especially oak. Found widely distributed in Fairbanks, Alaska.

9.8 *Hericium coralloides* (Scop.) Gray, *A Natural Arrangement of British Plants* (London) 1: 652 (1821)
(= *Hericium ramosum* (Bull.: Merat) Lebellier)
Mycologia 27: 367. 1935; Coker & Beers, Stip. Hydnums East. U. S. 14. pl. 8, 55, f. 1-3. 1951.
Tanana River bank, FP2A, M. M. Chen, 8/7/1990, specimen # AKMC 1014.
Edible. Delicious “Coral monkey head” mushroom, a super Chinese cuisine dish containing abundant protein, can also be used for medicinal purposes, anti-cancer. Similar to *Hericium erinaceus*. This species is already on the Chinese market. Basidiocarps fleshy, white when fresh, brownish when dry, consisting of a rooting base with several stout main branches from which short and more slender branchlets arise, the latter densely covered with pendent clusters of spines; spines terete, acute, 5-15 mm long; spores hyaline, smooth, subglobose, 5-6 x 4.5-5.5 μm, 1-guttulate. On cottonwood trunk with peripheral of white spruce forest willow ground. (27, 36)
(See M. M. Chen and Jo-Sing Yang, *Fungi Treasure* (2), p. 17-21)

9.9 *Galerina navcina* or *G. oiner.*
Tanana River bank, FP1A spot # 9, M. M. Chen, 8/7/1990, specimen # AKMC 1031.
Cap small 1-3 cm, grey to brown, surface covered with spines. Stipe 1 cm Gills brown, mid-distant, attached to stem. Cortinae: obvious gray-white small hairy scales. There are lots of this species.

9.10 *Hebeloma sp.*
Tanana River bank, FP2A, M. M. Chen, specimen # AKMC 0246, 0245A.
Cap fleshy, viscid, mostly glabrous; margin at first incurved; stipe central, fleshy, rarely fibrous, confluent with the cap; gills adnexed; spores ochraceous, elliptic to fusiform, smooth; growing on ground. (36)

References:

1. Anderson, J.P 1952. The Uredinales of Alaska and adjacent parts of Canada. Iowa State Coll. J Sci.26: 507-526.
2. Anonymous. 1960. Index of Plant Diseases in the United States. U. S. Dept. Agric. Handbook. No. 165. Washington, D. C. 531 P.
3. Arora, David.1986. Mushrooms Demystified. Ten Speed Press, Berkeley.
4. Bega, R.V. 1978. Diseases of Pacific Coast Conifers. U.S. Department of Agriculture, For. Serv., Washington D.C., Agric. Handbook. 521.
5. Bisby, G.R. 1938. The Fungi of Manitoba and Saskatchewan. National Research Council of Canada, Ottawa, Ont.
6. Bondarzew, A. 1953. Polyporaceae of the European Part of the USSR and the Caucasus. Leningrad.
7. Boyce, J.S. 1961. Forest pathology. McGraw-Hill, New York.
8. Cash, E. K.1953. A Checklist of Alaskan fungi. Pl. Dis. Repoter Suppl. 219:1-70.
9. Cash, E.K. (1953) The Plant Disease Reporter: Supplement 219. United States Department of Agriculture, Maryland.
10. Conners, I.L. 1967. An annotated index of plant diseases in Canada. Can. Dep. Agric., Ottawa, Ont. Publ. 1251.
11. Davidson, A.G.; Prentice, R.M. editors. 1967. Important forest insects and diseases of mutural concern to Canada, the United States and Mexico. Can. Dep. For. Rural Dev., Ottawa, Ont. Publ. 180.
12. Foster, R.E.; Wallis, G.W. 1969. Common tree diseases of British Columbia. Can. Dep. Fish. For., For. Branch, Ottawa, Ont. Publ. 1245.
13. Funk, A. 1981. Parasitic microfungi of western trees. Environ. Can., Can. For. Serv., Pac. For. Res. Cent., Victoria, B.C. Inf. Rep. BC-X-222.
14. Funk, A. 1985. Foliar fungi of Western trees. Can. For. Serv., Pac. For. Res. Cent., Victoria, B.C. Inf. Rep. BC-X-265.
15. Gilbertson, R. L. and L. Ryvarden. 1987. North American Polypores Volumes 1 and 2. Fungiflora, Oslo, Norway.
16. Hepting, G.H. 1971. Diseases of forest and shade trees of the United States. U.S. Dep. Agric., For. Serv., Washington, D.C. Agric. Handb. 389.
17. Hiratsuka, Y. 1977. Annotated checklist of tree and shrub diseases in the prairie provinces. Environ. Can., Can. For. Serv., North. For. Res. Cent., Edmonton, Alberta. Inf. Rep. NOR-X-178.
18. Hiratsuka, Y. 1977.; Powell, J.M. 1975. Pine stem rusts of Canada. Environ. Can., Can. For. Serv., Ottawa, Ont. Publ. 1329.
19. Holsten, E.H., Werner, R.A., Laurent, T.H. 1980. Insects and Diseases of Alaskan Forests. U.S. Government Printing Office, Washington, D.C.
20. Kobayasi, Y., Hiratsuka, N., Korf, R.P., Tubaki, K., Aoshima, K., Soneda, M., Sugiyama, J. 1967. Mycological Studies of the Alaskan Arctic. Annual Report of the Institute for Fermentation, Osaka.
21. Malhotra, S.S.; Blauel, R.A. 1980. Diagnosis of air pollutant and natural stress symptoms on forest vegetation in western Canada. Environ. Can., Can. For. Serv., North. For. Res. Cent., Edmonton, Alberta. Inf. Rep. NOR-X-228.
22. Manion, P.D. 1981. Tree disease concepts. Prentice-Hall Inc., Engelwood Cliffs, New Jersey.
23. Overholts, Lee O. 1953. The Polyporaceae of the United States, Alaska, and Canada. University of Michigan Press, Ann Arbor.
24. Partridge, A.D.; Miller, D.L. 1974. Major wood decays in the Inland Northwest. Idaho Res. Found. Inc., Nat. Resour. Ser. 3.
25. Peace, T.R. 1962. Pathology of trees and shrubs with special reference to Britain. Clarendon Press, Oxford.
26. Peterson, G.W.; Smith, R.S., Jr. 1975. Forestry nursery diseases in the United States. U.S. Dep. Agric., For. Serv., Washington, D.C. Agric. Handb. 470.
27. Phillips, Roger. 1991. Mushrooms of North America. Little, Brown and Company, Boston.

28. Riffle, J.W.; Peterson, G.W., technical coordinators. 1986. Diseases of trees in the Great Plains. U.S. Dep. Agric., For. Serv., Rocky Mt. For. Range Exp. Stn., Fort Collins, Colorado. Gen. Tech. Rep. RM-129.
29. Savile, D.B.O. 1962. Some fungal parasites of Onagraceae. Canada J. Bot. 40:1385-1398 hardwoods and develop a hydnaceous hymnophore at maturity.
30. Shaw, C.G. 1973. Host-fungus index for the Pacific Northwest. I. Hosts. Wash. Agric. Exp. Stn. Bull. 765.
31. Shaw, C.G. 1973. Host-fungus index for the Pacific Northwest. II. Hosts. Wash. Agric. Exp. Stn. Bull. 766.
32. Sinclair, W.A., Lyon, H.H., Johnson, W.T. 1987. Diseases of Trees and Shrubs. Cornell University Press, Ithaca.
33. Smith, W.H. 1970. Tree pathology. A short introduction. Academic Press, New York.
34. Sutherland, J.R.; van Eerden, E. 1980. Diseases of insect pest in British Columbia. B.C. Minist. For., Victoria, B.C.., and Envrion. Can., Can. For Serv., Pac. For. Res. Cent., Victoria, B.C., and Environ. Can., Can. For. Serv., Pac. For. Res. Cent., Victoria, B.C. Joint Rep. 12.
35. Tatter, T.A. 1978. Diseases of shade trees. Academic Press, New York.
36. Teng, S.C. 1996. Fungi of China. Mycotaxon, Ltd., Ithaca.
37. Viereck, Leslie A. 1972. Alaska Trees and Shrubs. United States Department of Agriculture, Washington, D.C.
38. Ziller, W.G. 1974. The tree rusts of western Canada. Environ. Can., Can. For. Serv., Ottawa, Ont Publ. 1329.

Reference Works In Chinese:

Chen, *Collection of Mushroom Prescriptions*, 2000.
The Macrofungi in China, Mao, 2001.

Index of Plants and Fungi:

Fig 1-1. Bank of the Tanana River

Fig 1-2. White Spruce forest

Fig 1-3. Spruce needle rust (*Chrysomyxa ledicola* lagh.)

Fig 1-4. *Sarcodon imbricatus* (L. ex Fr.) Karsten "White deer mushroom"

Fig 2-1a. *Clitocybe odora* (Fr.) Kummer

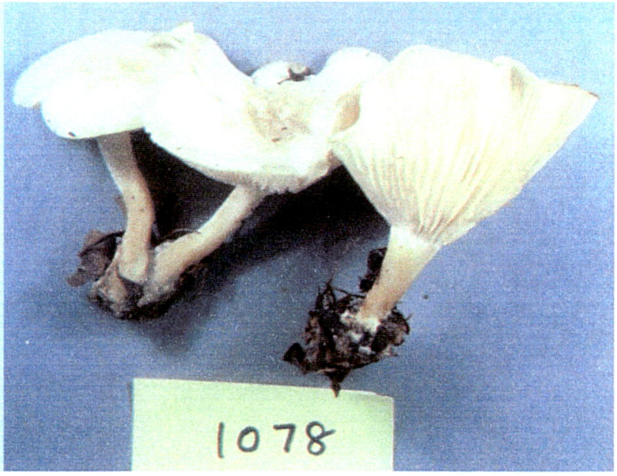

Fig 2-1b. *Clitocybe odora* (Fr.) Kummer

Fig 2-2. *Suillus luteus* (Fr.) S.F. Gray in young stage

Fig 2-3. *Amanita muscaria* var. *muscaria* (L. ex Fr.) Prs.

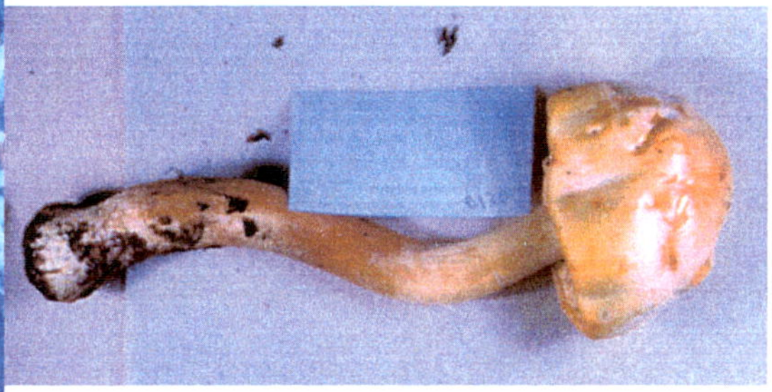

Fig 2-4. *Agaricus auqustus* Fr.

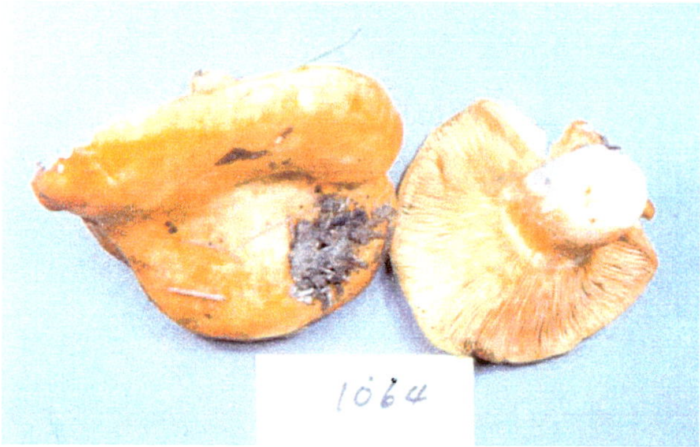

Fig 3-1. *Lactarius deliciosus* var. *deliciosus* (Fr.) S.F. Gray

Fig 3-2. *Phellinus tremulae* (bond.) Bond. & Boriss.

Fig 3-3. White spruce fired forest.

Fig 3-4. Quaking Aspen (*Populus tremuloides*)

Fig 4-1. *Hydnellum aurantiacum* (Fr.) Karsten

Fig 4-2. *Lycoperdon perlatum* Pers.

Fig 4-3. *Russula emetica* (Schaeff.: Fr.) Pers. ex S.F. Gray

Fig 4-4a. *Leccinum insigne* Smith, Thiers, & Watling

Fig 4-4b. *Leccinum insigne* Smith, Thiers, & Watling

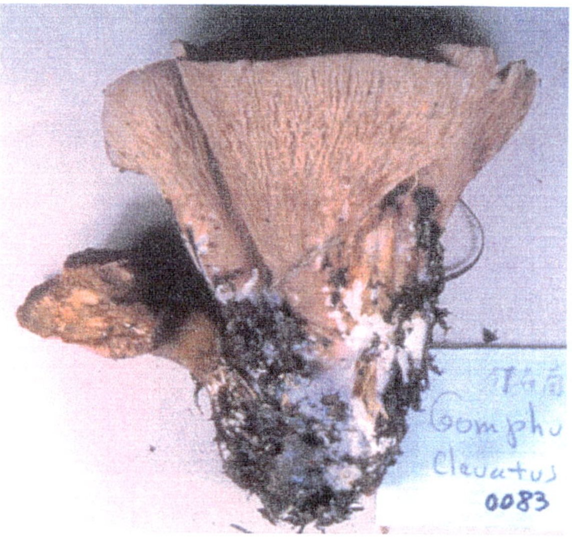

Fig 5-1. *Gomphus clavatus* (Fr.) S.F. Gray

Fig 5-2. *Panus rudis* Fr.

Fig 5-3. Fly Agaric

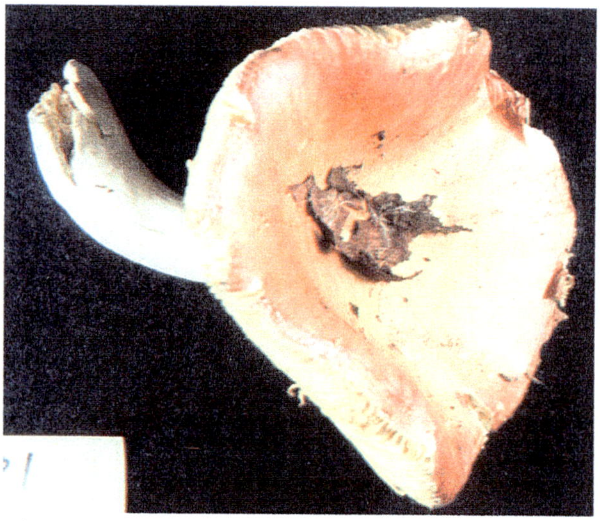

Fig 5-4. *Russula rosacea* (Bull.) Fr.

Fig 5-5. *Laccaria laccata*

New record for Alaska's *Populus*: *Cerrena unicolor*

Fig 6-1. Paper Birch forest (*Betula papyrifera*)

Fig 6-2. *Amanita* mushroom circle.

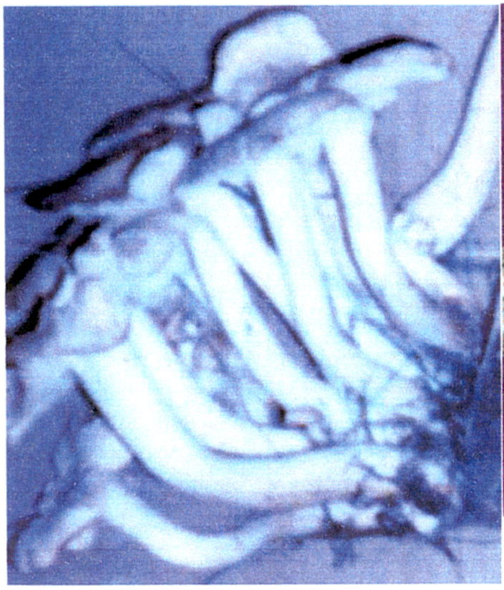

Fig 6-3. Fried Chicken Mushroom

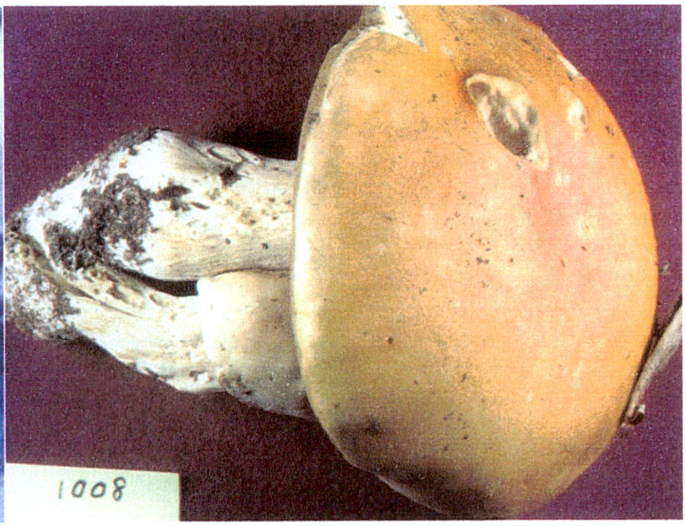

Fig 6-4. *Boletus edulis* Bull. Ex Fr.

Fig 7-1. Balsam poplar forest (*Populus balsamifera*)

Fig 7-2. *Cortinarius brunneus* circle

Fig 7-3. *Clitocybe aurantiace* (Wulf.) Stud

Fig 7-4. Coral monkey head

Fig 8-1. Black Spruce forest (*Picea mariana*)

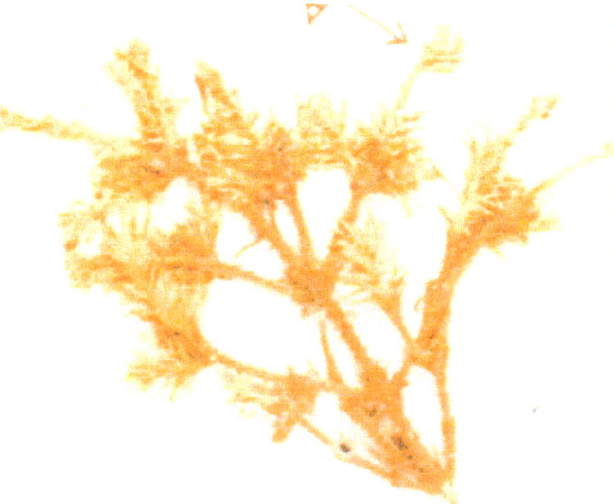

Fig 8-2. Witches' broom caused by *Chrysomyxa arctostaphyli*

Fig 8-3. *Leccinum alaska*

Fig 9-1. Willow (*Salix spp.*) on Tanana riverbank.

Fig 9-2. Powder mildew (*Uncinula salicis*)

Fig 9-4. Cortinarius subgroup. *Sericeocybe caninus* circle

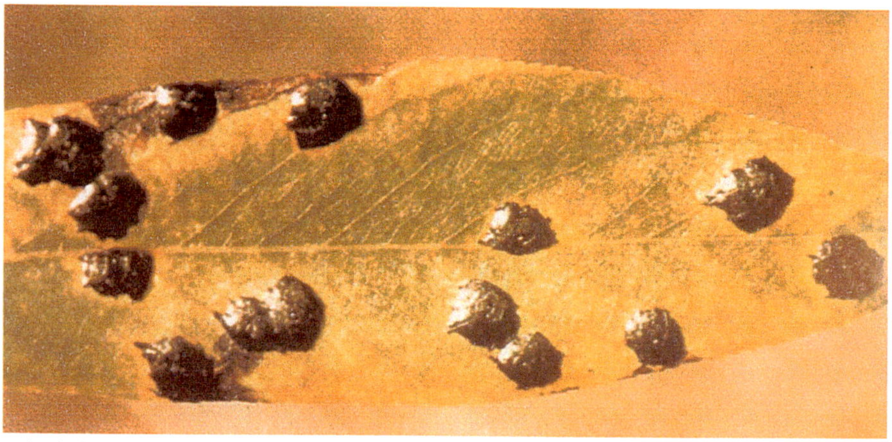

Fig 9-3. Willow leaf tar (*Rhytisma punctatum*)

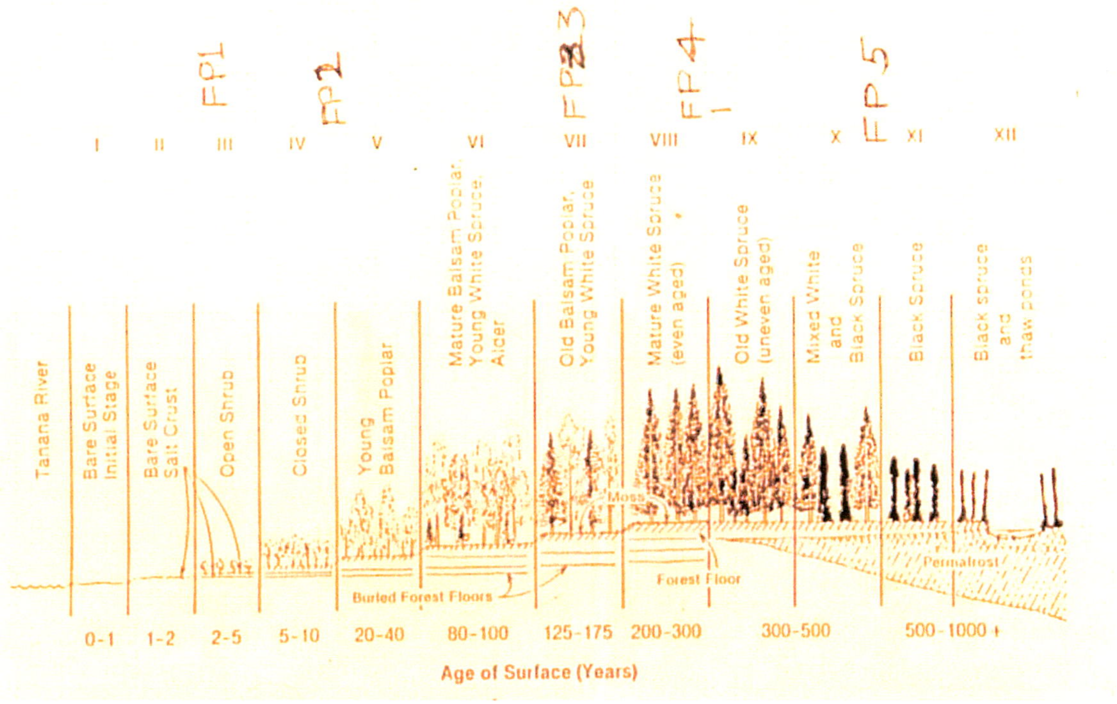

Main ecotype of forest spots on the Tanana River.

Fairbanks forest inland fungi team. (Second from left is Professor J. H. McBreath and the fourth is Leslie A. Viereck.)

Fungi Associated with Tanoak (*Lithocarpus densiflora* (H. & A) Rehd.) in California

Preface

In the last 10 years, tanoaks have suffered significant decline in the San Francisco Bay Area, as a result of increased development and a serious outbreak of sudden oak death, a newly emerging plant disease. A better understanding of biodiversity associated with tanoaks is therefore vital. As a mycologist and plant pathologist, I am specifically interested in the fungal component of this biodiversity.

This species list serves as a beginning point in a survey of fungal biodiversity associated with tanoaks. This list is a result of a broad literature review, summarizing much of the previously existing literature on tanoak-associate fungi, as well as previously unpublished specimen notes from the University of California Jepson Herbarium. Published sources include the *California Plants Disease Host Index* (Alex M. French, 1989), *Diseases of Trees and Shrubs* (Wayne, 1987), *Mushrooms Demystified* (David Arora, 1986), and *North American Polypores* (R. L. Gilbertson and L. Ryvarden, 1986). The specimen notes from Jepson Herbarium describe the fungal collections that date from 1920 to the present and represent the works of many historically significant mycologists.

This survey lists 100 species of tanoak-associated fungi and fungus-like oomycetes. This includes over 50 species of plant pathogenic fungi that affect tanoaks, such as *Armillaria* root disease, pine-oak rust, various canker diseases, powdery mildews, shoot molds, and other types of decay fungus. Other fungi include 18 species of mycorrhizal fungi that are directly associated with tanoaks, as well as saprotrophic fungi that are often found on dead wood and leaf litter produced by tanoaks. Special emphasis has been placed on tanoak-associated mushrooms that are important to humans, such as edible, medicinal, and toxic fungi. Thirty illustrations are attached.

It is my hope that we will eventually learn how to culture some of the mycorrhizal associates of tanoak, particularly those that have beneficial edible or medicinal properties. In this way, we can help restore tanoaks in areas where they have suffered heavy mortlity and at the same time provide a direct economic benefit from replanting these trees.

Contents

3 Decay

3.1 Wood decay

35) Datronia scutellata (Schwein.) Gilbn. & Ryvarden
36) Phellinus viticola (Schwein.)Donk
37) Crustomyces subabruptus (Bourd. And Galz.) Jülich
38) Hymenochaete tabacina (Sow.)Lév.
39) Lenzites betulina (Fr.) Fr.
40) Phellinus gilvus (Schwein.: Fr.) Pat.
41) Trametes cervina (Schwein.)Bres.
42) Trametes hirsuta (Wulf.: Fr.) Pilát
43) Trametes ochracea (Pers.) Gilbn. & Ryvarden
44) Phellinus ferreus (Pers.) Bourd. & Galz.
45) Poria vulgaris (Fr.) Quél.
46) Stereum hirsutum (Willd.: Fr.) S. F. Gray
47) Stereum fasciatum Schwein. forma zonatum
48) Porothelium fimbriatum (Pers.: Fr.) Fr.
49) Schizopora paradoxa (Schrad.: Fr.) Donk
50) Phanerochaete carnosa (Bert) Parm.
51) Peniophora ravenelii Cooke
52) Merulius tremellosus Schrad.
53) Bulgaria inquinans (Pers.: Fr.) Fr.

3.2 Thelephoraceae bark-rot

54) Peniophora incarnata (Pers.: Fr.) Karst.
55) Peniophora ravenelii Cooke
56) Ceriporia viridans (Berk. & Br.) Donk.

4 Mycorrhizal and edible mushrooms

4.1 Mycorrhizal mushrooms

57) Boletus aereus Bull: Fr.
58) Tricholoma magnivelare (Peck) Redhead
59) Tricholoma focale (Fr.) Ricken
60) Tricholoma terreum group
61) Cortinarius infractus (Pers.) Fr.
62) Cortinarius collinitus group
63) Cortinarius cotoneus group
64) Camarophyllus pratensis (Pers.: Fr.) Kumm.
65) Hygrophorus russula (Fr.) Quél.
66) Craterellus cornucopioides (L.) Pers.: Fr.
67) Cantharellus subalbidus Smith & Morse
68) Ramaria species
69) Hydnellum caeruleum (Hornem.) Karst.
70) Lactarius subvillosus Hesler & Sm.
71) Lactarius agrillaceifolius Hesler & Sm.
72) Russula cyanoxantha (Schaeff.) Fr.
73) Russula albonigra (Krombh.) Fr.
74) Gautieria parksiana Zell. & Dodge

4.2 Edible mushrooms

75)Hericium coralloides (Scop.) Gray
76)Pleurotus ostreatus (Jacq.: Fr.) Kumm.
77)Entoloma bloxamii (Berk. & Broome) Sacc.
78)Tremella mesenterica Retzius

5 Poisonous mushrooms

79) Agaricus hondensis Murrill
80) Omphalotus olivascens Bigelow, et al.

6 Saprotrophic on debris

6.1 Saprotrophic mushrooms

81) Clavariadelphus occidentalis Methven
82) Marasmius copelandii Peck
83) Marasmius rotula (Scop.: Fr.) Fr.
84) Setulipes quercophilus (Pouzar) Antonín
85) Macrotyphula juncea (Fr.) Berthier
86) Mycena capillaripes Peck

6.2 Saprotrophic microfungi

87) Coccomyces dentatus (Schm. & Kunze) Sacc.
88) Fusicolla foliicola Karst
89) Fusidium griseum Link.
90) Mollisia lithocarpi Cash
91) Monochaetia hysteriiformis (Berk. & Curt.) Guba
92) Protopeltis lithocarpi Miller & Bonar
93) Sphaerulina conflicta (Cooke) Barr
94) Aleurodiscus aurantius (Pers.: Fr.) Schröt.
95) Cystostereum pini-canadensis (Bourdot & Galzin) Chamuris
96) Pirex concentricus (Cooke & Ellis) Hjortstam & Ryvarden
97) *Mollisia cinerea* (Batsch.) Karst.
98) Bisporella citrina (Batsch: Fr.) Korf & Carp.
99)Rigidoprus terrestris (DC: Fr.) Ryvarden
100) Fusidium griseum L

1 PRIMARY PATHOGENS

1.1 *Rust*

There is general knowledge about tree rust disease such as pine-oak rust. These diseases are caused by a group of basidiomycetes of the order Uredinales that have evolved a complex life cycle not found in any other group of fungi. The obligatory parasitic life cycle usually involves two alternate plant hosts and five stages of the fungal life cycle. The scientific name in current usage for pine-oak rust is applied to the sexual stage in plants, such as *Quercus* and *Lithocarpus*.

1) *Cronartium quercuum* (Berk.) Miyabe ex Shirai, *Bot. Mag.*, Tokyo 13: 74 (1899)
(Fig. 1) Pine-oak rust
Collection notes and specimens: Marin Co. 1923, H.E. Parks; Humboldt Co. 1898, Blasdale; 1924, 1926 Parks, H.E. Parks & Tracy
On *Pinus spp.* II III and on oak or tanoak, there is sometimes heavy infection in Pine stems and branches. Pine-oak rust sexual stages are produced under the leaves of the tanoak. It commonly occurs in fall in California Coast Range in mixed pine/oak or pine/tanoak forests. Trees are easily infected.
Ref: 1, 2, 18.

1.2 *Powdery mildew*

This superficial growth often appears powdery. Most of these fungi penetrate and parasitize only the epidermal cell of their hosts, producing within the living host cells an absorbing structure (haustorium) that diverts water and nutrition to the fungus. The result is slow debilitation of the infected plant part. Symptoms include dwarfing, distortion, chlorosis, premature senescence, and browning of leaves, subnormal growth rate, blemished or aborted fruit, and depressed yields.

2) *Cystotheca lanestris* (Harkn.) Sacc.
(= *Sphaerotheca lanestris* Harkn.) Miyabe in ldeta.
(Fig. 2 & 3)
Collection notes and specimens: Gardner, et al 1972.
Oak powdery mildews.
Ref: 19, 20, 40 (pg. 14).

3) *Brasiliomyces trina* (Harkn.) R.Y. Zheng, *Mycotaxon* 19: 286 (1984)
(2) Collection notes and specimens: Yarwood & Gardner 1972.
New name of tanoak powdery mildew
Ref: 19, 40 (pg. 15).

4) *Microsphaera penicillata* (Wallr: Fr.) Lev.
(Fig. 4)
Powdery mildew disease.
Collection notes and specimens: Bay Area, Gardner, et al. 1972
Causes powdery mildew diseases of oak.
Ref: 16, 38 (p.134-136).

1.3 Canker

A necrotic lesion in the bark of the stem or root, often extending into the xylem; also, the scar left after shedding of bark tissues killed by localized disease or environmental injury. Some annual tanoak cankers enlarge only once and do so within an interval more brief than the growth cycle of the plant, usually less than one year. Other perennial cankers enlarge more than once during the year, while other diffuse cankers enlarge without a characteristic shape or a noticeable callus formation at the margins.

5) *Fusarium solani* (Mart.) Sacc., *Michelia* 2(7): 296 (1881)
(Fig. 5)
Root rot
Commonly call bark-rotting fungus. Causes cankers or twig dieback.
Ref: 18, (p. 118), 40, (p. 214-215).

6) *Diatrype disciformis* (Hoffm: Fr.) Fr. *Summa Veg. Scand.*, Section Post.: 385 (1849)
(Fig. 6)
Collection notes and specimens: Reaves Hts., 1931, Fawcett.
On bark.
Ref: 41.

7) *Diatrype stigma* (Hoffm: Fr.) Fr. *Summa Veg. Scand.,* Section Post.: 385 (1849)
Collection notes and specimens: Mill Valley, Marin Co., April 1925, H.E. Parks. Evidently a common and widespread fungus in California. Blight cankers and dieback on branches. Belongs to *Diatrypales*. See Janex-favre (Revue mycol. 42: 265, 1978) M. C. Janex-Favre. Ascospores formation in *Diatrype disciformis* (Hoffm.) Fries.
Ref: 40 (p. 140-144).

8) *Dendrothele candida* (Schwein.: Fr.) P.A. Lemke
(Fig. 7a & 7b)
Collection notes and specimens: Rattlesnake Camp, Mt. Tamalpais, Marin Co., 1926, H.E. Parks (coll. no. 3061); Humboldt Co., 1933, H.E. Parks & Tracy; Smith River, Del Norte Co., 1939 (coll. no. 6260).
This pathogen is a smooth patch and bark rot that colonizes and decomposes the dead, corky, outer layers of bark in living tree trunks. They are not known to cause harm to the trees, but the presence of their sporocarps on bark may be misinterpreted as indicating the presence of a canker or an extensive dead area with decaying wood beneath. White patches and bark patches are caused by this fungus, which belonging to the Aphyllophorales. The fruiting bodies are corticioid fungi that form smooth, more or less dislike, white fruiting bodies (basidiocarps) that may aggregate or coalesce into larger structures on bark.
Ref: 28, 40 (p. 168-169).

9) *Cryphonectria gyrosa* (Berk. & Broome) Sacc., *Syll. fung.* (Abellini) 17: 784 (1905)
(= *Endothia gyrosa* (Schwein.: Fr.) Fr. [as 'gyrosum'])
(Fig.8)
Collection notes and specimens: Green Valley Falls, Solano Co., Mar. 1935, L. Bonar.
The fungus infects broken branches or wound. In most circumstances it causes localized, slowly expanding cankers, but may girdle and kill branches and trunks of trees under stress.
Anamorphs are place in *Endothiella.*
Ref: 40 (p. 192).

Stipes RJ, Emert GH, Brown RD Jr. 1982. Differentiation of *Endothia gyrosa* and *Endothia parasitica* by disc electrophoresis of intramycelial enzymes and proteins. *Mycologia* 74:138-141. (Results corroborate morphological species concepts)
Walker J., Old KM, and Murry, DIL. 1985. *Endothia gyrosa* on Eucalyptus in Australia with notes on some other species of *Endothia* and *Cryphonectria*. *Mycotaxon* 23: 353-370. (Discussion of generic concepts, illustrations)

10) *Nectria coccinea* (Pers: Fr.) Fr., *Summa Veg. Scand.*, Section Post. 2: 388 (1849)
(Fig. 9a & 9b)
Collection notes and specimens: Marin Co. 1935, H.E. Parks.
Ref: 38 (p. 147), 40 (p 212-213)
Cotter HVT and Blanchard RO. 1981. Identification of the two *Nectria* taxa causing bole cankers on American beech. *Plant Disease* 65:332-334.

11) *Phytophthora ramorum* Werres, et al., *Mycological Research* 10: 1155 (2001)
Sudden oak death
Collection notes: Rizzo, et al. 2002
Ref: 47 (p. 1155-1165)
Rizzo DM, Garbelotto M, Davidson JM, Slaughter GW, and Koike ST. 2002. *Phytophthora ramorum* as the cause of extensive mortality of Quercus spp. and Lithocarpus densiflorus in California. *Plant-Disease* 86(3): 205-214.

1.4 Root disease

12) *Armillaria mellea* (Vahl: Fr.) P. Kumm.
(Fig. 10)
Collection notes and specimens: Raabe 1962, pg 45.
Armillaria root rot, also called shoestring root rot or, in some areas, mushroom root rot, is one of the best known and most damaging diseases of forest trees. *Armillaria* invades the bark and cambial region of roots and the root collar, killing roots and trees of all sizes. Some species or perhaps strains within species are virulent parasites, but others are opportunists that are selective on small or weak individuals such as those weakened by freezing, drought, or polluted air. *Armillaria* also colonize the declining root systems of trees felled or killed by other agents. The fungus can persist for decades in decaying wood in soil.
Ref: 3, 18 (p. 118), 40 (p. 308-312)
Raabe, R.D. 1962. Host list of the root rot fungus, *Armillaria mellea*. Hilgardia. 33(2):23-88.

2 OTHER PATHOGENS

Other parasites can live on either living or dead hosts and on various nutrient media; they are therefore called non-obligate parasites.

2.1 On stem

Canker, dieback, or sapwood rot. May be mistaken for wounds caused by sunburn or freezing temperature.

13) *Bonaria lithocarpi* (V.P.M. Miller & Bonar) Bat.

Collection notes and specimens: Inverness, Marin Co., March 1931, H.E. Parks (spec. no. 3573 – type specimen)

14) *Hypoxylon fuscum* (Pers: Fr.) Fr. *Summa veg. Scand.* (Sweden) 2: 384 (1849) [1846?]
Collection notes and specimens: Marin Co., L. Bonar.
Anthracnose or bark dry decay.
Ref: 18.

15) *Biscogniauxia mediterranea* (De Not.) O. Kuntze, *Revis. gen. pl.* (Leipzig) 2: 398 (1891)
(= *Nummularia clypeus* (Schwein.) Cooke)
(= *Hypoxylon mediterraneum* (De Not.) Ces. & De Not., *Comm. Soc. crittog. Ital.* 1(4): 202 (1863))
(Fig.11a & 11b)
Anthracnose and bark dry decay
Collection notes and specimens: Muir Woods, Marin Co., Feb 8, 1924, H.E. Parks (Coll. no. 7). Usually on the trunk and larger branches of oak. Belongs to Sphaeriales. *Biscogniaxia* was published to replace *Nummularia*, a later homonym.
Ref 18 (p. 188), 40 (p. 224)
Eckblad FE and Granmo A. 1978. *Nummularia* (Ascomycetes) in Norway. *Norwegian Journal of Botany* 25:69-75. 1978. (Describes this fungus as a species of *Biscogniauxia.* Key to two *Biscogniaxia spp.* Includes descriptions and illustrations).

16) *Hypoxylon thouarsianum* (Lév.) Lloyd, *Mycol. Writ.* 5: 26 (1919)
Collection notes and specimens: San Mateo Co., 1938, T. C. McCabe; Del Norte Co., 1944, H.E. Parks.

17) *Hypholoma capnoides* (Fr: Fr.) P. Kumm.
(= *Nematoloma capnoides* (Fr.) P. Karst., *Meddn Soc. Fauna Flora fenn.* 5: 61 (1880))
Collection notes and specimens: Muir Woods, Marin Co., 1938, T.T. McCabe; Harkness and P.C.F. Moore 1880.

18) *Phytophthora cinnamomi* Rands, *Meded. Inst. Plantenziekt.* 54: 1 (1922)
(Fig. 12)
Collection notes and specimens: unpublished records, 2001
Mycologia 79:56-63, 1979. Mycologia 72: 990, 1980.
Ref 38 (p. 178-180), 43.

19) *Steccherinum fimbriatum* (Pers.: Fr.) J. Eriks., *Symb. bot. upsal.* 16(1): 134 (1958)
(= *Odontia fimbriata* Pers., *Observ. mycol.* (Copenhagen) 1: 88 (1815))
(Fig. 13)
On dead wood sometimes causes the tanoak bark or twigs branches to decay. A shelf-like tooth fungus.
Collection notes and specimens: Mill Valley, Marin Co., April 7, 1925, H.E. Parks (Coll. no. 2707).
Ref: 3 (p. 612), 21

2.2 On twigs and branches

20) *Cyphellopsis anomala* (Pers.) Donk
(2) Collection notes and specimens: Del Norte Co., H.E. Parks, 1946

On old and dead wood, broken bark, twigs, and branches, decorticated twig, slash, logs, underside of old logs; associated with white rot.
Ref: Ginns J and Lefebvre MNL. 1993. Lignicolous corticioid fungi (Basidiomycota) of North Amerca: systematics, distribution, and ecology. St. Paul (MN): APS Press, 1993. (MSA Mycologia Memoir: 19).
Medd. Nedl. Myc. Ver. 18-20:128.

21) *Perrotia succinea* (W. Phillips) Dennis *Persoonia* 2: 82 (1962)
Collection notes and specimens: Redwood Park, San Mateo Co., 1924, H.E. Parks.
Belongs to Helotiales.
Ref 40 (p. 236)

22) *Mycosphaerella janus* (Berk. & M.A. Curtis) Petr.
(Fig. 14)
Collection notes and specimens: Mt. Tamalpais, Marin Co, 1922, H.E. Parks; Mt. Tamalpais, Marin Co, 1925, H.E. Parks; Inverness, Marin Co, 1926, H.E. Parks.
Ascomycete belonging to Dothideales. All have in common darkly pigmented somatic and fruit-body cells and feed saprotrophically. D.R. Reynolds lists 12 different names as synonyms, in 11 different genera.
Ref: Bull. Calif. Acad. Sci. 1:43.1884; Mycol. 44: 258.1952; Trans. Brit. Mycol Soc. 24: 282, 1940; Mycologia 63: 1173 (key)

2.3 *Shoot mold*

On leaves and smaller twigs, sooty molds may form a thin network of hyphae, a pellicle, a velutinous growth, or a pseudoparenchymatous crust. Hyphae of many sooty molds have a markedly mucilaginous outer wall, which absorbs water very readily. This wall acts as an adhesive, and undoubtedly helps maintain a moist leaf surface for longer period.

23) *Phaeosaccardinula anomala* (Cooke & Harkn.) V.P.M. Miller & Bonar, *University of Calif. Publ. Bot.* 29: 410 (1941)
Ref 33.

24) *Chaetasbolisia falcata* V.P.M. Miller & Bonar, *University of Calif. Publ. Bot.* 19: 413 (1941)
Sooty Mold
Collection notes and specimens: Inverness Ridge, Marin Co., Feb. 1932, L. Bonar; Muir Woods, Marin Co., 1963, H.E. Wise.
Quaderno No. 31:59-61, 1963
Ref: 18.

25) *Limacinula anomala* (Cooke & Harkn.) D.R. Reynolds
Collection notes and specimens: Inverness, Marin Co., H.E. Parks
On leaves.
Reynolds 1971 lists 12 different names as synonyms, in 11 different genera.
The sooty mold Ascomycete genus *Limacinula* makes this *Limacinula anomala* (Cke.&Harkn.)
Ref: Reynolds DR. 1971. Mycologia 63: 1186.

26) *Limacinia lithocarpi* V.P.M. Miller & Bonar

(= *Scolecobonaria lithocarpi* (V.P.M. Miller & Bonar) Batista)
Collection notes and specimens: Inverness Ridge, Marin Co., 1932, L. Bonar; Humboldt Co., 1931, H.E. Parks (coll. no. 3575)
On living leaves
Formerly, sooty mold phase was referred to as *Sclerobonaria* while the fruiting body was referred to as *Limacinia*. Usually with dark-colored hyphae, which produces brown to black colonies superficially on living plants.
Ref: Stanley J. Hughes. 1976. Mycologia 68: 693 1976.

2.4 On leaves

27) *Lachnum marginatum* (Cooke) Raitv., *Nizshie Rasteniya, Griby i Mokhoobraznye Sovetskogo Dal'nego Vostoka*, Griby. Vol. 2. Askomitsety. Erizifal'nye, Klavitsipital'nye, Gelotsial'nye (Leningrad): 301 (1991)
(= *Dasyscyphus marginatus* Cooke)
(= *Peziza marginata* Cooke, (1875))
Collection notes and specimens: Big Basin State Park, Santa Cruz Co., 1954, Bonar.
On dead attached leaves.

28) *Pseudomassaria polystigma* (Ellis & Everh.) Arx
Collection notes and specimens: Humboldt Co. 1930, H.E. Parks.
In spots on living and dead leaves of *Q. agrifolia* and in overwintered leaves of various *Quercus* species. Belongs to the order Sphaeriles.
Ref: 2
Margaret E. Barr. 1964. The Genus *Pseudomassaria* in North America. *Mycologia* 56: 841.

29) *Bonaria lithocarpi* (V.P.M. Miller & Bonar) Batista
(= *Protopeltis lithocarpi* V.P.M. Miller & Bonar)
Sooty molds
Collection notes and specimens: near Inverness, Marin Co., Mar. 1931, H.E. Parks, (coll. no. 3573).
Ref: 33

30) *Capnodium coffeae* Pat., *Bull. Soc. mycol. Fr.* 9: 150 (1893)
(Fig.16)
Collection notes and specimens: Inverness, Marin Co., 1932, L. Bonar.
March. Spec.
Ref 33 (p.122), 40 (p.30-31).

31) *Propolis quercifolia* (Cooke & Ellis) Sherwood, *Mycotaxon* 5: 327 (1977)
(misidentified as *Propolis farinosus* (Pers.) Fr.)
Collection notes and specimen: Nevada Co., 1964, McLaughlin & Lavares
Causes leaf spots.
Ref: 40 (p. 442).

32) *Pestalotiopsis montellica* (Sacc. & Voglino) T. Kobayashi, *Trans. Mycol. Soc. Japan* 15(4): 381 (1974)
(= *Pestalotia montellica* Sacc. & Voglino)
(=*Pestalotia castagnei* Desm.)
(Fig.17)

Collection notes and specimens: Muir Woods, Marin Co., 1923, L. Bonar; Smith River, Del Norte Co., March 1936, H.E. Parks; Weott, Humboldt Co., August 1949, L. Bonar; Butler Valley, Humboldt Co., 1949, J.P. Tracy.
Ref: 38 (P. 162).

33) *Phyllosticta quercus-ilicis* Sacc.
(Fig.18 a,b)
(2) Collection notes and specimens: Plumas Co. line, 1948, Quick.
Leaf spots.
Ref: 25 (p. 211), 38 (p. 176-177), 40 (p. 76-77)

2.5 On Sapwood Stain

34) *Chlorociboria aeruginosa* (Pers.) Seaver ex Ramamurthi, Korf & L.R. Batra, (1958)
Sapwood stain
(2) Collection notes and specimens: "San Francisco Bay region".
Ref: 2, 18.
Mycologia 28:391, 1936;Mycotaxon 1:200, 1974.

3 DECAY

3.1 Wood decay

35) *Datronia scutellata* (Schwein.) Gilb. & Ryvarden, *Mycotaxon* 22(2): 364 (1985)
Collection notes and specimens: Del Norte Co., 1946, H.E. Parks
Causes white rot of living or dead tanoaks.
Ref: 22.

36) *Phellinus viticola* (Schwein.) Donk
(Fig. 19)
Uniform white rot of dead wood of coniferous and hard wood.
Collection notes and specimens: John Brown Ranch, Saratoga, Santa Clara Co., Feb. 1921 H.E. Parks (coll. no. 1020)
Ref: 22.
Mycotaxon 9:53-85, 1979

37) *Crustomyces subabruptus* (Bourdot & Galzin) Jülich, *Persoonia* 10(1): 140 (1978)
(= *Odontia subabrupta* Bourdot & Galzin)
Collection notes and specimens: Del Norte Co., 1944, Park
Mushroom-like fruiting body with dentate gills.

38) *Hymenochaete tabacina* (Sowerby) Lév., *Annls Sci. Nat.*, Bot., sér. 3 5: 145 (1846)
(= *Hydnochaete tabacina* (Berk. & Curt.) Ryvarden *Mycotaxon* 15:441. 1982.)
Collection notes and specimens: Marin Co., 1940, L. Bonar.
A white rot in dead oakwood.
Ref: 3 (p. 606), 22.

39) *Lenzites betulina* (L.) Fr., *Epicrisis systematis mycologici* (Uppsala): 405 (1838)
Collection notes and specimens: Mendocino Co., Sept 17, 1969, Tom Tang; San Rafael, Marin Co., 1925, H.E. Parks; Del Norte Co., Feb. 1944, H.E. Parks

Mainly birch forest decay species.
Ref: 11, 22

40) *Phellinus gilvus* (Schwein.: Fr.) Pat., *Essai taxonomique*: 82 (1906)
(Fig 20)
Collection notes and specimens: Wunderlich Park, San Mateo Co., June 13, 1986, M.T. Seidl
Ref: 22.

41) *Trametes cervina* (Schwein.) Bres., *Annls mycol.* 1(1-2):81 (1903)
(2) Collection notes and specimens: Marin Co. 1941, W.B. Cooke
A white rot fungus.
Ref: *Mycologia* 53:487, 1961

42) *Trametes hirsuta* (Wulf: Fr.) Pilát, *Atlas des Champignons de l'Europe. Polyporaceae I* (Praha) 3:265 (1939)
(2) Collection notes and specimens: South Fork Smith River, Del Norte Co., 1949, W.B. & VG Cooke
Common broadleaf wood decay species
Ref: 22.

43) *Trametes ochracea* (Pers.) Gilb. & Ryvarden, *North American Polypores,* Vol. 2 *Megasporoporia - Wrightoporia* (Oslo): 752 (1987)
(= *P. betulinus* Fr.)
(= *Trametes pubescens*)
Collection notes and specimens: Muir Woods, Marin Co., 1926; San Rafael, Marin Co., Harkness & Moor
Causes heart rot of living trees.
In recent years the genus *Polyporus* has been defined in a narrow sense and many species have been transferred to other genera.
Ref: 22

44) *Phellinus ferreus* (Pers.) Bourdot & Galzin, *Hyménomycètes de France* (Sceaux): 627 (1928)
Collection notes and specimens: Marin Co., 1941
On wood.
Ref: 22, 26
USDA, Agricultural Research Service. 1960. *Index of plant diseases in the United States*. (Agriculture Handbook: 165) Washington (DC): USDA

45) *Poria vulgaris* (Fr.) Quél., *Syst. mycol.* (Lundae) 1: 381 (1821)
Collection notes and specimens: Muir Woods, Marin Co., March. 14, 1926, H.E. Parks (coll. no. 3078).
The taxonomic status of *Poria* is in question. Most species have been transferred to other genera. Lowe's (1963, 1966) works are still useful for species identification although the nomenclature is outdated.
Ref: 38 (p. 184-185).

46) *Stereum hirsutum* (Willd.: Fr.) S.F. Gray, *Epicrisis systematis mycologici* (Uppsala) 1: 549 (1938)
(= *Stereum rameale* (Schwein.) Burt)

(Fig. 21)
(2) Collection notes and specimens: Cataract Gulch, Mt. Tamapais, Marin Co., 1935, L. Bonar; Palo Alto
Found on many species of coniferous and hardwood trees in California.
Ref: 3 (p. 605-607).

47) *Stereum fasciatum* Schwein., *Epicrisis systematis mycologici* (Uppsala): 546 (1838) [1836] forma zonatum
(Fig. 22)
(2) Collection notes and specimens: Lake Lagunitas, Marin Co., Oct. 1925, H.E. Parks (coll. no. 2880)
Ref 3 (p. 606).
Welden AL. 1971. An essay on *Stereum*. *Mycologia* 63:790-799.

48) *Porothelium fimbriatum* (Pers.: Fr.) Fr.
(2) Collection notes and specimens: Del Norte Co., November 1947, H.E. Parks (coll. no. 7078).
Ref: *Mycologia* 49:682-685, 1957; Mycologia 53: 490, 1961.

49) *Schizopora paradoxa* (Schrad.: Fr.) Donk, *Persoonia* 5(1): 76 (1967)
(= *Rigidoporous terrestris* (DC.: Fr.))
Collection notes and specimens: In soil under *Lithocarpus densiflorus* and other oaks at Monte Rio, Sonoma Co. and Mt. Tamalpais, under redwood logs Humboldt Co.
The type collection and subsequent collections from the type region are always yellow. Other localities show the fungus as at first delicate pink or white and later turning yellow. A wood rot of dead trunk and logs.
Ref: 2, 22.

50) *Phanerochaete carnosa* (Burt) Parmasto, *Eesti NSV Teaduste Akadeemia Toimetised*, Bioloogia 16(4): 388 (1967)
(2) Collection notes and specimens: Phoenix Lake, Marin Co., Feb. 1926, H.E. Parks.
Ref: Ginns J and Lefebvre MNL. 1993. Lignicolous corticioid fungi (Basidiomycota) of North Amerca: systematics, distribution, and ecology. St. Paul (MN): APS Press, 1993. (MSA Mycologia Memoir: 19).
Mycologia Izv. Askad. Estonsk. Ssr. Biol. 16:383

51) *Peniophora ravenelii* Cooke, *Grevillea* 8(no. 45): 21 (1879)
(= *Phlebiopsis ravenelii* (Cooke) Hojort. 1987)
*Ref: 54
Ref: Ginns J and Lefebvre MNL. 1993. Lignicolous corticioid fungi (Basidiomycota) of North Amerca: systematics, distribution, and ecology. St. Paul (MN): APS Press, 1993. (MSA Mycologia Memoir: 19).

52) *Merulius tremellosus* Schrad., *Spicil. Fl. Germ.* 1: 139 (1794)
(= *Phlebia tremellosa* (Schrad.) Nakasone & Burds.)
(2) Collection notes and specimens: Cascade Gulch, Marin Co., 1947, H.E. Parks and L. Bonar.
Belongs to Corticiaceae. The Genus *Phlebia* is considered by forest pathologists to be of little importance, since they occur only as slash rots of economically important forest trees or mild pathogens of less valuable trees.
Ref.: *Mycologia* 48; 386 1956.

3.2 Thelephoraceae bark-rot

53) *Peniophora incarnata* (Pers.: Fr.) P. Karst., *Hedwigia* 28: 27 (1889)
Collection notes and specimens: San Mateo Co.

54*)* *Peniophora ravenelii* Cooke, *Grevillea* 8(no. 45): 21 (1879)
(= *Peniophora stratosa* Burt non Petch, hom. Illeg.)

Collection notes and specimens: A competiting saprobe prevents colonization by this species and tends to replace it in the root.
Ref: 40 (p. 316).
Ref: *Ann. Missouri Botanical Garden* 12: 333, 1925.

55*)* *Ceriporia viridans* (Berk. & Broome) Donk, *Meddel. Bot. Mus. Herb. Rijhs Universit. Utrecht.* 9: 171 (1933)
Wood rot of dead trunks and logs
Collection notes and specimens: Farr, et al. 1989 (Ref.: 17)
Ref: 1, 17.
Med. Bot. Mus. Univ. Utrecht 9.

56) *Bulgaria inquinans* (Pers.: Fr.) Fr., *Syst. mycol.* (Lundae) 2: 166 (1822)
(= *Phaeobulgaria inquinans* (Pers.: Fr.) Nannf.)
(Fig. 39)
Collection notes and specimens: Wunderlich park, San Mateo Co.; M.ST. Seidl (coll. no. 2806). Belongs to the Helotiales. On bark, branches, and logs of hardwoods, especially live oak or tanoak. Common in Santa Cruz Mountains, fall through spring.
Ref: 3 (p. 587-588), 31, 36, 42, 46.

4 MYCORRHIZAL AND EDIBLE MUSHROOMS

4.1 Mycorrhizal Mushrooms

These are some notable species form ectomycorrhizal associations with tanoak.

57) *Boletus aereus* Bull: Fr.
(Fig. 23)
Queen bolete
Collection notes and specimens: Arora 1986 (Ref.: 3).
Flesh: firm and white. Odor: pleasant. Taste: pleasant and very delicious, similar to *Boletus edulis*. Spores: spindle-shaped, smooth, 12-15x4-5n. Spore deposit: olive-brown. Habitat: under oak and tanoak. Uncommon in western North America. Season: Yunan, China, Oct.-Nov, California, Nov-Dec.
Edible and delicious mushroom. Forms ectomycorrhizae.
Ref: 3, 45.

58) *Tricholoma magnivelare* (Peck) Redhead, *Trans. Mycol. Soc. Japan* 25(1): 6 (1984)
(= *Armillaria ponderosa* Peck)
(Fig. 24)
American matsutake
Collection notes and specimens: Arora 1986 (Ref.: 3).

Flesh: firm; milky white. Odor: distinct pleasant, spicy odor. Taste: very delicious. Habitat: common in western North America. Under mixed hardwood, redwood, and other conifers in Pacific coast range mountain. One of the treasured mycorrhizal fungi. Cannot be cultivated yet, but much research on bring it into cultivation!
Ref 3.

59) *Tricholoma focale* (Fr.) Ricken
(= *Tricholoma zelleri*)
Collection notes and specimens: Arora 1986 (Ref.: 3).
Scattered to gregarious on ground in woods, northern North America. Extremely abundant under confers in the Pacific Northwest often in the same area as the *Tricholoma magnivelare*, but rather rare in the Bay Area and fruiting mainly in tanoak-madrone woods at higher elevation in the Coast Ranges, like *T. magnivelare*, in the late fall and early winter.
Ref: 3 (p. 188-189).

60) *Tricholoma terreum* group
(Fig. 32)
Collection notes and specimens: Arora 1986 (Ref.: 3).
Common mycorrhizal species in spruce forests and widely distributed in north temperate zone; also found with tanoak. Reportedly has anti-bacterial properties.
Ref: 3 (p. 182), 9.

61) *Cortinarius infractus* (Pers.) Fr., *Epicrisis systematis mycologici* (Uppsala): 261 (1838) [1836]
(Fig. 25)
Sooty-olive *Cortinartius*
Collection notes and specimens: Arora 1986 (Ref.: 3).
Spores: ellipsoid to ovoid, ochraceous, 7-9x5-6n, minutely roughened; Habitat: on ground in woods. Found in North America and Asia (Tibet and Qing-Hai, China).
Ref: 3, 9 (p. 147), 31.

62) *Cortinarius collinitus* group
(Fig. 27)
Collection notes and specimens: Arora 1986 (Ref.: 3).
Cap broadly conical orange-brown to rusty ochre, flesh pallid with a touch of violet. Odor and taste mild. Widespread in North America from September to October. This is a highly variable mushroom and seems to have often been muddled with other close species. It reportedly has anti-cancer properties and is a delicious mushroom.
Ref: 3 (p. 431-432), 31 (p. 258), 36.

63) *Cortinarius cotoneus* group
(Fig. 37)
Collection notes and specimens: Arora 1986 (Ref.: 3).
Cap olive-green or darker olive-brown, covered in minute fibrillose scales. Odor radishy, especially when crushed. Not edible.
Ref: 3 (p. 445), 36 (p.156).

64) *Camarophyllus pratensis* (Pers.: Fr.) P. Kumm., *Führ. Pilzk.* 117 (1871)
(= *Hygrophous pratensis* (Pers.: Fr.) Fr.)
Meadow waxy cap.

Collection notes and specimens: San Rafael, Marin Co., Harkness & Moore; Alameda; San Mateo Co., 1938, McCabe; Santa Cruz Co., Jan. 1939, McCabe; Crecent City, Del Norte Co., Dec. 1937, A.H. Smith; Alameda/Contra Costa Co. line, March 1938, McCabe; Mendocino Co., M.T. Seidl January 16, 1985 (coll. no. 367).
Common in winter under redwood and grassland, fall to early spring edible rate highly.
Ref: 3 (p. 110), 9, 36 (p. 63).

65) *Hygrophorus russula* (Fr.) Kauffman, *Publications Mich. geol. biol. Surv., Biol. Ser. 5* 26: 185 (1918)
(Fig. 26)
Collection notes and specimens: Arora 1986 (Ref.: 3).
Spores: ellipsoid, nonamyloid, 6-8x3-5n. Spore deposit: white. Habitat: scattered, gregarious, sometimes in fairy rings under oak and sometimes confers. Common and somewhat abundant in the eastern US. Fruiting body large, cap thick. Season: Nov-Feb in California.
Very delicious mushroom, widely distributed in North America.
Ref: 3, 9 (p. 113), 31 (p. 54 fig. 27), 36 (p. 73, fig.).

66) *Craterellus cornucopioides* (L.) Pers., *Mycol. eur.* (Erlanga) 2: 5 (1825)
(Fig. 29)
Horn of plenty or black chanterelle
Collection notes and specimens: Arora 1986 (Ref.: 3).
Common in Bay Area; scattered to gregarious; widely distributed to clustered in mixed hardwood and coniferous forests from midwinter to spring. Common ectomycorrhizal species under madrone and also live oak and tanoak. This is a very flavorful fungus; as David Arora said: it is one of his "five favorite flavorful flesh fungal fructifications."
Ref 3. (P.666 with color plate 182), 31, (p. 383, fig. 1077), 42.

67) *Cantharellus subalbidus* A.H. Smith & Morse
(Fig. 30)
White chanterelle
Collection notes and specimens: Arora 1986 (Ref.: 3).
Scattered in groups, on ground in mixed woods. Commonly found in the Pacific Northwest. Season, Sept.- Nov in northern California. Excellent edible mushroom and an ectomycorrizal with spruce and tanoak.
Ref: 9, 36, 45.

68) *Ramaria* species
Collection notes and specimens: Arora 1986 (Ref.: 3).
Fruiting bodies: slender, hollow, branching, compact, and almost parallel, divided near the tips. Flesh: fleshy-pliable, rubbery, drying to brittle and looking like translucent plastic. Odor: not distinctive
Taste: not distinctive. Edibility: unknown
Ref: 36 (p. 293, fig.)

69) *Hydnellum caeruleum* (Hornem.) P. Karst. [as 'coeruleum'], *Meddn Soc. Fauna Flora Fenn.* 5: 41 (1880)
Blue-Gray *Hydnellum*
Collection notes and specimens: Arora 1986 (Ref.: 3).

Solitary to gregarious or in fused clusters on the ground or in the woods; widely distributed. It is common in Bay Area, especially under oak, tanoak, and madrone, usually early in fall. It is not edible.
Ref: 3 (p. 625, fig.)

70) *Lactarius subvillosus* Hesler & A.H. Sm., *North American Species of Lactarius* (Ann Arbor): 278 (1979)
Collection notes and specimens: Arora 1986 (Ref.: 3).
Ref 3. (p. 66, 73).

71) *Lactarius argillaceifolius* Hesler & A.H. Sm., *North American Species of Lactarius* (Ann Arbor): 366 (1979)
Vulgar milk cap.
Collection notes and specimens: Arora 1986 (Ref.: 3).
Solitary, scattered, or in small groups in humus under hardwoods. Fairly common in Bay Area in late fall and winter. It is know only from the Pacific Coast, but variety *argillaceifollus* is widespread. Possibly poisonous; to be avoided.
Ref: 3 (p. 76)

72) *Russula cyanoxantha* (Schaeff.) Fr., *Epicrisis systematis mycologici* (Uppsala): 352 (1838) [1836]
(Fig. 28)
Variegated Russula
Collection notes and specimens: Arora 1986 (Ref.: 3).
Spores hyaline, white in deposit, subglobose, echinulate, 7-9x65-8 nm. Cystidia clavate-fusoid, 55-75x7-9um. Flesh: slightly greenish or unchanging in FeSo4.
Widespread throughout North America, scatted on ground in mixed tanoak/conifer forests. It is reported to have anti-cancer properties.
Ref 3 (P. 94-95 with color plate 12), 9 (p. 118), 31 (p. 344 fig. 967), 36.

73) *Russula albonigra* (Krombh.) Fr.
(Fig. 41)
Blackening Russula
Collection notes and specimens: Arora 1986 (Ref.: 3).
This mushroom is distinguished by its blackish staining reaction that takes place when it is cut or handled. The flesh of the whole mushroom is granulate and brittle and will easily crumble. The taste is mild to slightly acrid.
Ref: 3 (p. 89), 31 (p. 340, fig.951).

74) *Gautieria parksiana* Zeller & B.O. Dodge
Collection notes and specimens: Redwood park, San Mateo Co., 1924, Martha Watson.
A hypogeous ectomycorrhizal fungus, this and other hypogeous fungi are an important food source for squirrels and other burrowing animals.
Ref: 12 (p. 746)
Mycologia, 14: 196-7. 1922.

4.2 Edible Mushrooms

Several of the above-mentioned mycorrhizal mushrooms are also very good edibles, notably *Boletus edulis*, *Tricholoma magnivelare*, *Craterellus cornucopioides,* and *Cantharellus sublbidus*. The list below is of some other edible mushrooms found in association with tanoak.

75*)* *Hericium coralloides* (Scop.) Gray, *A Natural Arrangement of British Plants* (London) 1: 652 (1821)
(=*Hericium ramosum* (Bull.:Merat) Lebellier)
(Fig. 33)
Collection notes and specimens: Willits, Mendocino Co., 1934, Metcalf.
Fruiting body: annual, up to 40 cm broad, 20 cm tall, a loosely, branched structure arising from a short, tough, stalk, laterally attached to the substrate; individual branches slender, brittle; spines 3-8 mm long, pendant, arranged in rows; color: white when fresh, in age becoming cream, buff to buff-brown; flesh white, soft, except tough at the base; odor and taste mild. Spores 3.5-4.5 x 3.0-3.5 µm, nearly round to oval, smooth, amyloid; spores white in deposit. Habitat: Solitary or in small groups on hardwood logs; fruiting from after the start of the fall rains to mid-winter. Edibility: Edible and good.

Hericium coralloides is the most delicate of the three species that occur in our area. It has a loose, open branched structure, and relatively short spines which are arranged in rows except for the branch tips, which are tufted. *Hericium erinaceum,* more common than *H. coralloides,* also grows on hardwood logs and can be distinguished by an unbranched cushion-shaped base from which hang long, slender, white to cream-colored teeth. *Hericium abietis* has a fruiting body intermediate in structure between *H. coralloides* and *H. erinaceum.* It grows on conifer logs, sometimes forming impressive, large, white to cream-colored mass of fruiting bodies.
Ref: 3 (p. 615), 12 (p. 20-21), 31, (p. 424. fig. 1201), 36 (p. 279 with fig.).

76) *Pleurotus ostreatus* (Jacq.: Fr.) Quél., *Führer für Pilzfreunde* (Zwickau): 24, 104 (1871)
(Fig. 34a, 34b)
Oyster mushroom
Comments: *Pleurotus ostreatus* is believed to be a species complex. In the S.F. Bay Area, specimens can be found that vary from white and relatively thin-fleshed on oaks to thick fleshed, grey-brown shelves on cottonwood and willow. Whether these differences are environmentally induced or genetic is not clear, but most mycophagists prefer the large, thick-fleshed specimens collected from cottonwood. Usually in shelving masses or overlapping rows or columns on hardwood logs and stumps. Stupendous fruitings of several hundred pounds have been reported from a field where crushed coffee beans were dumped. It is easily cultivated on a wide variety of substrates, including compressed sawdust, shredded magazines, and presumably coffee grounds. If an oyster log is brought home from the wild and kept moist, it will produce crop regularly. It is a good edible mushroom and also has medicinal value.
Ref: 3 (p. 134-135 with fig.), 12 (p. 34), 31 (p. 66, fig. 61), 36.
Jura. Vosg. 1: 112. 1872.

77) *Entoloma bloxamii* (Berk. & Broome) Sacc. [as 'bloxami']
(misapplied name: *Entoloma madidum*)
(Fig. 36)
Midnight Blue Entoloma.
Collection notes and specimens: Arora 1986 (Ref.: 3).
Solitary, scattered, or gregarious on ground in woods; widely distributed but especially common along the west coast in the fall and winter. In the Bay Area two slightly different forms occur – one with conifers especially redwood, the other with tanoak and madrone. This is a good edible

mushroom. Arora points out that it has, "meaty flesh and good flavor. However, many fleshy *Entoloma* are toxic, so be sure of your identification!"
Ref: 3 (p. 243).

78) *Tremella mesenterica* Retzius, *Neues Magazin für die Botanik* 1: 111 (1794)
(= *lutescens* Pers.)
(Fig. 38)
Witch's Butter
Collection notes and specimens: San Mateo Co., 1987, Michelle Seildl.
Fruiting body: 1-7 cm broad, flabelliform to cerebriform, gelatinous, viscid to slippery when moist, hard and stiff when dry; pallid yellowish to yellowish-orange to orange in color, flesh gelatinous.
Spores: 7-18 X 6-14 µm, subglobose to elliptical, smooth, hyaline to pale yellowish. Basidia longitudinally septate. Habitat: Gregarious on wood, where it is parasitic on *Stereum* species.
Edibility: Edible.
Comments: *Tremella mesenterica* or Witches Butter is the name assigned to most collections of yellowish-orange jelly fungi in the S.F. Bay Area. *Dacrymyces palmatus*, however, is very similar and the two species cannot be reliably told apart without the use of a microscope. *Tremella mesenterica* has longitudinally septe basidium while the basidium in *Dacrymyces palmatus* resembles a tuning fork.
This fungus is a relative of *Tremella fuciformis*, which is cultivated in China as an edible mushroom and for its medicinal properties. Although *Tremella* species have little flavor of their own, they absorb the flavors of the foods that they are cooked with, and add an interesting texture.
Ref: 3 (p. 674), 31 (p. 516, fig. 1457)
Vet. Ak. Handl. 249. 1769.

5. POISONOUS MUSHROOMS

79) *Agaricus hondensis* Murrill, *Mycologia* 4: 296. 1912.
Felt-Ringed Agaricus
Collection notes and specimens: Arora 1986 (Ref.: 3).
Solitary to gregarious in forests, particularly where there are thick accumulations of fallen twigs and other debris, sometimes forming fairy rings. It is very common in central California. David Arora notes that it is, "Poisonous to many people, causing stomach distress, vomiting, etc. It is difficult to imagine a more delicious-looking mushroom, but it has an unpleasant, astringent-metallic taste even when cooked."
Ref 3. (p. 326).

80) *Omphalotus olivascens* H.E. Bigelow, O.K. Miller & Thiers, *Mycotaxon* 3:363-372-18, 1976
(Fig. 42)
Western Jack-O-Lantern Mushroom
Collection notes and specimens: Arora 1986 (Ref.: 3).
Pileus: 6-18 cm broad, convex, broadly convex at maturity; margin incurved at first, expanding and becoming wavy, upturned in age; surface smooth, moist dull orange to orange-brown, developing olive tones in age; flesh thin, pliant, same color as cap; odor and taste mild. Lamellae: decurrent, concolorous with cap or lighter, luminescent; veil absent. Stipe: 5-15 cm long, 1-4 cm thick, central to off-central, tapering downward, smooth, yellowish-olive, with brown stains at the base. Spores: 6.5-8 x 6-6.5 µm, globose to ovoid, smooth, inamyloid; spore depost cream to pale yellow. Habitat: Clustered at the base of hardwood stumps or from buried roots; most common with oaks, tanoaks, and other hardwoods. Fruits from late fall through mid-winter.
Edibility: toxic; causes severe gastrointestinal upsets.

Comments: Arora notes: "The Jack O'Lantern fungus is sometimes also called a False Chanterelle because of its yellowish color and decurrent gills. It can, however, be distinguished from the true chanterelle, *Cantharellus formosus*, by a combination of characters: *Cantharellus formosus* has ridges rather than true gills, never develops the olive tones of the Jack O'Lantern, and grows in oak duff, not on rotting wood. *Omphalotus olivascens* is interesting in that the fruiting bodies are luminescent, at least when fresh, though to appreciate this quality, it requires sitting for many minutes in a completely dark room before the greenish glow becomes visible. In fresh specimens this glow is sometimes bright enough to read a newspaper!"
Ref: 3 (p. 147-148, color plate 40 and 41), 4.

6 SAPROTROPHIC ON DEBRIS

6.1 Saprotrophic Mushrooms

81) *Clavariadelphus occidentalis* Methven, *Mycotaxon* 34(1): 169 (1989)
(= *Clavariadelphus pistillaris* (Fr.:L.) Donk)
(Fig. 31)
Collection notes and specimens: Arora 1986 (Ref.: 3).
Forming a large club, swollen at its apex; light yellow to deep ochre; flesh soft and spongy. Odor sickly, mushroomy. Taste mild to bitter, habitat gregarious to solitary in leaf litter, especially beech, found widespread in North America. Edible, but not good.
Ref: 36 (p. 291with fig.), 38 (p. 141).

82) *Marasmius copelandii* Peck
(misidentified as *Marasmius prasiosmus* (Fr.) Fr.)
(Fig. 40)
Collection notes and specimens: Woodside/Stanford University, San Mateo Co., E.B. Copeland (type spcimen); San Mateo Co., Feb. 1939, McCabe; Marin Co., Dec 1939, Miller.
Dead leaves in mixed forest of coast range. *Marasmius prasiosmus* is a European species of uncertain occurrence in North America, with whitish gills and smaller spores. Grows on leaf litter, occurring mainly in fields and also redwood forests, but also occurring with tanoak in mixed forests.
Ref: 3 (p. 207-208), 16.

83) *Marasmius rotula* (Scop.: Fr.) Fr., *Epicrisis systematis mycologici* (Uppsala): 385 (1838)
Picture NA P.76
Collection notes and specimens: Woodside, San Mateo Co., C. F. Baker
On leaf litter.
Ref: 3 (p. 203, 206), 16, 36.

84) *Setulipes quercophilus* (Pouzar) Antonín, *Česká Mykol.* 41(2): 86 (1987)
(= *Marasmius quercophilus* Pouzar)
Collection notes and specimens: San Mateo Co., 1931.
Pileus: 2-5 mm broad, convex, broadly so to plane in age, occasionally with the disc depressed; margin decurved, sometimes becoming plane, often sulfate; surface minutely pruinose, striate wrinkled to two-thirds the distance from the margin to the disc; color light-brown at the disc, pallid to cream-buff at the margin; context very thin, pallid; odor mild, taste: untried. Lamellae: adnexed, subdistant, moderately broad, whitish, lamellulae 1-2 seried. Stipe: 1-2.5 cm long, less than 1 mm thick, round, hair-like, equal, sometimes flexuous; surface at apex pallid to pale vinaceous-brown, sparsely pruinose, elsewhere glabrous, reddish-brown to dark-brown, instititious on leafy substrate,

scattered rhizomorphs near base. Spores 7.5-9.0 x 3-4.5 µm, ellipsoid to almond-shaped, smooth, nonamyloid, hyaline in KOH; spore deposit white.
Habitat: Solitary to gregarious on rotting hardwood leaves, notably species of oak (*Quercus*) and tanbark oak, (*Lithocarpus densiflora*); fruiting shortly after the fall rains.
Edibility Unknown; too small to have culinary value.
Ref: 16.

85) *Mycena capillaripes* Peck, *Ann. Rep. NY State Mus.* 41: 63. 1888
(Fig. 35)
Miniscule Mycena
Collection notes and specimens: Arora 1986 (Ref.: 3).
Pileus: 1-2 cm broad, convex, becoming bell-shaped, slightly knobbed or umbonate at maturity; margin entire to slightly scalloped; surface smooth, translucent-striate when moist, gray-brown, sometimes tinged pale vinaceous, fading to pale gray, then slightly furrowed; flesh thin, gray, unchanging; odor of bleach; taste mild. Lamellae: adnate, moderately broad, close, in age sub-distant, ashy-gray, edges pinkish. Stipe: 4-6 cm long, 1-2 mm thick, thin, fragile, hollow; equal or slightly enlarged at the base; apex faintly pruinose, otherwise smooth or polished, colored like the cap but the apex usually paler; veil absent. Spores 8-11 x 4-6.5 µm, smooth, elliptical, amyloid; spore deposit white. Habitat: on tanoak fallen leaves, most abundant in the mountains of Pacific Northwest and northern California. It fruits in mild, moist weather. fruiting in moist weather throughout the mushroom season. Edibility: Not edible.
Ref: 3 (p. 227).

6.2 Saprotrophic Microfungi

86) *Macrotyphula juncea* (Fr.) Berthier
Fairy Hair
Collection notes and specimens: Arora 1986 (Ref.: 3).
Scattered to gregarious in humus and leaf litter, on rotting twigs, etc.; widely distributed. It occurs in our area on oak and tanoak leaves and redwood needles, but it easy to overlook. It is fairly common in the fall and winter, especially along streams and in other dank places. David Arora said "Edibility: Utterly irrelevant - a couple hundred would be needed for a mouthful!"
Ref 3.(p. 636)

87) *Coccomyces dentatus* (J.C. Schmidt & Kunze) Sacc., *Michelia* 1(1): 59 (1877)
Collection notes and specimens: Santa Cruz Co., June 24, 1955, L. Bonar.
On leaves.
Ref: *Occasional papers of Farlow Herbarium* 15:46-49, 1980.

88) *Fusicolla foliicola* P. Karst
Collection notes and specimens: Alpine Lake, Marin Co., Dec. 31, 1963, L. Bonar.
On dead leaves.
Ref: 1.

89) *Fusidium griseum* Link.
(= *Cylindrocarpon griseum*)
Collection notes and specimens: near Alpine Dam, Marin Co., Feb. 6, 1960, I. Tavares
Leaf spots on dead leaves.
Ref 1 (p. 81).

90) *Mollisia lithocarpi* Cash, Sp. *Nov. Mycologia* 50: 647. 1958
Collection notes and specimens: Smith River, Del Norte Co., 1933; Orick-Orleans Grade, Humboldt Co., 1935, H.E. Parks; Big Basin State Park, Santa Cruz Co., 1954.
On dead leaves, frequently occur in leaf spots caused by various other fungi with which it is associated.
Ref: 1, 5 (p. 647).

91) *Monochaetia hysteriiformis* (Berk. & M.A. Curtis) Guba, *Monograph of Monochaetia and Pestalotia:* 37 (1961)
Collection notes and specimens: Marin Co., 1960, Bonar. On dead leaves, belongs to Coelomycetes Spp.
Ref: Guba, E. F. 1961. Monograph of *Monochaetia* and *Pestalotia.* Harvard University Press, Cambridge, MA 324 pp. (Key to 41 *Monochaetia spp.*, descriptions and illustrations)

92) *Protopeltis lithocarpi* V.P.M. Miller and Bonar
Collection notes and specimens: near Inverness, Marin Co., Mar. 1931, H.E. Parks (coll. no. 3573), type specimen.
On leaves.
Ref: 1 (p. 2434), 38 (p. 139).

93) *Sphaerulina conflicta* (Cooke) Barr
Collection notes and specimens: French AM, 1989. (Ref: 18).
Leaf spot .
Ref: 18, 26.

94) *Aleurodiscus aurantius* (Pers.: Fr.) J. Schröt.
Collection notes and specimens: Smith River, Del Norte Co., 26 March 1946, H.E. Parks (coll. no. 6895)
On dead wood and bark surface, smooth patch and bark rot.
Ref: 1 (p. 2083), 28, 40 (p. 168-169).

95) *Cystostereum pini-canadensis* subsp. *subabruptum* (Bourdot & Galzin) Chamuris, *Mycologia* 78(3): 385 (1986)
(= *Crustomyces pini-canadensis* subsp. *subabruptus* (Bourdot & Galzin) Ginns & M.N.L. Lefebvre, *Mycologia Mem.* (St. Paul) 19: 48 (1993))
Collection notes and specimens: Del Norte Co., 1944, H.E. Parks.
Ref: 1 (p. 81).

96) *Pirex concentricus* (Cooke & Ellis) Hjortstam & Ryvarden, in Hallenberg, Hjortstam & Ryvarden, *Mycotaxon* 24: 289 (1985)
(= *Phlebia concentrica* (Cooke & Ellis) Kropp & Nakasone)
Decay fungus.
Ref: 1, 6.
National fungus collections. U.S. Dept. Agric., Beltsville MD.

97) *Mollisia cinerea* (Batsch) P. Karst., *Mycol. Fenn.* (Helsinki) 1: 189 (1871).
Collection notes and specimens: Humboldt Co., 1940, H.E. Parks.; Del Norte Co., 1939, H.E. Parks.
Ref: *Nova Hedwigia* 23:49, 1972; *Kew Bull.* 1950: 171; *Mycologia* 25: 135, 26, 31: 3,1960-66;

Grevillea 5:36, 1876.

98) *Bisporella citrina* (Batsch: Fr.) Korf & S.E. Carp., *Mycotaxon* 1(1): 58 (1974)
(= *Helotium citrinum* (Hedw.) Fr.)
(= *Calycella citrina* (Hedw.) Boud.)
Collection notes and specimens: Muir Woods, Marin Co., 1925, H.E. Parks; Berkeley, Alameda Co., 1893, Blasdale; Mt. St. Helena, Napa Co., 1939, Wing & Spurrier.

99) *Rigidoprus terrestris* (DC: Fr.) Ryvarden
(? = *Byssoporia terrestris* (Pers.: Fr.) M.J. Larsen & Zak)
Collection notes and specimens: Monte Rio, Sonoma Co., and Mt. Tamalpais, Marin Co.
In soil under tanoak and other oaks. In the above collections, this fungus is always yellow. Other localities show the fungus as delicately pink or white and later yellow.
Ref: 44.

100) *Fusidium griseum* Link
(= *Cylindrocarpon griseum*)
Collection notes and specimens: Alpine Lake, Marin Co., Feb 6, 1960, I. Tavares.

References:

1. Anonymous. 1931-1970. California Fungi (Exsiccati) Nos. 11325
2. Anonymous. 1960. Index of plant diseases in United States U.S. Dept. Agri. Handbook. No.165. Washington, D.C. 531.
3. Arora, D. 1986. Mushrooms Demystified. Ten speed press, Berkeley
4. Bigelow, H.E., Miller, O.K. Jr., & Thiers, H.D. 1976. A new species of *Omphalotus*. Mycotaxon 3:363-372.
5. Bonar, Lee. 1962. *Stegopezizella balsameae* and *Gloeosporium balsameae*. BPI. National fungus collections. U.S. Dept. Agric.: Beltsville, MD.
6. Bretz, T.W. 1952. New hosts for the oak wilt fungus, *Chalara quercina* (Abstr.). Phytopathology 42(1): 3.
7. Bretz, T.W. 1955. Some additional native and exotic species of *Fagaceae* susceptible to oak wilt. Plant Dis. Rep. 39(6): 495-497.
8. Cash, E.K. 1958. Some New *Discomycetes* from California P. 647
9. Chen, M.M. & Mu, Z. et al. 1983. The Tibetan Fungi, The Chinese Academy of Sciences Press.
10. Chen, M.M. 1984-2001. Pine rust in California
11. Chen, M.M. 2001. The Inland Alaska Forest FUNGI (not yet published)
12. Chen, M.M. 2001. The Fungi Treasure.
13. Chen, M.M. 2001. Western Gall Rust in California 2001 (in press
14. Cooke, W.B. 1956. Mycologia 48: 368.
15. Cooke, W.B. 1957. The *Porotheleaceae: Porotheleum* Tai: P. 145: 78.
16. Desjardin, D.E. 1987. New and noteworthy marasmioid fungi from California. Mycologia 79: 123-134.
17. Farr, D.F., Bills, G.F., Chamuris, G.P., & Rossman, A.Y. 1989. Fungi on Plants and Products In United States.
18. French, A.M. 1989. California plant diseases host index.
19. Gardner, M.W., Yarwood, C.E., & Duafala, T. 1972. Oak mildews. Plant Dis. Rep. 56(4): 313-317.
20. Gardner, M.W., Yardwood, C.E., & Raabe, R.D. 1970. Unreported powdery mildews. IV. Dis Dis. Reporter 54: 399-402.

21. Geesteranus, R.A.M. 1974. Studies in The Genera Irpex and Steccherinum, Persoonia 7(4): 443-581.
22. Gilbertson, R.L., & Ryvarden, L. 1986. North American *Polypores*. Fungi flora-Oslo-Norway.
23. Guba, E.F. 1961. Monograph of *Monochaetia* and *Pestalotia* 342 P. Harvard Univ. Press: Cambridge.
24. Hepting, G.H. 1971. Diseases of forest and shade tree of the United States. U.S. Dep. Agric., Agric. Handb. 386: 226-227.
25. Horst, K.R. 1979. Westcott's Plant Disease Handbook (Fourth Edition) Van Nostrand Renhold Company.
26. Jong, S.C. 1971. North American Species of *Nummularia*. Mycologia 63: 862-876.
27. Lemke, P.A. 1964. The genus *Aleurodiscus* (senus striatco) in North America. Canada J. Bot. 42: 213-282.
28. Lemke, P.A. 1964. The genus *Aleurodiscus* (senus striatco) in North America. Canada J. Bot. 42: 723-768.
29. Lowe, J. L. 1957. *Polypoaceace* in North America.
30. Lowe, J.L., & Gilberson, R.L. 1961. Synopsis of The *Polyporaceae* of the Western United States and Canada. Mycologia 53: 474-511
31. Mao, X.L. 2000. The Macrofungi in China. He Nan Sciences and Technological Press
32. Marlatt, R.B. 1958. Isolation, inoculation, temperature relations and culture of a *Cercospora* pathogenic to Ficus elastica "decora". Pl. Dis reporter 54: 199-202.
33. Miller, V.M., & Bonar, L. 1941. A study of The *Perisporiacwae, Capnodiaceae*, and some other Sooty Molds from California. University of California Publications Botany 19 (12):405-428.
34. Neinhaus, F., & Yarwood, C.E. 1972. Transmission of virus from oak leaves fractionated with Sephadex. Phytopathology 62(3): 313-315.
35. Overholts, L.O. 1953. The *Polyporaceae* of the United States Alaska and Canada.
36. Phillips, R. 1991. Mushrooms of North America. Little, Brown and Company
37. Raabe, R.D. 1967. Plants resistant or susceptible *Armillaria* root rot Univ. California Agric. Ext. Ser. AXT-6, 6p.
38. Rossman, A.Y., Palm, M.E., & Spielman, L.J. 1990. A Literature Guide for the Identification of Plant Pathogenic Fungi. The American Phytopathological Society State Univ. Coll. For. Syracuse Univ., Tech. Publ. 80: 1-97 (Key to 68 species formerly placed in Fomes)
39. Sherwood, M.A. 1980. Taxonomic studies in the *Phacidiales*: The genus *Coccomyces (Rhytismataceae)*. Occas. Pap. Farlow Herb 15: 1-120.
40. Sinclair, W.A., Lyon, H.H., & Johnson, W.T. 1987. Diseases of Trees and Shrubs. Comstock publishing Associates, Cornell University Press.
41. Tai, F.L. 1979. Syllogu *Fungorum Sinicorum*. Science Press, Academia Sinica. Peking, China.
42. Teng, S.C. 1987. Fungi of China. Mycotaxon, LTD.: Ithaca, New York
43. Thorn, W.A., & Zentmyer, G.A. 1954. Hosts of *Phytophthora cinnanmomi* Rands Pl. Dis reporter 38: 47-52.
44. University Herbarium. 2001. University of California, Berkeley, Host Index Fungi of California (red book notes and black books) draft, file drawer
45. Wood, M. 2001. Fungi of The San Francisco Bay Area (Species Index and photo-Pictures)
46. Yu, Y. 2000. Collection of Mushrooms Prescription. The Agricultural Press
47. Werres, Sabine and Marwitz, Rainer 2001 *Phytophthora ramorum* sp. nov., a new pathogen on *Rhododendron* and *Viburnum* Mycological Research 105 (10): 1155-1165

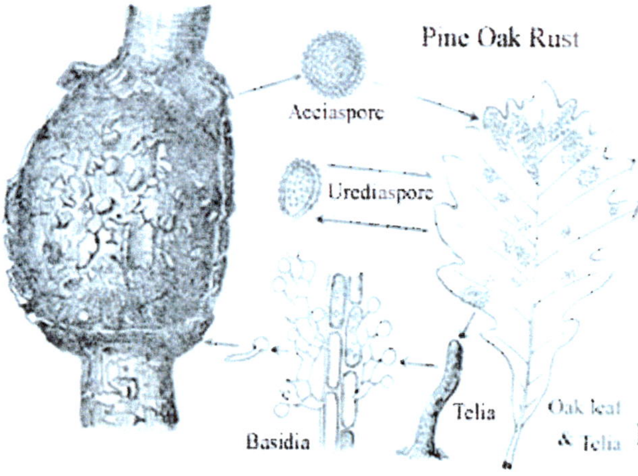

Fig 1. Cronartium quercuum (Berk.) Miyabe

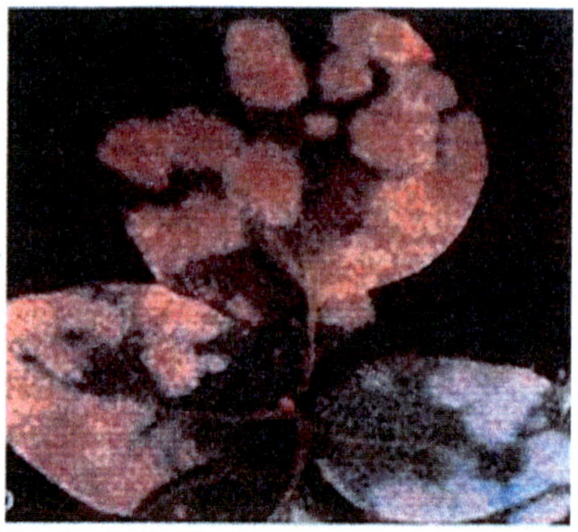

Fig 2. Sphaerotheca lanestris Harkn

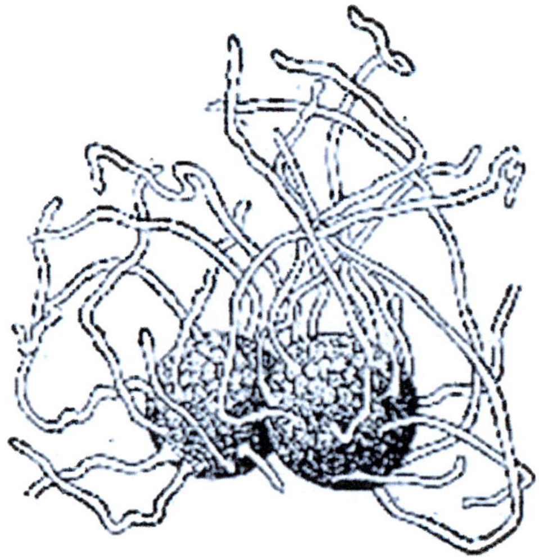

Fig 3. Sphaerotheca lanestris Harkn

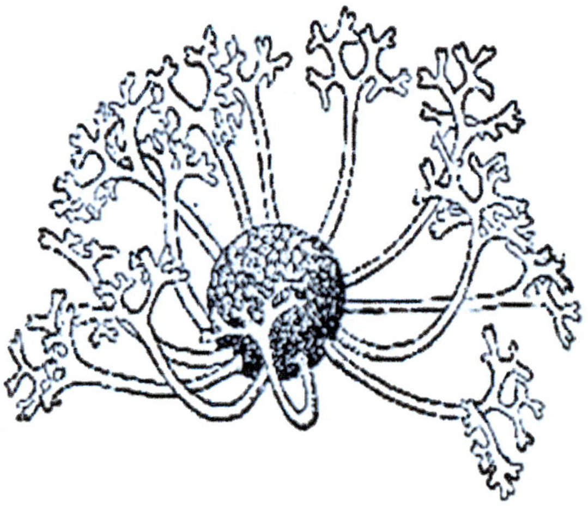

Fig 4. Microsphaera penicillata (Wallr.: Fr.)

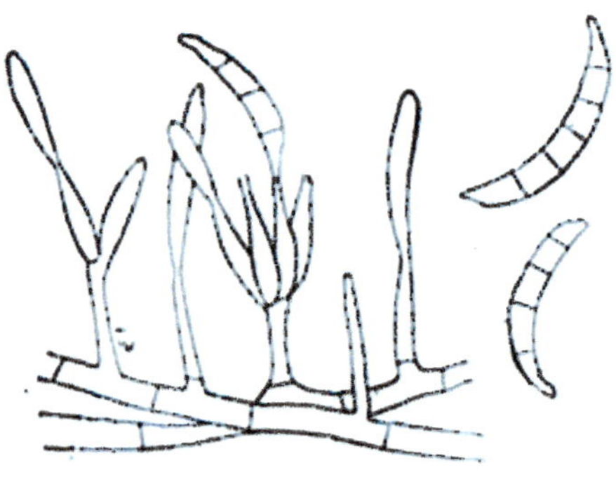

Fig 5. Fusarium

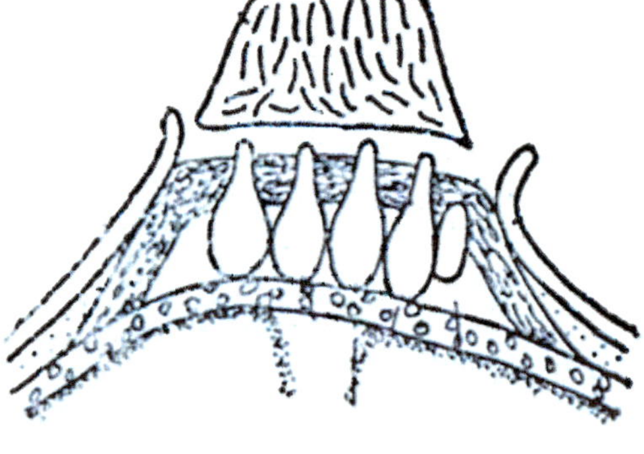

Fig 6. Diatrype disciformis (Hoffm.) Fr.

Fig. 7a Dendrothele candida (Schw.:Fr.) Lemke (Taken from Sinclair, 1987)

Fig. 7b Dendrothele candida (Schw.:Fr.) Lemke (Taken from Sinclair, 1987)

Fig. 8 Endothia gyrosa (Schw.) Fr.

Fig 9a. Nectria coccinea (Pers. ex Fr.)

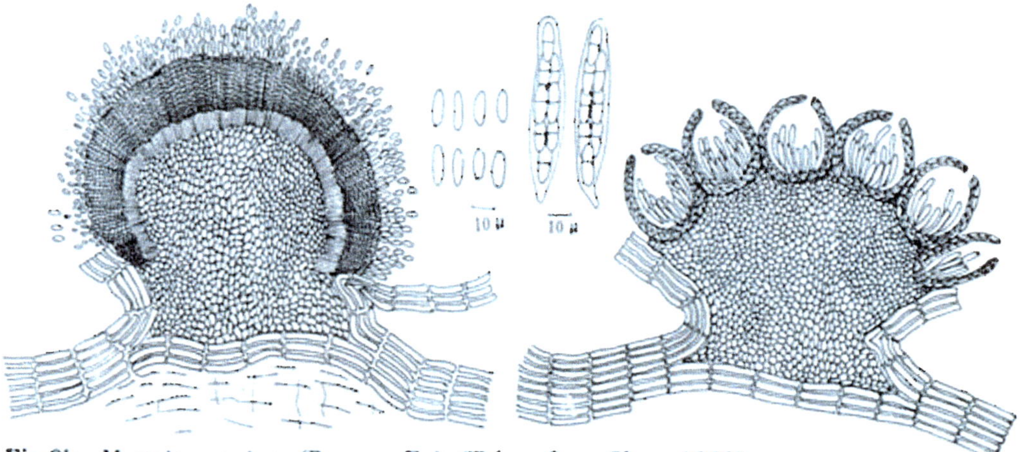

Fig 9b. *Nectria coccinea* (Pers. ex Fr.) (Taken from Shao, 1983)

Fig 10. *Armillaria mellea* (Vahl: Fr.)

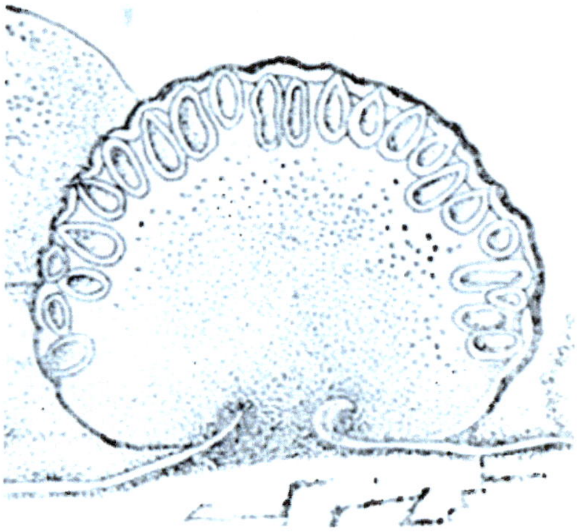

Fig 11a. *Hypoxylon mediterraneum* (De No.) J.H. Miller var. *microspora* Miller

Fig 11b. *Hypoxylon mediterraneum* (De Not.) J.H. Miller var. *microspora* Miller

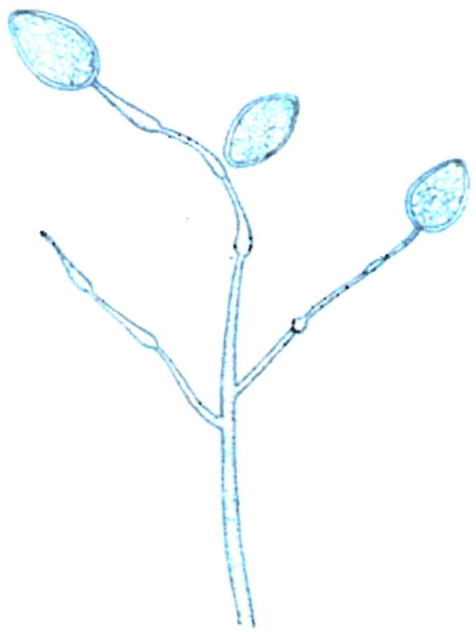

Fig 12. *Phytophthora*

Fig 13. *Steccherinum fimbriatum* (Pers.: Fr.) J. Eriksson (*Odontia fimbriata*) (From Arora, 1986)*

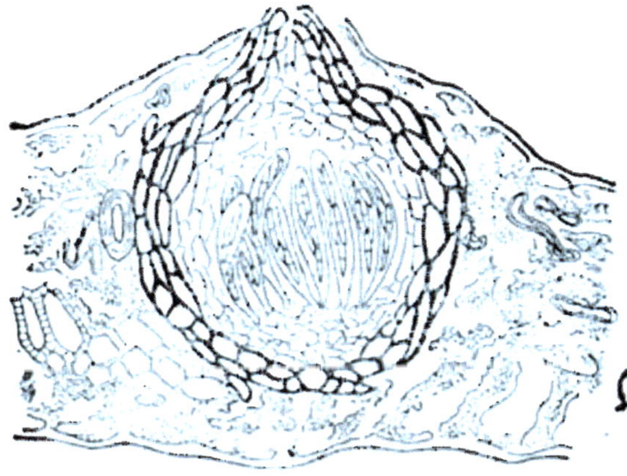

Fig 14. *Mycosphaerella janus* (Berk. & Curt.) Petrak, Sydowia 11: 340 1958.

Fig 16. *Capnodium coffeae* Pat. (From Teng, 1987)**

**Mushrooms Demystified* by David Arora
***Fungi of China* by Teng Shu Chün

Fig 17. *Pestalotia montellica* (Sacc. & Voglino) Tak. Kobayashi

Fig 18a. *Phyllosticta quercus* Sacc. & Speg. (Taken from Sinclair, 1987)

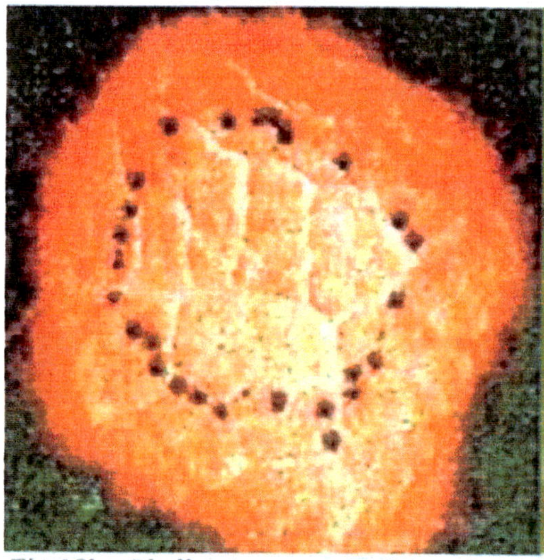

Fig 18b. *Phyllosticta quercus* (Sacc. & Speg.)

Fig 19. *Phellinus viticola* (Schw.) Donk (From Mao, 2001)

Fig 20. *Phellinus gilvus* (Schw.) Pat. (From Mao, 2001)

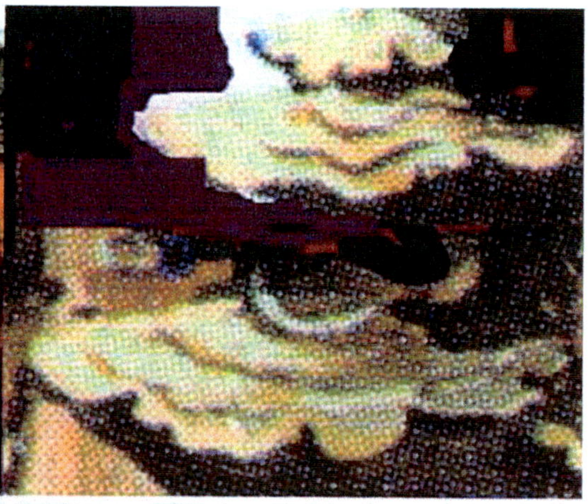

Fig 21. *Stereum hirsutum* (Willd.) S.F. Gray

Fig 22. *Stereum fasciatum* Schw. *forma zonatum* (From Mao, 2001)

Fig 23. *Boletus aereus* (Bull.: Fr)
(From Yang, 2003)

Fig 24. *Armillaria ponderosa* Peck
(From Yang, 2003)

Fig 25. *Cortinarius infractus* (Pers.) Fr.
(From Yang, 2003)

Fig 26. *Hygrophorus russula* (Fr.) Kauffman
(From Mao, 2001)

The Macrofungi in China by Mao Xiao Lan
Portraitist and mushroom printer Yang Chen
Diseases of Trees and Shrubs by Sinclair

Fig 27. *Cortinarius collinitus* group (From Mao, 2001)

Fig 28. *Russula cyanoxantha* (Schaeff.) Fr. (From Yang, 2003)

Fig 29. *Cratesellus cornucopiodes* (L.) Pers

Fig 30. *Cantharellus subalbidus* A. H. Smith & Morse (From Yang, 2003)

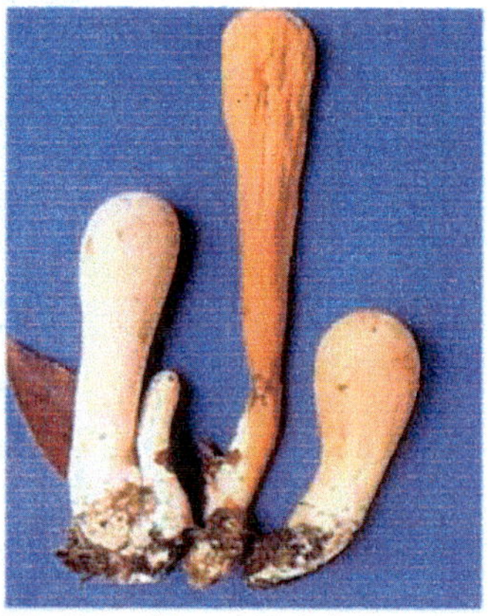

Fig 31. *Clavariadelphus occidentalis* Methvan

Fig 32. *Tricholma terreum*

Fig 33. *Hericium coralloides* (Scop.) Gray (From Abrams, 1982)

Fig 34a. *Pleurotus ostreatus* (Jacq.: Fr.) Quel (Oyster Mushroom)

Fig 34b. Oyster Mushroom 3015 Amycel grown in straw in milk carton.

Fig 35. *Mycena capillaris*
(From Yang, 2003)

Fig 36. *Entoloma bloxamii* (Beck & Broome) (From Yang, 2003)

Fig 37. *Cortinarius cotoneus*
(From Yang, 2003)

Fig 38. *Tremella mesenterica* Retzuis

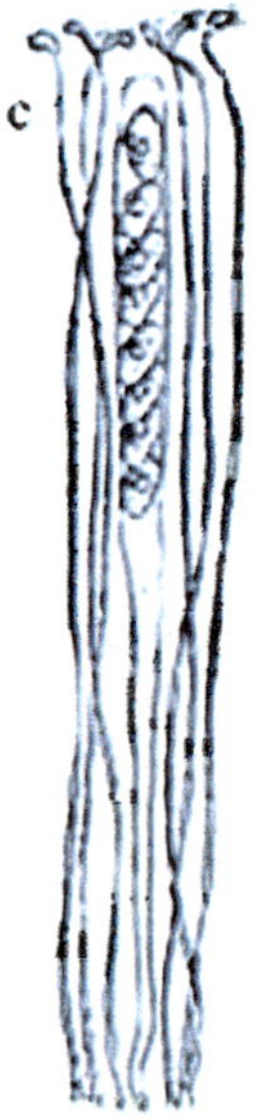

Fig 39. *Bulgaria inquinans* (Pers.: Hook)
a. ascocarps, b. ascospores, c. ascus & paraphyses

Fig 40. *Marasmius copelandii*
(From Yang, 2003)

Fig 41. *Russula albonigra* (Krombh.) Fr.
(From Yang, 2003)

Fig 42. *Omphalotus olivascens* H.E. Bigelow

Cultivation techniques for *Dictyophora, Polyporus umbellata,* and *Coprinus comatus*

ABSTRACT: Results of research on the bamboo long skirt mushroom (*Dictyophora indusiata.*), Zhu Ling (*Polyporus umbellata*) and drumstick mushroom (*Coprinus comatus*) are presented. The conditions for growing these three edible and medicinal species are best in China and California where the weather conditions are optimal for all three species. Detailed information on cultivation strains, spawn production and substrates, pH, temperature, humidity, and nutrition requirements for each species will be presented. Research has shown that *Dictyophora* cultivation requires special techniques: (1) inoculation of the tough mycelium with a "sharper applicable tool," and (2) careful management of the fruiting stage during which the basidiocarp (egg) breaks and cast skirts. Zhu Ling cultivation requires selected high-quality sclerotia and cultivated rhizomorph logs. New data show that *Coprinus comatus mycelium* grown on corn media rapidly produces spawn. Liquefaction of fruiting bodies at outdoor production can be prevented by effective growing measures.

1 *DICTYOPHORA* AND ITS CULTIVATION

Bamboo skirt mushrooms, *Dictyophora*, are basidiomycetes belonging to the family Phallaceae. There are many nicknames for these mushrooms in China, such as "ghost holding an umbrella," "mushroom Ginseng," "the veil mushroom," "the king of mushrooms," "the queen of mushrooms," "the flower mushroom," and "the king of medicinal mushrooms." In China, it has been used as a tonic to treat high blood pressure, tumors, and diseases of the kidneys, eyes and lungs (Liu 1978, Ying et al. 1987). An interesting characteristic of the bamboo mushroom is that it is a natural preservative for other foods. It also contains many proteins, about 15-18% of dry weight; different kinds of amino acids, about 16 types in all, including glutamic acid (Huang 1993) and high concentrations of riboflavin (vitamin B_2) (Huang 1993). Therefore, its nutritional value is very high. Bamboo skirt mushrooms are highly valued in Chinese cuisine for their flavor and fragrance, and they are often used for banquets in high-class restaurants.

In nature, *Dictyophora* species are found in the Sub-Alpine zone at elevations of 1000-2000 meters in *Pleioblastus* and *Sinocalamus* bamboo forests in Yunnan, Sichuan, and Guizhou. Cultivation began in 1968. Currently, *Dictyophora* species are grown in Fujian, Guizhou, Sichuan, Shandong and Liaoning (Liu 1978, Huang 1993, Fang & Yu 1996, Yao & Huang 1993, Lu et al. 1992).

1.1 *Dictyophora Desv.*

According to the Dictionary of Fungi and Names (Yu et al. 1986, Teng 1996), Dictyophora comprises nine species, four of which are known to be edible: *D. indusiata* (Vent.) Fischer, *D. duplicata* (Bose) Fischer, *D. echinovolvata* Zang and *D. merulina* Berk. Of these, *D. indusiata* and *D. echinovolvata* are frequently cultivated (The Edible Mushrooms Institute, The Academy of Shanghai Agriculture 1991, Huang 1993, Chen 1987, Fang & Yu 1996), although only *D. indusiata* is available in quantity.

1.2 Dictyophora indusiata *(Vent.) Fischer. Morphological description* (Fig. 1)

Receptacle 12-20 cm. high, sheathed at the base by a whitish volva 3-5 cm. in diameter. Pileus campanulate, 3-5 cm. long and broad, strongly reticulate, covered with olive-citrine slimy gleba of slightly unpleasant odor, apex truncate and perforate. Veil white, extended below the cap for more than 10 cm., made up of slender tubular threads, with polygonal meshes 5-10 mm in diameter. Stem white, hollow, with cellular-spongy walls, tapering gradually upward, 2-3 cm thick near the base. Spores 3-3.5 X 1.5-2 mm. Distributed in Yunnan, Hainan, Jiangxi, Guangdong. (Teng 1996).

1.3 *Long Skirt mushroom cultivation* (Dictyophora indusiata) (Fig. 2)

Spawn culture production goes through three steps: 1) isolation of the mother culture, 2) original culture expansion, and 3) spawn grain propagation. Keeping the culture sterile is very important; everything should be kept clean, including the laboratory, clothes, tools, and especially the cultivator.

1.3.1 *Mother culture*

Use the spore isolation with general guidelines for spore collection; place the receptive body (basidiocarp bud) hand-up in the spore collection glass container; incubate at 22°C until the sheathed volvo develops and breaks, the pileus becomes campanulate, long and broad, and the central stem becomes covered with slimy gleba. Put a drop of glutinous slimy gleba into sterile water, make a spore suspension, then inoculate in the special PDA medium*. When the spores germinate into mycelium, they are ready for the mother culture.

Since the entire long skirt bamboo mushroom is composed of compressed mycelia, the basidiocarp buds represent the best locations for tissue isolation. Select a few clean living basidiocarp bud specimens, split the bud into two parts to expose interior tissue by cutting tough mycelium with "Lu's sharper applicable tool" (Lu et al. 1992); excising a piece of tissue for transfer into the special PDA dish as mother culture.

*Special PDA media menu: bamboo root pieces 100g, fert roots 100g, pine needles 5-10 g, potato flour 100g , dextrose 20 g, agar 18-20g, and water, 1,000 ml (Huang 1993).

1.3.2 *Original culture and spawn cultivation*

Inoculation of sterilized grain from original PDA dish culture and inoculation of grain from grain master jars. Use the same cultivation method and a grain ingredient: 75% bamboo substrates, some times use reed instead of bamboo (Chen 1997, Liu et al. 1997), 18% bran, 3% soy bean powder, 1% limestone, 1% sugar. The mixture should have an 65% water content, and a pH of 6.5 (Mei et al. 1997). Mix the compost then add the supplement; the compost at filling should release some moisture when firmly squeezed. Fill the mason jars with spawn grain with the lids making an imperfect seal to allow some air exchange during commercial spawn makers' autoclave sterilization. 2-3 original special PDA dishes per 500 ml-grain master mason jar inoculation. After that, place jars at 18°C during the first week, at 20°C for the second week, and 23°C for the third and fourth weeks. After this, incubate at 15-29°C for 60 days cultivation (Fang & Yu 1996).

1.3.3 *Preparation of spawn*

Recently research has shown (Mei et al. 1997) that a high-quality spawn culture can be produced using the following method: A substrate of 68% mixed hardwood sawdust, 5% bamboo leaves (or reeds,

Phragmites communis) and 5% needles (add 2 ml boiling water), 18% bran, 3% soybean litters, 2% sugar, and 1% limestone is autoclaved in plastic bags (240x120x.4mm) and inoculated with the contents of one spawn culture mason jar per bag. The bags are incubated at 15-29°C for 80 days.

1.3.4 *Outdoor "Mushroom Qi Bed" production*

This is a popular field cultivation method used by Chinese Farmers. A "Mushroom Qi Bed" consists of well-drained rectangular areas (100 x 400 x 30 cm) on which the mushrooms are grown. These plots are separated by ridges for growing mushrooms, 0.5% limestone water is used on a substrate of bamboo litter or reeds or wood chips that has been sterilized for 48 hours. The substrate should be well drained but retain 65% moisture content. For each square meter use 25 kg substrate; pile it 80-100 cm. wide with 10 cm.-high ridges on the Qi bed. Spread one layer of spawn culture then cover with 5 cm. of substrate. The ratio of substrate to spawn is 15:1. Then spread another layer of spawn. (Fig. 3)Repeat the process until there are three layers of spawn. The final, top layer should consist of a 5 cm.-thick layer of composted bamboo (or reed) litter. Maintain consistent humidity and aeration with plastic covers in the outdoor bed.

1.3.5 *The management techniques*

To achieve good fruiting with this method, the fruiting should occur 40-60 days after spawning. The temperature for mycelial growth is 28-33°C. An additional fifteen days are needed at 20-24°C for the mycelium to reach physical maturation. Then the surface of the substrate is full of mycelium, in mass primordium formation. At this time the relative humidity should be 85%-90%, and the water content of the substrate should be 60%-65%. During the period of about 30 days between breaking–basidio carp (egg) to skirts casting, air circulation is required and a temperature of 18-25°C (no higher than 25°C). Basidiocarp buds (the eggs) gradually develop into an ovaliform basidiocarp. As soon as the top membrane becomes thin, the mushrooms will all break the basidiocarps for the next 1-2 days. Then, in several hours the fruiting bodies mature and the long skirts spread, (Fig. 4) to be harvested immediately (Mei et al. 1997).

After harvesting, only the stem (not the skirt or the receptacle) is suitable for eating.

2 CULTIVATION OF POLYPORUS UMBELLATA

Zhu Ling, *Polyporus umbellate* (Pers.) Fr., is a basidiomycete fungus belonging to the Polyporaceae family (Fig. 5a). The sclerotium of Zhu Ling is medicinally unique. In traditional Chinese medicine, it has been used mainly as a diuretic, but recently it has been shown to be beneficial in treating lung cancer and leukemia (Liu 1978, Ying 1987).

Zhu Ling grows on honey fungi (*Armelliaria mellea*) rhizomorph for nutrients, leading to a unique cultivation process (Huang 1993, Yao & Huang 1993).

2.1 *Selected high-quality sclerotia*

Zhu ling cultivation uses sclerotium as spawn, the exposed fruiting body borne out of underground sclerotia. Sclerotium is irregularly shaped while the fruiting bodies are rounded, clarinet-shaped, the surface having a swollen appearance and bearing small scales and fine strands. The exterior is colored black to brown with internal color being near white to light yellow. After drying, the tissue is firm, woody, and cork-like when squeezed (Fig. 5b). The tissue side of the fruiting body is white but after exposure to air, it becomes light brown. Zhu Ling sclerotia have three varieties of shapes which Chinese farmers describe, graphically, as pig scat, chicken scat, and horse scat. It is better to select the pig and horse scat appearing sclerotia for sclerotia spawn. Use young sclerotia, which are 1 to 2 years old. The color should be yellow-green and the consistency soft when squeezed (Tian & Jian 1998).

2.2 *Honey fungi logs cultivation*

The spawn plug method is used for the cultivation of *Armellaria mellea* rhizomorph in logs, in the similar way, as traditional Shiitake are cultivated on logs. Alder logs are appropriate for rhizomorph substrate material. 1 x 2 cm sized chip sections are chopped for chip cultures, and then inoculated and incubated at 25°C for one month. When the *Armelliaria mellea* mycelium is fully-grown on hardwood chips, it is inoculated into each "scale of fish" spot of log. Each log is about 50 cm long and 8-10 cm. in diameter with water content of 70% and receives 10 spots inoculation. The logs are laid in a sand-lined pit 50 cm. deep, 70 cm. high and 70 cm. wide, with sufficient moisture and minimal air exchange. After 2 months of incubation at normal room temperature, the *Armelliaria mellea* rhizomorph will have fully colonized the exterior of the log providing excellent nutrients for Zhu ling production (Liu 1978, Tain & Jian 1998).

2.3 *Zhu Ling cultivation*

First a humus rich cultivation bed of 50 cm wide and 50 cm deep is prepared (in China this bed is referred to as a "den"). A 5 cm layer of tree leaves lines the bottom of the den and on this is placed a layer of fully colonized Honey Fungus logs that have been prepared by the above method. Above and in contact with this layer is another layer of new un-inoculated hardwood logs and the spaces in between are filled up with fresh tree leaves and chips and conditions are kept moist. The important point is that the sclerotia must reach to Honey Fungus logs in exposed positions during inoculation and cultivation process. Because the strong *Armelliaria mellea* mycelia rhizomorph sustained providing nutrients for new sclerotia grown on logs under the wood chips under the humus soil, keep conditions moist. After four or five more log layers are added, a 10 cm mixed layer of wood chips and humus soil are placed on top of the den. For cultivation of sclerotia from germination to growth until production, the new sclerotia require a process of 2-3 years (Tain & Jain 1998, Yao & Huang 1993).

3 CULTIVATION OF THE SHAGGY MANE MUSHROOM FOR FOOD AND MEDICINE

Coprinus is a genus of the basidiomycete fungi belonging to the family Coprinaceae.

3.1 *Nutritional value*

A variety of *Coprinus comatus* (Muell. Ex Fr.) Gray var. *ovatus*, also known as the "White Chicken leg mushroom," is an edible wild mushroom which is common in many parts of China, appearing often after rain from late spring through autumn. It has long been prized as a food with fine color, texture and flavor and as a valuable medicinal resource. Traditionally, the white chicken leg mushroom has been used by Chinese physicians to calm the mind and to treat the spleen, stomach and digestive system. Modern chemical analysis (Ying et al. 1987) revealed that white Chicken leg mushroom contains 20 kinds of amino acids (among them 8 kinds from human beings). The mushroom cap contains aspartyl, asparagines and glutamine. The stem is a source of glutamine, glyceric acid, threonine, ß-aminobutyric acid, isoleucine, and lysine. The Chinese Medical Fungi Illustrated Monograph reports that compounds extracted from white Chicken leg mushroom were used to treat mice with malignant tumors and appeared to have a significant effect in restraining the development of the cancers. Results also suggests that white Chicken leg mushroom contains compounds which can lower blood sugar in mice and may have a potential for the treatment of diabetes. Many generations of Chinese farmers have cultivated the white Chicken leg mushroom and in recent decades growers in the USA, the Netherlands, France, Germany, Italy, and Japan have been successful in the commercial production of *Coprinus* from which a variety of food and medicinal products are made. (Liu 1978, Ying et. Al. 1987)

3.2 Coprinus comatus *var.* ovatus *cultivation process*

The ground-arch-roof mushroom room is recommended as the best environment for cultivation of winter crops (Fig. 6a). A suitable temperature should be kept during harvesting winter fruiting-bodies and should be maintained inside the growing chamber while cold outdoor temperature provides the best conditions for preserving fresh mushrooms while they are being transported to market.

3.2.1 *Mother culture*

Research shows (Liu et al. 1998) that the best mother culture media is multiple PDA with wheat. It is composed of potato flour 200 g, glucose 20 g, 1% KH_2PO_4, 0.5% $KHPO_4$, 0.5% $MgSO_4$, peptone 3 g and agar 20 g. The standard method requires 25-27°C for seven to nine days. Currently, the Shan Dong Chicken leg 9653, Gui Zhou, CC100, Jiang Shu 963, Hunan CC168 are the most productive strains (Luo & Qian 1999).

3.2.2 *Original and spawn culture cultivation*

Use PDA to expand the mother culture on the petri dish. The spawn grain ingredients consist of cottonseed casings (75%), sugar (1%), limestone (2%), and water to about 65% approximately. The pH level should be neutral after inoculation onto the spawn media into 500-ml mason jars. Place the sterilized grain-filled jars in the sterile room until ready for inoculation (Luo & Qian 1999, Luo 1997). Also research suggests (Luo 1997) that when using a large size grain such as corn as inoculation grain, a hole should be made at the center of the jar's cover with a woody pencil, because this gives a much easier inoculation process, when filling the jar from top to bottom.

3.2.3 *Cultivation techniques*

Use the standard method to inoculate. Incubation requires 23-25°C for 20 to 30 days. Then mycelium will colonize the entire medium. Plastic Bags of 18 x 35 cm are used. Cottonseed meal (100 kg) is considered the best filling substrate with a supplement of Urea (0.5 kg), phosphorous fertilizer (2.0 kg), limestone (1-2 kg), and water (150 kg). Use a bag filling machine. Seal bags by tying up both ends. Standard sterilization: autoclave for 10-12 hours. Inoculate both sides. Each jar of spawn grain will be enough for forty bags. Incubate twenty-five to thirty days in the temperature range of 24-26°C for spawn run. (Zhu 1998, Mei 1997). Once white mycelium has thoroughly filled the bags, they are ready for ground cultivation.

Inside a ground-arch mushroom cover; prepare a den of 30 cm. deep in sandy rich soil. After removing the plastic bags, distribute the mycelium "logs" 30 per square meter leaving a 2 cm. space between each one. Completely fill and cover the den with the clean sandy soil. This species grows well under soil. Maintain temperature within 16-22°C and 85-95% humidity (Liu et al. 1999). Chicken leg mushroom fruiting body requires both moisture and oxygen. Ground level windows built into the ground-arch mushroom cover aid ventilation. Mushroom fruiting buds will then emerge from the ground soil. Fruiting bodies quickly form within 7-10 days (Liu et al. 1999) (Fig. 6b), depending on moisture and air circulation. Mushrooms are ready for harvest at 70-80% maturation. If they are left to mature further, "ink" will start leaking from the fruit body and the mushroom will decompose rapidly. This process of cultivation is a prevent-liquefaction technique.

REFERENCES

Chen, H. 1997. The Utilization of Reeds (*Phragmites communis*) for *Dictyophora indusiata* cultivation. *Edible Fungi* 5: 26.

Chen, S.Y. 1987. *Encyclopedia of Edible Mushrooms Cultivation.* Beijing: Chinese Agriculture Publisher House. pp. 450-459.

Fang, S.C. & Yu, P.S. 1996. *The Dictyophora High Yield Integrate Cultivation with Orchard Field. Edible Fungi*: 29.

Huang, J.P. 1993. *Mushroom Cultivation.* Hunan: Hunan Sciences and Technology Publisher House. pp. 206-221.

Lin, X., Zhou, X. W. & Tao, Y. F. 1998. The Preliminary Report on Selection of Stock Culture Medium of *Coprinus comatus. Edible Fungi of China* 2: 22-24.

Liu B. 1978. *Medicinal Fungi of China.* Tai Yuan, Shan Xi the People's Publish House. pp.252, pp. 111-114, pp. 177-182.

Liu, R.B., Wei, C.Y. & Xie, L.G. 1997. The Model of Integrates Cultivation of Bamboo Mushroom and Soybean Crops. *Edible Fungi* 1:33.

Liu, W.B., Hu, Z.M. & Wang, C.H. 1999. The Study on *Commercial Cultivation Techniques of Coprinus Commatus. Edible Fungi of China* 2: 32-33.

Lu, Z.Z., Chen, L.G. & Xie, B.G. 1992. *The 200 Questions of Edible Mushroom on Cultivation.* Beijing. Chinese Agricultural Publisher House.

Luo, T.X. 1997. The CC100 *Coprinus comatus* Charaters and the Techniques of Cultivation Spawn Rapidly. *Edible Fungi* 4:16.

Luo, T.X. & Qian, Z.M. 1999. The Key Techniques of the CC100 *Coprinus comatus* Cultivation. *Edible Fungi* 4: 14-15.

Mei, L.P., Wu, M.F. & Wu, C.M. 1997. *The Outdoor cultivation Tehcniques of Dictyophora. Edible Fungi* 5: 27-28.

Teng, S.C. 1996. *Fungi of China.* Ithaca, New York: Mycotaxon LTD. p. 489.

The Edible Mushroom Institute, The Academy of Shanghai Agriculture. 1991. *The Flora of Edible Mushrooms of China.* Beijing: Chinese Forestry Publisher House. pp. 271-274.

Tian, M.L. & Jiang, J.C. 1998. The Techniques Cultivation of *Polyporus umbellate. Edible Fungi of China* 1:22.

Yao, Z.F. & Huang, Y. 1993. *The Cultivation Technique of Common Chinese Medicinal Herbs.* 1989. Beijing: Jing Duang Publisher House. pp. 545 550.

Ying, J.Z. et al. 1987. *Icons of medicinal fungi from China.* Beijing. Science Press. pp.572-575.

Yu, Y.N. & Zhen, R.Y. et al. 1986. *The Dictionary of Fungi and names.* Sciences Press. Beijing. p. 210.

Zhu, J.B. 1998. The Cultivation Techniques of *Coprinus comatus* on Shanghai Nan-Hui County. 1998. *Edible Fungi* 3:32.

From
Science and Cultivation of Edible Fungi, vol 2. 2000. Rotterdam, Netherlands: A.A. Balkema. pp. 543-548

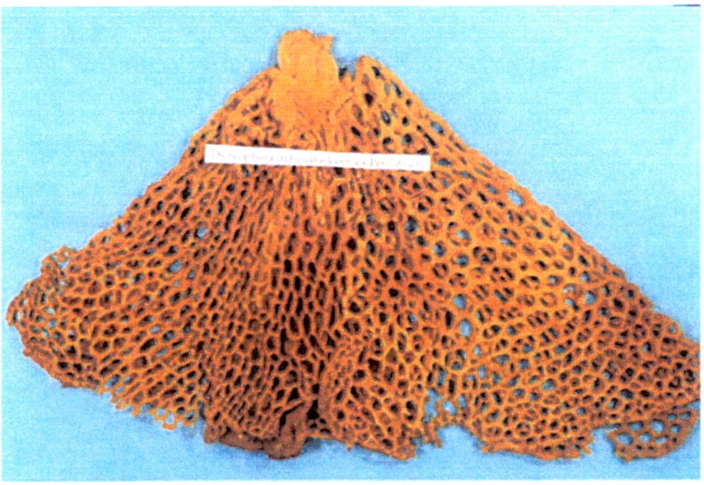

Fig 1. *Dictyophora indusiata* specimen.

Fig 2. *Dictyophora indusiata* (Huang Jian Ping).

Fig 3. Layers of spawn.

Fig 4. Skirts casting.

Fig 5a. Zhu Ling Fruiting Body (Shanghai Edible Fungi Institute)

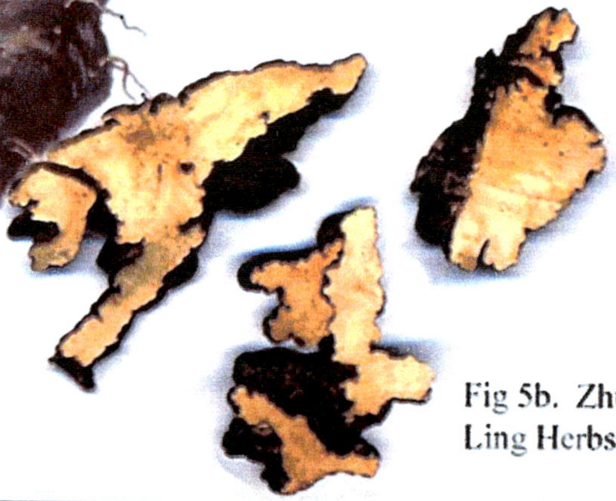

Fig 5b. Zhu Ling Herbs

Fig 6a. A ground-arch-roof room

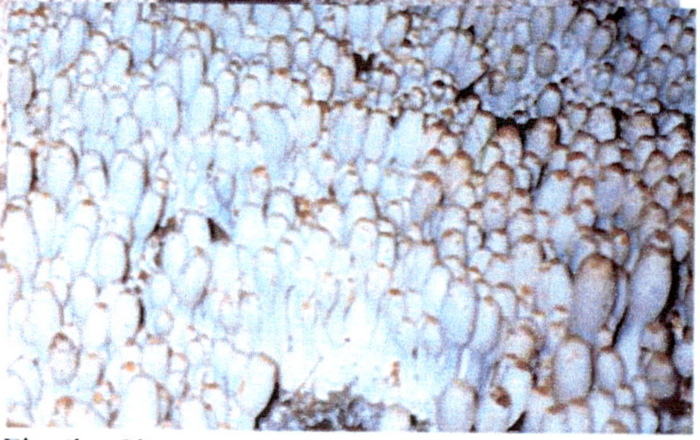

Fig 6b. Shaggy mane mushroom (Xu Zhou, China)

“白色金子”- 持续发展

(American Edible Mushroom Sustainable Tendencies)

人类赖以生存之空间，科学家们称之为生物圈（Biosphere），而今由于经济，社会或工业过速及不合理运作，损害生物圈有效持续发展，造成水，空气，食物污染，于是这一与人类生命和生活品质息息相关的问题已提到当前重要议事日程，这个问题不仅广为东西方人民大众之关注，而且对国际的科学家提出了严峻的挑战，这个课题是“地球如何持续发展”。

一，蘑菇文化和市场

中国是农业历史最悠久的国家，美国仅两百年历史，不论东西方历史文化不同，但人民大众都有食菇和栽培菇类历史经验，美国食菌原产品种类虽少，但单产高，品质好，产业化管理科学；中国几千年人民伟大实践选育出丰富多样栽培种，真菌产品用于生物防治，药物，化妆品，尤以新的加工食，药品种日新月异，国内市场畅通，逐渐进入国际市场。传统的西方文化认为野生种产生经济价值有限，必须大力发展孟达尔遗传学以促进增产，增加新品种及菇类产业化。但是由于 DNA 实验工程昂贵，日前出现的新品种仍是从野生中选出的；鉴于近年来美国民众掀起学习中医传统热潮，随之中草药包括菌药很快进入了华人集聚之大城市如旧金山，洛杉矶，纽约，西雅图等市场；美国是个法制国家，中药材正在全力争取它们站脚之合法地位，据知近期即公布草，菌药进口规定；目前药菌备受关注，大小生物制药公司利用其先进设备，生物高科技生产新生物药制品，试图找出抗爱滋病，抗癌以及用于心血管，疑难病之新药物，这些都是世界商人们投资生物技术之焦点。

至于大家还有另一重点是用真菌清理环境，美国环保署及犹他大学专家们所进行之“蘑菇是 DDT，PCBs，Dioxon 之杀手”就是一个重要研究成果，但在环境保护，废物处理研究项目上需求品目繁多，利用真菌为资源广为政府及大小环保公司重视。

二，精致农业

品种繁多的东方食用菌的消费者多为各大城市的“中国城菜场”以及各个万人以上人口之城镇之中国餐馆，无数美国人都以有生吃到一种新品种蘑菇为自豪，他们说“中餐佳肴东方菇味最鲜美”；普遍的中国餐馆使用干香菇，罐头草菇及黑毛木耳，近十年这几种菇为中国餐馆生意带来了数以亿计的盈利，但美国餐馆，法式，意大利及其他欧洲餐馆及高档西方大饭店仅用少量新鲜菇类如扣子菇，大钟菇，香菇。目前社会上流行素食主义，吃素的人多了，鲜蘑菇大受欢迎，仅知我所在之柏克莱加大里 38%女学生素食，18%男学生素食，对蘑菇美誉为“荤中之素，素中之荤”的说法颇为推崇。

去年世界三百个城市生活品质比较，旧金山名列首矛，该区称旧金山海湾地区，位于北美西海岸加利福尼亚中部，是典型地中海气候，产有红杉和花旗松又是世界著名产菇地区，可食菌根产量极富，每当十月雨季来临各式各样奇光异彩之蘑菇出土，给海湾大众带来别致风味和艺术享受和美食，加州农业中果蔬渝牧为全美首屈一指，而该地之精致农业- 蘑菇小农及大菌种公司，产业化蘑菇厂正如雨后春笋般建立起来。

三，濒危遗传基因库保护

人们对菇类是如此之热爱，但传统文化中（不论东西方）均认知野生质地好，因此随着人们的需求“采野菇”逐渐成为一门商业，据奥立岗州农产品林产品中价值最高收入为菇类，最新出版之书籍称它为 Non-Timber Forest Products，中国称它为林副产品，此概念渐已在西方流行起来；然而一些事故发生了，大约是一九九二年前后日人以高价收购北美西海岸“松茸”时，发生了乱采，采光乃至相互抢采和枪杀等命案，无独有偶就在中国由于冬虫夏草之传奇故事，被像炒作股票式“冬虫夏草”一磅售价六百美元，商品经包装真假难分，使这一高原特有种采售过度，为此不少生态学家和真菌学家提出保护“白色金子”警告，要求自然保护和永续利用那些濒危遗传基因资源；首先联合真菌团体向政府要求经济及政策支持，以下是他们一些做法：

(1) 以美西北真菌协会，旧金山真菌学会为主持“保持永续性采集野生菇类”研究，并制定菇类执照法（由政府部门制定管理），无采集执照者于国有森林，公园，地区不得采野厌

(2) 与林业，生态学部门单位大力提倡研究菌根菌，发展菌根菌生产工厂，使用飞机散布菌根菌，特别针对主要针叶林地区及新造林地区。

(3) 学者著书立说以提高真菌在生态环境中之地位；近期提出 No Fungi, No Future 之口号；九零年密西根真菌学界发表了震撼全美最大新闻- 一株世界最大菌类；1995 年泰勒发表于真菌学报中之文章- 早泥炭纪的菌根化石研究；九零年代一本巨著- “真菌群落她在生态系统中的组织和作用”由六十个大学六十位生态学和真菌学教授合作写成。以及一九九七年由美国农部及柏克莱加大帕门教授和卡帕拉教授合写的“持续发展中之真菌学”，这些著作均围绕着以持续发展为中心解决人口增加及资源减少的矛盾。

四，迎接信息时代来临

近二十年来经济飞速发展，各国生物学界著书立说，用科技使经济发展步入正轨，我国改革开放政策成功推行，使菇类发展迅速并逐步走向国际市场，其中信息是前导，近期在电脑网路上出现了一个十分新和丰富信息库这就是 Mycology Resources，它拥有世界最新的真菌教学，研究及普及的家庭专页（Home Page）其中与食菌有关的信息最多如有关采集，学会，新闻期刊，研究室，遗传工程信息，分类争议以及真菌演替系统教学等。现仅举几例：

(1) 由瑞典 Eric Danell (E-mail: Eric Danell@mykopat.slu.se)创立的“可食菌根蘑菇”专题，其中包括 Boletus,Cantharellus, Lactarius, Tricholoma, Tuber, Terfezia 等和其他块菌，对以上“”正在研究者共一百十三位专家，分布于科威特，美国，新西兰，法国，加拿大，英国，瑞士，波多黎哥，以色列，意大利，比利时，澳大利亚，西班牙，波兰，日本，立陶宛，葡萄牙，挪威，泰国，芬兰等二十一国，Eric Danell 本人为瑞典农业大学森林真菌学和病理学教授。

(2) 由 Tom Volk 美国威州大学-La Crosse 分枝助教授创立有关 Armillaria 种分类体系，他是个充满热情刻苦奋斗之后起之秀，他的电子信件地址为*。

(3) 由 Davis W. Fischer 创立的“蘑菇问题之真答案”专题包括蘑菇生态学，再生产，可食性，毒菇，菇名，贡布茶，栽培，未来资源，菇著作，美国菇类学会和俱乐部等；电子信件地址为 Basidium@aol.com。以上电脑网路信息有用与否需常查看，不少商业启发性的信息，趣味性，多为义务宣传教育，增加人们对真菌认识，增长见识以提高人民素质，对开放中的中国进入国际市场是需要这些信息的。本人专页在九所 University of California 拥有的森林实验室专栏内，请查 mmchen@nature.berkeley.edu。

五，药菌或其他菌种专家咨询

世界新的生物制品是当前经济发展之重点，特别在医学，农学中，这类巨大投资和收益的产品必须有真菌学家参与评估，评估者获得知识市场认可，有时由大学或公司推荐，有时为同行推举，获得聘请，在合同任务上签字，以确保产品的基本价值，其中包括产品价值及生态服务价值的评估，评估者最基本做到：

(1) 每一品种具体经济值（野生和栽培）。

(2) 评估生物量，产物作 DNA 分子分析，天然产品的化学结构。

(3) 产品价值不仅现实价，对未来之价值评估。

(4) 其中包括对环境正负影响，污染危险度之评估。

除药菌需确保质量第一外，传统原产品种须列为濒危生物名单加以保护。

目前金针菇产业化正在美国西部兴起，正被西方市场纳入各类欧美食品文化中，有关专家正从金针菇产业化中学习改进资询内容及方法。所以一个新产品不能靠传说哪个市场需要，哪个不可能发展，要有法律的，政策的，社会和经济的以及科学技术各方面详尽分析，最后供给经济学家及政策决策者讨论，以完善生产前之评估过程。

六，国际蘑菇研究-教育合作交流

蘑菇在真菌学中是最具生命力的一类，她随着人类文明进步概念的普及逐步认同大家来自一个地球村，我们分享地球村的环境和食物；各国都有蘑菇研究，教育事业和专家，他们获得政府政策上和经费上支持（美国近二分之一为私营的机构），他们负有科技重任，要求彼此应加强生产管理及技术合作交流，尤其国际交流崭获尤多。喜闻我国适度减缓大型基本建设，而将经费用于加强投入科学技术领域，这是英明的，因为持续发展农业的动力需要科技推动，我国菇类发展在近十几年对经济发展，对扶贫作出举世瞩目贡献，但面临新时代来临问题也相当多，所以蘑菇科技倍受重视，国家和地方封闭不利于经济发展，国际间合作交流彼此取长补短，各种科技信息在世界电脑网路上，不分国界，不分民族在迅速流动着，国际的蘑菇科技及市场信息将会起着前导作用；有时蘑菇是一个地区经济发展的种子，这种现象不仅在发展中国家，在我担任柏克莱加州大学蘑菇学课程中不少学生来自中南美洲及南非等。就是发达国家也逐步利用蘑菇生产的优势发展经济，比如今年二月份美国奥立岗州经济发展部拨款支持在美发展东方菇类，首先从品种多样化和市场需要与西方据垄断地位的扣子菇（Agaricus）在经济上作比较研究，在大学设置长短期基础蘑菇课，以及小型试验新品种，建立食菌国际品种产值和市场研究。

北美洲蘑菇学会订立明年二月在旧金山真菌学会地址召开，该会为世界最大真菌学会，拥有千余菇类爱好者，为加强中美人民友谊在旧金山这个盛产菇类地方，我热烈迎接祖国朋友来访，交流学术，增进友谊，记得十年前曾在柏克莱加州大学植病系欢迎由戚佩坤教授率领的第一个中国真菌学界代表团，当时本人主要作林病及国际检疫研究，但从最近十年来已全力投入食菌的教学和研究工作，在这方面我们有更多的共同语言了。祝大会圆满成功。写于美国加尼福利亚州大学，一九九七，七，二十。

The characteristic edible mushrooms: China and America

ABSTRACT: A rich and large variety of mushroom species can be found in the forests of the California Pacific coast range and Sierra Nevada Mountains. Four main edible species are prevalent throughout the Pacific coast: *Morcella esculenta, Boletus edulis, Cantharellus cibarius,* and *Tricholoma magnivelare.* San Francisco itself has a unique history in mushroom mycology with up to over 1,000 wild species growing in the Bay Area. The Mycological Society of San Francisco is an educational and scientific organization for the public that promotes forays and fairs so that all people can share information and knowledge of mushrooms. The cultivation mushroom industry in the United States traditional started with the *Agaricus bisporus* mushroom and has begun to diversify their products with the *Lentinus edodes* and *Pleurotus* species mushroom. In the United States, Amycel and Sylvan are businesses that serve throughout America and also internationally. The Chinese have a 2,000-year history in cuisine and medicinal fungi that exceeds 20-30 species for cultivation and hundreds of species used for specific pharmaceutical needs. Over the many years the Chinese have developed unique agroforest techniques in cultivating the mushrooms. Presently the Chinese produce 40 million tons of mushrooms, which is 60% of the world's production. As the abundance of diverse mushroom species become more well known throughout the world, humans will come to enjoy the natural food and medicinal properties of these fungi treasures.

A rich and large variety of mushroom species can be found in the forests of the California Pacific coast range and Sierra Nevada Mountains especially during the months of October to April. There are already 950 names in Mycological Society of San Francisco database of recent fungus fair collections. Up to 294 species can be collected for a single fair in the San Francisco Bay Area alone. Four main species are prevalent throughout the Pacific coast: *Morcella esculenta, Boletus edulis, Cantharellus cibarius,* and *Tricholoma magnivelare.*

San Francisco has a unique history in mycology. The Mycological Society of San Francisco [MSSF] was founded in 1950 as an educational and scientific organization for the public. The MSSF has a very large membership including amateurs and even professional mycologists. The group has activities all year long such as forays and mushrooms fairs. The MSSF works to promote and preserve cultural traditions of mushroom collecting. The library includes various handbooks, guidelines and identification books such as "Mushroom Demystified" by Bay Area's own David Arora from Santa Cruz. This wonderful book contains 3,500 some species and is a good and easy reference for the average person. Other books that extend mushroom knowledge to the public are *Mushrooms of Northeastern North America* by Alan Bessete, *Mushrooms and other Fungi of the Midcontinental United States* by D. Huffman, L. Tiffany, and G. Knaphus, *Mushrooms of Colorado and the Southern Rocky Mountains* by Vera Stucky Evenson, *Mushrooms and Truffles of the Southwest United States* by Jack S. States, *Texas Mushrooms: A Field Guide* by Susan and Van Metzler, *Mushrooms of Western Canada* by Helen Schalkwijk-Braendsen (*John Lennie Mycena New, Nov. 2001*).

The button mushroom *Agaricus bisporus* in the U.S. has been industrialized to both the East and the West and it is a large commodity. The East, particularly at Pennsylvania State University, is focused on the compost research and education with their mushroom science PhD program. In the West, the largest mushroom company is in Monterey Bay, California, which produces many button mushrooms for the country. Over the past twenty years, the shiitake (香菇 *Lentinus edodes*) and oyster mushrooms (平菇) have also been introduced to the industry. However, previously, only the button mushroom was known as the typical mushroom. Now Americans have become more familiar with different mushrooms and are using them more in their foods. Amycel, the largest mushroom spawn business in the west coast also produces shiitake and oyster along with their original button mushrooms. Sylvan, the east coast and international industry also produces other mushrooms. The shiitake mushroom is the other most demanded mushroom because 95% of dried shiitake are used in Chinese restaurants. Majority of the fresh shiitake (新鲜香菇) are used in other high-grade western restaurants.

More than 1,800 years ago, during the East Han Dynasty in China, people began to discover and use wild fungi for their unique medicinal properties. Today, Americans know many of the precious edible fungi, but most know little about the medicinal fungi. These medicinal fungi are irreplaceable gifts of nature to people.

The Chinese know of 930 edible mushroom and 90 species can be cultivated to fruiting bodies and among these 20-30 species can be produced commercially (Mao 2000). The shiitake is 61% of the world's total mushroom production. In Huang Shan County, China, there are 230,000 farmers directly working year round on the cultivation of about 10 species. The labor is cheap and work is abundant. During 1950-1960 there was a period of technique reform. Many agricultural institutions such as Shanghai Academy of agriculture, edible research institute first started research. Also Professor Yang Xin Mei (杨新美教授) of Hua Zhong Agricultural University worked on the *Tremella* liquid inoculation spawn. Professor Yang is the author of *The Science of Chinese Edible Mushroom Cultivation,* which is among several important monographs that he wrote of mushroom cultivation. It was the basis of later developments of cultivation and Yang and his senior students, Prof. Luo Xin Chang and Prof. Lin Fang Cai, were recognized at the genetic and biogeography of mushroom sciences works (Luo 2000). In the 1960's the button mushroom was developed in suburban Shanghai and was popular among foreigners buyers. During the 1970's the shiitake and the wood ear were reformed to grow in sawdust substrate in plastic bags. This was a major contribution because millions of bags could be used as opposed to the logs, which were more expensive. Fu Jian Province began to use machinery and reduced their manual labor. With faster production the oyster mushroom became more common. From the 1980's to the present day, 40 million tons of mushroom are being produced in China, which is 60% of the world's production (The Edible Fungi Association of China 中国食用菌协会). Mushrooms are the sixth in rank of agriculture production of China. Shiitake production has development from the south to the northern part of China. Many more wild species are being studied and coming into commercial species for mass production. Among them: *Aurcularia polytricha, Flammulina velutipes, Coprinus comatus, Dictyophora indusiata ,Grifola frondosa, Pleurotus ferula Agrocybe cylindracea, Cordyceps militaris* and *Pholiota cylindracea*. They also developed potential at inner China marketing. Recently published Chinese books on mushrooms are: *The flora of edible mushrooms 中国食用菌志* by the Shanghai Academy of Agriculture, editors including many scholars such as Liu Bo (刘波)，Zang Mu (藏穆) (1991) etc., and *Edible Fungi Encyclopedia* (1993) by Huang Nian Lai (黄年来)

Mushroom cultivation of China is a very specific agroforestry system. These mushrooms are grown side by side in equal amounts along with other agriculture so to balance the needs of each product. For example rice is grown with wood ear or button mushroom and corn is grown with oyster. Because they are grown together to supplement each other there is a 10-30% increase in production. Many agroforestry articles are published by editor Zhang Guan Ya (主编张光亚) in the journal, *Edible Fungi of China (中国食用菌*) and in editions by the Shanghai edible research institute journal, *Edible Fungi (食用菌).*

Mushrooms are not only used in cooked foods but also in drinks and pharmaceutical products. In *Collection of Mushroom Prescriptions* (1999) (蕈菌医方集成), Chen Shi Yu (陈士瑜) writes of the

2,000-year Chinese medicinal history, including uses of 297 species with Latin names. Chen Shi Yu studied thousands of ancient Chinese medical texts written by generations of Chinese medical doctors to glean 3,840 prescriptions that utilize medicinal fungi such as *Ganoderma lucidum* (灵芝).

Coming into the new millennium, humans can hope to use fungi even more widely for food and medicine. To achieve our lofty goal of increasing mushroom cultivation for production of food, we must work hard on such aspects as acceptance of a greater variety of mushrooms by public, standards of production (quality control), biodiversity of species in nature, sustainability of natural resources, and optimization of quality of spawn. Mushroom production can be non-commercial as well as industrialized. A free exchange between science and technology as well as improved marketing practices are essential to modernization of the production of edible mushrooms.

REFERENCES

Mao, X.L. 2000. Review and prospects of edible mushroom species resources in China. Edible Fungi of China 19: 9-13.

The edible fungi association of China. 2000. Half the century of Chinese mushroom industry. Edible Fungi of China 19: 1-13.

Luo, X.C. 2000. Prospects of utilization of development of new technology in mushrooms. Edible Fungi of China 19: 18-21.

He, Y.S. 1994. The Chinese Xiang Gu mushroom (*Lentinus edodes*). Shanghai Sciences and Technology Press. pp. 239.

Yao, S.X., & Zhang, S.C. 1997. The Cultivation new technology of Hua Gu (*Lentinus edodes*). Second Title: The deamic, technological skills managemement and Marketing of The Xiang Gu (*Lentinus edodes*) cultivate as Haul Gu.

Taken from: International Shiitake Mushroom Conference. 2002. Biyang, China.

Four main edible species are prevalentthroughout the Pacific coast : *Armillaria ponderosa* (new name: ***Tricholoma magnivelae***), ***Boletus edulis,,Morcella esculenta, Cantharellus cibarius.***

Armillaria ponderosa

Boletus edulis

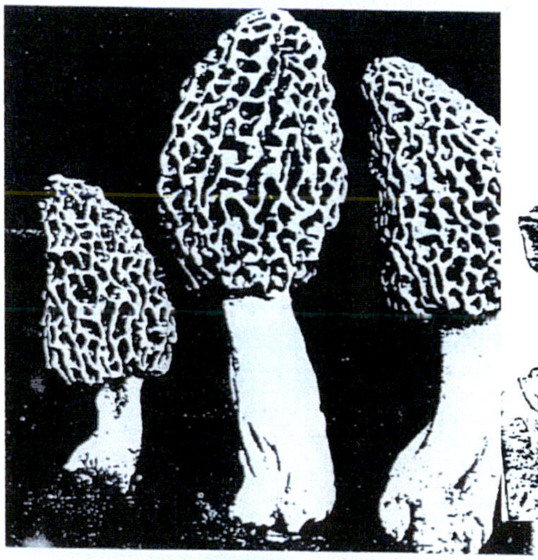

Morcella esculenta

Cantharellus cibarius

真菌珍品

Shiitake *(Lentinus edodes)*

Oyster (*Pleurotus ostretus*)

Lin Zhi (*Ganoderma lucidum*)

Maitake (*Grif ola f rondosa*)

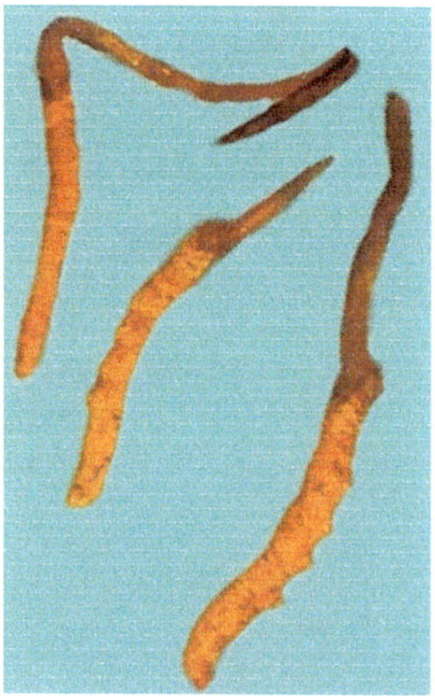

Winner worm summer f ungus (Cordy ceps sinensis)

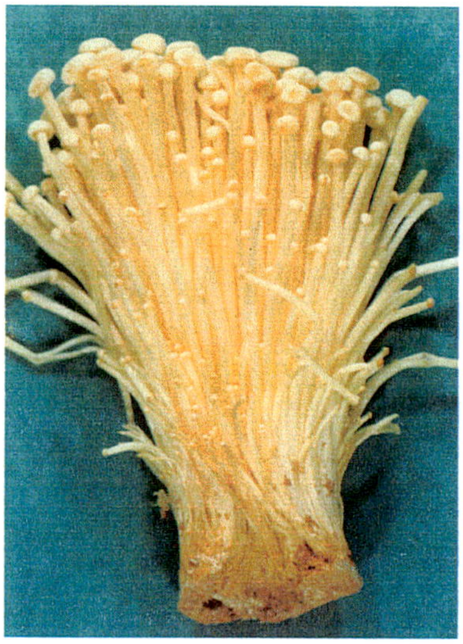

Enoki (*Flammulina velutip es*)

STUDENT WORKING ON MUSHROOM ISOLATION

Black truffle (*Tuber melanosporum*)

Fresh mushroom at the market

Button mushroom (*Agaricus bisporus*)

Portabella (*Agaricus bisporus*)

Chanterelle (*Cantharellus cibarius*)

Index of Polyporaceae and Melampsoceae of Tibet

At the Program of Geological and Ecological Studies at **Qinghai-Xizang** (Tibet) Plateau

The Tibet Plateau, known as the "world's roof", is a treasury of information on natural sciences. Yet, over the long years this plateau was approached by only a few scientists and it remained enshrouded in mystery. Not content with what has been known, scientists both in China and abroad have long cherished the desire to explore into the unknown and to bring this hidden treasury of information to light.

A serious effort was started in China in the early fifties to study the plateau, and in the years that followed, altogether seven large-scale scientific expeditions to Xizang were organized by Academia Sinica, with more than 1,400 scientific workers participating. The effort was amply rewarded. Not only were many significant findings made but also research on a wide range of disciplines was done. As a result, knowledge about the plateau attained a new level.

The following is a list of two families of Tibetan fungi, Polyporaceae and Melampsoceae, with citations where each species can be found (1975-1976). The citations are derived from a book on Tibetan Fungi, for which I was the main contributing author. In addition, I have included an appendix of Basidiomycotina using the AINSWORTH (1966) system.

1. *Polyporaceae*

Amauroderma

Amauroderma rugosum (Bl. & Nees) Bres., Ann. Myc. 8: 256, 1912;---*Polyporus rugosus* Bl. & Nees, Nova Acta Acad. Caes. Leop. 3: 21. 1826.

Bjerkandera

Bjerkandera adusta (Widdl. ex Fr.) Karst., Soc. Fauna Flora Fenn. Meddel. 5: 38. 1879;---*Polyporus adustus* Willd. ex Fr., Syst. Myc. 1: 363. 1821.

***Coltricia* Mich. ex Gray.**

Coltricia montagnei (Fr.) Boud., Hymen. France 1: 631. 1927;---*Polyporus montagnei* Fr., Ann. Sci. Nat. Bot. 2, 5: 341. 1836;---*Xanthochrous montagnei* Quel., Jura Vosg. 1: 252. 1872.

Coltricia perennis (L. ex Fr.) Murr., Journ. Mycol. 9: 91. 1903;---*Polyporus perennis* L. ex Fr., Syst. Myc. 1: 350. 1821.

Coltricia schweinitzii (Fr.) Cunn., Dep. Sci. Ind. Res. Pl. Dis. Div. Bull. 77: 7. 1948;---*Polyporus schweinitzii* Fr., Syst. Myc. 1: 351. 1821.

***Cryptoporus* (Pk.) Hubbard.**

Cryptoporus volvatus (Pk.) Schear, Bull. Torr. Bot. Club. 29: 450. 1902;---*Polyporus volvatus* Peck, Rept. N. Y. St. Mus. 27: 98. 1875.

***Daedalea* Pers. em. Pat.**

Daedalea borealis (Fr.) Quel., Enchir. Fung. P. 184. 1886;---*Polyporus borealis* Fr., Syst. Myc. 1:366. 1821.

Daedalea quercina (L.) Fr., Syst. Myc. 1: 333. 1821;---*Agaricus quercina* L., Sp.Pl.P.1176. 1753.

***Daedaleopsis* Schroet**

Daedaleopsis tricolor (Bull. ex Fr.) Bond. et Sing., Ann. Myc. 39: 64. 1941;---*Lenzites tricolor* (Bull.) Fr., EPicr. Syst. Myc. P. 406. 1836.

***Favolus* Fr. em Ames.**

Favolus alveolarius (DC. ex Fr.) Quel., Ench. Fung. P. 185. 1886;---*Merulius alveolarius* DC., Flore Fr. 4: 43. 1815;---*Cantharellus alveolaris* DC. ex Fr., Syst. Myc. 1: 322. 1821.

Favolus arcularius (Batsch ex Fr.) Ames, Ann. Myc. 11: 241. 1913;--- *Polyporus arcularius* Fr., Syst. Myc. 1: 342. 1821.

Favolus sp. aff. ***multiplex*** Lev., Ann. Sci. Nat. Ser. 3. Bot. 12: 203. 1844.

Favolus squamosus (Huds. ex Fr.) Ames., Ann. Myc. 11: 241. 1913;---*Polyporus squamosus* Huds. ex Fr., Syst. Myc. 1: 342. 1821.

***Fomes* (Fr.) Gill**

Fomes annosus (Fr.) Cke., Crev. 14: 20. 1885;---*Polyporus annosus* Fr., Syst. Myc. 1: 373. 1821.

Fomes conatus (Weinm.) Gill., Champ. Fung. P. 684. 1878;---*Polyporus connatus* Weinm. Florae Ross., P. 332. 1836.

Fomes marginatus (Fr. ex Fr.) Gill., Champ. Fr. P. 683. 1878;---*Polyporus marginatus* Fr., Syst. Myc. 1: 372. 1821.

Fomes marmoratus (Berk et Curt.) Cooke., Grevillea 14: 18, 1885;---*Polyporus marmoratus* Berk. et Curt., Am. Acad. Arts Sci. Proc. 4: 122. 1860.

Fomes officinalis (Vill. ex Fr.) Faull., Roy. Canad. Ins. Trans. 11: 185. 1916;---*Fomitopsis officinalis* (Vill ex Fr.) Bond. et Sing., Ann. Myc. 39: 55. 1941.

Fomes pinicola (Sw. ex Fr.) Cke. Grev. 14: 17. 1885;---*Fomitopsis pinicola* (Fr.) Karst., Bidr. Kanned. Finl. Nat. Folk. 68: 306. 1889;---*Polyporus pinicola* Fr., Syst. Myc. 1: 372. 1821.

Fomes roseus (Alb. et Schw. ex Fr.) Cke., Grev. 14: 19. 1885;---*Polyporus roseus* Fr., Syst. Myc. 1: 372. 1821.

***Ganoderma* Karst.**

Ganoderma applanatum (Pers. ex Wallr.) Pat., Bull. Soc. Myc. Fr. 5: 67. 1889;---*Fomes applanatus* (Pers. ex Wall.) Gill., Champ. Fr. 1: 686. 1878; --*Polyporus applanatus* (Pers.) Wallr., Fl. Crypt. Germ. 2: 591. 1833.

Ganoderma applanatum (Pers. ex Gray) Pat. var. ***gibbosum*** (Bl. et Nees.) Teng ( )p. 450. 1963.

Ganoderma cochlear (Bl. Et Nees.) Bres., Hedw 51: 313. 1912;---*Polyporus cochlear* Bl. et Nees. in Lloyd; Myc. Writ. 3. 1906.

Ganoderma lobatum (Schw.) Atk., Ann. Myc. 6: 190. 1908.

Ganoderma lucidum (Leyss. ex Fr.) Karst., Rev. Mycol. 3 (9): 17. 1881;---*Boletus lucidus* Leyss. Fl., Halensis p. 300. 1783;---Polyporus lucidus Leyss. ex Fr., Syst. Myc. 1:353. 1821.

Ganoderma oroflavum (Lloyd.) Teng ( ) p. 449. 1963;---*Fomes oroflavus* Lloyd, Synopsis Gen. Fomes p. 262. 1915.

Ganoderma sessile Murr., Bull. Torrey Bot. Club 29: 604. 1902;---*Polyporus polychromus* Copeland, Ann Mycol. 2: 507. 1904;---*Ganoderma subperforatum* Atk., Bot. Gaz. 46: 337. 1908.

Ganoderma tsugae Murr., Bull. Torr. Bot. Club. 29: 601. 1902.

***Gloeophyllum* Karst.**

Gloeophyllum abietinum (Bull. Ex Fr.) Karst., Hattsv. 2: 80. 1879;---*Daedalea abietina* Bull. ex Fr., Syst. Myc. 1: 334. 1821.

Gloeophyllum imponens (Ces.) Teng, ( ) P. 532. 1964;---*Daedalea imponens* Ces., Atti. Acc. Sci. fisich. Mat. Napoli 8: 7. 1879.

Gloeophyllum saepiarium (Wulf. Ex Fr.) Karst., Hattsv. 2: 80. 1879;---*Lenzites saepiaria* (Wulf. ex Fr.) Fr., Epicr. Myc. P. 407. 1838;---*Daedalea saepiaria* Wulf. ex Fr., Syst. Myc. 1:333. 1821.

Gloeophyllum straitum (Sw. ex Fr.) Murr., Bull. Torr. Bot. Club 32: 370. 1905;---*Agaricus striatus* Swartz., Prodr. P. 148. 1788;---*Daedalea striata* Fr., Syst. Myc. 1: 334. 1821.

Gloeophyllum subferrugineum (Berk.) Bond. et Sing., Ann. Myc. 39. 64. 1941---*Lenzites subferruginea* Berk., Hooker's London J. Bot. 6: 134. 1854.

Gloeophyllum trabeum (Pers. ex Fr.) Murr., N. Am. Fl. 9: 129. 1908;---*Agaricus trabeum* Pers., Syn. Meth. Fung. P. 24. 1801;---*Daedalea trabea* Fr., Syst. Myc. 1: 335. 1821.

***Hirschioporus* Donk em. Bond. et Sing.**

Hirschioporus abietinus (Dicks. ex Fr.) Donk, Medd. Bot. Mus. Utrecht. 9: 168. 1933;---*Boletus abietinus* Dicks, Pl. Cr. Br. 3: 21. 1793;---*Polyporus abietinus* (Dicks.) Fr., Syst. Myc. 1: 370. 1821.

Hirschioporus fusco-violaceus (Schrad. ex Fr.) Donk, Med. Bot. Mus. Utrecht. 9: 169. 1933;---*Hydnum fusco-violaceus* Schrad., Sp. Fl. Germ. P. 180. 1794.

Hirschioporus laricinus (Karst.) Teramoto in Chiba & Teramoto, Bull. Tokyo. Univ. Forest. 43: 31. 1952;---*Lenzites laricina* Karst., Acta Soc. Faun. Fl. Fenn. 27 (4): 4. 1905.

Hirschioporus pergamenus (Fr.) Bond. et Sing., Ann. Myc. 39: 63. 1941;---*Polyporus pergamenus* Fr., Epicr. Syst. Myc. P. 480. 1838; *Polystictus pergamenus* Fr., Nov. Symb. Myc. P. 69. 1851.

Hirschioporus sector (Ehrenb. ex Fr.) Teng, ( ), p. 485. 1963;---*Boletus sector* Ehrenb., Horae Phys. Berol. P. 86. 1820;---*Polyporus sector* Fr., Syst. Myc. 1: 505. 1821.

***Lenzites* Fr. em. Donk.**

Lenzites betulina (L. ex Fr.) Fr., Epicr. Syst. Myc. P. 405. 1836;---*Agaricus betulinus* L., Sp. Pl. p. 1176. 1753;---*Lenzites ochraceus* Lloyd., Myc. Notes 66: 1130. 1922.

Lenzites sp. aff. ***Cyclogramma*** Pat., Bull. Soc. Myc. Franc. 23: 73. 1907.

Lenzites extimia Berk. in Hook., Journ Bot. 6: 134 1854;---F L Balfour-Browne, Bull. Brit. Mus. (Nat. Hist.) 1: 191. 1955.

Lenzites flaccida (Bull.) Fr. Epicr. Syst. Myc. P. 406. 1838;---*Agaricus flaccidus* Bull., Champ. P. 394. 1788

Lenzites palisoti Fr., Syst. Myc. 1: 335. 1821.

Lenzites platyphylla Lev., Champ. Exot. P. 179. 1855;---Sacc., Syll. Fung. 5: 645. 1887.

***Piptoporus* Karst.**

Piptoporus betulinus (Bull. ex Fr.) Karst., Rev. Myc. 3 (9): 17. 1881;---*Polyporus betulinus* Bull. ex Fr., Syst. Myc. 1: 358. 1821;---*Boletus betulinus* Bull., Herb Fr. Pl. p. 312. 1786.

***Polyporus* (Mich.) Fr. ex Fr.**

Polyporus brumalis Pers. ex Fr., Syst. Myc. 1: 348. 1821;---*Boletus brumalis* Pers., Mag. Bot. (Neues) 1: 107. 1794.

Polyporus confluens (Alb. et Schw.) Fr., Syst. Myc. 1: 355. 1821;---*Boletus confluens* Alb. et Schw., Cons. Fung. Leipz. P. 244. 1805;---*Polyporus aurantius* Trog Fl. P. 354. 1832.

Polyporus dispansus Lloyd., Stip Polyp. Add. P. 192. 1912.

Polyporus dryophilus Berk., Lond Journ. Bot. 6: 321. 1847.

Polyporus elegans (Bull.) Fr., Epicr. Myc. P. 440. 1838;---*Boletus elegans* Bull., Herb. Fr. Pl. 46. 1780.

Polyporus flabelliformis Klotzsch, Linnaea 8: 483. 1833.---*Microporus flabelliformis* (Klotz. ex Fr.) Kuntze, Rev. Gen Pl. 3 (2): 496. 1898.

Polyporus frondosus (Dicks.) Fr., Syst. Myc. 1: 355 1821;---*Boletus frondosus* Dicks. Pl. Crypt. Brit. 1: 18 1785;---*Caloporus frondosus* Quel., Fl Myc. P. 406. 1888.

Polyporus leucospongia Cooke et Harkness, Grevillea 11: 106 1883;---*Spongiporus leucospongia* (Cooke et Harkness) Murr., Bull. Torrey Bot. Club 32: 474 1905.

Polyporus montanus (Quel) Ferry., Rev. Myc. 19: 144. 1897;---*Cerioporus montanus* Quel., Fl. Myc. P. 408. 1889.

Polyporus picipes Fr., Syst. Myc. 1: 353. 1832;---*Polyporus varius* Grev., Crypt. Fl. 4: 202. 1824;---*Melanopus varius* Bourdot et Galzin, Hym Fr. p 527. 1874.

Polyporus resinosus (Schrad.) Fr., Syst. Myc., 1: 361. 1821;---*Boletus resinosus* Schrad., Sp. Flora Germ. P. 171. 1794.

Polyporus tomentosus Fr., Syst. Myc 1: 351. 1821;---*Coltricia tomentosa* (Fr.) Murr., Bull. Torrey Bot. Club 31: 346. 1904.

Polyporus triqueter Secr., Mycogr. Suisse. 3: 101. 1833.

Polyporus zonatus Nees ex Fr., Syst. Myc. 1: 368. 1821;---*Boletus zonatus* Nees., Syn. Pilze, p. 221. 1817.

***Polystictus* Fr.**

Polystictus meleagris Berk., Chall. n. 157. in Sacc., Syll. Fung. 6: 227. 1888.

Polystictus microloma Lev., Champ. Exot. 183. 1855;---Sacc., Syll. Fung. 6: 221. 1888.

Polystictus murinus Lev., Ann. Sci. Nat. Bot. 3: 185. 1844;---Sacc., Syll. Fung. 6: 226. 1888.

Polystictus pinsitus Fr., Elench. Fung. 1: 95. 1828.

Polystictus polyzonus (Pers.) Fr., Epicr. Syst. Myc. P. 477. 1838;---*Coriolus polyzonus* (Pers.) Imaz., Bull. Tokyo. Mus. 6: 80. 1943.

Polystictus unicolor (Bull. ex Fr.) Lloyd., Myc. Writ. Lett. 63: 6. 1916;---*Daedalea unicolor* (Bull.) Fr., Syst. Myc 1: 336. 1821;---*Boletus unicolor* Bull., Herb. France. Pl. 408. 1788;---*Trametes unicolor* (Bull. ex Fr.) Cke., Austr. Fung. P. 162. 1892.

Polystictus velutinus (Fr.) Cooke, Grevillea 59: 83. 1886;---*Polyporus velutinus* Fr., Syst. Myc. 1: 368. 1821.

Polystictus versicolor (L.) Fr., Syst. Myc. 1: 363. 1832;---*Coriolus versicolor* (Fr.) Quel., Ench p. 175 1886;---*Boletus versicolor* L., Sp. Pl. 2: 1176. 1753.

Polystictus villosus (Sw) Fr, Syst. Myc. 1: 344 1821;---*Boletus villosus* Sw., Fl. Ind. 1933.

Polystictus xanthopus Fr., Nov. Symb Myc. P. 58. 1851.

***Poria* Pers. ex Gray**

Poria barbaeformis Berk. et Curt., North Amer. Fung. P. 144. 1860.---Bakshi: Indian Polyporaceae. P. 159. 1971.

***Pyropolyporus* Murr.**

Pyropolyporus fastuosus (Lev.) Murr., Bull. Torr. Bot. Club 34: 479. 1907;---*Fomes fastuosus* (Lev.) Cooke, Grevillea 14: 18. 1885;---*Polyporus fastuosus* Lev., Ann. Sci. Nat. Bot. 3, 2: 190. 1844.

Pyropolyporus fomentarius (L. ex Fr.) Teng, ( ). P. 763. 1963;---*Boletus fomentarius* L., Sp. Pl. p. 1176. 1753;---*Polyporus fomentarius* Fr., Syst. Myc. 1: 374. 1821;---*Fomes formentarius* (L. ex Fr.) Fr., Summa Veg. Scand. 2: 321. 1849.

Pyropolyporus rimosus (Berk.) Teng. ( ) p. 470. 1964;---*Polyporus rimosus* Berk., Lond. Jour. Bot. 4: 54. 1854;---*Phellinus rimosus* (Berk.) Pil., Ann. Myc. 38: 80. 1940.

***Trametes* Fr.**

Trametes abietina (Fr.) Pilat., Atl. Champ. Eur. 3: 273. 1936;---*Polyporus abietimus* Fr., Syst. Myc. 1: 370. 1821.

Trametes carbonaria (Berk. et Curt.) Overh., Mycologia 23: 126. 1931;---*Hexagonia carbonaria* Berk. et Curt., Grevillea 1: 68. 1872;---*Trametes sequoiae* Copeland., Ann. Mycol. 2: 507. 1904.

Trametes cervina Lloyd, Myc. Not. 35: 470. 1910.

Trametes cinnabarina (Jacq.) Fr., Summa Veg. Scand. P. 323. 1849;---*Polyporus cinnabarinus* Jacq. ex Fr., in Syst. Myc. 1: 371. 1821;---*Boletus cinnabarius* Jacq., Flora Aust. 4: 2. 1776.

Trametes confragosa Lloyd, Myc. Writ. Letter n. 42: 15, 1913.

Trametes corrugata (Pers.) Bres., Hedwigia 51: 316. 1912;---*Polyporus corrugatus* Pers., Gaud. Bot. Freyc. Voyage Uranie, p. 172. 1826;---*Polystictus persoonii* Cooke, Grevillea 14: 85. 1885;-*Daedalea sanguinea* Klotzsch, Linnaea. 8: 481. 1833.
Trametes cubensis (Mont.) Sacc., Syll. Fung. 9: 198. 1891;---*Polyporus cubensis* Mont., Ann, Sci. Nat. Bot. 2: 364. 1837.
Trametes devexa Berk., Journ. Linn. Soc. 13: 165. 1873.
Trametes gallica Fr., Hym. Eur. p. 582. 1874.
Trametes gibbosa (Pers.) Fr., Epicr. Syst. Myc. p. 492. 1838;---*Daedalea gibbosa* Pers., Syn. Meth. Fung. P. 501. 1801;---*Pseudotrametes gibbosa* Pers. ex Singer, Mycologia 36: 68. 1944.
Trametes hirsute (Wulf. ex Fr.) Pilat., Atl. Champ. Eur. 3. Poyp. 1:265. 1936;---*Polyporus hirsutus* Wulf. ex Fr., Syst. Myc. 1: 367. 1821.
Trametes insularis Murr., Bull. Torrey Bot. Club. 35: 405. 1908.
Trametes lactinea Berkl, Ann. Mag. Nat. Hist. 10: 373. 1843.
Trametes malicola Berk. et Curt., Journ. Acad. Nat. Hist. Phili. 3: 209. 1956.
Trametes meyenii klotzsh, Nov. Act. Nat. Cur. Suppl. p. 236. 1843;---*Polyporus myenii* Klotzsch *in* Bakshi, Indian Polyp. p. 170. 1971.
Trametes orientalis (Yasuda) Imaz., Bull. Tokyo Sci. Mus. 6: 73. 1943;---*Polystictus orientalis* Yasuda., Bot. Mag. Tokyo. 36: 155. 1922.
Trametes pubescens (Schum. Ex Fr.) Pilat., Atlas Champ. Eur. 3, Polyp. 1: 268. 1939;---*Tyromyces pubescens* (Schw. ex Fr.) Imaz., Bull. Tokyo. Sci. Mus. 6: 84. 1943.
Trametes serialis Fr., Hym. Eur. p. 585. 1874;---*Polyporus serialis* Fries, Syst. Myc. 1: 370. 1821;---*Coriolellus serialis* (Fr.) Murr., N. Am. Flora 9: 29. 1907.
Trametes suavelens (L.) Fr., Epicr. Syst. Myc. p. 491. 1838;---*Boletus suaveolens* L., Sp. Plant. Ed. 1: 1177. 1753.
Trametes trogii Berk., Trog-schwiez. Schw. 2: 52. 1850.

***Tyromyces* Karst.**
Tyromyces floriformis (Bres.) Bond. et Sing., Ann. Myc. 39: 51. 1941;---*Polyporus floriformis* Bres., Fung. Trid. 1: 61. 1884.
Tyromyces lacteus (Fr.) Murr., N. Am. Flora 9: 36. 1907;---*Polyporus lacteus* Fr., Syst. Myc. 1: 359. 1821.
Tyromyces sulphureus (Bull. ex Fr.) Donk., Med. Bot. Mus. Utrecht. 9: 145. 1933;---*Polyporus sulphureus* Fr., Syst. Myc. 1: 357. 1821;---*Laetiporus sulphureus* (Bull. ex Fr.) Bond. et Sing., Ann. Myc. 19: 51. 1941.

***Xanthochrous* Pat.**
Xanthochrous hispidus (Bull. ex Fr.) Pat., Ess. Tax. Hym. p. 98. 1900;---*Polyporus hispidus* Bull. ex Fr., Syst. Myc. 1: 362. 1821.

2. *Melampsoraceae*

***Aecidium* Pers.**
Aecidium akebiae P. Henn., Hedwigia 34: 154. 1900.
Aecidium cimicifugatum Schw., Trans. Am. Philos. Soc. 4: 293. 1832.
Aecidium hydrangeae P. Henn. In Engl. Bot. Jahrb. 37: 159. 1905.
Aecidium mori Barcl., Journ. Asiatic Soc. Bengal, 50: 226. 1891.
Aecidium paeoniae Kon., Fungi Rossiae Exs. No. 177. 1898.
Aecidium pusatillae Tranz., Travaux Musee Bot. L' Acad. Imper. Sc. St.-Petersbourg Liver. 7: 115. 1909.

Aecidium rhododendri Barcl., Sci. Mem. Med. Off. Arm Ind. 6: 71. 1891.
Aecidium senecionis-scandentis Sawada, Journ. Taihoku Soc. Agr. For 7: 40. 1943.
Aecidium sino-rhododendri Wilson, Notes Bot. Gard. Edinburgh. 12: 261. 1921.

Blastospora **Diet.**
Blastospora itoana Togashi et Onuma, Bot. Mag. Tokyo 45: 6. 1931.

Caeoma
Caeoma makinoi Kus., Bot. Mag. Tokyo. 17: 15. 1903.

Chrysomyxa **Unger**
Chrysomyxa expansa Diet., Engl. Bot. Jahrb. 28: 287. 1900.
Chrysomyxa himalense Barclay, Sci. Mem. Med. Army Ind. 5: 83. 1890.
Chrysomyxa ledi (Alb. et Schw) de Bary. Var. ***rhododendri*** (de Bary) Savile, Can. Journ. Bot. 33: 491, 1955;---*Chrysomyxa rhododendri* (DC) de Bary, Bot. Zeitung, 37: 809. 1879;---*Uredo rhododendriI* DC., Fl. Fr. 6: 86. 1815.
Chrysomyxa stilbae Wang, Chen et Guo, Acta Microbiol. Sinica 20(1): 16, 1980.

Coleopuccinia **Pat.**
Coleopuccinia sinensis Pat., Rev. Myc. 11: 35. 1889.

Coleosporium **Lev.**
Coleosporium asterum (Diet.) Syd., Ann. Myc. 12: 107. 1914;---*Stichospora asterum* Diet., Engl. Bot. Jahrb. 27: 566. 1899.
Coleosporium brevius Barcl. Journ Asiat. Soc. Bengal. 59: 89. 1890;---*Caeoma clematidis* Thum., Myc. Univers. P. 539-1875
Coleosporium complanatum Barcl. Ured. Simla 101. t. v. f. 1. 3. 8. 1891.
Coleosporium geranii Pat., Rev. Myc. 12: 135. 1890.
Coleosporium heteropappi (P. Henn.) Tranz., Conspectus Uredinalium URSS, P.378. 1939;---*Uredo heteropappi* P. Henn., Bot. Jahrb. 34: 597. 1904.
Coleosporium ligulariae Thuem., Bull. Soc. Imp. Nat. Moscou 52: 140. 1877.
Coleosporium lonicerae Wang et Wei, Acta Microbiol. Sinica 20(1): 17-18, 1980.
Coleosporium pedicularidis Tai, Farlowia 3: 100. 1974.
Coleosporium perillae Syd. Hedw. 38: 141. 1899.
Coleosporium saussureae Thum, Beitr. Pilzfl. Sibir. 4: 212. 1880.
Coleosporium zangmui Wang et Wei Acta Microbiol. Sinica 20(1): 17, 1980.

Cronartium **Fr.**
Cronartium flaccidium (Alb. Et Schw.) Wint., Pilze Deut. 1: 236. 1884.
Cronartium himalayense Bagchee, Indian Forest Rec. Bot. ser. 18: 14. 1933;--- Peters. Rept. Tottori Myc. Inst. Jap. 10: 217. 1973.
Cronartium quercuum Miyabe in Shirai, Bot. Mag. Tokyo 13: 74. 1899.
Cronartium ribicola Fisch. In Rab., Fung. Eur. No. 1595. Hedw. 11: 182. 1872. Diet. in E. et P., Nat. Pfl. 1: 79. 1897.

Gymnosporangium **Hedw. f.**
Gymnosporangium clavaraeforme (Jacq.) DC., Fl. Franc. 2: 217. 1805.
Gymnosporangium japonicum Syd., Hedw. 38: 141. 1899.
Gymnosporangium nipponicum Yamada in Hiratsuka f. Mem. Tottori Agr. Coll. 3: 143. 1935.

Hyalopsora **Magn.**

Hyalopsora japonica Diet., Ann. Myc. 12:84. 1914.

***Melampsora* Cast.**
Melampsora coleosporioides Diet., Engl. Bot. Jahrbucher 32: 50. 1902.
Melampsora euphorbia-dulcis Otth. Mitl Mat. Ces. Bern., P. 70. 1869.
Melampsora larici-capraearum Kleb., Forstl. Nat. Zeits. P. 469. 1897.
Melampsora larici-tremulae Kleb. Ztschr. Pflanzenkr. 9: 146. 1899.
Melampsora ribesi-purpureae Kleb., Pringzh. Jahrb. Wissenschaftl. Bot. 35: 667. 1901.
Melampsora salicis-viminalis Wang et Guo, Acta Microbiol. Sinica 20 (1): 18, 1980.
Melampsora stellerae Teich, ( ). CP. ( ) 19(24): 181. 1934.

***Melampsoridium* Kleb.**
Melampsoridium betulinum (Desm.) Kleb., Zeits. Pfl.-Kr. 9: 21. 1899.

***Monosporidium* Barclay**
Monosporidium andrachnis Barclay, Journ. Asiatic Soc. Bengal 56: 371. 1887.

***Nyssopsora* Arth.**
Nyssopsora koelreuteriae (Syd.) Tranz., Journ. Soc. Bot. Russie 8: 132. 1925;--- *Triphragmium koelreuteriae* Syd. Ann. Myc. 55: 211. 1913.

Peridermium
Peridermium ephedrae Cke., Indian Forest. 3: 95. 1877;---*Accidium ephedrae*
Peridermium sinensis Wang et Guo, Acta Microbiol. Sinica 20 (1): 22-23, 1980.
Peridermium yunshae Wang et Guo Acta Microbiol. Sinica 20 (1): 22, 1980.

***Phakopsora* Diet.**
Phakopsora ampelopsidis Diet. et Syd., Hedw. 37: 217. 1898.---Uredo vitis Thu m. Pilze Weinst. P. 182. 1878.

***Phragmidium* Link.**
Phragmidium potentillae (Pers.) Karst., Myc. Fenn. 4: 49. 1879;---*Puccinia potentillae* Pers., Syn. Meth. Fung. P. 229. 1801.
Phragmidium rosae-rugosae Kasai, Trans. Sapporo Nat. Hist. Soc. 3: 38. 1910.
Phragmidium rubi (Pers.) Winter in Rabenh., Kr. Fl. Deutschl., 1 (1): 230. 1884;---*Uredo rubi-fruiticosi* Pers., Syn. Fung. P. 218. 1801;---*Puccinia mucrassatum* Link, Spec. Plant., 2: 85. 1825.
Phragmidium shengezhense Chen et Chen, Act. Phytopath. 9: 27. 1979.
Phragmidium zamonense Chen et Chen, Act. Phytopath. 9: 28. 1979.

***Pileolaria* Cast.**
Pileolaria klugkistiana (Diet.) Diet., Ann. Myc. 19: 301. 1921;---*Uromyces klugistiana* Dietel, Engl. Bot. Jahrb. 27: 570. 1900.

***Puccinia* Pers.**
Puccinia akebiae Wang et Wei, Acta Mircobiol. Sinica 20 (1): 18-19, 1980.
Puccinia angelicae Fuck., Gahrb. Nass. Wer. Nat. 23-24: 52, 1872.
Puccinia *angelicae* P. Henn. Var. ***centellae*** Chen, Acta Microbiol. Sinica 20(1): 20, 1980.
Puccinia asteris Duby, Bot. Gall. 2: 888. 1830;---*Leptopuccinia asteris* Sydow, Ann. Mycol. 20: 120. 1922.
Puccinia atrofusca (Dudl. et Thom.) Holw., Journ. Myc. 10: 288. 1904; *Uromyces atrofuscus* Dudl. et Thom., Journ. Myc. 10: 55. 1904.

Puccinia bolleyana Sacc., Syll. Fung. 9: 303. 1891.
Puccinia caricis Rebent., Fl. Neomarch. P. 356. 1804.
Puccinia caricis-brunnear Diet., Ann. Myc. 4: 306. 1906.
Puccinia chelonopsidis Balfour-Browne, Bull. Brit. Mus. (Nat. Hist.) 1: 207. 1955.
Puccinia circaeae Pers., Disp. Meth. Fung., P. 39. 1797.
Puccinia clintoniae-udensis Bud. var. ***tibetica*** Chen Acta Microbiol. Sinica 20(1): 21, 1980.
Puccinia cnici-oleracei Pers. ex Desm. Cat. Pl. Onnris. P. 24. 1823;---*Puccinia asteris* Duby, Bot. Gall. 2: 888. 1830;---*Puccinia artemisicola* P. et H. Syd., Monogr. Ured. 1: 4. 1902;---*Puccinia Phramidioides* Liou et Wang, Contr. Inst. Bot. Acad. Peiping 2: 158. 1934.
Puccinia coronata Cda., Ic. Fung. 1: 6, 1837.
Puccinia deyeuxiae-scabrescentis Wang et Wei, Acta Microbiol. Sinica 20 (1): 20, 1980.
Puccinia elymi Westend., Bull. Acad. Brux. 182: 408. 1851;---*Rostropia elymi* (West.) Lagerh., Journ. Bot. Fr. 3: 188. 1898.
Puccinia gentianae Rorhl., Deuta. Fl. Ed. 2, 3 (2): 131. 1813;---*Uredo gentianae* Str., Ann. Wett. Ges. 2: 102. 1810.
Puccinia haleniae Arth. et Holw., Bull. Geol. Nat. Hist. Surv. Minn. 3: 30. 1887.
Puccinia helianthii Schw., Sc. Nat. Ges. Leipzig 1: 73. 1822.
Puccinia heterocoloris Chen, Acta Microbiol. Sinica 20 (1): 22, 1980.
Puccinia heucherae (Schw.) Diet., Ber. Deutsch. Bot. Ges. 9: 42. 1891.
Puccinia himalensis (Barc.) Diet., Hedw. 29: 259. 1890;---E. J. Butler et Al., Pung. India P. 102. 1960.
Puccinia kusanoi Diet., Bot. Jahrb. 27: 568. 1899.
Puccinia lakanensis Chen et Wei, Act. Phytopath, Sinica 9 (1): 26. 1979.
Puccinia leveillei Mont., Gay. Fl. Chil. 8: 41. 1852.
Puccinia menthae Pers., Syn. Meth. 1: 227. 1801;---*Uredo menthae* Pers., Syn. Fung. P. 220. 1801;---*Dicaeoma menthae* S. F. Gray, Nat. Arr. Brit. Pl. 1: 542. 1821.
Puccinia nepalensis Barcl.et Diet. Syd. Hedw. P. 265. 1891;---Syd. Monogr. Ured. 1:578.
Puccinia paludosa Plowright, Monogr. Brit. Ured. Ustil. P. 174. 1889;---*Puccinia pedicularis* Lagh., Tromso Mus. Aarsh. 17: 68. 1895.
Puccinia pimpinellae Mart., Fl. Mosq. 2: 226. 1817.
Puccinia poae-nemoralis Otth., Mitth. Nat. Ges. Bern, P. 113. 1871;---*Puccinia poae-sudeticae* Joerst., Nytt. Mag. Nat. 70: 325. 1932.
Puccinia recondite Rob. ex Desm., Bull. Soc. Bot. France 4: 798. 1857;---*Puccinia hordei* Fckl., Jahrb. Nass. Ver. Nat. 16: 27. 1873;---*Puccinia rubigo-vera* Wint., Rabh. Krypt-Fl. 1: 217. 1881.
Puccinia silvatica Schroter in Cohn's Beitr. 3: 68. 1879;---*Aecidium taraxaci* Ktze et Sch., Mycol. Heft. 1: 85. 1817;---*Caeoma compositarum* Link., Sp. Pl. 2: 50. 1825.
Puccinia smilacis Schw., Schr. Nat. Ges. Lepzig 1: 72. 1822.
Puccinia sorghi Schw., Trans. Am. Phil. Soc. 2 (4): 259. 1832;--*Uredo zeae* Desm., Ann. Sci Nat. 13: 182. 1840.
Puccinia striiformis West., Not. 7: 10. 1861;---*Puccinia glumatum* (Schmedt) Eriks. et Henn., K. Landtbr. Akad. Handl. Tidskr. 33: 169. 1894.
Puccinia taylorii Balfour-Browne, Bull. Brit. Mus. (Nat. Hist.) 1: 209. 1955.
Puccinia vaginatae Juel., Bot. Not. P. 56. 1893.
Puccinia violae-reniformis Wang et Wei, Acta Microbiol. Sinica 20 (1): 19, 1980
Puccinia yokogurae P. Henn., Hedw. 42: 107 1903.

***Pucciniastrum* Otth.**
Pucciniastrum potentillae Komarov in Jacz., Fung. Rossiae Exsicc. No. 327. 1899;- Hedwigia 14:128. 1900;---N. Hiratsuka, Monogra. Pucciniastreae P. 237. 1936.

***Ravenelia* Berk.**
Ravenelia indigoferae Tranz., in Dietel, Hedwigia 33: 369. 1894.

***Roestelia* Rebent.**

Roestelia levis (Crowell) Kern, Taxon. Acc. Gymnosporangium P. 86. 1972;---*Gymnosporangium leve* Crowell, Journ. Arnold Arb. 17: 50. 1936.

Roestelia sikangensis (Petr.) Joerstad, Archiv. Bot. 4: 363, 1959;---*Gymnosporangium sikangense* Petr., Act. Hort. Gothob. 17: 120. 1949.

***Thekopsora* Magn.**

Thekopsora areolata (Fr.) Magn., Sitzungaber d. Ges. Nat. Freunde Berl. P. 58. 1875;---*Aecidium strobilinum* Reess Abh. Nat. Ges. Hall. 11:105. 1869;---*Pucciniastrum aereobilinum* Otth., Mitt. Nat. Ges. Bern. P. 85. 1863.

***Tranzschelia* Arth.**

Tranzchelia pruni-spinosae (Pers.) Diet. Ann. Myc. 20: 31. 1922.

Uredo

Uredo clemensiae (Arth. et Cumm.) Hiratsuka f., Transact. Myc. Soc Japan 5: 4. 1957;--*Pucciniastrum clemensiae* Arth. et Cumm. Philipp. Journ. Sci. 51: 479. 1936.

Uredo myriactidis Sumdaram, Indian Phytopath. 14: 203. 1962.

Uredo piptanthus Chen, Act. Phytopath. Sin. 12: 27. 1982.

Uredo pseudocystopteridis Wang et Wei, Acta Microbiol. Sinica 20 (1): 26-27, 1980.

Uredo rhododendronis Wang et Guo, Acta Microbiol. Sinica 20(1): 23, 1980.

Uromyces

Uromyces appendiculatus (Pers.) Unger. Enfl. Bodens P. 216. 1936;---*Uromyces phaseoli* (Rebent.) Wint., Hedw. 19: 37. 1880.

Uromyces hedysari-obscuri (DC.) Lev. Orbigny. Dict. Univ. Hist. Nat. 12: 786. 1849;---*Puccinia hedysari-obscuri* DC., Syn. Pl. Gall P. 46. 1806.

Uromyces lapponicus Lagerh., Bot. Notiser. 7: 274. 1890.

Uromyces polygoni-avicularlae (Pers.) Karst., Mycol. Fenn. 4: 12. 1879;---*Puccinia polygoni-avicularis* Pers., Syn. Meth. Fung. P. 227. 1801.

Uromyces viciae-fabae (Pers.) Schroet., Hedwigia 14: 161. 1875;---*Uredo viciae-fabae* Pars., Syn. Fung. P. 221. 1801;---*Uromyces fabae* de Bary, Ann. Sci. Nat. Bot. 4 (20): 80. 1863.

Index of forest diseases and Insects of Siberia and the Soviet Far East
(Pine, Spruce, fir and Larch)

Pests and Pathogens on Coniferous Trees of the Eastern Soviet Union

I. Siberian Region

Main Species of Bark Beetles (Scoytidae)

A. Larch Forests

Larix gmelini (Rupr.) Rupr.

Dryocoetes Baicalicus Reitt.
Ips acuminatus Eichh.
I. duplicatus Sahlb.
I. sexdentatus Boern.
I. subelongatus Motsch.
I. typographus *Lin.*
Orthotomicus starki Spess.
Pityogenes baicalicus Egg.
P. chalcographus L.
P.irkutensis Egg.
Pityophthorus micrographus L.
Polygraphus sachalinensis Egg.
Scolytus morawitzi Sem.
Trypodendron lineatus Ol.

Larix sibirica Ledeb.

Ips acuminatus Eichh.
I. duplicatus Sahlb.
I. sexdentatus Boern.
I. subelongatus Motsch.
I. typographus Lin.
Orthotomicus starki Spess.
Pityogenes baicalicus Egg.
P. chalcographus L.
Pityophthorus micrographus L.
Polygraphus sachalinensis Egg.
Trypodendron lineatum Ol.

B. Spruce and Fir Forests

Picea obovata Ledeb.
Abies sibirica Ledeb.

Carphoborus tephlouchovi Spess.
Cryphalus abietis Ratz.
C. saltuaris Wse.
Dendroctonus micans Kug.
Dryocoetes autographus Ratz.
Hylastes cunicularius Er.
Hylurgops glabratus Zett.
Ips acuminatus Eichh.
I. duplicatus Sahlb.
I. typographus L.
Orhotomicus laricis F.
O. starki Spess.
O. suturalis Gyll.
Phthorophloeus spinulosus Rey
Pityogenes bidentatus Hbst.
P. chalcographus L.
P. quadridens Hart.
Pityophthorus morosovi Spess.
P. traegardhi Spess.
Polygraphus poligraphus L.
P. punctifrons Thoms.
P. subopacus Thoms.
Trypodendron lineatum Ol.
Xylechinus pilosus Ratz.

C. Pine Forests

Pinus sibirica Ledeb.

Dryocoetes autographus Ratz.
Hylastes opacus Er.
H. opacus Er.
Hylurgops glabratus Zett.
H. palliatus Gyll.
Ips duplicatus Sahlb.
I. sexdentatus Boern.
Orthotomicus golovjankoi Pjat.
O. laricis F.
O. proximus Eichh.
O. suturalis Gyll.
Pityogenes bidentatus Hbst.
P. chalcographus L.

P. quadridens Hart.
Pityophthorus micrographus L.
Polygraphus subopacus Thoms.
Trypodendron lineatum Ol.

D. Scotch Pine Forests

Pinus sylvestris L.

Blastophagus minor Hart.
B. piniperda L.'*Carphoborus cholodkovskyi* Spess.
Dendroctonus micans Kug.
Hylastes ater Pyk.
H. opacus Er.
Hylurgops glabratus Zett.
H. spessivtzevi Egg.
Ips acuminatus Eichh.
I. sexdentatus Boern.
Orthotomicus laricis F.
O. proximus Eichh.
O. suturalis Gyll.
Pityogenes bidentatus Hbst.
P. chalcographus L.
P. irkutensis Egg.
P. quadridens Hart.
Polygraphus poligraphus L.
Trypodendron lineatum Ol.

Wood Borers (Cerambycidae)

A. Tundra Larch Forests

Larix gmelini (Rupr.) Rupr.

Acanthocinus carinulatus Gebl.
Acmaeps pratensis Laich.
A. septentrionis Thoms.
A. smaragdula F.
Anoplodera sequensi Reitt.
Asemum striatum L.
Callidium coriaceum Payk.
Judolia sexmaculata L.
Monochamus impluviatus Motsch.
M. sutor L.
M. urussovi Fisch.
Pogonocherus fasciculatus Deg.
Tetropium gracilicorne Reitt.

B. Larch Forests

Larix sibirica Ledeb.

Acanthocinus carinulatus Gebl.
Acmaeops septentrionis Thoms.
A. smaragdula F.
Anoplodera sequensi *Reitt.*
A. variicornis *Dalm.*
Asemum striatum L.
Callidium chlorizans Sols.
C. violaceum L.
Clytus arietoides Reitt.
Cornumutila quadrivittata Gebl.
Gaurotes virginea L.
Judolia sexmaculata L.
Leptura arcuata Panz.
Monochamus impluviatus Motsch.
M. salutarius Gebl.
M. sutor L.
M. urussovi Fisch.
Nivellia extensa Gebl.
Pogonocherus fasciculatus Deg.
Rhagium inquisitor L.
*Strangalia attenuata*L.
Tetropium gracilicorne Reitt.
Xylotrechus altaicus Gebl.

C. Spruce and Fir Forests

Picea obovata Ledeb.
Abies sibirica Ledeb

Acanthocinus griseus F.
Acmaeops pratensis Laich.
A. septentrionis Thoms.
Anoplodera sequensi Reitt.
Arhopalus rusticus L.
Asemum striatum L.
Clytus arietoides Reitt.
Evodinus borealis Gyll.
Judolia sexmaculata L.

Molochus minor L.
Monochamus salutarius Gebl.
M. sutor L.
M. urussovi Fisch.
Pogonocherus fasciculatus Deg.
Pronocera brevicollis Gebl.
Rhagium inquisitor L.
Saperda interrupta Gebl.
Spondylis buprestoides L.
Strangalia attenuata L.
Tetropium castaneum L.

D. Pine Forests

Pinus sibirica Ledeb.

Acmaeops angusticollis Gebl.
A. septentrionis Thoms.
A. smaragdula F.
Anoplodera rufiventris Gebl.
A. rubra L.
A. sequensi Reitt.
Arthopalus rusticus L.
Asemum striatum L.
Callidium coriaceum Payk.
Clytus arietoides Reitt.
Evodinus borealis Gyll.
Monochamus salutarius Gebl.
M. sutor L.
M. urussovi Fisch.
Pogonocherus fasciculatus Deg.
Rhagium inquisitor L.
Tetropium castaneum L.
Tragosoma depsarium L.

E. Scotch Pine Forests

Pinus sylvestris L.

Acanthocinus aedilis L.
A. griseus F.
Acmaeops marginata F.
Anoplodera rubra L.
A. virens L.
Arhopalus rusticus L.
A. tristis F.
Asemum striatum L.
Callidium violaceum L.
Clytus arietoides Reitt.
Evodinus borealis Gyll.
Guarotes virginea L.
Judolia sexmaculata L.
Monochamus galloprovincialis Ol.
Pachyta quadrimaculata L.
Pogonocherus fasciculatus Deg.
P. ovatus Goeze
Pronocera brevicollis Gebl.
Rhagium inquisitor L.
Spondylis buprestoides L.
Tragosoma depsarium L.

Flatheaded Borers (Buprestidae)

A. Pine Forests

Pinus sylvestris L.

Phaeops cyanea

Weevils (Curculionidae)

A. Larch Forests

Hylobius abietis L.
H. moria

B. Spruce and Fir Forests

Pissodes cembrae Motschulaky

C. Pine Forests

Pissodes cembrae Motschulaky

Wood Wasps (Siricidae)
(Siberia and Far East)

A. Tundra Low Density Forests

Paururus noctilio F.
Urocerus gigas L.

B. Spruce, Fir, and Pine Forests

Paururus ermak Sem.
P. juvencus L.
P. mongolorum Sem. et Guss.
P. noctilio F.
Tremex satanas Sem.
Urocerus antennatus Marl.
U. gigas L.
Xeris spectrum L.
Xoanon mysta Sem.

C. Fir Forests

Paururus ermak Sem.
P. juvencus L.
P. noctilio F.
Tremex satanas Sem.
Urocerus gigas L.
Xeris spectrum L.

D. Larch Forests

Paururus ermak Sem.
P. juvencus L.
P. mongolorum Sem. et Guss.
P. noctilio F.
Urocerus antennatus Marl.
U. gigas L.
U. umbra Sem.
Xeris spectrum L.

E. Scotch Pine Forests

Paururus junencus L.
P. noctilio F.
Urocerus gigas L.
U. tardigradus Ced.

F. Coniferous and Broad-Leaf Larch Forests

Paururus ermak Sem.
P. juvencus L.
P. mongolorum Sem. et Guss.
P. noctilio F.
Urocerus antennatus Marl.
U. gigas L.
Xeris spectrum L.
Xiphydria eborata Knw.
Xoanon matsumurae Roh.
X. mysta Sem.

II. Seacoast Forests Region (Far East)
Main Species of Bark Beetles (Scolytidae):

A. Larch Forests

Larix gmelini (Rupr.) Rupr.

Cryphaus latus Egg.
Dryocoetes baicalicus Reitt.
D. hectographus Reitt.
D. rugicollis Egg.
Ips acuminatus Eichh.
I. duplicatus Saheb.
I. sexdentatus *Boern.*
Orthotomicus laricis Fabr.
O. suturalis Gyll
Pityogenes chalcographus L.
Polygraphus sachalinensis Egg.
Trypodendron lineatum Ol.

Larix olgensis Henry

Cryphalus latus Egg.
Dryocoetes baicalicus Reitt.
D. hectographus Reitt.
D. rugicollis Egg.
Ips acuminatus Eichh.
I. duplicatus Saheb.
I. sexdentatus Boern.
Orthotomicus laricis Fabr.
O. suturalis Gyll.
Pityogenes chalocographus L.
Trypodendron lineatum Ol.

B. Spruce and Fir Forests

Picea jezoensis Carr.

Blastophagus puellus Reitt.
Dryocotes hectographus Reitt.
D. rugicollis Egg.
Hylourgops glabratus Zett
H. palliates gyll.
Ips acuminatus Eichh.
I. sexdentatus Boern.
I. typographus L.
O. golovjankoi Pjat.
O. laricis Fabr.
O. suturalis Gyll.
Piyogenes chalcographus L.
Polygraphus jezoensis Niis.
P. punctifrons Thoms.
P. sachalinensis Egg.
P. subopacus Thoms.
Trypodendron lineatum Ol.
T. proximum Niis.
Xylechinus pilosus Ratz.

Picea koraiensis Nakai

Dryocoetes hetographusi Reitt.
D. rugicollis Egg.
Hylurgops palliates Gyll.
Ips acuminatus Eichh.
I. sexdentatus Boern.
I. subelongatus Motsch.
I. typographus L.
Orthotomicus golovjankoi Pjat.
O. laricis Fabr.
O. suturalis Gyll.
Pityogenes chalcographus L.
Polygraphus jezoensis Niis.
P. punctifrons Thoms.
P. sachalinensis Egg.
Scolytus morawitzi Sem.
Trypodendron lineatum Ol.
T. proximum Niis.

Abies holophylla Maxim.

Dryocoetes hectogrphus Reitt.
D. rugicollis Egg.
D. striatus Egg.
Ips duplicatus Saheb.
Hylurgops palliates Gyll.
Orthotomicus golovjankoi Pjat.
Pityogenes chalcographus L.
Polygraphus proximus Blandf.
P. schalinensis Egg.
Trypodendron lineatum Ol.

Abies nephrolepis (Trautv.) Maxim.

Dryocoetes hectographus Reitt.
D. rugicollis Egg.
D. striatus Egg.
Hylurgops palliatus Gyll.
Orthotomicus golovjankoi Pjat.
O. laricis Fabr.
Pityogenes chalcographus L.
Polygraphus proximus Blandf.
P. sachalinensis Egg.
Trypodendron lineatum Ol.

C. Pine Forests

Pinus koraiensis Sieb. et Zucc.

Blastophagus pilifer Spess.
Dryocoetes hectographus Reitt.
Hylastes parallelus Chapusis
H. plumbeus Blandf.
Hylugops imitator Reitt.
H. interstitialis Chap.
H. spessivtzevi Egg.
Ips acuminatus Eichh.
I. sexdentatus Boern.

I. typographus L.
Orthotomicus golovjankoi Pjat.
O. laricis Fabr.
O. proximus Eichh.
O. suturalis Gyll.
Pityogenes chalcographus L.
Trypodendron lineatum Ol.

Pinus sylvestris L.
P. stylvestris mongolica Litv.

Blastophagus pilifer Spess.
Dryocoetes hectogrphus Reitt.
Hylastes attenuatus K.
H. interstitialis Chap.
Ips acuminatus Eichh.
I. sexdentatus Boern.
I. typographus L.
Orthotomicus laricis Fabr.
O. suturalis Gyll.
Pityogenes chalcographus L.
Polygraphus sachalinensis Egg.
Trypodendron lineatum Ol.

Wood Borers (Cerambycidae)

A. Larch Forests

Larix gmelini (Rupr.) Rupr.

Acanthocinus aedilis L.
A. carinulatus Gebl.
Arhopalus rusticus L.
Asemum striatum L.
Callidium aeneum Deg.
C. ciolaceum L.
Cyrtoclytus capra Germ.
Monochamus salutarius Gebl.
M. urussovi Fisch.
Pogonocherus fasciculatus Deg.
Rhagium inquisitor L.
Tetropium castaneum L.
T. gracilicorne Reitt.
Xylotrechus altaicus Gebl.

B. Spruce and Fir Forests

Picea jezoensis Carr.

Acanthocinus aedilis L.
A. carinulatus Gebl.
A. griseus F.
Arhopalus rusticus L.
Asemum striatum L.
Callidium violaceum L.
Crytoclytus capra Germ.
Monochamus salutarius Gebl.
M. sutor L.
M. urussovi Fisch.
Pogonocherus fasciculatus Deg.
Rhagium inquisitor L.
Semanotus undatus L.
Tetropium castaneum L.
T. gracilicorne Reitt.

Picea koraiensis Nakai

Acanthocinus aedilis L.
A. carinulatus Gebl.
A. griseus *F.*
Arhopalus rusticus L.
Asemum striatum L.
Callidium violaceum L.
Cyrtoclytus capra Germ.
Monochamus salutarius Gebl.
M. sutor L.
M. urussovi Fisch.
Pogonocherus fasciculatus Deg.
Rhagium inquisitor L.
Tetropium castaneum L.

Abies holophylla Maxim.

Acanthocinus aedilis L.
A. carinulatus Geb.
Arhopalus rusticus L.
Asemum striatum L.
Callidium violaceum L.
Monochamus salutarius Gebl.
M. sutor L.
M. urussovi Fisch.
Pogonocherus fasciculatus Deg.

Rhagium inquisitor L.
Tetropium castaneum L.
T. gracilicrne Reitt.

Abies nephrolepis (Trautv.) Maxim.

Acanthocinus aedilis L.
A. carinulatus Gebl.
A. rusticus L.
Asemum striatum L.
Callidium violaceum L.
Monochamus salutarius Gebl.
M. sutor L.
M. urussovi Fisch.
Tetropium castaneum L.
T. gracilicorne Reitt.

C. Pine Forests

Pinus koraiensis Sieb. et Zucc.

Acanthocinus aedilis L.
Arhopalus rusticus L.
Asemum striatum L.
Callidium violaceum L.
Crytoclytus capra Germ.
Monochamus salutarius Gebl.
M. urussovi Fisch.
Pogonocherus fasciculatus
Rhagium inquisitor L.
Tetropium castaneum L.
T. Gracilicorne Reitt.

Pinus sylvestris L.

Acanthocinus aedilis L.
A. carinulatus Gebl.
Arhopalus rusticus L.
Asemum striatum L.
Callidium violaceum L.
Crytoclytus capra Germ.
Monochamus salutarius Gebl.
M. sutori L.
M. urussovi Fisch.
Pogonocherus fasciculatus
Rhagium inquisitor L.
Tetropium castaneum L.

Flatheaded Borers (Buperstidae)

A. Larch Forests

Larix gmelini (Rupr.) Rupr.

Ancylocheira sibirica Fleisch.
A. strigosa Gebl.
Anthaxia quadripunctata L.
A. reticulata Motsch.
Chrysobothris chrysostigma L.
Melanophila acuminata Deg.
Phaenops guttulata Gebl.

B. Spruce and Fir Forests

Picea jezoensis Carr.

Ancylocheira sibirica Fleisch.
A. strigosa Gebl.
Anthaxia quadripunctata L.
A. reticulata Motsch.
Chrysobothris chrysostigma L.
Melanophila acuminata Deg.
Phaenops guitulata Gebl.

Picea koraiensis Nakai

Ancylocheira sibirica Fleisch.
A. strigosa Gebl.
Anthaxia quadripunctata L.
A. reticulata Motsch.
Chrysobothris chrysostigma L.
Melanophilia acuminata Deg.
Phaenops guttulata Gebl.
Abies holophylla Maxim.
Anthaxia quadripunctata L.
A. reticulata Motsch.
Chrysobothris chrysostigma L.
Melanophila acuminata Deg.

Abies nephrolepis (Trautv.) Maxim.

Anthaxia quadripunctata L.

A. reticulata Motsch.
Chrysobothris chrysostigma L.

C. Pine Forests

Pinus koraiensis Sieb. et Zucc.

Ancylocheira sibirica Fleisch.
A. strigosa Gebl.
Anthaxia quadripunctata L.
A. reticulata Motsch.
Chrysobothris chrysostigma L.
Melanophila acuminata Deg.

Pinus sylvestris var. *mangolica* Litv.

Ancylocheira sibirica Fleisch.
A. strigosa Gebl.
Anthaxia quadripunctata L.
A. reticulata Motsch.
Chrysobthris chrysostigma L.
Melanophila acuminata Deg.

Weevils (Curculionidae)

A. Larch Forests

Larix gmelini (Rupr.) Rupr.

Hylobius albosparsus Boh.
Pissodes gyllenhali Gyll.

B. Spruce and Fir Forests

Picea jezoensis Carr.

Cryptorrhynchus electus Roel.
Hylobius albosparsus Boh.
H. haroldi Faust.
H. piceus Deg.
H. pinastri Gyll.
Pissodes gyllenhali Gyll.
Sipalinus gigas F.

Picea koraiensis Nakai

Cryptorrhynchus electus Roel.
Hylobius albosparsus Boh.
H. haroldi Faust.
H. piceus Deg.
H. pinastri Gyll.
Pissodes gyllenhali Gyll.

Abies holophylla Maxim.

Niphades variegatus Roel.
Sipalinus gigas F.
Melandryidae spp.

Abies nephrolepsis (Trautv.) Maxim.

Niphades variegatus Roel.

C. Pine Forests

Pinus koraiensis Sieb. et Zucc.

Cryptorrhynchus electus Roel.
Hylobius albosparsus Boh.
H. haroldi Faust
H. pinastri Gyll.
Pissodes gyllenhali Gyll.
Sipalinus giagas F.

Pinus sylvestris var. *mangolica* Litv.

Hylobius albosparsus Boh.
Pissodes gyllenhali Gyll.

Blazed Tree Borer (Melandryidae)

A. Larch Forests

Larix gmelini (Rupr.) Rupr.

Serropalpus barbatus L.

Wood Wasps (Siricidae)

A. Larch Forests

Larix gmelinii (Rupr.) Rupr.

Paururus ermak Sem.
P. juvencus L.
Urocerus antennatus Marl.
U. gigas L.
Xoanon mysta Sem.

B. Spruce and Fir Forests

Picea jeoensis Carr.

Paururus ermak Sem.
P. juvencus L.
Urocerus antennatus Marl.
U. gigas L.

Picea koraiensis Nakai
Paururus ermak Sem.
P. juvencus L.
Urocerus antennatus Marl.
U. gigas L.
Xoanon mysta Sem.

Abies holophylla Maxim.

Paururus ermak Sem.
P. juvencus L.
Urocerus antennatus Marl.
Xoanon mysta Sem.

Abies nephrolepis (Trautv.) Maxim.

Paururus ermak Sem.
P. juvencus L.
Urocerus antennatus Marl.
U. gigas L.
Xoanon mysta Sem.

C. Pine Forests

Pinus koraiensis Sieb. et Zucc.

Paururus ermak Sem.
P. juvencus L.
Urocerus antennatus Marl.
U. gigas L.
Xoanon mysta Sem.

Pinus sylvestris var. *mangolica* Litv.

Paururus ermak Sem.
P. juvencus L.
Urocerus antennatus Marl.
U. gigas L.
Xoanon mysta Sem.

Pathogens
(Pathogens of Coniferous Trees in the Eastern Soviet Union)

A. Larch (*Larix* spp.)

Hypodermella laricis
Meria laricis
Melampsoridium betulinum
Melampsora larici-capraearum
Melampsora larici-epitea
Melampsora larici-populina
Melampsora larici-tremulae
Melampsora poulnea
Hymenochaete abietina
Lachnellula willkommii
Phacidiophycnis pseudotsugae
Columnocystis abietina

Fomitopsis officinalis
Fomitopsis pinicola
Ganoderma lucidum
Haplopilus fibrillosus
Hirschioporus abietinus
Hymenochaete tabacina
Ischnoderma resinosum
Laetiporus sulphreus
Phellinusignarius
Phellinus pini
Phellinus torulosis
Pholiota destruens
Polyporus squamosus
Armillaria spp.
Heterobasidion annosum
Inonotus circinatus
Inonotus heinrichii
Inonotus tomentosus
Phaeolus schweinitzii
Polyporus osseus

B. Spruce (*Picea* spp.)

Lachnellula calyciformis
Lachnellula subsilissimus
Ascocalyx spp.
Pezicula spp.
Lophodermium macrosporum
Chrysomyxa ledi-ledi
Chrysomyxa ledi-rhododendri
Chrysomyxa woroninii
Chysomyxa pirolata
Pucciniastrum areolatum
Heterobasidion annosum
Phaeolus schweinitzii
Inonotus tomentosus
Phellinus pini
Phellinus chrysoloma
Phellinus weirii
Armillaria spp.

C. Fir (*Abies* spp.)

Lachnellula spp.
Pucciniastrum goepptertianum
Pucciniastrum epilobi
Uredinopsis spp.
Melampsora spp.
Melampsorella spp.
Ophiostoma spp.
Heterobasidion annosum
Phaeolus schweinitzii
Inonotus tomentosus
Phellinus pini
Phellinus chrysoloma
Armillaria spp.
Phellinus weirii

D. Pine (*Pinus* spp.)

Cronartium flaccidum
Cronartium spp.
Lachnellula spp.
Ophiostoma spp.
Heterobasidion annosum
Bursaphelenchus spp.
Phaeolus schweinitzii
Inonotus tomentosus
Phellinus pini
Armillaria spp.
Phellinus weirii
Phellinus torulosus

E. Scotch Pine (*Pinus sylvestrisi)*

Lagenidiales

Lagenidium pygmaeum Zopf. (dust coat)
Phytophthora cactorum Lev. and Cohn (seedling)
Ph. cinnamomi Kands. (seedling)
Phythium aphanidermatum (seedling)
P. debaryanum Hesse (seedling)
P. irregulare Buism. (seedling)
P. pyrilobum Trow. (seedling)
P. ultimum Trow. (seedling)
P. torulosum F. (seedling)

Mucorales

Thamnidium elegans Link

Eurotiales

Elaphomyces cervinus (Pers.) Schrot (seed)
Ceratocystis (Ophiostomas) (lumber)
C. minor (Hedge.) Hunt (lumber)

C. piceae (Muxh.) Bakshi (lumber)
C. pini (lumber)
C. pilifera (lumber)

Sphaeriales

Herpotrichia jiniperi (Duby.) Petrak. (needle) (lumber)
H. nigra Karst. (lumber)
Niesslia vermicularia Zer. (branch)
N. pusilla (Fr.) Sch. (needle dryed, branch)
Spharia pinastri Fr. (branch)

Xylariales

Coniochaeta malacotriha (Niessl.) Trav. (branch, lumber)
Rosellinia helena (Fr.) Sch. (Root)
R. obliquata Wint. (cone's sclaes)
Hypoxylon diathrauston Rehm. (branch)

Allantosphacriales

Calosphaeria abietis Krieger (bark)
C. ligniaria (Grev.) Mass. (lumber, bark)
Diaporthe eres Nits. (bark)
Valsa collicula (Worm) Cke. (bark)
V. decumbens (Sch.) Nits. (bark)
V. pini (Alb. et Schw.) Fr. (bark)
V. superficiales Fr. (bark)
Valsella abietis (Rostr.) Munk. (branch0dried)

Melanosporarales

Melanospora chionea (Fr.) Cda (bark and rotted wood)

Hyporcreales

Calonectria cucurbitula (Fr.) Sacc. (bark and rotted wood)
Gibberella suabinetti (Mont.) Sacc. (seedling)
Hypocrea alutaceae (Pers. ex Fr.) Ces. (needle and seed)
Nectria cinnabarina (Tode) ex Fr. (branch)
N. cucurbitula (Tode) ex Fr. (branch)
N. viridescens Booth (branch)
Ophionectria scolecospora Bref. (needle, branch)

Pezizales

Phizina undulata Fr. (root)
Desmazierella acicola Lib. (needle)
Discina perlata (Fr.) Fr. (trunk)
Peziza calycina Fr. (trunk)
P. resinae Fr. (trunk)
Pseudoplectania melaena Fr. (branch)

Phacidiales

Coccophacidium pini (Schw. ex Fr.) Rehm. (branch)
Hypodermella empla Fr. (needle)
H. arcuata Dark (needle)
H. cerina D. (needle)
H. concolar L. (needle)
Hypodermella limitata (needle)
H. montana F. (needle)
H. pedatum D. (needle)
H. sulcigena (Rostr.) Tubeuf (needle)
Lasiostictis fimbriata (Schw.) Baumber (needle)
Lophodermium brachysporum Rostr. (needle)
L. durilabrum Darker (needle)
L. nitens (Darker) needle
L. pinastri (Sch.) Chev. (needle)
Phacidium convexum Dearn (needle)
P. infestans Karst. (needle)
P. planum Davis (needle)
Pseudographis pinicola (Nyl.) Rehm (bark)

Ostropales

Naemacclus niveus Sacc. (needle)
Stictis fimbriata Schw. (cone)

Helotiales

Mitrula pusilla (Nees.) Fr. (branch)

Sclerotinia graminearum Elen.
Orbilia chryzocoma (Bull.) Sacc. (branch)
Cenangium abietis (Pers.) Duby
C. acicolum Rehm
C. atropurpureum Cash
Cenangium ferruginosum Fr.
C. pinicola (Reb.) Karst. (branch)
Crumenula abietina Lagerb. (branch)
C. sororia Darst.
Dermatea pini Phill. et Harkn. (branch)
Tympanis buchsii Rehm. (branch)
T. confusa Nyl. Conn. (branch)
T. hypopodia Ny. Conn. (branch)
T.pinastri Tul. (branch)
Dasyscypha agassizzi (Berk ex Curt.) Sacc. (branch)
D. arida Sacc. (branch)
D. calyciformis (Willd.) Rehm.
D. ellisiana (Rehm) Sacc. (branch)
D. oblongospora Hahn ex Ayers. (branch)
D. pini (Brunch.) Hahn ex Ayers. (branch)
D. pulverulentus (Lib.) Sacc.
Lachnellula calycina Sacc. (branch)
L. chrysophthalma (Pers.) Karst. (branch)
L. flavorirens (Bres.) Dennis (branch)
L. fuscosanguonea (Rehm) Dennis (branch)
L. pini (Brunch.) Dennis (branch and stem)
L. pseudofarinacea Dennis (branch)
Pezizella lythri Sacc.
P. minuta Decern.
Phialea acuum (Alb. et Schw.) Rehm.
Biatorella resinae (Fr.) Mudd.
Pragmopora amphibolaI Massal.
Scleroderris lagerbergii Germ.
Tryblidiopsis pinastri (Pers.) Karst. (branch)

Dothideales

Physalospora obtusa (Schw.) Cke.
Phaeocryptopus pinastri (Ell and Sacc.) Petz.
Scirrhia acicola (Dearn) Siggers
Scirrhia pini Funk.
Scorias spongiosa (Schw.) Er.
Cucurbitaria pithyophila (Fr.) de N. (branch)
Botryosphaeria ribis Gross. (branch)

Capnodiales

Capnodium pini Berk. et Curt. (branch)

Hysteriales

Hypoderma brachysporum (Rostr.) Tubeuf.
H. conigenum Cooke
H. desmazierii Duby
H. pallidula Br.
H. pinicolaI Brunch
Hypoderma saccatum Dark. (branch)
Hysterium contortum Ditt. (branch)
H. crispum Fr. (branch)
H. elatinum Fr. (branch)
Hysterographium nova Caesariense (ell.) Roum.
Lophium mytilinum Pers. x Fr. (branch)

Aphyllophorales

Aleruodiscus amorphus (Pers.) Rab.
A. polygonius (Pers.) H. et L.
Amylostereum areolatum Fr. Boidin (bark, lumber)
Athelia galzinii (Bourd.) Donk.
Cavulicium macconii (Burt) John Erikss et Boid ex Parm. (branch)
Corticium byssinum (Karst.) Mass.
C. centrifugum (Lev.) Bres. (log)
C. evolvens Fr. (branch)
C. laeve Br. (lumber)
C. mutabile Bres. (lumber)
C. ochroleucum Bres. (lumber)
C. pelliculare Karst. C. pertenue Karst.
C. sulphureum Fr.
C. terrigenum Bres. (lumber)
C. teutoburgense Brinkm. (lumber)
Cytidia albo-melea (Bond.)
Gloecystidium alutaceum (Sch.) bourd. et Galz. (lumber)

Gloecystidium inaequale H. et L. (bark, lumber)
G. ochraceum (Fr.) Litsch. (bark, lumber)
G. sphaerospora (H. et L.) Bourd. et Galz.
Glocoporus amorphus f. molluscus (Fr.) Killern.
Gl. dichrous (Fr.) Bres.
Hyphodonti arguta Erikss.
H. subalutaceae (Karst.) Erikss.
Metulodontia cremeo-alutacea Parm.
Peniophora agrillaceae Bres.(branch)
P. cremea Bres.
P. flavoferruginea (Karst.) Ltisch.
P. gigantea (Fr.) Mass
P. serialis (Fr.) H. et L.
P. subalutacea (Karst.) H. et L. (stem)
P. velutina (Fr.) Cooke (stem)
Phlebia gigantea
Phlebiella candidissima (Schw.) Bond. et Sing.
Trechispora candissima (Schw.) Bond. et Sing.
Stereum abietinum (Pres. ex Fr.) Epicr.
S. pini (Fr.) Fr.
S. rugisporum (Ell. et Ev.) Burt.
S. sanguinolentum Alb. et Schw.
Botryobasidium bortryosum (Bres.) Jo rikss.
B. subcoronatum (Hohn) Donk.
Sarcodon fuligineo-albus (Fr.) Quel.
S. imbrieatum (Fr.) Karst.
S. laevigatum (Fr.) Quel.
Thelephora fibriata Schw.
T. aciniata
T. terrestris Ehrenb.
Tomentella isabellina (Fr.) H. et L.
T. ochracea Fr.
T. subfusca (Karst.) H. et L.
Clavaria afflata Lagger.
C. apiculata Fr.
C. purpurea Fr.
Clavariadelphus ligula (Fr.) Donk.
C. truncatus (Quel.) Donk.
Mucronella calva (Fr.) Fr. (lumber)
M. subtilis Karst. (bark, lumber)
Pistillaria fusifornmis Kauf.
P. paradoxa (Karst.) Corner
Pterula multifida Fr.
Typhula abietina Cornder
Kavinia bourdotii (Bres.) John Erikss. (lumber)
K. himantia (Schw.) John Erikss. (lumber)
Lentaria delicatal (Fr.) Corner (lumber)
L. epichnoa (Fr.) Corner (lumber)
L. micheneri (Berk. et Curt.) Corner (lumber)
Lentaria soluta (Karst.) Pil. (lumber)
L. virgata (Fr.) Corner (lumber)
Ramaria apiculata var. *compacta* (Bourd. et Gatz.) Corner
R. crispula (Fr.) Quel. (branch)
R. flaccida (Fr.) Ricken
R. invalii (Cott. et Wakef.) Donk.
Auriscalpium vulgare (Fr.) Karsk.
Hydnum auriscalpium Fr. (lumber)
H. niveum Fr.
H. repandum Fr.
H. tomentosum Sch.
Odonitia ambigua Karst.
O. arguta (Fr.) Quel.
O. bicolor Alb. et Schw.
O. floccosa (Erikss.) Nicol. (lumber)
O. fusco-atra (Fr.) Bres. (branch)
O. grisea Bres.
O. hydnoides (Cook. et Massae) Hohn (bark, lumber)
O. lactea Karst.
O. papillosa Karst.
O. queletii Bourd. et Galz. (branch)
O. soloewskii Jack. (lumber)
Radulum byssinum Bres. (lumber)
R. orbiculare Fr. (branch, stem)
R. pendulum Fr. (branch, stem)
R. quercinum (branch, stem)
R. spathulatum Bres.(lumber)
Xylodon candidum (Her.) Bourd.
Merulius aureus Fr. (branch)
M. himantioides Fr. (lumber)
M. molluscus Fr. (lumber)
M. pinastri (Fr.) Burt. (lumber)
Meruliporia taxicola (Pers.) Bond. et Sing. (branch)
Serpula lacrymans (Wulf. et Fr.) Bond.
S. minor (Fr.) Bond.
S. pinastri (Fr.) Bond.

S. silvester (R. Falck) Bond.
Abortiporus borealis (Fr.) Sing.
Amylocytis lapponicus (Rom.) Bond. et Sing. (bark, lumber)
Amyloporia lenis (Karst.) Bond. et Sing.
A. xantha (Fr.) Bond. et Sing. (lumber)
Bjerkandera adusta (Willd. ex Fr.) Karst.
B. fumosa (Pers. ex Fr.) Karst. (lumber)
Ceraporia taxicol (Pers.) E Kom. (lumber)
Chaetoporellus aurens (Peck.) Bond. (stem)
C. radulus (Pers.) Bond. et Sing. (lumber)
C. rixosus (Karst.) Bond. et Sing.
C. subacidus (Peck) Bond. et Sing. (lumber)
Coniophora arida (Fr.) Karst. (branch)
C. cerebella (Pers.) Sch.
C. puteana(Schum. ex Fr.) Karst. (lumber)
Coniophorella byssoidea Fr. (lumber)
C. olivaceae Karst. (branch)
Coniophorella umbrina (Alb. et Schw.) Bres. (branch)
Coriolellus anceps (Pec.) Parm. (lumber)
C. flavescens (Bres.) Bond. et Sing. (lumber)
C. serialis (Fr.) Murr.
C. squalens (Karst.) Bond. et Sing.
C. subsinuosus (Fr.) Bond. et Sing.
Coriolus cervinus (Schw.) Bond. (lumber)
C. hoehnelii (Bres.) Bond. et Sing.
C. sinuosus (Fr.) Bond. et Sing.
C. vaporarius (Fr.) Bond. et Sing. (lumber)
C. subsinuosus(Fr.) Bond. et Sing. (lumber)
Fibuloporia bomoycina (Fr.) Bond. et Sing.
F. mollusca (Pers.) Bond. et Sing.
F. reticulata Pers. Bond. (lumber)
F. vaillantii (Dc. ex Fr.) Bond. e Sing. (stem, lumber)
F. unita var. *multistratosa* Pil. (lumber)
fomitopsis annosa (Fr.) Karst.
F. crassa (Karst.) Bond. (stem)
F. officinalis (Vill.) Bond. et Sing. (stem)
F. pinicola (Schw. ex Fr.) Karst.
F. rosea (Alb. et Schw. ex Fr.) Karst
F. stellae (Pil.) Bond.
F. subrosea (Weir.) Bond. et Sing. (lumber)
Funalia trogii (Berke.) Bond. et Sing. (lumber)
Gloeophyllum odoratum (Fr.) Jmaz. (lumber)
G. sepiarium (Fr.) Karst. (lumber)
G. trabeum (Fr.) Murr. (lumber)
Hapalopilus aurantiacus (Rostr.) Bond. et Sing.
H. fibrillosus (Karst.) Bond. et Sing.
H. nidulanus (Fr.) Karst. (branch)
Jrpex lacteus (Fr.) (stem)
Laetiporus suolphureus (Fr.) Bond. et Sing. (stem)
H. ochraceo-lateritius (Bond.) Bond. et Sing. (lumber)
Hirschioprus abietinus (Fr.) Donk. (lumber)
H. fusco-violaceus (Her.) ex Fr. Donk. (lumber)
Osmoporus odoratus (Wulf.) Sing. (lumber)
O. protractus (Fr.) Bond.
Oxyporus ravidus (Fr.) Bond. et Sing.
O. pearsonii (Pil.) E. Kom.
Podoporia sanguinolanta (Alb. et Schw.) Hohn
P. vitrea (Fr.) Donk.
Polyporus picipes (Fr.) Karst. (stem)
Polystictus circinatus (Fr.) Karst.
P. circinatus var. *trgueter* Bres.
P. tomentosus (Fr.) Karst. (stem)
Poria placenta (Fr.) Cke.
P. vulgaris (Fr.) Cke.
P. weirii Cke.
Trametes heteromorpha (Fr.) Bres. (lumber)
Tyromyces albellus (Peck.) Bond. et Sing.
T. albidus (Sch. ex Secr.) Murr. (lumber)
T. caesius (Sch. ex Fr.) Murr. (lumber)

T. cinerascens (Bres.) Bond. et Sing.
Tyromyces destructor (Schrad.) Bond. et Sing.
T. erubescens (Fr.) Bond. et Sing.
T. floriformis (Quel.) Bond. et Sing. (bark, lumber)
T. fragilis (Fr.) Donk.
T. kymatodes Donk (stem)
T. lacteus (Fr.) Murr. (stem)
T. leucomalleus Murr.
T. mollis (Fr.) Karst.
T. resupintus (B. ex Pil) Bond. et G.
T. semipleatus (Peck.) Murr.
T. semisupinus (Berk. et Kurt.) Murr.
T. sericeo-mollis (Ram.) Bond. et Sing.
T. stipticus (Fr.) Coll. et Ponz.
T. tephroleucus(Fr.) Donk.
T. trabeus (Rost.) Bourd et Jalz.
T. undosus (Peck) Murr. (stem)
Ganoderma applanatum (Pers. ex Wallr.)
Hymenochaete fuliginosa (Pers.) Bres.
Inonotus hispidus (Bull ex Fr.) Karst. (stem)
I. radiatus (Sow. ex Fr.) Karst.
Ischnoderma resinosum (Fr.) Karst. (stem, lumber)
Phaeolus schweinitzii (Fr.) Pat.
Phellinus contiguus (Pers.) Bourd. et Galz. (stem, lumber)
P. demidoffii (Lev.) Bond. et Sing. (stem, branch)
P. hartigii (All. et Sch.) Bond.
P. isabellinus (Fr.) Bourd. et Galz. (stem)
Phellinus nigrolimitatus (Rom.) Bourd. et Galz.
P. pini (Thore et Fr.) Pil.
P. pini Til. var *tipicus* Pil. *f. pithyusa* Negr.
P. pini var. *abietis* (Karst.) Pil. (branch, stem)
P. pini Pil. var. *abietis* Karst. *f. caucasicus* Nigr.
P. pini var. *pini* (Thore et Fr.) Pil. (stem)
Cyphella vernalis Weinm. (bark, lumber)
C. digitalis Alb. et Schw.
C. griseo-pallida Weinm.
Schizophyllum commune Fr. (stem)

Agaricales

Armillariella mellea (Fr.) Karst.
Catathelasma imperiale (Fr.) Sing.
Clitocybe auantiaca (Fr.) Stud.
Collybia dryophila (Fr.) Kumm.
C. maculata (Fr.) Kumm.
Lentinus lepideus (Fr.) Fr.
L. sulcatus Berk.
L. squamosus H.
L. vulpinus (Fr.) Fr.
Lepista nuda (Fr.) Cke.
Tricholoma flavovirens (Fr.) Lund.
T. poentosum (Fr.) Quel.
Tricholomopsis rutilans (Fr.) Sing.
Pholiota adiposa (Fr.
P. flammans (Fr.) Kumm.
Stropharia aeropharia (Fr.) Fr.
Cortinarius violaceus (Fr.) Fr.
Paxillus atrotomentosus (Fr.) Fr. (stem, lumber)
P. acheruntius Fr.
P. involutus (Fr.) Fr. (stem, lumber)
P. panuoides (Fr.) Fr.
Gomphidius rutilus (Fr.) Lund. et Nant.
Boletus edulis f. pinicola (Vitt.) Vassilk.
Leceinum percandidum (Vassilk.) Watling
Suillus bovinus (Fr.) O. Kuntze
S. granuiatus (Fr.) O. Kuntze
S. luteus (Fr.) S. F. Gray
S. piperatus (Fr.) O. Kuntze
Russula aurata Fr.
R. decolorans (Fr.) Fr.

Tulasnellales

Tulasnella araeosa Bourd. et Galz.
T. fuscoviolaceae Bres.
T. violaceae (Johan, Olsen) Juel.

Dacrymycetales

Arrhytidia involuta (Schw.) Coker
Calocer cornea (Fr.) Fr.
C. visoca (Pers.) Fr.
Cerinomyces altaicus Parm.

C. canadensis Jacks et Martin
C. crustulinus ((Bourd.) et Gats) Martin
Ditiola brunnea (Martin) Kennedy
D. nuda Berk. et Br.
Dacrymyces chrysocous (Fr.) Tul.
D. dictyosporus Martin
D. deliquescens (Merat) Duby
D. estonicus Raitv.
D. ovisporus Bref.
D. palmatus Schw.) Bres.
D. tortus Fr.
Guepiniopsis merulinus (Pers.) Pat.

Tremellales

Ditangium cerasi (Tul.) Cost. et Duf.
Exidia pithya Fr.
E. saccharina Fr.
E. testaceae Raitv.
Exidiopsis calcea (Pers.) Wells.
E. fugacissima (Broust. et Galz) Sac. et Trott
Protodontia piceicola (Kuhnl.) Martin
Pseudohydnum gelatinosum (Fr.) Karst.
Stypella papillata Moller
Tremella encephata (Willd.) Pers. (branch)
T. foliaceae Fr.
T. translucens Gordon

Auriculariales

Septobasidium linderi Couch
S. pinicola Snell.

Uredinales

Coleosporium apocynaceum Cke.
C. campanulae (Pers.) Lev.
C. crowellii Cumm.
C. euphrasiae (Schum) Wint.
C. helianthi Arth.
C. inconsipicuum Hedge et Long.
C. inulae (Kze.) Rabenh.
C. ipomoeae (Sch.) Arth.
C. laciniariae Arth.
C. melampyri (Rebent) Karst.
C. petasitis (Dc.) Lev.
C. pinicola Arth.
C. pulsatillae (Str.) Lev.
C. rhinanthacearum Lev.
C. senecionis Kickx.
C. solidaginis (Sch.) Thuem.
C. sonchi (Str.) Lev.
C. sonchi-arvensis (Pers.) Lev.
C. terebinthinaceae (Sch.) Arth.
C. tissilaginis (Pers.) Lev.
C. vernoniae Berk. et Curt.
Cronartium cerebrum Hedge et Long.
C. coleosporioides Hedge et Long. (branch)
C. compotniae Arth. (branch)
Cronartium flaccidum (Alb. et Schw.) Wint. (branch, stem)
C. himalayense W.
C. quercus Schrot f. sp. fusiforme Sch.
C. ribicola (Lasch.) Fisch. v. Waldh. (stem, branch)
C. strobilinum Hedge et Hohn.
Endocronartium harknessii Hir.
Melampsora pinitorquea (Fr.) Rostr. (branch, stem)
Peridermium comptoniae (Link. Chev.
P. fusiforme Chev. (branch, stem)
P. kurilense (Link.) Chev. (branch, stem)
P. montezumae Cummis sp. nov. (branch)
P. cerebrum Chev. (branch, stem)
P. pini Lev. et Keb. (branch)
P. pyriforme L. (branch, stem)
P. stalactiforme L. (branch, stem)

Moniliales

Aspergillus flavus Link.
A. glaucus Link.
A. herbariorum F.
A. niger V. Tiegh.
A. wentii Wehm.
Botrutis cinerea Pers. ex Fr.
Fusoma pinii Harting
Helicomyees condidus Sacc. (branch_
Penicillium coryophilum Dietr.
P. glaucum Link.
P. luteum Jukal
Trichoderma viride var. *kirhanense* (lumber)

Krap. Pol. Sizoa
Trichothecium roseum Link.
Verticicladiella sp.
Verticillum albo atrum Rke. et Berth.
V. terrestre Pke. et Berth.
Alternaria alternata (Fr.) Keissl.
A. humicola Qud. (lumber)
A. tenuis Nees. emius Neerg.
Cercospora pinidensiflorae
Cladosporium herbarum Link. ex r.
Helicosporium phaeosporum (Fres.) Sacc.
Nigrospora gallarum (Nol.) Potl.
Phialophora fastigiata (Lager. et Melin) Conant. (lumber)
Pullularia pullulans (De-By) Berkhout (lumber)
Rhinocladiella atrovierns Nannf. (lumber)
Sporodesmium cladosporioides Cda. (lumber)
Stachybotrys macrocarpa L. (lumber) (lumber)
Trichoporium hetermorphum Nannf.)lumber)
Leptographium lundbergii Lagerh. (lumber)
Aegerita torulos Sacc.
Bactridium flavum Kze. (branch)
Exoporium pyrosporum Hohn et Melin (branch)
Fusarium bulbigenum W.
F. latertium Nees f. pini Hepting
F. martii App. et Woll.
F. oxysporum Sch. var. aurantiacum (Dk.) Wr.
F. sporotrichioides Sherb.
Tuberculina maxima Rostr. (branch)

Melanconiles

Cryptosporium lunasporum Linder
C. pinicola Linder
Cylindrosporium acicola Bres.
Gloeosporium pineae Bub.
G. pini Oud.
Monochaetia pinicola Dearn.
Pestalotia funerea Desm.
P. partigii Tub.
P. peregrina Ell. et Martin
P. truncata var. *lignicola* Grove
Phragmotrichum chailletii Kze.
Stilbospora pinicola Berk.
Truncatella truncata (Lev.) Stey

Sphaeropsidales

Zythia cucurbitula Jacz. (branch)
Z. resinae (Ehr.) Karst. (branch)
Brunchorstia destruens Erikss.
B. pinea (Karst.) Hohn
Leptothyrium pinastri Karst.
L. stenosporum Dearn.
Leptostroma pinastri Desm.
Coniothyrium dispersellum Farst. (branch)
C. pini Wudem.
Cytospora curreyi Sacc. (branch)
C. kazachstanica Sch. (branch)
C. kunzei Sacc.
C. pinastri Fr.
Diplodia conigena Desm.
Diplodia megalospora Berk. et Curt. (branch)
D. natalensis P. Evans (branch)
D. pinea Kickx.
D. sapinea (branch)
D. thujeae West. (branch)
Diplodiella crustaceae Karst. (branch)
D. pini-silverstris All. (branch)
D. pityophila Sacc. et Penz.
Dothistroma pini Herb.
D. septospora Hulb.
Haplosporella pini Fk. (branch)
Helicomyces candidius Sacc.
Hendersonia acicola Munch et Tub. (branch)
H. folicola (Berk.) Fekl
H. pini (branch)
H. strobilina Curr.
H. thujae Died. (branch)
Hendersonula pini Died. (branch)
H. pinicola Dearn. (lumber)
Hermiscium antiquum (Cda.) Sacc. (lumber)
Phoma acicola (Lev.) Sacc.
P. bacteriophilla Pk. (branch)
P. cembrae Karst.

P. douglasii Oud. (branch)
P. eguttulata Karst.
P. geniculata Sacc. (branch)
P. harknessii Sacc. (branch)
P. inopinata Oud. (branch)
Phoma juniperi (Desm.) Sacc.
P. piciana Karst.
P. pinastrella Sacc.
P. pinastri (Oud.) Sac. (branch)
P. pinicola Sacc.
P. strobiligena Desm.
Phomopsis conorum Isacc.) Died. (branch)
P. occulta (Sacc.) Trav.
P. strobi Syd. (branch)
Rhisosphaerella pini Maubl.
Sclerophoma pini Gucev. sp. n.
S. pithya v. Hohn (branch) (branch)
S.pithyophila (Corda) Hohn.
S. pityella (Sacc.(Hohn
Septoria acuum Oudem.
S. pinicola Dearn.
S. spadicea Pat.
Spharonaema aciculare Fr. (branch)
S. piliferum Sacc. (bark, lumber)
S. pithyum Sacc. (stem, branch)
Sphaeropsis ellissii Sacc. (bark, lumber)
S. maorum Pk. (branch)
Biatoridina pinastri Golov. et Zxhzedr. (branch, stem)
Discula brunneo-tingens H. Meyer (lumber)
D. pinicola (namn.) Petr. var. mammosa Lagerh. (lumber)
Discula rubra H. Meyer (lumber)
Dothichiza ferruginosa Sacc. (branch)
D. kazachstanica Sch.
Patellina caesia Ell. et Stansf.
Pseudopatellina conigena V. Hohn
Rhizochonia endophytica var.-*filicata* var *nov.* (branch)
R. globularis sp. nov. (branch)
R. hiemalis K. (branch)
R. solani Kuhn (branch)

Bacteria

Erwinia multivora Sez. Parf. (stem, root)
Pseudomonas halepensis L. (stem, root)

P. pini Vuil.

Angiospermae

Visum austriacum Wiesb. (branch)
V. sp. (branch, branch)
Arceuthobium pusilum K. (branch)
A. americanum L. (branch)

Viroae

Tobacco mottl virus
Tobacco ringspot virus

Wood Decay and Canker Diseases of *Abies Sibirica* Ledb.

Lechnellula calyciformis (Fr.) Dharne. = *Dasychypha clyciformis* Rehm.
Scleroderris sp.
Lophodermium nerviseguum (D.S.) Rehm.
Herpotrichia nigra Hart.
Aleurodiscus amorphus (Fr.) Schroet.
Calyptospora goeppertiana Kuehn.
Melamsorella caryophyllasearum Schr. =*M. cerastii* (Pers.) Wint.
Pucciniastrum spilobii Otth.
Bactrodesmium obligum Sutton var. sattonnii Hughesctwhite
Cirrenalia donnae Sutton
Capnobotrys neesii Hughes.
Seiridium abietium (Ell. et Ev.) Sutton
Toxosporium camptospermum (Pk) Maublanc
Micropera pinastri Sacc.
Zythiostroma pinastri Karst.
Phoma abietella-sibirica Schw.
Sclerophoma pithiophila (Cda) Hohn.
Rhizosphaera pini (Corda) Maubl.

Phellinus hartigii
Armillariella mellea
Heterobasidion annosus
Laetiporus sulphureus
Phaeolus schweinitzii
Fomitopsis pinicola
Ganoderma pplanatum
Fomes fomentarius
Gloeophyllum sepiarium
Schizophyllum commune
Stereum sanguinolentum

Stem Insects in Larch and Pine, and Their Location in the Log

Larch

Acanthocinus griseus (phloem)
Acmaeops septentrirnis (phloem)
Anoplodera variicornis (feed on withered tree)
Asemum amurense (phloem, xylem)
Callidium chlorizans (stem)
C. violacaum (stem)
Dryocoetes baicalicus (stem phloem)
Hylobius abietis (seedling collar and young stem phloem)
H. albosparsus (seedling collar and young stem phloem)
Ips acuminatus (branch phloem)
I. subelongatus (stem phloem)
Malanophila guttulata (stem)
Melanophil acuminata (stem)
Monochamus salutarius (xylem, phloem)
M. sutor (xylem, phloem)
M. urussovi (xylem, phloem)
Pityogenus chatcographus (stem thick branch)
Rhagium (phloem)
Sirex ermak (stem xylem)
Tetropium castaneum (xylem, phloem)
T. gracilicornis (xylem, phloem)
Urocerus gigas taiganus (stem xylem)
Xeriss pectrums pectrum (stem xylem)
Xyloterus lineatus (stem xylem)

Pine

Acanthocinus aedilis (xylem, phloem)
A. griseus (phloem)
Acmaeops septentrirnis (phloem)
Anthaxis quadripunctata (stem)
Arhopalus rusticus (xylem, phloem)
Asemum amurense (xylem, phloem)
Blastophagus minor (stem phloem, shoot)
B. piniperda (stem phloem, shoot)
Burprestis sibirica (stem)
Callidium chlorizans (stem)
Chryocothris sauccedanea (stem)
Dendroctonus micons (trunk phloem)
Hylastes angustatus (stem phloem)
Hylobius albosparsus (seedling collar and young stem phloem)
H. abietis haroldi (seedling collar and young stem phloem)
Ips acuminatus (stem phloem)
I. sexdentatus (stem phloem)
Magdalis (shoot tip)
Melanophila guttulata (stem)
Melanophila acuminata (stem)

Major Trees of California Forests
加利福尼亚州主要森林树种名单

Common Name	Latin Name	Chinese Name
Conifers		
Pines	*Pinus spp.*	松属
Bishop pine	*Pinus muricata*	加州沼松
Jeffrey pine	*Pinus jeffreyi*	黑材松
Knobcone pine	Pinus atenuata	
Limber pine	Pinus flexilis	柔松
Lodgepole pine	Pinus contorta var murrayana	扭叶松
Monterey pine	Pinus radiata	辐射松
Ponderosa pine	Pinus ponderosa	西黄松
Singleleaf pinyon pine	Pinus monophyla	单针松
Sugar pine	Pinus lambertiana	糖松
True Firs	Abies spp.	真冷杉类
Red fir	Abies magnifica	加州红冷杉
White fir	Abies concolor	白冷杉
Douglas-fir	Pseudotsuga menziesii	花旗松
Giant sequoia	Sequoiadendron giganteum	巨杉
Incense-dedar	Libocedrus decurrens	
Port-Orford-cedar	Chamaecyparis lawsoniana	美国扁柏
Sitka spruce	Picea sitchensis	西特呵云杉
Hardwoods	Quercus spp.	橡树
California black oak	Quercus kelloggii	加州黑橡树
Coast live oak	Quercus agrifolia	常绿橡树
Other, California natives		
California laurel	Umbellularia californica	加州桂
Elm	Ulmus spp.	榆树
Pacific madrone	Arbutus enziesii	浆果鹃
Tanoak	Lithocarpus densiflorus	密花石柯
Willow	Salix spp.	柳树

Major Diseases of California Forests **加州主要森林病害名单**

Disease	Latin	Chinese name
Cankers		
Diplodia blight of pines	Diplodia pinea (Sphaeropsis sapinea)	松叶疫病
Pine pitch canker	Fusarium subglutinans f.	松流脂病

	sp. Pini	
Phomopsis canker	Diaporthe lokoyae	间座壳溃疡
Dwarf Mistletoes		
Western dwarf mistletoe	Arceuthobium camplopodum	西槲寄生
Foliage diseases		
Elytroderma needle disease	Elytroderma deformans	针叶病
Fir needle cast	Lirula abietis-concoloris	冷杉针叶病
Fusarium hypocotyls rot	Fusarium spp.	镰刀菌
Leaf spot of arbutus	Coccomyces arbutifolius	
Powdery mildew on live oaks	Sphaerotheca lanestris	橡树白粉病
Root diseases		
Annosus root disease	Heterobasidion annosum	针叶根朽病
Black stain root disease	Leptographium wagneri	针叶树黑变病
Charcoal root rot	Macrophomina phaseolina	炭垠体病
Laminated root disease	Phellinus weirii	层根付朽病
Phytophthora root rot	Phytophthora spp.	疫霉烂根病
Port-Orford-cedar root Disease	Phytophthora lateralis	美国扁柏疫霉病
Rusts		
Western gall rust	Peridermium harknessi	西方锈瘤
White pine blister rust	Cronartium ribicola	白松疱锈病
True mistletoes		
Leafy mistletoe	Phoradendron spp.	真槲寄生

Major Insects of California Forests **加利福尼亚州主要森林害虫名单**

Common Name	Latin Name	Chinese Name
Bark beetles		
California fivespined ips	Ips paraconfusus	波辣孔夫子达小蠹
Cedar bark beetles	Phloeosinus sp.	肤小蠹
Douglas-fir beetle	Dendroctonus pseudotsugae	花旗松大小蠹
Douglas-fir engraver	Scolytus unispinosus	花旗松皮小蠹
Fir engraver	Scolytus ventralis	冷杉皮小蠹
Jeffrey pine beetle	Dendroctonus jeffreyi	光背大小蠹虫
Monterey pine ips	Ips mexicanus	墨西哥齿小蠹
Mountain pine beetle	Dendroctonus ponderosae	山松大小蠹
Pine engravers	Ips spp.	齿小蠹类
Red turpentine beetle	Dendroctonus valens	红脂大小蠹
Western pine beetle	Dendroctonus brevicomis	西 松 大 小 蠹
Defoliators		

California oakworm	Phryganidia californica	加州橡蛾
California budworm	Choristoneura carnana California	加州卷蛾
Douglas-fir tussock moth	Orgyia pseudotsugata	花旗松古毒蛾
Elm leaf beetle	Xanthogaleruca luteola	榆叶甲
Fruittree leafroller	Archips argyrospilus	果树黄卷蛾
Gypsy moth	Lymantria dispar	午毒蛾
Lodgepole pine needleminer	Coleotechnites milleri	针叶麦蛾
Modoc budworm	Choristoneura retiniana	麦达克卷蛾
Pine needleminer	Coleotechnites sp.	松麦蛾
Silverspotted tiger moth	Halisidota argentata	银点灯蛾
Tent caterpillar	Malacosoma sp.	天幕毛虫
Western tent caterpillar	Malacosoma californicum	加州天幕毛虫
White fir needleminer	Epinotia meritana	白冷杉小卷叶蛾
Tree regeneration insects		
Black vine weevil	Otiorhynchus sulcatus	黑葡萄耳象
Branch and twig beetles	Carphoborus pinicolens	
	Pityophthorus tuberculatus	星坑小蠹
	Pityophthorus conifertus	
	Pityophthorus jeffreyi	
	Piyophthorus sp.	
Douglas-fir reproduction weevil	Cylindrocopturus furnissi	花旗松再生象
Gouty pitch midge	Cecidomyia piniinopsis	古特脂瘿蚊
Pine needle sheathminer	Zelleria haimbachi	松梢巢蛾
Pine reproduction weevil	Cylindrocopturus eatoni	松细枝象
Flatheaded fir borer	Melanophila drummondi	扁头冷杉吉丁虫
Roundheaded fir borer	Tetropium abietus	冷杉幽天牛
Africanized honey bee	Apis mellifera scutellata	非洲意蜂
Asia longhorned borer	Anaphlophora glabripennis	亚洲长角天牛

Index of Forest Diseases and Insects in North America

(Taken from "http://www.na.fs.fed.us/spfo/pubs/fidl.htm, accessed July 27, 2002)

List of Forest Diseases of North America

Annosus Root Disease of Western Conifers
Heterobasidion annosum
Annosus Root Rot in Eastern Conifers
Heterobasidion annosum
Anthracnose Diseases of Eastern Hardwoods
Apiognomonia quercina
Apiognomonia tiliae
Apiognomonia veneta
Asteroma inconspicuum
Asteroma microspermum
Colletotrichum gloeosporioides
Cryptocline betularum
Discula umbrinella
Glomerella cingulata
Gnomonia caryae
Gnomonia leptosyla
Kabatiella apocrypta
Armillaria Root Disease
Armillaria mellea
Atropellis Canker of Pines
Atropellis piniphila
Beech Bark Disease
Nectria coccinea
Black Stain Root Disease of Conifers
Laptographium wageneri
Brown-Spot Needle Blight of Pines
Scirrhia acicola
Butt Rot of Southern Hardwoods
Ganoderma lucidus
Canker-Rots in Southern Hardwoods
Irpex mollis
Polyporus hispidus
Poria laevigatta
Poria spiculosa
Cankers on Western Quaking Aspen
Ceratocystis fimbriata
Cryptosphaeria populina
Cytospora chrysosperma
Encoelia pruinosa
Hypoxylon mammatum

Chestnut Blight
Cryphonectria parasitica
Comandra Blister Rust
Cronartium comandrae
Cytospora Canker of True Firs
Cytospora abietis
Diplodia Blight of Pines
Diplodia pinea
Decay and Discoloration of Aspen
Armillariel1ea
Collybia velutipes
Fomes igniarius
Ganoderma applanatum
Gleocystidium karstenii
Libertella sp
Peniophora polygonia
Pholiota aurivella
Pholiota spectabilis
Pholiota squarrosu
Pleurotus ostreatus
Polyporus dryophilus
Polyporus polygonia
Radulum casearium
Trechispora raduloides
Decay and Discoloration of Sugar Maple
Dothistroma Needle Blight of Pines
Dothistroma pini
Dry Face of Naval Stores Pines
Douglas-fir Dwarf Mistletoe
Arceuthobium douglasii
Fir Dwarf Mistletoe
Arceuthobium spp.
Dwarf Mistletoe of Ponderosa Pine in the Southwest
Arceuthobuim vaginatum
Dwarf Mistletoe of Sugar Pine
Arceuthobium californicum
Eastern Dwarf Mistletoe on Black Spruce
Arceuthobium pusillum
Eastern Gall Rust
Cronartium quercuum
Eutypella Canker of Maple
Eutypella parasitica
Elytroderma Disease of Ponderosa Pine
Elytroderma deformans
Melampsorella caryophyllacearum
Fusiform Rust of Southern Pines
Cronartium fusiforme
Gray Pine Dwarf

Arecuthobium campylopodum
Arceuthobium occidentale
Heart Rots of Appalachian Hardwoods
Armillaria mellea
Hydnum septentrionale
Inonotus glomeratus
Phellinus pini
Heart Rots of Balsam Fir
Ceriporiopsis rivulosa
Echinodontium tinctorium
Fomitopsis officinalis
Fomitopsis pinicola
Ganoderma applanatum
Hericium abietis
Laetiporus sulphureus
Perenniporia subacida
Phellinus pini
Postia sericeomollis
Stereum sanguinolentum
Veluticeps fimbriata
Heart Rots of Central Region Hardwoods
Heart Rots of Douglas-fir
Fomitopsis pinicola
Polyporus schweinitzii
Heart Rots of Engelmann Spruce and Subalpine Fir in the Central Rocky Mountains
Columnocystis abietina
Coniophora puteana
Coriolellus serialis
Echinodontium sulcatum
Fomitopsis piniola
Haematostereum sanguinolenta
Phellinus nigrolimitatus
Phellinus pini
Pholiota alnicola
Polyporus tomentosus
Heart Rots of Incense-Cedar
Fomes pini
Polyporus amarus
Polyporus schweinitzii
Heart Rots of Red and White Firs
Armillaria mellea
Echinodontium tinctorium
Fomes annosus
Fomes officinalis
Fomes pinicola
Ganoderma sp
Pholiota adiposa
Polyporus dryadeus
Polyporus schweinitzii
Polyporus sulphureus
Heart Rots of Redwood
Poria albipellucida
Poria sequoiae
Heart Rots of Western Hemlock
Armillaria mellea
Echinodontium tinctorium
Fomes annosus
Fomes pini
Fomes pinicola
Fomes robustus
Pholiota adiposa
Polyporus schweinitzii
Polyporus sulphureus
Poria subacida
Stereum abietinum
Stereum sanguinolentum
Hemlock Dwarf Mistletoe
Arceuthobium tsugense
Hypoxylon Canker of Aspen
Hypoxylon mammatum
Laminated Root Rot of Douglas-fir
Phellinus weirii
Laminated Root Rot of Western Conifers
Phellinus weirii
Larch Dwarf Mistletoe
Arceuthobium laricis
Limber Pine Dwarf Mistletoe
Arceuthobium cyanocarpum
Littleleaf Disease
Phytophthora cinnamomi
Pythium spp.
Lodgepole Pine Dwarf Mistletoe
Arceuthobium americanum
Mistletoes on Hardwoods in the United States

Phoradendron californicum
Phoradendron flavescens
Phoradendron rubrum
Phoradendron tomentosum
Phoradendron villosum
Nectria Canker of Hardwoods
Nectria galligena
Needle Cast of Southern Pines
Lophodermium pinastri
Needle Discolorations of Western Larch
Coleophora laricella
Hypodermella laricis
Meria laricis
Nursery Diseases of Southern Hardwoods
Botrytis cinerea
Cytospora spp.

Diaporthe lokoyae
Dothistroma pini
Macrophomina phaseoli
Peridermium harknessii
Pratylenchus and Meloidogyne spp
Pythium, Rhizoctonia, Phytophthora, and Fusarium
Rosellinia herpotrichoides
Sirococcus strobilinus
Xiphinema bakeri
Nursery Diseases of Southern Pines
Cronartium quiercuum
Fusarium moniliforme
Fusarium spp.
Fusarium spp.
Meloidodera spp.
Meloidogyne spp.
Phytophthora spp.
Pratylenchus spp.
Pythium spp.
Rhizoctonia solani
Scirrhia acicola
Sclerotium bataticola
Nursery Diseases of Western Conifers
Botrytis cinerea
Cytospora spp.
Dothistroma pini
Macrophomina phaseoli
Meloidogyne spp.
Peridermium harknessii
Phomopsis lokoyae
Phytopthora spp.
Pratylenchus spp.
Pythium spp.
Rosellinia herpotrichoides
Sirococcus strobilinus
Xiphinema bakeri
Oak Decline
Agrilus bilineatus
Armillaria mellea
Oak Wilt
Ceratocystis fagacearum
Phomopsis Blight of Junipers
Phomopsis juniperovora
Phoradendron on Conifers
Phoradendron boleanum
Phoradendron capitellatum
Phoradendron densum
Phoradendron hawksworthii
Phoradendron juniperinum
Pinyon Pine Dwarf Mistletoe
Arceuthobium divariatum
Pitch Canker of Southern Pines
Fusarium moniliforme
Pole Blight of Western White Pine
Port-Orford-Cedar Root Disease
Phytophthora lateralis
Red Rot of Ponderosa Pine
Polyporus anceps
Rust-red Stringy Rot
Echinodontium tinctorium
Sapstreak Disease of Sugar Maple
Ceratocystis coerulescens
Scleroderris Canker of Northern Conifers
Gremmeniella abietina
Sirococcus Shoot Blight
Sirococcus strobilinus
Southern Cone Rust
Cronartium strobilinum
Strumella Canker of Oaks
Conoplea globosa
Sweetfern Rust on Hard Pines
Cronartium comptoniae

Sweetgum Blight
Botryosphaeria ribis
Walnut Anthracnose
Gnomonia leptostyla
Western Dwarf Mistletoe on Ponderosa Pine
Arceuthobuim vaginatum

Western Gall Rust on Hard Pines
Endocronartium harknessii
White Pine Blister Rust
Cronartium ribicola
White Trunk Rot of Hardwoods
Fomes igniarius

List of Forest Insect Pests in North America

Ambrosia Beetles
Platypus spp.
Arizona Five-Spined Ips
Ips lecontei
Bagworm
Thyridopteryx ephemeraeformis
Balsam Wooly Aphid
Adelges piceae

Black-Headed Budworm in Western United States
Acleris gloverana
Black Pineleaf Scale
Nuculaspis californica
Black Turpentine Beetle
Dendroctonus terebrans
Boxelder Bug
Boisea trivittatus
Bronze Birch Borer
Agrilus anxius
California Five-Spined Ips
Ips paraconfusus
California Flatheaded Borer
Melanophila californica
California Oakworm
Phryganidia californica
Carpenterworm
Cossula magnifica
Columbian Timber Beetle
Corthylus columbianus
Douglas-fir Beetle
Dendroctonus pseudotsuqae
Douglas-Fir Tussock Moth
Orgyia pseudotsugata
Eastern Pineshoot Borer
Eucosma gloriola

Eastern Subterranean Termite
Reticulitermes flavipes
European Pine Sawfly
Neodiprion sertifer
Elm Sawfly
Cimbex Americana
Elm Spanworm
Ennomos subsignarius
European Pine Shoot Moth
Rhyacionia buoliana
Fir Engraver
Scolytus ventralis
Fir Tree Borer
Melanophila drummondi
Forest Tent Caterpillar
Malacosoma disstria
Gouty Pitch Midge
Cecidomyia resinicola
The Green-Striped Mapleworm
Anisota rubicunda
Gypsy Moth
Lymantria dispar
The Hemlock Borer
Melanophila fulvoguttata
Hemlock Sawfly
Neodiprion tsugae
Hemlock Scale
Fiorinia Externa
Introduced Pine Sawfly
Diprion similis
Ips Bark Beetles in the South
Ips avulsus
Ips calligraphus calligraphus
Ips grandicollis
Jack Pine Budworm
Choristoneura pinus
Jack Pine Sawfly
Neodiprion pratti
Jeffrey Pine Beetle
Dendroctonus jeffreyi
Larch Casebearer in Western Larch
Coleophora laricella
Larch Sawfly
Pristiphora erichsonii
Large Aspen Tortrix
Choristoneura conflictana
Loblolly Pine Sawfly
Neodiprion taedae
The Locust Borer
Megacyllene robiniae
Lodgepole Needle Miner
Coleotechnites miller
Monterey Pine Ips
Ips mexicanus
Mountain Pine Beetle
Dendroctonus ponderosae
Nantucket Pine Tip Moth
Rhyacionia frustrana
Pales Weevil
Hylobius pales
Pandora Moth
Caloradia pandora
Pine Butterfly
Neophasia menapia
Pine Engraver in the Western United States
Ips pini
Pine Looper
Caripeta spp.
Lambdina spp.
Pine Needle-sheath Miner
Zelleria haimbachi
Pine Tortoise Scale
Toumeyella numismaticum
Pine Reproduction Weevil

Cylindrocopturus eatoni
Pine Root Collar Weevil
Hylobius radicis
Pine Sawfly
Neodiprion spp.
Pinyon Needle Scale
Matsucoccus acalyptus

Pinyon Sawfly
Neodiprion edulicolus
Ponderosa Pine Tip Moth
Rhyacionia spp.
Poplar-and-Willow Borer
Cryptorhynchus lapathi
Redheaded Pine Sawfly
Neodiprion lecontei
Redhumped Oakworm
Symmerista canicosta
Red Oak Borer
Enaphalodes rufulus
Red Pine Scale
Matsucoccus resinosae
Red Turpentine Beetle
Dendroctonus valens
Roundheaded Pine Beetle
Dendroctonus adjuntus
Saddled Prominent
Heterocampa guttivitta
Saratoga Spittlebug
Aphrophora saratogensis
Silver Fir Beetle
Pseudohylesinus sericeus
Sitka Spruce Weevil
Pissodes strobi
Six-Spined Engraver Beetle
Ips calligraphus
Slash Pine Seedworm
Cydia anaranjada
Southern Pine Beetle
Dendroctonus frontalis
Southwestern Pine Tip Moth
Rhyacionia neomexicana
Spear-Marked Black Moth
Rheumaptera hastata
Spruce Beetle
Dendroctonus rufipennis
Spruce Budworm
Choristoneura fumiferana
Spruce Budworm in the Eastern United States
Choristoneura fumiferana
Sugar Maple Borer
Glycobius speciosus

Sugar Pine Cone Beetle
Conophthorus lambertianae
Texas Leaf-Cutting Ant
Atta texana
Tuliptree Scale
Toumeyella liriodendri
Two-lined Chestnut Borer
Agrilus bilineatus
Variable Oakleaf Caterpillar
Heterocampa manteo
Walkingstick
Diapheromera femorata
Walnut Caterpillar
Datana integerrima
Western Pine Beetle
Dendroctonus brevicomis
Western Spruce Budworm
Choristoneura occidentalis
Western Tent Caterpillar
Malacosoma californicum
Western Tussock Moth
Orgyia vetustu
White Fir Needle Miner
Epinotia meritana
White Grubs in Forest Tree Nurseries and Plantations
Phyllophaga spp.
White-Pine Cone Beetle
Conopthorus coniperda
White Pine Weevil
Pissodes strobi
White-Spotted Sawyer
Monochamus scutellatus
Yellow-Headed Spruce Sawfly
Pikonema alaskensis
Yellow-Poplar Weevil
Odontopus calceatus
Zimmerman Pine Moth
Dioryctiria zimmermani

Index of Forest Tree Pathogens in China

Abelia spp.

Erysiphe abeliae Zheng et Chen (Powder mildew) Sichuan
Pyropolyporus pectinatus (Kl.) Murr. Var. jasmini
(Quel.) Bres. (Decay) Hebei Guangdong

Abies spp.

Cucurbitaria pithyophila (Schmidt et Kunze ex Fr.) Ces.et de Not. (Dieback) Sichuan
Cytospora curreyi sacc. (Canker) Sichuan
Fomes fomentarius (l. ex Fr.) Kickx. (Heartwood mix pocket decay) Jilin
Fusarium spp. (Damping off) Sichuan
Gloeophyllum abietinum (Bull. ex. Fr.) Karst. (Decay) Gansu
(=*Lenzites abietina* (Bull.) Fr.)
Gloeophyllum saepiarium (Wulf. ex. Fr.) Karst. (Decay) Gansu
(=Lenzites saepiaria (Wulf.) Fr.)
Hericium coralloids (scop. ex Fr.) Pers. ex Gray (White rot) Tibet Sichuan
Heterobasidion annosus (Fr.) Bref. (Pine root rot) Sichuan
Inonotus dryadeus (Pers. ex Fr.) Murr. (Decay) Sichuan
Inonotus hispidus (Bull. ex Fr.) Karst. (Leaf tip rot) Jilin
Lachnellula agassizii Berk. et Curt. (Canker) Sichuan
(=Dasyscypha agassizii (Berk. et Curt.) Sacc.)
Lachnellula caliciodes DC. (Canker) Jilin Heilongjiang
(=Dasyscypha caliciodies (DC.) Sacc.)
Laetiporus sulphureus (Bull. ex Fr. Bond.) et Sing. (Trunk brown cube rot) Sichuan
(=Polyporus sulphureus (Bull.) Fr.)
(=Tyromyces sulphureus (Bull. ex Fr) Donk)
Lophodermium nervisequium (DC. ex Fr.) Rehm (Leaf spot) Jilin
Lophodermium pinastri (Schrad. ex Fr.) Chev. (Falling needles) Jilin
Melampsorella cerastii (Mart.) Wint. (Leaf rust) Jilin
(=Fomes igniarius (L. et Fr.) Kickx)
Nectria cinnabarina (Tode) Fr. (N. Canker) Jilin
Phaeolus schweinitzii (Fr.) Pat. (Trunk brown cube rot) Hebei Jilin Heilongjiang Xinjiang Sichuan Yunnan
(=Fomes igniarius (L. et Fr.) Kickx)
Phellinus hartigii (Allesch. et Schabl.) Imaz. (Sapwood brown rot) Sichuan Yunnan Heilongjiang
Phellinus igniarius (L. ex Fr.) Quel. (Heartwood white rot) Jilin
Phellinus pini var. abietis Karst. (White pocket rot) Sichuan Yunnan
(=Fomes pini var. abietis (Karst.) Ovesh.)
Pseudoplectania nigrella (Pers.) Fuck. Sichuan
Pythium spp. (Damping off) Sichuan
Rhizoctonia spp. (Damping off) Sichuan
Stereum hirsutum (Willd.) Fr. (Decay) Hunan

***Abies chensiensis* Van Tiegh.**

Fomitopsis pinicola (Sw. ex Fr.) Karst. (Brown cube rot) Gansu
Hirschioporus abietinus (Dicks. ex Fr.) Donk (Decay) Gansu
Phellinus pini (Thore ex Fr.) Ames var. abietis Karst. (White pocket rot) Shanxi Gansu
Pucciniastrum sp. (Leaf rust) Shanxi

***Abies delavayi* Franch.**

Gloeophyllum subferrugineum (Berk.) Bond. et Sing. (Decay) Gansu
Hirschioporus abietinus (Dicks. ex Fr.) Donk (Decay) Gansu
Lophodermium nerviesquium (DC. ex Fr.) Rehm (Needle spot) Sichuan
Phellinus pini (Thore ex Fr.) Ames var. abietis Karst. (White pocket rot) Gansu

***Abies fabri (Mast.)* Craib**

Cryptodrema substygium (Berk. et Br.) Imaz. (Decay) Tibet
Fomitopsis pinicola (Sw. ex Fr.) Karst. (Brown cube rot) Tibet
Hericium coralloides (Scop. ex Fr.) Pers. ex Gray (Decay) Tibet
Heterobasidion annosus (Fr.)Bref. (Pine root rot) Tibet
(=Fomes annosus (Fr.) Cooke.)
(=Fomitopsis annosa (Fr.) Karst.)
Laetiporus sulphureus (Bull. ex Fr.) Bond. ex Sing. (Trunk brown cube rot) Sichuan
Melampsora abietis-capracarum Tub. (Leaf rust) Sichuan
Phaeolus chweintzii (Fr.) Pat. (Trunk brown cube rot) Sichuan
Phellinus hartigii (Allesch. et schnabl) Imaz. (Sapwook brown rot) Sichuan Yunnan
Phellinus pini (Thore ex Fr.) Ames var. abietis Karst. (White pocket rot) Tibet
Polyporus frondosus (Dicks.) Fr. (Decay) Tibet
Comprehensive (Hidden rot) Sichuan

***Abies fargesii* Franch.**

Fomitopsis pinicola (Sw. ex Fr.) Karst. (Brown cube rot) Shanxi Gansu
Hirschioporus abietinus (Dicks. ex Fr.) Donk (Decay) Gansu
(=Polyporus abietinus (Dicks.) Fr.)
Phaeocrytopus abietis Naum. (Black mold) Shanxi
Phellinus pini (Thore ex Fr.) Ames var. abietis Karst. (White pocket rot) Shanxi
(=Trametes abietis Sacc.)

***Abies faxoniana* Rehd. et Wils.**

Ganoderma applanatum (Pers.) Pat. (Brown rot) Sichuan
Heterobasidion annosus (Fr.) Bref. (Pine root rot) Sichuan
Inonotus dryadeus (Pers. ex Fr.) Murr. (Brown rot) Sichuan
Phaeocryptus abietis Naum. (Black mold) Yunnan
Phaeolus schweinitzii (Fr.) Pat. (Trunk brown cube rot) Sichuan
Phellinus hartigii (Allesct. et schnabl.) Imaz. (Sapwood brown rot) Sichuan
Phellinus pini (Thore ex Fr.) Ames var. abietis Karst. (White pocket rot) Sichuan
Polyporus dichrous Fr. (Brown rot) Sichuan

***Abies forrestii* C.C. Rogers**

Melampsorella cerastii (Mart.) Wint. (Leaf rust) Yunnan
Phaeolus schweinitzii (Fr.) Pat. (Trunk brown cube rot) Sichuan

***Abies georgei* Orr.**

Fomitopsis pinicola (Sw. ex Fr.) Karst. (Brown cube rot) Yunnan Sichuan
Gloeophyllum abietinum (Bull. ex Fr.) Karst. (Decay) Sichuan
Heterobasidion annosus (Fr.) Bref. (Pine root rot) Sichuan Yunnan
Inonotus dryadeus (Pers. ex Fr.) Murr. (Decay) Sichuan
Laetiporus sulphureus (Bull. ex Fr.) Bond. et Sing. (Trunk brown cube rot) Sichuan
Phaeolus schweinitzii (Fr.) Pat. (Trunk brown cube rot) Sichuan Yunnan
(=Polyporus schweinitzii Fr.)
Phellinus hartigii (Allesch. ex schnabl.) Imaz. (Sapwood brown rot) Yunnan Sichuan
Phellinus pini (Thore ex Fr.) Ames var. abietis Karst. (White pocket rot) Yunnan Sichuan
Comprehensive (Hidden rot) Sichuan

***Abies georgei var. smithii (Viguie et Gaussen)* Cheng et L. K. Fu**

Fomitopsis pinicola (Sw. ex Fr.) Karst. (Brown cube rot) Tibet
(=Fomes pinicola (Sw. ex Fr.) Cooke.)

Inonotus dryadeus (Pers. ex Fr.) Murr. (Brown rot) Tibet
Laetiporus sulphureus (Bull. ex Fr.) Bond. et Sing. (Trunk brown cube rot) Tibet

***Abies holophylla* Maxim.**
Coriolus hirsutus (Wulf. ex Fr.) Quél. (Heartwood white rot) Jilin Heilongjiang
Fomitopsis pinicola (Sw. ex Fr.) Karst. (Brown cube rot) Jilin
Ganoderma applanatum (Pers.) Pat. (Decay) Jilin
Phellinus pini (Thore ex Fr.) Ames var. abietis Karst. (White pocket rot) Jilin

***Abies nephrolepis* (Trautv.) Maxim.**
Phellinus hartigii (Allesch. et schnabl.) Imaz. (Sapwood brown rot) Heilongjiang

***Abies recurvata* Mast.**
Hirschioporus abietinus (Dicks. ex Fr.) Donk (Decay) Gansu
Phellihus pini (Thore ex Fr.) Ames var. abietis Karst. (White pocket rot) Gansu

***Abies sibirica* Ledeb.**
Calyptospora goeppertiana Kuehn (Leaf rust) Xinjiang
Fomitopsis pinicola (Sw. ex Fr.) Karst. (Brown cube rot) Xinjiang
Lophodermium nerviesquium Rehm (Needle spot) Xinjiang
Melampsorella cerastii (Mart.) Wint. (Leaf rust) Xinjiang
Comprehensive (Hidden rot) Xinjiang

***Abies spectabilis* (D. Don) Spach.**
Microporus xanthopus (Fr.) pat. (Decay) Tibet
Phaeolus schweinitzii (Fr.) Pat. (Trunk brown cube decay) Tibet
Phellinus hartigii (Allesch. et Schnabl) Imaz. (Sapwood brown rot) Tibet
Phellinus pini (Thore ex Fr.) Ames var. abietis Karst. (White pocket rot) Tibet

***Abrus mollis* Hance**
Ravenelia ornata Syd. (Leaf rust) Guangdong

***Abrus precatorius* L.**
Ravenelia ornate Syd. (Leaf rust) Taiwan

***Acacia auriculaeformis* A. Cunn. Ex Benth.**
Oiduim sp. (Powder mildew) Guangdong

***Acacia confusa* Merr.**
Cephaleuros virescens Kunze (Algae) Taiwan Guangdong Fujian Guangxi
Cercospora acaciae-confusae Saw. (Leaf spot) Taiwan
Coriolus polyzonus (Pers.) Imaz. (Decay) Taiwan
Corticium salmonicolor Berk. et Br. (Damping off) Taiwan
Ganoderma applanatum (Pers.) Pat. (Decay) Taiwan
Ganoderma tropicum (Jungh.) Bres. (Decay) Taiwan
Hexagonia heteropora (Mont.) Imaz. (Decay) Taiwan
Loranthus parasiticus (Linn.) Merr. (Mistletoe) Guangxi
Meliola koae Stev. (Black mildew) Taiwan Guangxi
(=Meliola acaciae-confusae Saw.)
Phaeosaccardinula javanica (Zimmm.) Yamam. (Black mildew) Taiwan
(=Capnodium javanica Zimm.)
Phellinus robustus (Karst.) Bourd. Et Galz. (Sapwood brown rot) Taiwan
Poliotelium hyalospora (Saw.) Mains. (Leaf rust) Guangdong Guangxi Fujian Taiwan
Septobasidium acaciae Saw. (Plaster) Taiwan

***Acacia farnesiana* Willd.**
Uredo formosana (Syd.) Tai (Leaf rust) Taiwan

Acanthopanax spp.
Meliola araliicola Yamam. (Black mildew) Guangxi

***Acanthopanax divaricatus* (Sieb. Et Zucc.) Seem.**
Mycospharella araliae C. Harkn. (Leaf spot) Hebei

***Acanthopanax gracilistylus* W. W. Smith**

Aecidium acanthopanacis Diet. (Leaf rust) Jiangsu

***Acanthopanax sessiliflorus* (Rupr. Et Maxim.) Seem.**

Ascochyta acanthopanacis (Syd.) P.K.Chi (Leaf spot) Jilin

***Acanthopanax trifoliatus* (L.) Merr.**

Meliola acanthopanacis Yamam. (Black mildew) Taiwan

Acer spp.

Armillariella mellea (Vahl ex Fr.) Karst. (Root decay) Dongbei

Botryodiplodia acerina Ell. et Ev. (Canker) Liaoning

Cercospora acerina Hart. (Leaf spot) Huabei

Cuscuta japonica Choisy (Dodder) Zhejiang

Eriophyes aceris Hodgkiss (Leaf felt spot) Liaoning

Eriophyes macrochelus-eriobius Nal. (Leaf felt spot) Heilongjiang

Favolus squamosus (Huds. ex Fr.) Ames (Decay) Dongbei

Fomes fomentarius (L. ex Fr.) Kickx. (Brown rot) Dongbei

(= Pyropolyporus fomentarius (L. ex Fr.) Teng.)

Inonotus hispidus (Bull. ex Fr.) Karst. (Leaf tip rot) Dongbei

Laetiporus sulphureus (Bull. ex Fr.) Bond. et Sing. (Trunk brown cube rot) Dongbei

Oxyporus populinus (Schum. ex Fr.) Donk (Decay) Jilin

(=Fomes connatus (Fr.) Gill.)

Phellinus igniarius (L. ex Fr.) Quél. (White pocket rot) Gansu

Phyllosticta negundinis Sacc. Et Speg. (White leaf spot) Heilongjiang

Phyllosticta platanoidis Sacc. (Leaf spot) Liaoning Heilongjiang

Phytophthora cactorum (Leb et Cohn.) Schröt. (Bleeding canker) Taiwan

Poria lurida Bres. (Decay) Hebei

Rhytisma acerinum (Pers.) Fr. (Tar spot) Henan Zhejiang Jiangsu Fujian Jilin Liaoning Sichuan

(Asexual:Melasmia acerina Lév.)

Rhytisma punctatum (Pers.) Fr. (Tar spot) Liaoning Heilongjiang Sichuan Anhui Hunan Ningxia

(Asexual:Melasmia punctata Sacc. Et Roum.)

Sawadaea bicornis (Wallr. ex Fr.) Homma (Powder mildew) Liaoning Jilin Heilongjiang Sichuan Jiangsu

(=Uncinula aceris (DC.) Sacc.)

Sawadaea tulasnei (Fuck.) Homma (Powder mildew) Henan Hebei

(=Uninula tulasnei Fuck.)

(=Uncinula aceris (DC.) Sacc. Var. tulasnei (Fuck.) Salm.)

Steccherinum septentrionale (Fr.) Bank. (Decay) Jilin Heilongjiang Hebei

Taphrina sp. (Leaf blister) Heilongjiang

Trametes malicola Berk. et Curt. (Decay) Tibet

Tyromyces fissilis (Berk. et Curt.) Murr. (Decay) Jilin Heilongjiang Hebei Shanxi

Physical (Physical belt disease) Liaoning Heilongjiang

Bacterial (Leaf spot) Gansu

***Acer buergerianum* Miq.**

Alternaria negundinicola (Ell. et Barth.) Joly Henan

Cuscuta japonica Choisy (Dodder) Zhejiang

Dinemasporium acerinum Peck (Dieback) Jiangsu

Ganoderma applanatum (Pers.) Pat. (Decay) Guangdong

Melasmia acerina Lév. (Tar spot) Hunana

Melasmia punctata Sacc. et Roum. (Tar spot) Anhui Zhejiang Jiangsu Jiangxi

Schizothyrium annuliforme Syd. et Butl. Henan Zhejiang Anhui Jiangsu

Uncinula duncoides Zheng et Chen (Powder mildew) Jiangsu
Uncinula nankinensis Tai (Powder mildew) Zhejiang Anhui Jiangsu

***Acer catalpifolium* Rehd.**
Sawadaea bicornis Wallr. ex Fr. Homma (Powder mildew) Sichuan
Sawadaea polyfida (Wei) Zheng et Chen (Powder mildew) Sichuan
(=Uncinula polyfida Wei)

***Acer caudatum* Wall.**
Sawadaea bomiensis Zheng et Chen (Powder mildew) Tibet

***Acer caudatum var. georgei* Diels.**
Sawadaea bicornis (Wwallr. ex. Fr.) Homma (Powder mildew) Yunnan

***Acer caudatum var. prattii* Rehd.**
Melasmia punctata Sacc. et Roum. (Tar spot) Sichuan

***Acer davidii* Franch.**
Melampsoridium caeris Jϕrst. (Leaf rust) Sichuan
Rhytisma acerinum (Pers.) Fr. (Tar spot) Shanxi

***Acer ginnala* Maxim.**
Rhytisma acerinum (Pers.) Fr. (Tar spot) Shanxi
Rhytisma punctatum (Pers.) Fr. (Tar spot) Liaoning Jilin Heilongjiang Hubei

***Acer henryi* Pax**
Melasmia acerina Lév. (Tar spot) Henan Shanxi
Rhytisma punctatum (Pers.) Fr. (Tar spot) Henan

***Acer mandshuricum* Haxim.**
Fomes fomentarius (L. ex Fr.) Kickx. (Brown rot) Jilin
Oxyporus populinus (Schum. ex Fr.) Donk (Decay) Jilin

***Acer maximowiczii* Pax**
Eriophyes sp (Leaf felt spot) Shanxi
Rhytisma acerinum (Pers.) Fr. (Laequer spot) Shanxi

***Acer mono* Maxim.**
Botryodiplodia acerina Earst. ex Ev. Liaoning
Fomes fomentarium (L. ex Fr.) Kickx. (Brown spot) Jilin
Inonotus radiatus (Sow. ex Fr.) Karst. Var. licentii Pilát (Decay) Jilin
Oxyporus populinus (Schum. ex Fr.) Donk (Decay) Jilin
Phyllosticta minima (Berk. et Curt.) Ell. et Ev. (Leaf spot) Shanxi
Rhytisma punctatum (Pers.) Fr. (Tar spot) Dongbei Jilin
Sawadaea bicornis (Wallr. ex Fr.) Homma (Powder mildew) Heilongjiang

***Acer mono var. platonoides* Miyabe**
Phyllosticta platanoidis Sacc. (Leaf spot) Jilin
Sawadaea bicornis (Wallr. ex Fr.) Homma (Powder mildew) Liaoning
Sawadaea tulasnei (Fuck.) Homma (Powder mildew) Liaoning

***Acer negundo* L.**
Alternaria negundinicola (Ell. et Barth.) Joly (Black spot) Henan
Cercospora acerina Hart. (Leaf spot) Hebei
Cuscuta monogyna Vhal (Dodder) Xinjiang
Meliola aceris Yamam. (Black mildew) Neimeng
Phyllosticta arida Earle (Leaf spot) Jilin
Phyllosticta negundicola Sacc. (Leaf spot) Liaoning Jilin Heilongjiang Neimeng
Sawadaea bicornis (Wallr. ex Fr.) Homma (Powder mildew) Jilin Liaoning
Sawadaeanegundinis Homma (Powder mildew) Jilin
(=Unicinula negundinis (Homma) Tai)
Septoria negundinis Ell. et Ev. (Leaf spot) Jilin
Septoria samarae Peck. (Powder mildew) Jilin Liaoning Heilongjiang

Verticillium albo-atrum Reinke et Berth. (Wilt) Jilin Xinjiang
Verticillium dahliae Kleb. (Wilt) Xinjiang

***Acer oblongum* Wall.**
Meliola aceris Yamam. (Black mildew) Taiwan
Rhytisma acerinum (Pers.) Fr. (Tar spot) Hunan
Uncinula aduncoides Zheng et Chen (Powder mildew) Hunan
(=Uncinula sinensis Tai et Wei)

***Acer palmatum* Thunb.**
Macrophomina phaeseoli (Maubl.) Ashby (Trunk decay) Jiangsu

***Acer pilosum* Maxim.**
Coriolus versicolor (L. ex Fr.) Quél. (Decay) Gansu

***Acer pseudosieboldianum* (Pax) Kom.**
Mycosphaerella alarum Ell. et Halst. (Leaf spot) Liaoning
Rhytisma punctatum (Pers.) Fr. (Tar spot) Liaoning Dongbei

***Acer rubescens* Hay.**
Pucciniastrum hikosanense Hirats. f. (Leaf rust) Henan Taiwan

***Acer saccharium* Marsch.**
Phyllosticta negundinis Sacc. (White leaf spot) Neimeng
Septoria Saccharina Ell. et Ev. (Leaf dry spot) Heilongjiang Jiangsu
Septoria seminalis Sacc. (Leaf dry spot) Jiangsu

***Acer semenovii* Regel et Header**
Cuscuta monogyna Vhal (Dodder) Xinjiang
Cylindrosporium aceris-oblusati Bub. Xinjiang
Cytospora annulata Ell. et Ev. (Canker) Xinjiang

***Acer sinense* Pax**
Melasmia acerina Lév. (Laequer spot) Hunan

***Acer tegmentosum* Maxim.**
Coriolus hirsutus (Wulf. ex Fr.) Quel. (Decay) Jilin Heilongjiang
(=Trametes hirsutus (Wulf. ex Fr.) Pilát)
Eriophyes macrochelus-eriobius Nal. (Leaf felt spot) Jilin
Oxyporus populinus (Schum. ex Fr.) Donk (Decay) Jilin
Septobasidium bogoriense Pat. (Plaster) Heilongjiang

***Acer tetramerum var. betulifolium* (Maxim) Rehd.**
Melasmia punctata Sacc. et Roum. (Laequer spot) Gansu
Rhytisma acerinum (Pers.) Fr. (Laequer spot) Hubei

***Acer Truncatum* Bunge**
Asteromella platanoidis (Sacc.) Petr. Shanxi
Rhytisma punctatum (Pers.) Fr. (Tar spot) Dongbei
Sawadaea tulasnei (Fuck.) Homma (Powder mildew) Henan Beijing

***Acer ukurunduense* Trautv. et Mey.**
Rhytisma punctatum (Pers.) Fr. (Tar spot) Taiwan
Sawadaea bicornis (Wallr. ex Fr.) Homma (Powder mildew) Heilongjiang

***Achras sapota* L.**
Aithaloderma clavatispora Syd. (Black mildew) Taiwan
Capnodium walteri Ssacc. (Sooty mold) Taiwan
Hypocapnodium setosum (Zimm.) Speg. (Sooty mold) Taiwan
Phaeosaccardinula javanica (Zimm.) Yamam. (Sooty mold) Taiwan
Triposporiopsis spinigera (Höhn.) Yamam. (Sooty mold) Taiwan

Actinidia spp.
Cercospora iteodaphnes (Thüm.) Sacc. (Leaf spot) Guizhou
Phyllactinia actinidiae-formosanae Saw. (Powder mildew) Anhui Guizhou

Phyllactinia actinidiae-latifoliae Saw. (Powder mildew) Guangxi
Pseudomonas syringae van Hall pv. actinidia (Canker) Sichuan Shandong Hunan Beijing Shanxi

***Actinidia arguta (Sieb. et Zucc.)* Planch. et Miq.**
Monilia fructigena Pers. (Sclerotium) Jilin

***Actinidia chinensis* Planch.**
Phyllactinia actinidiae-formosanae Saw. (Powder mildew) Guangxi
Phyllactinia actinidiae-latifoliae Saw. (Powder mildew) Hunan

***Actinidia formosana* Hayata**
Phyllactinia actinidiae-formosanae Saw. (Powder mildew) Taiwan
Pucciniastrum acinidae Hirats. f. (Leaf rust)

***Actinidia kolomikta* (Rupr. et Maxim.) Maxim.**
Uncinula necator (Schw.) Burr. (Powder mildew)
(Asexual: Oidium tuckeri Berk.) Dongbei Jilin

Actinodaphne spp.
Meliola actinodaphnes Hansf. (Sooty mold) Guangxi
Xenostele Echinacea (Berk.) Syd. (Leaf rust) Sichuan

***Actinodaphne mushaensis* (Hayata) Hayata**
Appendiculella kiraiensis (Yamam.) Hansf. (Black mildew) Taiwan
Armatella formosana Yamam. (Black mildew) Taiwan

***Actinodaphne pedicelata* Hayata**
Phellinus williamsii (Murr.) Pat. (Decay) Taiwan

***Adina rubella* (Sieb. et Zucc.) Hance**
Loranthus parasiticus (Linn.) Merr. (Mistletoe) Province not determined

Aesculus sp.
Botryodiplodia aesculina Pass.
Glomerella cingulata (Stonem.) Spauld. et Schrenk (Anthracnose) Province not determined

***Aesculus swlsonii* Rehd.**
Sawadaea aesculi Zheng et Chen (Powder mildew) Sichuan

Ailanthus spp.
Phyllactinia ailanthi (Golov. et Bunk.) Yu (Powder mildew) Shandong
(=Phyllactinia corylea (Pers.) Karst.)
(=Phyllactinia suffulta (Rebent.) Sacc. f. ailanthi Golov. et. Bunk.)
Schizophyllum multifidum (Batsch) Fr. (Decay) Shanxi
Stereum purpureum (Pers.) Fr. (Decay) Shanxi
Uncinula delavayi Pat. (Powder mildew) Yunnan

***Ailauthus altissima* (Mill.) Swingle.**
Capnodium elongatum Berk et Desm. (Sooty mold) Shanxi
Cercospora glandulosa Ell. et Kell. (Cycle spot) Neimeng Shanxi
Coriolus hirsutus (Wulf ex Fr.) Quél. (White rot) Shanxi
Funalia hispida (Bagl.) Pat. (White rot) Shanxi
Haplosporella ailanthi Ell. et. Ev. (Leaf spot) Hebei
Macrophomina phaesoli (Maubl.) Ashby (Stem rot) Anhui
Nyssopora cedrelae (Hori) Tranz. (Leaf rust) Hubei Hunan Guangxi Guangdong Jiangxi Anhui Shandong Taiwan
Phyllactinia ailanthi (Golov. et Bunk.) Yu (Powder mildew) Liaoning Heilongjiang Anhui Hubei Jiangsu Henan Gansu Hebei Jiangxi Hunan Sichuan Shanxi Shanxi Shandong Fujian
Phyllosticta ailnathi Sacc. (Leaf spot) Henan Hebei
Rhizoctonia solani Kühn. (Seedling damping off) Henan
Schizophyllum commune Fr. (Sapwood white rot) Shanxi
Uncinula delavayi Pat. (Powder mildew) Jiangsu Sichuan

Akebia spp.
Myiocopron smilacis (de Not.) Sacc. (Mold spot) Zhejiang Hubei
***Akebia quinata* (Thunb.) Decne.**
Aecidium akebiae P. Henn. (Leaf rust) Jiangsu Zhejiang Anhui Fujian Gansu
Cuscuta japonica Choisy (Dodder) Zhejiang
Microsphaera akebiae Sawada. (Powder mildew) Jiangsu Sichuan
Puccinia akebiae Wang et Wei (Leaf rust) Tibet

***Akebia trifoliate* (Thunb.) Koidz.**
Aecidium akebiae P. Henn. (Leaf rust) Jiangsu Zhejiang Jiangxi Hunan
Microsphaera akebiae Sawada (Powder mildew) Henan
Alangium sp.
Phyllactinia alangii Yu et Lai (Powder mildew) Sichuan
Septoria taiana Syd. (Leaf spot) Sichuan
Typhulochaeta alangii Yu et Lai (Powder mildew) Sichuan
***Alangium chinense* (Lour.) Harms**
Phyllactinia alangii Yu et Lai (Powder mildew) Sichuan Jiangsu Zhejiang Anhui Jiangxi
Septoria taiana Syd. (Leaf dry spot) Jiangsu
***Alangium platanifolium* (Sieb. et Zucc.) Harms**
Cuscuta japonica Choisy (Cuseata high plant mistletoe) Zhejiang
Loranthus parasiticus (Linn.) Merr. (Mistletoe) Guangxi
Phyllactinia alangii Yu et Lai (Powder mildew) Jiangsu Sichuan
Septoria acerina Speg. (Leaf spot) Anhui
Albizia spp.
Inonotus hispidus (Bull. ex Fr.) Karst. (Leaf tip rot) Hebei
(=Xanthochrous hispidus (Bull. ex Fr.) Pat.)
Lenzites shichiana (Teng et Ling) Teng (Decay) Hebei
Oxyporus populinus (Schum. ex Fr.) Donk (Decay) Dongbei Guangdong Jiangxi Guizhou
Ravenelia japonica Diet. et Syd. (Leaf rust) Dongbei Hebei Jiangxi Guizhou
Ravenelia sessilis Berk. (Leaf rust) Guangdong Hainandao
***Albizia chinensis* (Osbeck) Merr.**
Meloidogyne incognita (Koforid et White) chitwood (root nematodes) Guangdong
Ravenelia sessilis Berk. (Leaf rust) Guangdong
***Albizia julibrissin* Durazz.**
Coriolus versicolor (L. ex Fr.) Quél. (Decay) Shanxi
Cuscuta japonica Choisy (Dodder) Zhejiang
Hydnum diversidens Fr. (Decay) Shanxi
Irpex sp. (Decay) Shandong
Loranthus parasiticus (Linn.) Merr. (Mistletoe) Guangxi
Nectria cinnabarina (Tode) Fr. (Canker) Tibet Shanxi
(Asexual : Tubercularia vulgaris Tode)
Ravenelia japonica Diet. et Syd. (Leaf rust) Sichuan Henan Taiwan
Schizophyllum commune Fr. (Decay) Shanxi
Sphaerophragmium acaciae (Cooke) Magn. (Leaf rust) Guangdong Guangxi
***Albizia kalkora* (Roxb.) Prain**
Cephaleuros virescens Kunze (Algae)
Ravenelia japonica Diet. et Syd. (Leaf rust) Sichuan Shandong Jiangsu Hunan
***Albizia lebbek* (L.) Benth.**
Colletotrichum gloeosporiodes penz. (Anthracnose) Guangdong
(=Colletotrichum lebbek (Syd.) Petr.)
Phoma lebbek Saw. (Dieback) Taiwan

Sphaerophragmium acaciae (Cooke) Magn. (Leaf rust) Guangdong Guangxi Fujian Taiwan

***Albizia odoratissma* (L.f.) Benth.**

Meloidogyne incognita (Koforid et White) Chitwood (root nematodes) Guangdong

***Albizia procera* (Roxb.) Benth.**

Meloidogyne incognita (Koforid et White) Chitwood (root nematodes) Guangdong
Ravenelia sessilis Berk. (Leaf rust) Guangdong

***Albizia yunnanensis* Fr.**

Ravenelia japonica Diet. et Syd. (Leaf rust) Yunnan

Aleurites spp.

Cerccospora aleuritdis Miyake (Black leaf spot) Sichuan Guangxi
Coniothyrium aleuritis Teng (Leaf spot) Jiangsu
Melampsora aleuritidis Cumm. (Leaf rust) Guizhou

***Aleurites cordata* R. Br.**

Phyllactinia aleuritidis Yu et Lai (Powder mildew) Sichuan
(=Phyllactinia corylea (Pers.) Karst.)
Cercospora aleuritidis Miyake (Black leaf spot) Sichuan Hunan Yunnan

***Aleurites fordii* Hemsl.**

Botryosphaeria ribis (Tode) Grossenb. et Duggar Sichuan
Cercospora aleuritidis Miyake (Black leaf spot) Sichuan Henan Anhui Guangdong Guangxi Jiangsu Fujian Guizhou Shanxi Jiangxi Hunan Yunnan Hubei
Elytranthe ampullacea Don. (Mistletoe) Guangxi
Fusarium oxysporum Schlecht. (Tung oiltree wilt) Sichuan Zhejiang Anhui Guangdong Guangxi Hunan Jiangxi
Loranthus chinensis DC. (Mistletoe) Guangxi
Loranthus yadoriki Sieb. et Zucc. (Mistletoe) Guizhou
Macrophoma sp. (Black Stiff fruit) Guangxi
Phoma sp. (Brown Stiff fruit) Guangxi
Schizophyllum commune Fr. (Decay) Hunan
Septobasidium bogoriense Pat. (Felt fungus)
Uncinula miyabei (Salm.) Sacc. et Syd. (Powder mildew) Sichuan
Uncinula nishidana Homma (Powder mildew) Sichuan
Viscum angulatum Heyne (Mistletoe) Guangxi
Viscum articulstrum Birm. f. (Mistletoe) Sichuan Guangxi Guizhou

***Aleurites Montana* (Lour) Wils.**

Cercospora aleuritidis Miyake (Blight) Sichuan Hunan Guangdong
Colletotrichum sp. (Anthracnose) Guangxi Guangdong
Pestalotia sp. (Brown leaf spot) Hunan Guangdong

Alniphyllum sp.

Coriolus hirsutus (Wulf. ex Fr.) Quél. (Decay) Jiangxi

***Alniphyllum pterospermum* Matsum.**

Meliola alniphylli Yamam. (Black mildew) Taiwan

Alnus spp.

Antennaria pannosa Berk. (Black mildew) Yunnan
Melasmia alni Syd. (Black mole) Guangdong
Phyllosticta bellunensis Mart. (Leaf spot) Liaoning
Plasmodiophora alni (Woron.) Möll. (Root swell) Yunnan

***Alnus cremastogyne* Burkill**

Loranthus yadoriki Sieb. et Zucc. (Camphor mistletoe) Hunan
Melampsoridium alni (Thüm) Diet. (Leaf rust) Sichuan
Phyllactinia alni Yu et Han (Powder mildew) Sichuan
(=Phyllactinia fraxini (DC.) Homma)

Alnus formosana **(Burkill) Makino**
Hypocapnodium setosum (Zimm.) Speg. (Black mildew) Taiwan
Phyllactinia alni Yu et Han (Powder mildew) Taiwan

Alnus japonica **(Thunb.) Steud**
Coriolus versicolor (L. ex Fr.) Quél. (Decay) Shandong
Eriophyes brevitarsus Fr. (Leaf felt spot) Liaoning
Fomes fomentarius (L. ex Fr.) Kickx (White rot) Sichuan
Hirschioporus laricinus (Karst.) Teram. (Decay) Shandong
Melasmia alni Syd. (Blacktar spot) Guangdong
Melampsoridium alni (Thüm) Diet. (Leaf rust) Liaoning
Phyllactinia alni Yu et Han (Powder mildew) Shandong
Phyllosticta bellunensis Martin (Leaf spot) Liaoning
Uncinula miyabei (Salm) Sacc. et Syd. (Powder mildew) Liaoning
Septoria alnifolia Ell. et Ev. (Leaf spot) Liaoning

Alnus mandshurica **(Call) Hand-Mazz.**
Fomitopsis scutellata (Schw.) Bond. et Sing. (Decay) Neimong
Lenzites betulina (L.) Fr. (Decay) Neimong
Schizophyllum commune Fr. (Decay) Neimong

Alnus nepalensis **D. Don**
Phyllactinia alni Yu et Han (Powder mildew) Yunnan

Alnus sibirica **Fisch. ex Trucz.**
Coriolus hirsutus (Wulf. ex Fr.) Quél. (Decay) Heilongjiang
Ganoderma applanatum (Pers.) Pat. (Decay) Dongbei
Melampsoridium hiratsukanum Ito ex Hirats. (Leaf rust) Dongbei
Phellinus igniarius (L. ex Fr.) Quél. (Decay) Dongbei
Phyllactinia alni Yu et Han (Powder mildew) Heilongjiang
Phyllosticta bellunensis Martin (Leaf spot) Liaoning
Septoria alnifolia Ell. et Ev. (Leaf spot) Liaoning Henan

Amelanchier sp.
Coleopucinia sinensis Pat. (Leaf rust) Yunnan

Amorpha fruticosa **L.**
Haplosporella amorphae (Ell. et Barth) Togashi (Leaf spot) Hebei
Physical belt disease Neimong Ningxia
Bacterial leaf spot Neimong

Ampelopsis spp.
Phyllactinia ampelopsidis Yu et Lai (Powder mildew) Jiangxi
Physopella ampelopsidis (Diet. et Syd.) Cumm. Et Ramechar (Leaf rust) Anhui Jiangxi Shanxi Fujian
Pucciniostele hashiokai (Hirats. f.) Cumm. (Leaf rust) Guizhou

Ampelopsis brevipedunculata **(Maxim.) Koehne.**
Physopella ampelopsidis (Diet. Et Syd.) Cumm. Et Ramachar (Leaf rust) Anhui Zhejiang Henan Taiwan
Plasmopara viticola (Berk. et Curt.) Berl. et de Toni (Frost mold) Sichuan
Septoria alnifolia Ell. et Ev. (Leaf spot) Henan
Septoria ampelopsidis-heterophyllae Miura (Leaf spot) Liaoning

Ampelopsis acntoniensis **(Hook. et Arn.) Planch.**
Pucciniostele hashiokai (Hirats. f.) Cumm. (Leaf rust) Taiwan

Ampelopsis humulifolia **Bunge**
Phyllactinia ampelopsidis Yu et Lai (Powder mildew) Hebei Taiwan
(=Phyllactinia corylea (Pers.) Karst.)
Physopella ameplopsidis (Diet. et Syd.) Cumm. et Ramachar (Leaf rust) Taiwan

Septoria ampelopsidis-heterophyllae Miura (Leaf spot) Liaoning Hebei

***Ampelopsis japonica* (Thunb.) Makino**

Physopella ampelopsidis (Diet. et Syd.) Cumm. et Ramachar (Leaf rust) Anhui

***Anthocephalus chinensis* (Lamk.) Rich ex Walp.**

Phyllosticta sp. (Leaf spot) Guangxi

***Aphanathe aspera* planch.**

Cuscuta japonica Choisy (Dodder) Zhejiang

Aralia spp.

Cercospora araliae P. Henn. (Leaf spot) Anhui Guangxi Jiangxi Guizhou

***Aralia chinensis* L.**

Aecidium araliae Saw. (Leaf rust) Henan Taiwan
Cercospora araliae P. Henn. (Leaf spot) Liaoning Taiwan Zhejiang
Cladosporium araliae Saw. (Black leaf spot) Taiwan
Meliola araliicola Yamam. (Black mildew) Taiwan
Nyssopsora asiatica Lütj. (Leaf rust) Jilin Dongbei

***Aralia dasyphylla* Miq.**

Cercospora araliae P. Henn. (Leaf spot) Anhui Guangxi Guizhou Jiangxi

***Aralia decaisneana* Hance**

Aecidium araliae Saw. (Leaf rust) Taiwan
Cercospra araliae P. Henn (Leaf spot) Taiwan
(=Cercosporiopsis araliae (Henn) Miura)
Meliola araliicola Yamam. (Sooty mold) Taiwan

***Aralia elata* (Mip.) Seem.**

Aecidium araliae Saw. (Leaf rust) Shanxi Taiwan
Cercospora araliae P. Henn. (Leaf spot) Liaoning
Nyssopsora asiatica Lütj. (Leaf rust) Jilin Heilongjiang
Triphragmium clavellosum var. asiatica Komarov (Leaf rust) Heilongjiang Jilin

***Areca catechu* L.**

Brachysporium arecae (Berk. et Br.) Sacc. Taiwan
Gloeosporium catechu Syd. (Anthracnose) Taiwan
Guignardia arecae Sacc. (Leaf spot) Taiwan

***Artocarpus heterophyllus* Lam.**

Physalospora rhodina Berk. et Curt. (Fruit rot) Guangdong Guangxi
(Asexual : Diplodia natalensis Evans)
Rhizopus artocarpi Racib. (Soft rot) Guangdong Guangxi
(=Rhizopus stolonifer (Ehrenb. ex Fr.) Vuill.)

***Atraphaxis pyrifolia* Bunge**

Puccinia platypoda Syd. (Leaf rust) Xinjiang
Trichocladia atraphaxis Golov. (Powder mildew) Xinjiang

Bambusa spp.

Aithaloderma bambusinum Petr. (Black mildew) Zhejiang
Balansia take (Miyake) Hara (Witches' broom) Guangdong Zhejiang
Cladosporium graminum Link Guangdong
Coniosporium bambusae (Thüm. et Bolle) Sacc. Guangdong
Coniosporium shiraianum (Syd.) Bub. Guangdong
Didymella eumorpha (Berk. et Curt.) Sacc. Guangdong
Endodothella bambusae (Rabenh.) Thesiss. et Syd. Guangdong
Eutypella bambusina Penz. et Sacc. Guangdong
Konradia bambusina Racib. Guangdong Guangxi Fujian
Melanconium bambusinum Speg. Fujian
Meliola bambusae Pat. (Black mildew) Guangdong

Meliola tenella Pat. var. atalantiae (Pat.) Hansf. (Black mildew) Guangdong
Myriangium haraeanum Tai et Wei Fujian
Phragmocarpella japonica Hara Fujian
Phyllachora orbicular Rehm (Black mole) Zhejiang Jiangxi Yunnan
Phyllachora shiraiana Syd. (Black mole) Jiangsu Guangdong Fujian Guizhou
Phyllachora sinensis Sacc. (Black mole) Jiangsu Zhejiang Jiangxi Hunan Fujian Guangdong Sichuan Yunnan
Puccinia kwanhsienensis Tei (Leaf rust) Sichuan
Puccinia longicornis Pat. et Har. (Leaf rust) Zhejiang Fujian
Puccinia phyllostachydis Kus. (Leaf rust) Yunnan
Schlerotium fumigatum Nakta ex Hara (Schlerotium) Taiwan
Septocytella bambusina Syd. (Black mildew) Zhejiang Hunan
Shiraia bumbusicaola P. Henn. (Shiraia disease) Henan
Stereostratum corticiodides (Berk et Br.) Magn. (Leaf rust) Jiangsu Zhejiang Henan Guangdong Sichuan Guizhou Shandong Anhui
Tetraploa aristata Berk. et Br. Guangdong
Trematosphaerella bambusae (Miyake et Hara) Tai Jiangsu
(=Leptosphaerella bambusae Miyake et Hara)
Ureda ignava Arth. (Leaf rust) Guangdong

***Bambusa multiplex* (Lour.) Raeusch.**
Linochora howardii Syd. Guangdong
Myriangium haraeanum Tai et Wei (Rounded leaf spot) Henan
Phyllachora orbicular Rehm (Black mole) Guangdong
Prosthemiella bambusina Syd. Guangdong

***Bambusa stenostachya* Hack.**
Dasturella divina (Syd.) Mundk. et Khesw. (Leaf rust) Taiwan
Phellinus torulosus (Pers.) Bourd. et Galz. (Decay) Taiwan
Phyllachora maculans (Karst.) Theiss et Syd. (Black mole) Guangdong

***Bambusa textilis* McClure**
Fusarium moniliforme Sheld (Stem and trunk decay) Guangdong Guangxi

***Bambusa vulgaris* Schrad.**
Colletotrichum septorioides Sacc. (Anthracnose) Sichuan Guangdong
Coniosporium shiraianum (Syd.) Bubak Guangxi
Dasturella divia (Syd.) Mundk. et Khesw. (Leaf rust) Taiwan

Berberis spp.
Microsphaera berberidis (DC. ex Mer.) Lév. (Powder mildew) Henan Hebei Gansu Neimeng Xinjiang
(=Erysiphe berberidis DC.)
(=Microsphaera berberidis Lév.)
Puccinia graminis Pers. (Leaf rust) Hebei Henan Shanxi Shanxi Jiangsu Gansu Xinjiang Guangxi Neimeng Fujian Sichuan

***Berberis amurensis* Rupr.**
Microsphaera berberidis (DC. ex Mér) Lév. (Powder mildew) Liaoning Hebei Beijing Sichuan
Microsphaera multaappendicis C.Y. Zhao et Yu (Powder mildew) Xinjiang
Puccinia graminis pers. (Leaf rust) Jilin Heilongjiang Shanxi Qinghai
Puccinia pygmaea Erikss. (Leaf rust) Jilin

***Berberis amurensis var. japonica* (Regel) Rehed.**
Puccinia graminis Pers. (Leaf rust) Heilongjiang

***Berberis chinensis* Poir.**
Microsphaera berberidis (DC. ex Mév.) Lév. (Powder mildew) Jilin Heilongjiang

***Berberis circumserrata* Schneid.**

Microspaera berberidis (DC. ex Mér.) Lév. (Powder mildew) Gansu

***Berberis dasystachya* Maxim.**

Microsphaera berberidicola Tai (Powder mildew) Henan

***Berberis delavayi* Schneid.**

Puccinia graminis Pers. (Leaf rust) Sichuan

***Berberis diaphana* Maxim.**

Microsphaera berberidis (DC. ex Mér.) Lév. (Powder mildew) Sichuan

Microsphaera sichuanica Yu (Powder mildew) Sichuan

***Berberis dielsii* Fedde**

Puccinia graminis Pers. (Leaf rust) Henan

***Berberis gilgiana* Fedde.**

Puccinia graminis Pers. (Leaf rust) Henan Shanxi

***Berberis heteropoda* Schneid.**

Microsphaera berberidis (DC. ex Mér.) Lév. var. dimorpha Yu et C. Y. Zhao (Powder mildew) Xinjiang

Otthia amelanchieris Karst. Xinjiang

Puccinia arrhenatheri (Kleb.) Erikss (Leaf rust) Xinjiang

Puccinia graminis Pers. (Leaf rust) Xinjiang

Septoria berberdis Niessl. (Leaf spot) Xinjiang

***Berberis kawakamii* Hayata**

Uredo clemensiae (Arth. et Cumm.) Hirats. f. (Leaf rust) Taiwan

***Berberis morrisonensis* Hayata**

Aecidium niitakense Hirats. (Leaf rust) Taiwan

***Berberis poiretti* Schneid.**

Aecidium berberidis Pers. (Leaf rust) Liaoning

Melasmia berberidis Thüm. et Wint. (Black mole) Liaoning

Microsphaera berberidis (DC. ex Mér.) Lév. (Powder mildew) Liaoning Jilin Hebei

Puccinia culmicola Diet. (Leaf rust) Jilin Hebei Henan Anhui

Puccinia graminis Pers. (Leaf rust) Yunnan Hebei

***Berberis silva-taroucana* Schneid.**

Puccinia gramanis Pers. (Leaf rust) Sichuan

***Berberis thunbergii* DC.**

Puccinia graminis Pers. (Leaf rust) Shanxi Yunnan

***Berberis virgetorum* Schneid.**

Puccinia graminis pers. (Leaf rust) Jiangxi

***Berberis vulagris* L.**

Microsphaera berberidis (DC. ex Mér.) Lév. (Powder mildew) Xinjiang

Puccinia graminis Pers. (Leaf rust) Shanxi Xinjiang

Betula spp.

Armillariella mellea (Vahl ex Fr.) Karst. (Root rot) Dongbei Huabei Sichuan Gansu Yunnan

Asterosporium hoffmanii Fr. Yunnan

Bjerkandera adusta (Willd. ex Fr.) Karst. (Decay) Liaoning Heilongjiang

Cerrena unicolor (Bull. ex Fr.) Murr. (Decay) Hebei Xinjiang

Coriolus hirsutus (Wulf ex Fr.) Quél. (Decay) Heilongjiang Shanxi Hebei Sichuan

Cytospora personata Fr. (Canker) Xinjiang

Daedaleopsis confragosa (Bolt. ex Fr.) Schrot. (Decay) Heilongjiang

(=Daedalea confragosa Fr.)

Daldinia concentrica (Bolt.) Des. et de Not (Decay) Xinjiang

Eriophyes rudis longisetosus Nal. (Leaf felt spot) Xinjiang

Eriophyes rudis typicus Nal. (Leaf felt spot) Xinjiang

Fomes fomentarius (L. ex Fr.) Kickx. (Brown rot) Heilongjiang Shanxi Sichuan Tibet Gansu Xinjiang Hebei
Fomes rufolaccatus Lloyd. (Decay) Xinjiang
Fomitopsis pinicola (Sw. ex Fr.) Karst. (Brown cube rot) Heilongjiang Xinjiang
Ganoderma applanatum (Pers.) Pat. (Decay) Heilongjiang Sichuan Yunnan Xinjiang
Gloeophyllum saepiarium (Wulf. ex Fr.) Karst. (Decay) Shanxi Heilongjiang
Gloeosporum betularium Ell. et Mart. (Anthracnose) Hebei
Gloeosporium betulinum West. (Anthracnose) Xinjiang
Hypoxylon multiforme Fr. (Decay) Xinjang
Inonotus obliquus (Fr.) pil. (Decay) Xinjiang
Inonotus rheades (Pers.) Pilát (Decay) Heilongjiang
Irpex lacteus Fr. (Decay) Sichuan Heilongjiang
Laetiporus sulphureus (Bull. ex Fr.) Bonp. et Sing (Trunk brown rot) Sichaun Heilongjiang
Lenzites betulina (L.) Fr. (Decay) Liaoning Heilongjiang
Lenzits tricolor (Bull.) Fr. (Decay) Tibet Heilongjiang
Libertella betulina Desm. Xinjiang Heilongjiang
Melampsoridium betulinum (Dasm.) Kelb. (Leaf rust) Hebei Sichuan Xinjiang Heilongjiang Jilin
Microsphaera betulae Magn. (Powder mildew) Sichuan Xinjiang
Nectria cinnabarina (Tode) Fr. (Canker) Dongbei Sichuan Xinjiang
Nectria cocinea (pers.) Fr. (Canker) Sichuan
Nectria ditissima Tul. (Canker) Sichuan
Nectria galligena Bres. (Dieback) Liaoning
Oxyporus populinus (Schum. ex Fr.) Donk (White rot) Sichuan
Phellinus igniarius (L. ex Fr.) Quél. (Heartwood white sponge rot) Jilin Shanxi Hebei Henan Dongbei Neimeng Shanxi Tibet Sichuan Gansu Ningxia Qinghai Xinjiang Yunnan
Phellinus robustus (Karst.) Bourd. et Galz. (Sapwood brown rot) Xinjiang
(=Fomes robustus Karst.)
Phyllactinia alni Yu et Han (Powder mildew) Xinjiang Sichuan
Piptoporus betulinus (Bull. ex Fr.) Karst (Brown cube rot) Jilin Heilongjiang Henan Shanxi Dongbei Neimeng Gansu Sichuan Tibet
(=Polyporus betulinus Fr.)
Podosphaera erıneophila Naum. (Powder mildew) Xinjiang
Pycnoporus cinnabarinus (Jaq.) Karst. (Decay) Neimeng Xinjiang Qinghai
(=Trametes cinnabarina (Jacq.) Fr.)
Sclerotinia betulae Woron. (Seed blight) Xinjiang
Septoria betulae Westend. (Leaf spot) Xinjiang
Stereum fasciatum Schw. (Decay) Xinjiang
Stereum pubescens Burt. (Decay)
Stromatinia betulae (Woron.) Naum. Xinjiang

***Betula albo-sinensis* Burk.**

Lenzites betulina (L.) Fr. (Decay) Shanxi
Melampsoridium betulinum (Desm.) Kleb. (Leaf rust) Hebei
Stereum fasciatum Schw. (Decay) Shanxi

***Betula chinensis* Maxim.**

Dothidella betulina (Fr.) Sacc. (Dieback) Liaoning
Euryachora betulina (Fr.) Schröt. (Dieback) Liaoning
Favolus squamosus (Huds. ex Fr.) Ames (Decay) Heilongjiang
(=Polyporus squamosus (Huds.) Fr.)
Fomes fomentarius (L. ex Fr.) Kickx. (Decay) Shanxi Gansu
Phellinus igniarius (L. ex Fr.) Quél. (Heartwood white sponge rot) Shanxi

Piptoporus betulinus (Bull ex Fr.) Karst. (Brown rot) Shanxi Gansu
Septoria chinensis Miura (Leaf spot) Liaoning

***Betula dahurica* Pall.**
Eriophyes sp. (Leaf felt spot) Neimeng
Gloeosporium sp. (Peony anthracnose) Neimeng
Melampsoridium sp. (Leaf rust) Neimeng
Phyllactinia alni Yu et Han (Powder mildew) Liaoning

***Betula ermanii* Cham**
Melampsoridium betulinum (Desm.) Kleb (Leaf rust) Jilin

***Betula japonica* Sieb. et Winkl.**
Fomitopsis pinicola (Sw. ex Fr.) Karst (Brown cube rot) Gansu
Piptorus betulinus (Bull. ex Fr.) (Brown rot) Gansu
Pycnoporus cinnabarinus (Jacq.) Karst. (Decay) Qinghai

***Betula luminiera* H. Winkl.**
Phyllactinia alni Yu et Han (Powder mildew) Guizhou

***Betula pendula* Roth.**
Eriophyes rudis longissetosus Nal. (Leaf felt spot) Xinjiang
Eriophyes rudis typicus Nal. (Leaf felt spot) Xinjiang
Fomes fomentarius (L. ex Fr.) Kickx. (Decay) Xinjiang Heilongjiang
Fomes rufolaccatus Lloyd (Decay) Xinjiang
Fomitopsis pinicola (Sw. ex Fr.) Karst. (Brown rot) Xinjiang
Ganoderma applanatum (Pers.) Pat. (Decay) Heilongjiang
Gloeosporium betulinum Westend. (Anthracnose) Hebei Xinjiang
Melampsoridium betulinum (Desm.) Kleb. (Leaf rust) Xinjiang
Microsphaera betulae Magn. (Powder mildew) Xinjiang
Phellinus igniarius (L. ex Fr.) Quél. (Heartwood white sponge rot) Heilongjiang Xinjiang
Phyllactinia alni Yu et Han (Powder mildew) Xinjiang
Piptoporus betulinus (Bull. ex Fr.) Karst. (Brown rot) Xinjiang
Podosphaera erineophila Naum. (Powder mildew) Xinjiang

***Betula platyphylla* Suk.**
Cerrena unicolor (Bull. ex Fr.) Murr. (White rot) Shanxi Neimeng
(=Coriolus unicolor (Bull. ex Fr.) Pat.
Coriolus hirsutus (Wulf. ex Fr.) Quél. (White rot) Neimeng
Coriolus versicolor (L. ex Fr.) Quél. (White rot) Neimeng
Cytospora betulina Ehrenb. (Canker) Neimeng
Favolus squamosus (Huds. ex Fr.) Ames (Decay) Heilongjiang
Fomes fomentarius (L. ex Fr.) Karst. (Decay) Shanxi Neimeng
Fomitopsis pinicola (Sw. ex Fr.) Karst. (Brown cube rot) Neimeng
Fumago vagans pers. (Black mildew) Liaoning
Hirschioporus pargamenus (Fr.) Bond. et Sing. (White rot) Neimeng
Lenzites betulina (L.) Fr. (White rot) Shanxi Neimeng
Melampsoridium betulinum (Desm.) Kleb. (Leaf rust) Heilongjiang Neimeng
Piptoporus betulinus (Bull ex Fr.) Karst. (Brown rot) Shanxi Dongbei Neimeng Tibet
Pycnoporus cinnabarinus (Jacq.) Karst. (Decay) Neimeng
Schizophyllum commune Fr. (Decay) Neimeng
Stereum hirsutum (willd.) Fr. (Decay) Shanxi Neimeng
Trametes suaveolens (L.) Fr. (White rot) Neimeng

***Betula tianschanica* Rupr.**
Cerrena unicolor (Bull. ex Fr.) Murr. (Decay) Xinjiang
(=Daedelea unicolor Fr.)
Coriolus fibula (Fr.) Quél (Decay) Xinjiang

(=Polystictus fibula (Sow.) Fr.

Coriolus versicolor (L. ex Fr.) Quél. (Decay) Xinjiang
Eriophyes rudis longisetosus Nal. (Leaf felt spot) Xinjiang
Eriophyes rudis typicus Nal. (Leaf felt spot) Xinjiang
Fomes fomentarius (L. ex Fr.) Kickx. (Decay) Xinjiang
Fomes rufolaccatus Lloyd (Decay) Xinjiang
Fomitopsis pinicola (Sw ex Fr.) Karst. (Brown cube rot) Xinjiang
Fumago sp. (Black mildew) Xinjiang
Ganoderma applanatum (Pers. ex Gray) Pat. (Decay) Xinjiang
Hypoxypora personata Fr. Xinjiang
Inonotus obliquus (Fr.) Pit. (Decay) Xinjiang
Lenzites tricolor (Bull.) Fr. (Decay) Xinjiang
Melampsoridium betulinum (Desm.) Kleb. (Leaf rust) Xinjiang
Microsphaera betulae Magn. (Powder mildew) Xinjiang
Nectria sp. (Dieback) Xinjiang
Phellinus igniarius (L. ex Fr.) Quél (White rot) Xinjiang
Phellinus robustus (Karst.) Bourd. et Galz. (Decay) Xinjiang
Phyllactinia alni Yu et Han (Powder mildew) Xinjiang
Podosphaera eriophila Naum. (Powder mildew) Xinjiang
Polyporellus varius (Pers ex Fr.) Karst. (Decay) Xinjiang

(=Polyporus varius (Pers.) Fr.

Pycnoporus cinnabarinus (Jacq.) Karst. (Decay) Xinjiang
Septoria betulae West. (Leaf spot) Xinjiang
Stereum fasciatum Schw. (Decay) Xinjiang
stromatinia betulae (Woron.)Naum.
Trametes suaveolens (L.) Fr. (Decay) Xinjiang

***Betula utilis* D. Don**

Fomes fomentarius (L. ex Fr.) Kickx. (Brown rot) Tibet
Ganoderma applanatum (Pers.) Pat. (Decay) Tibet
Melampsoridium betulium (Desm.) Kleb. (Leaf rust) Tibet
Phellinus igniarius (L. ex Fr.) Quél (Heartwood white sponge rot) Tibet
Piptoporus betulinus (Bull ex Fr.) Karst. (Brown rot) Tibet

***Bischoffia javanica* Bl.**

(=Bischoffia trifoliata Hook f.)

Aithaloderma clavatispora Syd. Taiwan
Cercospora bischofiae Yamam. (Leaf spot) Taiwan
Pestalotia adusta Ell. et Ev. (Leaf spot) Taiwan
Phellinus williamsii (Murr.) Pat. (Decay) Taiwan
Uncinula bischofiae Wei (Powder mildew) Sichuan

***Brachystachyum densiflorum* (Rendle) Keng**

(=Arundinaria densiflora Rendle.)

Balansia take (Miyake) Hara (Black leaf spot) Zhejiang
Coccostroma arundinariae (Hara) Teng (Black leaf spot) Zhejiang

(=Coccostromopsis arundinariae (Hara) Teng)

Cuscuta japonica Choisy (Cuseata high plant mistletoe) Zhejiang
Phyllachora phyllostachydis Hara (Black swell) Zhejiang
Shiraria bambusicola P. Henn. (Bamboo swell) Sichuan
Stereostratum corticioides (Berk. et Br.) Magn. (Leaf rust) Zhejiang

Broussonetia spp.

Cercospora broussonetiae Chupp et Linder (Leaf spot) Guangxi
Phyllactinia brossonetiae-kaempferi Saw. (Powder mildew) Shanxi Hunan Shanxi

Septobasidium bogoriense Pat. (Plaster) Jiangsu Zhejiang Guangxi

***Broussonetia papyrifera* (L.) Vent.**

Loranthus parasiticus (Linn.) Merr (Mistletoe) Guangxi
Myxosporella miniata Sacc. Jiangxu
Nothopatella chinensis Miyake (Dieback) Hebei
Phakopsora fici-erectae Ito et Otani (Leaf rust) Jiangsu Taiwan
Phomopsis broussonetiae (Sacc.) Diet. (Stem spot) Jiangsu
Phyllactinia brossonetiae-kaempferi Saw. (Powder mildew) Jiangsu Jiangxi Henan Sichuan Shanxi Shanxi
Septobasidium bogoriense Pat. (Plaster) Taiwan
Septogloeummore Briosi et Cav (Brown leaf spot) Jiangsu
Uredo broussonetiae Saw. (Leaf rust) Taiwan

***Burretiodendron hsienmu* Chun et How**

Cercospora sp. (Leaf spot) Guangxi
Phyllosticta sp. (Leaf spot) Guangxi

Buxus sp.

Meliola buxicola Doidge (Black mildew) Yunnan

***Buxus sinica* (Rehd. et wits) cheng.**

Gloeosporium pachybasium Sacc. (Leaf spot) Shanxi
Macrophoma candollei (B. et Br.) Berl. et Vogl. (Leaf spot) Shanxi

***Cajanus cajan* (L.) Millsp.**

Cercospora instabilis Rangel (Leaf spot) Guangxi
Colletotrichum gloeosporioides Penz. (Anthracnose) Yunnan
(=Colletotrichum cajani Rangel)
Oidium sp. (Powder mildew) Yunnan
Sclerotinia sclerotiorum (Lib.) de Bary (Sclerotiorum) Guangxi Sichuan
Uromyces dolicholi Arth. (Leaf rust) Guangdong Taiwan

Calligonum spp.

Leptothyrium bornmiilleri Magn. (Dieback) Xinjiang
leveillula polygonacearum Golov. (Powder mildew) Xinjiang
Trichocladia atraphaxis Golov. (Powder mildew) Xinjiang

***Calophyllum inophyllum* L.**

Pestalotia calabae Westend. (Leaf spot) Guangdong

Camellia spp.

Asterina camelliae Syd. (Black mildew) Yunnan
Elytranthe fordii (Hance) Merr. (Elytranthe mistletoe) Guangxi
Phaeosaccardinula javanica (Zimm.) Yamam. (Black mildew) Taiwan

***Camellia japonica* L.**

Cornyeum camelliae Mass. (Leaf spot) Jilin
Glomerella cingulata (Stonem.) Spauld et Schrenk (Anthracnose) Jilin Areas along the Yangtze River
(Asexual: Colletotrichum gloeosporiodes Penz.)
(=Collectotrichum camelliae Mass.)
Leoptothyrium camelliae P. Henn. (Leaf spot) Sichuan
Pestalotia guepini Desm. (Cycle spot) Sichuan
Phyllosticta theae Speschn. (Leaf spot) Sichuan

***Camellia oleifera* Abel**

Agricodochium camelliae Liu. Wei et Fan (Soft rot) Jiangxi Hunan
Capnodium sp. (Smoke mold) Jiangxi Hunan Guangxi
Cassytha filiformis L. (Cassytha mistletoe) Guangxi
Cercospora sp. (Leaf spot) Hunan Jiangxi Anhui

Cephaleuros virescens Kunze (Algae) Henan Anhui Guangdong Guangxi Hunan Yunnan Fujian Taiwan Guizhou
Coccochorina hottai Hara (Canker) Henan
Corticium sp. (Camellia half witches' broom) Zhejiang Hunan Jiangxi Guangxi Guangdong
Cuscuta sp. (Cuseata high plant mistletoe) Jiangxi
Deuterophoma sp. (Leaf spot) Guangxi
Diplodia sp. (Leaf spot) Guangxi
Exobasidium gracile (Shirai) Syd. (Camellia leaf gall) Zhejiang Henan Jiangxi Hunan Guangxi Anhui Taiwan Guizhou Sichuan Guangdong Fujian
Glomerella cingulata (Stonem.) Spauld et Schrenk (Anthracnose) Jiangsu Zhejiang Henan Shanxi Hunan Guangxi Sichuan Anhui Guizhou Jiangxi
(=Guignardia camelliae (Cooke) Butl.)
Loranthaus chinensis DC. (Mistletoe) Guangxi
Loranthus parasiticus (Linn.) Merr. (Mistletoe) Zhejiang Jiangxi Hunan Guangxi Sichuan
Loranthus sampsoni Hance (Mistletoe) Jiangxi Guangxi Hunan Guizhou
Marothecium camelliae
Loranthus yadoriki sing. et Zucc. (Camphor mistletoe) Guangxi Hunan Hubei Fujian Sichuan Guangdong
Meliola camelliae (Catt.) Sacc. (Black mildew) Jiangsu Guangxi Anhui Hunan Sichuan Guizhou
Neocapnodium sp. (Black mildew) Zhejiang Hunan Jiangxi Anhui Sichuan Fujian Guizhou Yunnan
Pestalotia theae Saw. (Gray leaf spot) Henan Guangxi
Phyllosticta theaefolia Hara (Gray leaf spot) Anhui Hunan Guangxi
Physalospora sp. (Shoot blight) Hunan Guangdong
Piggotia sp. (Brown leaf spot) Hunan
Sclerotium rolsii Sacc. (Southern blight) Zhejiang Jiangsu Hunan Anhui Sichuan
Septoria sp. (Gray leaf spot) Hunan Anhui Guizhou
Sphaceloma sp. (Scab) Guangdong Guangxi
Viscum sp. (True mistletoe) Jiangxi

***Camellia reticulata* Lindl. f. simplex sealy**

Pestalotia guepini Desm. (Leaf spot) Jilin
(=Pestalotia puttemansii P. Henn.)
Verticillium albo-atrum Reinke. Et Berth. (Wilt) Jilin

***Camellia sasanqua* Thunb.**

Aithaloderma clavatispora Syd. Taiwan
Colletotrichum gloeosporioides Penz. (Anthracnose) Hunan
(=gloeosporium thear-sinensis Miyake)
Exobasidium gracile (Schirai) Syd. (Leaf galls) Anhui Zhejiang Fujian Taiwan Guangxi Guangdong Hunan Jiangxi
Pestalotia guepini Desm. (Leaf spot) Jiangsu
Phaeosaccardinula javanica (Zimm.) Yamam. (Black mildew) Taiwan

***Camellia sinensis* O. Ktze.**

E. vexans Mass (Camellia leaf galls) Hunan Jiangxi Zhejiang Fujian
E. reticulatum Ito et Saw. (Camellia leaf galls) Hunan Jiangxi Zhejiang Fujian

***Camptotheca acuminata* Decne.**

Cercospora camptothecae Tai (Leaf spot) Sichuan
Phyllactinia corylea (Pers.) Karst. (Powder mildew) Sichuan

Caragana spp.

Cucurbitaria sinica teng (Dieback) Hebei Qinghai
Mazzania tranzschelii (Woronich.) Teng (Caragana gall) Gansu

Melasmia caraganae Jacz. (Black mole) Hebei
Septoria caraganae P. Henn. (Leaf spot) Xinjiang
Uromyces genistae-tinctoriae (Pers.) wint. (Leaf rust) Neimeng Xinjiang
Uromyces laburni (DC.) Otth (Leaf rust) Hebei Xinjiang

***Caragana arborescense* (Amm.) Lam.**
Fumago vagans Pers. (Black mildew) Liaoning Jilin
Microsphaera caraganae Magn. (Powder mildew) Liaoning
Microsphaera grossulariae (Wallr.) Lév. (Powder mildew) Jilin

***Caragana fructescens* (L.) Medic.**
Uromyces laburni (DC.) Otth (Leaf rust) Dongbei Hebei

***Caragana frutex var. latifolia* Schneid.**
Melasmia caraganae Jacz. (Black mole) Hebei
Uromyces laburni (DC.) Otth (Leaf rust) Hebei

***Caragana jubata* (Pall.) Poir.**
Septoria subiniae Pat. (Leaf spot) Tibet

***Caragana manshurica* Kom.**
Microsphaera grosulariae (Wallr.) Lév (Powder mildew) Heilongjiang

***Caragana microphylla* (Pall) Lam.**
Erysiphe sp. (Powder mildew) Ningxia
Fumago sp. (Black mildew) Ningxia
Inonotus hisipidus (Bull Carpinus spp.. ex Fr.) Karst. (Leaf tip rot)
Mazzantia tranzschelii (Woronich.) Teng (Caragana gall) Ningxia
Microsphaera caraganae Magn. (Powder mildew) Neimeng
Microsphaera pseudolonicerae (Salm.) Blum. (Powder mildew) Sichuan
Petrakomyces caraganicola Pai et Hou (Black mole) Neimeng
Uromyces genistae-tinctoriae (Pers.) Winter. (Leaf rust) Neimeng Ningxia
Uromyces laburni (DC.) Otth (Leaf rust) Dongbei Hebei
Virus Ningxia

***Caragana pygmaea* (L.) DC.**
Melasmia caraganae Jacz. (Black mole) Hebei

***Caragana rosea* Turcz.**
Uromyces laburni (DC.) Otth (Leaf rust) Dongbei Hebei

***Caragana sinica* (Buchoz.) Rehd.**
Melasmia caraganae Juzz. (Black mole) Hebei
Microsphaera grossulariae (Wallr.) Lév. (Powder mildew) Hebei
Microsphaera longissima M. Y. Li (Powder mildew) Beijing
Uromyces laburni (DC.) Otth (Leaf rust) Dongbei Hebei Taiwan

***Caragan sophorifolia* Ben.**
Uromyces laburni (DC.) Otth (Leaf rust) Dongbei
Melampsoridium carpini (Fuck.) Diek (Leaf rust) Guizhou
Phyllactinia guttata (Wallr. ex Fr.) (Powder mildew)
(=Phyllactinia corylea (Pers.) Karst.)
(=Phyllactinia corylea Karst. em. Salm.)
(=Phyllactinia guttata Wallr.)
(=Phellactinia suffulata (Rebent.) Sacc.) Guizhou

Carpinus turczaninowii Hance
Nectria cinnabarina (Tode) Fr. (Canker)
Taphrina carpini (Rostr.) Juhans. Shanxi
Uncinula verniciferae P. Henn. (Powder mildew) Hebei
Uromyces genistae-tinctoriae (Pers.) Wint. (Leaf rust) Neimeng

***Cassia siamea* Lam**

Meliola aethiops Sacc. (Black mildew) Taiwan

Castanea spp.

Diplodia castaneae Sacc. (Leaf spot) Fujian

Endothia parasitica (Murr.) P. J. et H. W. Anders. (Chestnut blight) Hebei Guangxi Jiangsu Beijing Liaoning

Phyllactinia roboris (Gachet) Blum. (Powder mildew) Yunnan

(=Erysiphe roboria Gachet)

(=Erysiphe quercus Merat)

(=Phyllactinia corylea (Pers.) Karst.)

(=Phyllactinia quercus (Merat) Homma)

(=Phyllactinia suffulta (Rebent.) Sacc. f. quercina Jacz.)

Pucciniastrum castaneae Diet. (Leaf rust) Jiangxi Anhui Guizhou

***Castanea crenata* Sieb. et zucc.**

Coniothyrium tirolensis Bub. (Leaf spot) Liaoning

Endothia parasitica (Murr.) P. J. et H. W. Anders. (Chestnut blight) Liaoning

Limacinia chenii Saw. et Yamam. (Black mildew) Taiwan

Neocapnodium tanakae (Shirai et Hara) Yamam. (Black mildew) Taiwan

(=Capnodium tanakae Shirai et Hara)

Pucciniastrum castaneae Diet. (Leaf rust) Taiwan

Scorias communis Yamam. (Black mildew) Taiwan

Triposporiopsis spinigera (Höhn.) Yamam. (Leaf spot) Taiwan

***Castanea henryi* (Skan) Rend. et Wils.**

Cronartium quercuum (Bonk.) Miyabe (Leaf rust) Anhui

Pucciniastrum castaneae Diet. (Leaf rust) Hunan

***Castanea mollissima* Bl.**

Actinopelte japonica Sacc. (Leaf spot) Liaoning

Armillariella mellea (Vahl ex Fr.) Karst. (Root rot) Jilin

Ascochyta quercus Sacc. et Speg. (Gray leaf spot) Shanxi

Botryodiplodia theogromae Pat. Hebei

Cephaleuros virescens Kunze (Algae) Guangdong

Coryneum sp. (Dieback) Shandong

Coryneum kunzei Corda var. castaneae Sacc. et Roum. (Leaf spot) Jiangsu

Cronartium quercuum (Berk.) Miyabe (Leaf rust) Jiangxi Zhejiang Henan Shanxi Anhui Sichuan Yunnan

Elytranthe bibracteolata (Hance) le. comte var. sinensis le Comte (Elytranthe mistletoe) Guangxi

Elytranthe fordii (Hance) Merr. (Maple mistletoe) Guangxi

Endothia parasitica (Murr.) P. J. et H. W. Anders. (Cheastnut blight) Hebei Jiangsu Zhejiang Henan Guangdong Hunan Jiangxi Anhui Shanxi Shandong Liaoning

Glomerella cingulata (Stonem.) Spauld. et Schrenk (Anthracnose)

Hypoxylon howeianum Peck (White rot) Shanxi

Laetiporus sulphureus (Bull. ex Fr.) Bond. et Sing (Brown rot) Shanxi

Loranthus parasiticus (Linn.) Merr. (Mistletoe) Guangxi

Loranthus yadoriki Sieb et zucc. (Loranthus mistletoe) Guangxi Fujian Hunan Hubei Zhejiang Sichuan Shanxi

Macrophoma kawatsukai Hara (Leaf spot) Sichuan

Macrophomina phaseoli (Maubl.) Ashby (Stem rot) Jiangsu

Melanconis monodia Tul. (Dieback) Guangdong

Meliola sp. (Black mildew) Liaoning

Microsphaera alni (Wallr.) Salm. (Powder mildew) Liaoning Jilin Hebei Jiangsu Henan Jiangxi Sichuan Guizhou Yunnan

Monochaetia pachyspora Bub. (Leaf spot) Henan Shanxi Guangxi
Monochaetia turgida (Atk.) Sacc. (Leaf spot) Liaoning
Penicillium sp. (Green mold) Henan
Phyllactinia roboris (Gachet) Blum. (Powder mildew) Liaoning Jilin Heilongjiang Hebei Jiangsu Zhejiang Shandong Anhui Guangxi Guangdong Henan Sichuan Guizhou Yunnan Jiangxi
Phyllosticta castaneae Ell. et Ev. (Leaf spot) Jilin
Phyllosticta maculiformis (Pers.) Sacc. (Leaf spot) Henan Guangdong
Pucciniastrum castaneae Diet. (Leaf rust) Henan Guangdong Hunan Sichuan Yunnan Guangxi
Taphrina rubro-brunnea (Peck.) Sacc. (Leaf eviel) Shanxi
Trichothecium roseum (Bull.) Link (Red powder mildew) Henan
Valsa sordida Nits. (Canker) Dongbei
(Asexual : Cytospora chrysosperma (Pers.) Fr.)
Virus (Leaf eviel) Liaoning
Viscum japonicum Thunb. (Mistletoe) Fujian

***Castanea seguinii* Dode**
Phyllactinia roboris (Gachet) Blum. (Powder mildew) Jiangsu
Pucciniastrum castaneae Diet. (Leaf rust) Hunan

Castanopsis spp.
Erysiphe fagacearum Zheng et Chen (Powder mildew) Yunnan Sichuan
Trametes kusanoana Imaz. (Brown pocket rot) Hainandao
Typhulochaeta japonica Ito et Hara (Powder mildew) Sichuan

***Castanopsis concolor* Rehd et Wils.**
Erysiphe fagacearum Zheng et Chen (Powder mildew) Yunnan Sichuan

***Castanopsis delavayi* Franch.**
Erysiphe fagacearum Zheng et Chen (Powder mildew) Yunnan

***Castanopsis hystrix* A. DC.**
Meliola castanopsina Yamam. (Black mildew) Taiwan
Meliola quercina Pat. (Black mildew) Yunnan
Phellinus robustus (Karst.) Bourd et Galz. (Sapwood brown rot) Hunan

***Castanopsis kaweakamii* Hay.**
Meliola shiiae Yamam. (Black mildew) Taiwan

***Castanopsis sclerophylla* (Lindl.) Schottky**
Chaetothyrium sinense Teng (Black mildew) Zhejiang

***Castanopsis stipitat* (Hayata) Kaneh et Hatus.**
Meliola castanopsina Yamam. (Black mildew) Taiwan
Meliola fusispora Yamam. (Black mildew) Taiwan
Meliola shiiae Yamam. (Black mildew) Taiwan

***Castanopsis subacuminata* Hayata**
Meliola subacuminata Yamam. (Black mildew) Taiwan

***Castanopsis taiwaniana* Hayata**
Meliola taiwaniana Yamam. (Black mildew) Taiwan

***Castanopsis tibetana* Hance**
Irenina castanopsis Hansf. Guizhou
Meliola castanopsidis Hansf. (Black mildew) Guizhou
Pseuderiospora castanopsidis Keissl. (Black mildew) Hunan

Casuarina spp.
Pseudomonas solanacearum E. F. Smith (Casuarina wilt) Fujian

***Casuarina equisetifolia* L.**
Cassytha filiformis L. (Cassytha mistletoe) Guangdong

Pseudomonas solanacearum E. F. Smith (southern bacterial wilt) Guangdong Guangxi Fujian

Catalpa spp.

Cerrena unicolor (Bull. ex Fr.) Murr. (Decay) Jilin Heilongjiang

Coriolus hirsutus (Wulf. ex Fr.) Quel. (Decay) Hebei Xinjiang

***Catalpa bungei* C. A. Mey**

Cercospora catalpae wint. (Leaf spot) Henan

Cytospora sp. (Canker) Xinjiang

Phoma sp. (Damping off) Henan

Sclerotium rolfsii Sacc. (Southern blight) Henan

***Catalpa ovata* Don**

Alternaria catalpae (Ell. et Mart.) Joly (Leaf spot) Jilin Henan

Cercospora catalpae Wint. (Leaf spot) Jiangsu Henan Ningxia

Macrosporium catalpae Ell. et Mart. (Leaf spot) Henan

Meloidogyne marioni Goodey (Ocen root-knot nematode) Henan

Phyllactinia corylea (Pers.) Karst. (Powder mildew) Henan Ningxia

Phyllosticta catalpae Ell.et Martin (Leaf spot) Jilin

Schizophyllum commune Fr. (Decay) Tibet

Sclerotium rolfsii Sacc. (Southern blight)

Sphaerotheca catalpae Wang ex Z. Y. Zhao (Powder mildew) Shanxi Xinjiang

***Catalpa speciosa* warder**

Diplodia catalpae Speg. (Leaf spot) Shandong

Phyllactinia corylea (Pers.) Karst. (Powder mildew) Jiangsu

Sphaerotheca catalpae Wang ex Z. Y. Zhao (Powder mildew) Xinjiang

Trametes hirsute (Walf. ex Fr.) Pil. (Decay) Xinjiang

***Celastrus orbiculatus* Thunb.**

Mycosphaerella sp. (Leaf spot) Liaoning

Phyllosticta sp. (Leaf blight) Liaoning

Uncinula sengokui Salm. (Powder mildew) Liaoning Fujian

Celtis spp.

Cercospora profuse H. et P. Syd. (Leaf spot) Huabei

Pleochaeta shiraiana (P. Henn.) Kimbr et Korf (Powder mildew) Shanxi Jiangxi Fujian Guizhou Henan Jiangsu

(=Unicinula shiraiana P. Henn.)

(=Uncinulopsis shiraiana (P. Henn.) Hara

(=Uncinulopsis polychaeta (Berk. et Curt.) Homma emend. Wei)

Uncinula clintonii Peck (Powder mildew) Jiangsu Yunnan

***Celtis biondii* Pamp.**

Uncinula clintonii Peck (Powder mildew) Sichuan

***Celtis bungeana* Bl.**

Coniothyrium celtidicola Miura (Gray leaf spot) Liaoning

Coriolus versicolor (L. ex Fr.) Quél. (Decay) Jiangxi

Phyllactinia corylea (Pers. Karst. (Powder mildew) Hebei

Pleochaeta shiraiana (P. Henn.) Kimbr. et Korf (Powder mildew) Henan

Uncinula clintonii Peck (Powder mildew) Henan

***Celtis koraiensis* Nakai**

Coniothecium album Miura (White leaf spot) Liaoning

***Celtis sinensis* Pers.**

Cercospora spegazzinii Sacc. (Leaf spot) Guangdong

Neocapnodium tanakae (Shirai et Hara) Yamam. (Black mildew) Taiwan

Pleochaeta shiraiana (H. Penn) Kimbr. et Korf (Powder mildew) Shanxi Jiangxi Guangxi Fujian Sichuan Yunnan Jiangsu Zhejiang Taiwan Guizhou

Pseudoperonospora celtidis (Waite) Wils. (Downy mildew) Taiwan
Uncinula clintonii Peck (Powder mildew) Jiangsu Zhejiang Guangxi Fujian Sichuan Yunnan Taiwan

***Centella asiatica* (L.) Urban**
Puccinia centellae M. M. Chen & Wang (Leaf rust) Tibet

***Cephalotaxus fortunei* Hook. f.**
Pestalotia sp. (Leaf spot) Shanxi

***Cephalotaxus sinensis* (Rehd. et wils.) Li**
Pestalotia sp. (Leaf spot) Shanxi

***Cercis chinensis* Bunge**
Cercospora chionea Ell. et Ev. (Leaf spot) Jiangsu Henan Anhui
Phoma cercidicola P. Henn. (Leaf spot) Henan

Chamaecyparis sp.
Macrophomina phaseoli (Maubl.) Ashby (Stem decay) Jiangsu

***Chosenia macrodepis* (Turcz.) Kom.**
Coriolus hirsutus (Wulf. ex Fr.) Quél. (Decay) Neimeng
Funalia trogii (Berk.) Bond. et Sing. (Decay)
Stereum hirsutum (Willd.) Fr. (Decay) Neimeng

***Cimicifuga heracleifolia* Kom.**
Coleosporium cimicifugatum Thüm. (Leaf rust) Liaoning

Cinnamomum spp.
Albugo tragopogonis (Pers.) Gray (White rust) Sichuan
Exobasidium sawadae Yamada (Leaf gall) Guangdong Guangxi (=Elaeodema cinnamomi Syd.)
Glomerella cingulata (Stonem.) Spauld. et Schrenk (Anthracnose) Sichuan
Phyllactinia roboris (Gachet) Blum. (Powder mildew) Yunnan
Phyllosticta cirsii Desm. (Leaf spot) Sichuan
Phyllosticta nobilis Thüm. (Leaf spot) Sichuan
Puccinia cinnamomi Tai (Leaf rust) Sichuan
Puccinia cinnamomicola Cumm. (Leaf rust) Guangxi
Puccinia obtegens (Link) Tul. (Leaf rust) Sichuan
Pucciniastrum castaneae Diet. (Leaf rust) Anhui Jiangxi Yunnan
Sawadaea bicornis (Wallr. ex Fr.) Homma (Powder mildew)
Sawadaea polyfida (Wei) Zheng et Chen (Powder mildew) Sichuan (=Uncinula polyfida Wei)
Trametes hirsute (Wulf. ex Fr.) Pilát (Decay) Sichuan

***Cinnamomum appelianum* Hand-Mazz.**
Capnodium footii Berk. et Desm. (Black mildew) Hunan

***Cinnamomum camphora* (L.) Presl.**
Asterina cinnamomi Syd. Sichuan
Capnodium sp. (sooty mold) Jiangxi
Cephaleuros virescens Kunze (Algal spot) Jiangxi Guangdong Guangxi Taiwan Sichuan Fujian
Corticium centrifugam (Lév) Bres. (Fisfeye fruit rot of Cinnamomum) Taiwan (Asexual: Scleroium rolfsii Sacc.)
Corticium sasakii (Shirai) Matsum (White butt rot) Taiwan
Elytranthe bibracteolata (Hance) le Comte var. sinensis le Comte (Elytranthe mistletoe) Guangxi
Eriophyes sp. (Leaf felt spot) Guangdong Jiangxi Fujian Guangxi
Exobasidium sawadae Yamada (Leaf gall) Guangdong Guangxi Hunan Sichuan Taiwan

Glomerella cingulata (Stonem.) Spauld. et Schrenk (Anthracnose) Guangdong Guangxi Taiwan Hunan Shanxi Henan

Hypochnus cinnamomi Saw Taiwan

Inonotus triqueter (Alb. et Schw. et Fr.) Karst. (Decay) Tibet

Loranthus yadoriki Sieb. et Zucc. (Mistletoe)

Meliola zigzag Berk. et Curt. (Black mildew) Guangxi Guangdong Zhejiang Fujian Hunan Sichuan Shanxi Hubei

Phaeolus schweinitzii (Fr.) Pat. (Brown cube rot) Tibet

Phellinus linteus (Berk. et Curt.) Teng (Decay) Tibet

Phellinus williamsii (Murr.) Pat. (Decay) Taiwan

Pucciniastrum castaneae Diet. (Leaf rust) Fujian

Sclerotium cinnamomi Saw. (Sclerotia) Taiwan

Septobasidium albidum Pat. (Felt fungus) Fujian Taiwan

***Cinnamomum camphora var. nominale* Hayta**

Cercospora cinnamomi. Saw. et Karst. (Leaf spot) Taiwan

Erysiphe cichoracearum DC. (Powder mildew) Taiwan (Asexual :Oidium tabaci Thüm.)

***Cinnamomum cassia* Neos et. Eberm**

Botryodiplodia theobromae Pat. (Gray leaf spot) Guangdong Guangxi

***Cinnamomum cassia* Presl**

Cephaleuros virescens Kunze (Algal spot) Guangdong

Exobasidium sawadae Yamada (Leaf gall) Guangdong Guangxi

***Cinnamomum glanduliferum* (Wall.) Nees**

Exobasidium sawadae Yamada forma brunnea (Keissl) Tai (Leaf gall) Hunan

Glomerella cingulata (Stonem.) Spauld. et Schrenk. (Anthracnose) Guangdong Guangxi

Cinnamomum humanni (Nees.) Bl.

Exobasidium sawadae Yamada. (Leaf gall) Guangdong Guangxi

Phyllosticta sp. (Leaf dry spot) Guangdong

***Cinnamomum jensenianum* Hand-Mazz.**

Chlorocyphella aeruginascens (Karst.) Keissl. Hunan

***Cinnamomum lourerii* Nees.**

Aecidium cinnamomi Racib. (Leaf rust) Taiwan

***Cinnamomum micranthum* Hayata**

Exobasidium sawadae Yamada (Leaf gall) Taiwan

***Cinnamomum pedunculatum* Nees**

Exobasidium sawadae Yamada (Leaf gall) Guangdong

***Cinnamomum pseudo-lourerii* Hayata**

Armatella longispora Yamam. (Black mildew) Taiwan

***Cinnamomum randaiensis* Hayata**

Armatella longispora Yamam. (Black mildew) Taiwan

***Cladrastis platycarpa (Maxim)* Mak.**

Loranthus maclurei Merr. (Yellowwood mistletoe) Guangdong Guangxi Fujian Guizhou

***Cleidiocarpon cavaleriei (Lévl.)* Airy-Shaw**

(=Sinopimelodendron kwanysiense Tsiang)

Pseudomonas solanacearum E. F> Smith (Southern bacterial wilt) Guangdong Guangxi

Clematis sp.

Puccinia lakanensis M. M. Chen & Wang (Leaf rust) Tibet

***Cintonia udensis* Trautv et Mey.**

Puccinia clintoniae-udensis Bub. var. tibetica M. M. Chen (Leaf rust) Tibet

Cornus spp.

Pestalotia corni Allesch. (Leaf spot) Hebei

Phellinus igniarius (L. ex Fr.) Ouél. (White rot) Hebei
Ramularia angustissima Sacc. Hebei
Septoria cornicola Desm. (Leaf spot) Zhejiang

Cornus controversa **Hemsl**

Cercospora cornicola Tray et Earle (Leaf spot) Shandong Tibet
Coriolus versicolor (L. et Fr.) Quél (Decay) Tibet

Cornus officinalis **Sieb et Zucc.**

Septobasidium bogoriense Pat. (Plaster) Zhejiang

Corylus spp.

Cususta japonica Choicy (Cuseata high plant mistletoe) Liaoning
Fumago vagans Pers. (Black mildew) Liaoning
Ghnomoniella coryli (Batsch) Sacc. (Black mildew) Liaoning Shanxi
Mamiania coryli (Batsch ex Fr.) Ces ex de Not. (Black leaf spot) Liaoning Zhejiang Shanxi Anhui
Microsphaera variabilis Yu (Powder mildew) Liaoning Sichuan
(=Microsphaera alni (Wallr.) Salm.)
(=Macrosphaera coryli (Jacz.) Golov.)
Phyllactinia guttata (Wallr. ex Fr.) Lév. em. Yu (Powder mildew) Jiangxi
Phyllosticta corylaria Sacc. (Leaf spot) Liaoning
Taphrina coryli Nishida. (Leaf eviel) Liaoning

Corylus heterophylla **Fisch**

Ascochyta cortyli Sacc. (Leaf spot) Jilin
Botrytis cinerea Pers. (Gray leaf spot) Jilin
Fumago vagans. Pers. (Black mildew) Jilin
Gnomoniella coryli (Batsch) Sacc. (Anthracnose) Jilin Heilongjiang Shanxi
Mamiania coryli (Batsch ex Fr.) Ces et de Not. (Black leaf spot) Heilongjiang Dongbei Hebei Neimeng
Mamiania fimbriata (Pers.) Ces. et de Not (Black leaf spot) Henan
Microsphaera variabilis Yu (Powder mildew) Liaoning Jilin Hebei Henan Shanxi Jiangxi Sichuan Neimeng
Mycosphaerella corylina Karst. (Leaf spot) Hebei
Phyllactinia guttata (Wallr. ex Fr.) Lév. em. Yu (Powder mildew) Jilin Hebei
Phyllosticta corylaria Sacc. (Leaf spot) Jilin
Pucciniastrum coryli Kom. (Leaf rust) Liaoning Jilin

Corylus heterophylla var. sutchuenensis **Franch.**

Ghomoniella coryli (Batsch) Sacc. (Anthracnose) Shanxi
Microsphaera variabilis Yu (Powder mildew) Sichuan

Corylus mandshurica **Maxim. et Rupr.**

Cercospora corylina Ray (Leaf spot) Jilin
Coriolus versicolor (L. ex Fr.) Quél. (Decay) Heilongjiang
Microsophaera variabilis Yu (Powder mildew) Sichuan
Pucciniastrum coryli Kom. (Leaf rust)
Septoria corylina Peck. (Leaf spot) Jilin

Corylus yunnanensis **(Franch.) A. Camus**

Microsphaera verruculosa Yu et Lai (Powder mildew) Sichuan

Cotinus coggygria **Scop.**

Pileolaria cotini-coggygriae Tai et Cheo (Leaf rust) Henan
Uncinula verniciferae P. Henn. (Powder mildew) Hebei Henan Shanxi

Cotinus coggygria var. cinerea **Engl.**

Pileolaria cotini-coggygriae Tai et Cheo (Leaf rust) Hebei Henan Shanxi
Uncinula verniciferae P. Henn. (Powder mildew) Hebei

***Cotinus coggygria var. pubescens* Engl.**

Uncinula vernifciferae P. Henn. (Powder mildew) Sichuan

Cotoneaster spp.

Coleopuccinia sinensis Pat. (Leaf rust) Gansu Sichuan Yunnan Tibet

Entomosporium mespili (DC.) Sacc. (Leaf spot) Xinjiang

Gymnosporangium clavariiforme (Jacg.) DC. (Leaf rust) Yunnan

Gymnosporangium fusisporum Ed. Fischer (Leaf rust) Xinjiang

Monilia fructigena Pers. (Fruit blight) Xinjiang

Roestelia nanwutaiana (Tai et Cheo) Jϕrst. (Leaf rust) Shanxi

Roestelia sikangensis (Petr.) Jϕrst. (Leaf rust) Xinjiang

***Cotoneaster acutifolia* Turcz.**

Coleopuccinia sinensis Pat. (Leaf rust) Tibet Sichuan

Rosetelia nanwutaiana (Tai et Cheo) Jϕorst. (Leaf rust) Shanxi

Roestelia sikangensis (Petr.) Jϕrst. (Leaf rust) Tibet Sichuan

***Cotoneaster adpressus* Bois**

Roestelia sikangensis (Petr.) Jϕrst. (Leaf rust) Sichuan

***Cotoneaster ambigena* Rehd. et wils.**

Gymnosporangium confusum Plowr. (Leaf rust) Gansu

***Cotoneaster franchetii* Bios**

Coleopuccinia kunmingensis Tai (Leaf rust) Yunnan

***Cotoneaster integerrima* Medius.**

Coleopuccinia sinensis Pat. (Leaf rust) Neimeng

Eriophyes sp. (Leaf felt spot) Neimeng

***Cotoneaster microphylla* wall.**

Coleopuccinia kunmingensis Tai (Leaf rust) Yunnan

***Cotoneaster morrisonensis* Hayata**

Aecidium dichrocephalae P. Henn. (Leaf rust) Taiwan

***Cotoneaster multiflora* Bunge**

Roestelia nanwutaiana (Tai et Cheo) Jϕrst. (Leaf rust) Shanxi

***Cotoneaster rubens* W. W. Smith**

Gymnosporangium confusum Plowr. (Leaf rust) Yunnan

***Cotoneaster soongoricus* (Regel ex Herd) Popov**

Roestolia nanwutaiana (Tia et Cheo) Jϕrst. (Leaf rust) Xinjiang

***Cotoneaster tenuipes* Rehd. et wils.**

Coleopuccinia sinensis Pat. (Leaf rust) Sichuan

Crataegus spp.

Gymnosporangium clavariiforme (Jacq.) DC. (Leaf rust) Jiangsu Shanxi

Gymnosporangium haraeanum Syd. (Leaf rust) Hebei Jiangsu Xinjiang

Phyllactinia pyri (Cast.) Homma (Powder mildew) Xinjiang

(=Erysiphe pyri Cast.)

(=Phyllactinia corylea (Pers.) Karst.)

(=Phyllactinia guttata f. pyri Fuck.)

(=Phyllactinia sulffulta Sascc. f. pyri Rehm.)

(=Phyllactinia pyri-serotiniae Saw.)

Phyllosticta crataegicola Sacc. (Leaf spot) Liaoning

Podosphaera oxyacanthae (DC.) de Bary (Powder mildew) Liaoning

Roestelia magna (Crowell) Iϕrst. (Leaf rust) Huazhong

Truncospora truncatospora (Lloyd) Ito (Sapwood rot) Hebei

***Crataegus altaica* (Lour.) Lange**

Coriolus versicolor (L ex Fr.) Quél (Decay) Xinjiang

***Crataegus pinnatifida* Bunge**

Ascochyta crataegicola Allesch. (Leaf spot) Jilin
Coniothyrium tirolensis Bub. (Leaf spot) Jilin
Coryneum crataegicola Miura (Leaf spot) Liaoning Jilin Sichuan
Coryneum microstictum Berk. et Br. (Leaf spot) Liaoning
Eriophyes sp. (Leaf felt spot) Xinjiang
Eriophyes piri var. crataegi (Leaf felt spot) Neimeng
Gymnosporangium clavariiforme (Jacq.) DC. (Leaf rust) Hebei
Gymnosporangium haraenum Syd. (Leaf rust) Jilin Hebei Henan Jiangsu Shanxi Hubei Xinjiang Sichuan
Macrophoma kawatsukai Hara (Leaf spot) Sichuan
Penicillium frequentans Westl. Jiangsu
Phellinus pomaceus (Pers. ex Gray) Quél. (Decay) Liaoning Jilin Jiangsu Zhejiang Hebei Shanxi Henan Xinjiang Sichuan Yunnan Ningxia
Phyllactinia pyri (Cast.) Homma (Powder mildew) Yunnan Xinjiang
Phyllosticta crataegicola Sacc. (Leaf spot) Liaoning Jilin Hebei Shandong
Phyllosticta solitaria Ell. et Ev. (Leaf spot) jilin
Podosphaera oxyacanthae (DC.) de Bary (Powder mildew) Liaoning Jilin Hebei Shandong Jiangxi Guizhou Yunnan
Sphaeropsis demersa (Bon.) Sacc. Jilin

***Crataegus purpurea* Bosc. ex DC.**

Podosphaera oxycanthae (DC.) de Bary (Powder mildew) Liaoning

***Crataegus songarica* K. Koch.**

Fomes fomentarius (L. ex Fr.) Kickx. (Decay) Xinjiang
Funalia gallica (Fr.) Pat. (Decay) Xinjiang
(=Trametes gallica Fr.)
Gymnosporangium confusum Plowr. (Leaf rust) Xinjiang
Phellinus pomaceus (Pers. ex Gray) Quél. (Brown rot) Xinjiang
Phyllactinia pyri (Cast.) Homma (Powder mildew) Xinjiang
Phyllosticta michailouskoensis Elenk et Ohl (Leaf spot) Xinjiang
Septoria crataegi Kickx. (Leaf dry spot) Xinjiang

***Cryptomeria japonica* (L. F.) D. Don**

Cercospora sequoiae Ell. et Ev. (Cryptomeria blight) Jiangsu Zhejiang Jiangxi Taiwan Fujian Sichuan Anhui
(=Cercospora cryptomeriae Shirai)
Corticium centrifugum (Lév.) Bres (Sclerotium) Taiwan
Macrophomina phaseoli (Maubl.) Ashby (Stem decay) Jiangsu
Pestalotia foedans Sacc. et Ell. (Leaf spot) Henan Jiangsu
(=Pestalotia shiraiana P. Henn.)
Phoma cryptomeriae kassai (Leaf spot) Zhejiang Jiangsu
Sclerotiopsis sp. (Blight)

Cudrania spp.

Phyllachora cudrani P. Henn. (Black mole) Taiwan
Phyllactinia moricola (P. Henn.) Homma (Powder mildew) Sichuan
(=Phyllactinia moricola (P. Henn.) Saw.)
(=Phyllactinia suffulta (Rebent.) Sacc. var. moricola P. Henn.)
(=Phyllactinia corylea (Pers.) Karst.)
Systremma natans (Tode) Theiss. et Syd. Anhui
Uncinula clintonii Peck (Powder mildew) Guangxi
Uncinula miyabei (Salm.) Sacc. et Syd. (Powder mildew) Sichuan
Uredo sinensis (Syd.) Trott. (Leaf rust) Guangdong

***Cudrania cochinchinensis* (Lour.) Kudo et Masam.**

Hymenopsis cudraniae Mass. Taiwan

***Cudrania cochinchinensis var. gerontogea* Masam.**

Hymenopsis cudraniae Mass. Taiwan

Uredo cudraniae Saw. (Leaf rust) Taiwan

***Cudrania tricuspidata* (Carr.) Bur.**

Cercospora vanierae Chupp et Linder (Leaf spot) Guizhou

Tubercularia vulgaris Tode (Dieback) Yunnan

(Asexual :Nectria cinnabarina (Tode) Fr.)

Uncinula aspera Doidge var. clavulata Zheng et Chen (Powder mildew) Sichuan Yunnan

Uncinula clintonii Peck (Powder mildew) Guangxi

Uncinula pseudoehretieae Aheng et Chen (Powder mildew) Guangxi

Uncinula yaanensis Tao et Li (Powder mildew) Sichuan

***Cunninghamia lanceolata* (Lamb.) Hook**

Botryosphaeria cunninghamiae Huang (Branch wilt) Hubei Fujian

Cassytha filiformis L. (Cassytha Mistletoe) Guangdong Guangxi

Cephaleuros parasiticus Karst. (Algae) Guangxi

Cephaleuros virescens Kunze (Algae) Henan Hunan Fujian

Cercospora sequoiae Ell et Ev. (Blight) Anhui

Coriolus hirsutus (Wulf. ex Fr.) Quél. (Decay) Zhongnan

Discosia sp. (Blight) Sichuan

Discosia artocreas (Tode) Fr. (Blight) Henan

Elytranthe bibracteolata (Hance) le Comte (Elytranthe mistletoe) Guangdong Guangxi

Fusarium spp. (Damping off) Guangdong Guangxi Sichuan Anhui Hunan Henan

Glomerella cingulata (Stonem.) Spauld. et Schrenk (Anthracnose) Jiangsu Zhejiang Guangdong Jiangxi Hunan Fujian Anhui Hubei Henan Guizhou

Gnomonia sp. (Cedar leaf spot) Guangdong

Gyrophana acrymans (Wulf.) Pat. (Trunk decay) Sichuan

Helicobasidium purpureum (Tul.) Pat. (Violet root rot) Jiangsu Zhejiang Henan Anhui Guangdong Sichuan

Hypoderma cunninghamiae Teng (Needle cast) Shanxi Henan Hunan Fujian Yunnan Sichuan Guangdong

Hypoderma desmazieri Duby (Needle cast) Jiangxi

Lophodermium uncinatum Dark. (Needle cast) Jiangsu Zhejiang Jilin Henan Fujian Hunan Guangxi Anhui Guizhou Jiangxi Sichuan

Macrophomina phaseoli (Maubl.) Ashby (Stem decay) Henan

Macrosporium helminthosporioides (Corda) Ssacc. et Trav. (leaf spot) Jiangsu Hunan

Myiocopron sp. (Blight) Henan

Pestalotia apiculata Huang (Shoot blight) Fujian

Pestalotia foedans Sacc. et Ell. (Leaf blight) Jiangsu Jiangxi Hunan Guangdong Guangxi Sichuan Yunnan Fujian Guizhou

Phoma sp. (Leaf spot) Hubei Sichuan

Phomopsis sp. (Shoot blight) Jiangsu Zhejiang

Phyllosticta cryptomeriae Kawamura (blight) Anhui Guangdong Hunan

Phytophthora sp. (Cunninghamia shoot blight) Sichuan Anhui

Pseudomonas cinninghamiae Nanjing F.P.I.C. et al. (Bacterial leaf spot) Jiangsu Zhejiang Henan Jiangxi Guangdong Hunan Fujian Sichuan Guizhou Anhui Guangxi

Pythium spp. (Damping off) Henan Sichuan Guangdong Guangxi

Pythium ulfimum Tyow (Root rot) Sichuan

Rhizoctonia solani Kühn (Damping off) Henan Anhui Guangdong Guangxi Sichuan Hunan
Robillarda sp. Sichuan Henan
Sclerotium rolfsii Sacc. (Sclerotium) Anhui Jiangxi
Physical : (Physical blight) Anhui Zhejiang Jiangxi Fujian Hunan Hubei Guangdong Sichuan

Cupressus sp.
Gymnosporangium cunninghamianum Barcl. (Leaf rust) Yunnan

***Cupressus duclouxiana* Hickel**
Gymnosporangium cunninghamianum Barcl. (Leaf rust) Yunnan
Valsa sp. (Canker) Sichuan
Fusicoccum sp. (Tip canker of Cupressus) Sichuan

***Cupressus funebris* Endl.**
Alternaria sp. (Damping off) Sichuan
Ascochyta sp (Leaf spot) Sichuan
Fusarium sp. (Damping off) Sichuan
Fusicoccum sp. (Tip canker of Cupressus) Sichuan
Gymnosporangium tsingchenensis Wei (Leaf rust) Zhejiang Sichuan
Macrophoma sp. (Leaf rust) Sichuan
Macrophomina phaseoli (Maubl.) Ashby (Stem rot) Jiangsu Guizhou
Pestalotia sp. (Leaf spot) Sichuan
Phellinus hartigii (Allesch. et Schnabl) Imaz. (Sapwood brown rot) Tibet
Phoma sp. (Leaf spot) Sichuan
Robillards sp. Sichuan

***Cycas revoluta* Thunb.**
Ascochyta cycadina Scalia (Leaf spot) Jilin
Coniothyrium olivaceum Bon. (Leaf spot) Sichuan
Pestalotia cycadis Allesch. (Leaf spot) Jiangsu

***Cycas taiwaniana* Carr.**
Zygosporium oscheoides Mont. forma euonymi Bianchi Taiwan

***Cyclobalanopsis glauca* (Thunb.) Oerst.**
(=Quercus glauca Thunb.)
Cronartium quercuum (Berk.) Miyabe (Pine oak gall rust) Jiangsu
Cryptoderis quercina (Teng) Teng Zhejiang
Erysiphe fagacearum Zheng et Chen (Powder mildew) Zhejiang Anhui
Erysiphe gracilis Zheng et Chen (Powder mildew) Zhejiang
Hadronema orbiculare Syd. Guizhou
Hypocapnodium setosum (Zimm.) Speg. Taiwan
Loranthus yadoriki Sieb. et Zucc. (Mistletoe) Zhejiang Fujian Guangdong Guangxi Hunan Hubei Sichuan Shanxi
Meliola cyclobalanopsina Yamam. (Black mildew) Taiwan
Meliola taityuensis Yamam. (Black mildew) Taiwan
Microsphaera alphitoides Griff. et Maubl. (Powder mildew) Jiangsu Gansu
Phaeosaccardinula javanica (Zimm.) Yamam. (Black mildew) Taiwan
Sphaerotheca lanestris Harkn. (Powder mildew) Taiwan
Sphaerotheca wrightii (Berk. et Curt.) Höhn. (Powder mildew) Jiangsu Zhejiang Fujian Taiwan Hunan
Triposporiopsis spinigera (Höhn.) Yamam. (Black mildew) Taiwan

Dalbergia spp.
Maravalia achroa (Syd.) Arth. et Cumm. (Leaf rust) Sichuan Guizhou Yunnan
Meliola bantamensis Hansf. (Black mildew) Yunnan

***Dalbergia balansae* Prain.**

Pestalotia adusta Ell. et Ev. (Leaf spot) Guizhou
Phyllachora dalbergiicola P. Henn. (Black tar spot) Guangdong Guangxi

***Dalbergia hupeana* Hance**

Macrophoma dalbergiicola Teng (Leaf spot) Jiangsu
Maravalia achroa (Syd.) Arth. et Cumm. (Leaf rust) Jiangsu
Phyllachora dalbergiicola P. Henn. (Black tar spot) Jiangsu Zhejiang Guangdong Guangxi
Phyllachora dalbergiicola Syd. (Leaf spot) Zhejiang

***Dalbergia sisso* Roxb.**

Maravalia achroa (Syd.) Arth. et Cumm. (Leaf rust) Taiwan
Phellinus williamsii (Murr.) Pat. (Decay) Taiwan

Dandrocalamus laliflorus

Fusarium semilectum (Nectria ditisima) (Wilt)

Diospyros spp.

Coricolus versicolor (L. ex Fr.) Quél. (Decay) Fujian
Phyllactinia kakicola Saw. (Powder mildew) Anhui

***Diospyros kaki* L.**

Aithaloderma clavatispora Syd. (Black mildew) Taiwan
Botryodiplodia theobromae Pat. Guangdong
Botrytis cinerea Pers. (Botrytis blight) Sichuan
(Sexual : Sclerotinia fuckeliana (de Bary) Fuck.)
Cassytha filiformis L. (Cassytha mistletoe) Guangxi
Cephaleuros parasiticus Karst. (Algal) Guangdong
Cephaleuros virescens kunze (Algal) Guangdong
Cercospora kaki Ell. et Ev. (leaf spot) Liaoning Hebei Shanxi Henan Jiangsu Shandong Neimeng Fujian Anhui Guangxi Yunnan Sichuan Taiwan Hunan Hubei
Chaetoscorias vulgare Yamam. (Black mildew) Taiwan
Colletotrichum gloeosporioides Penz. (Anthracnose) Hebei Henan Jiangsu Zhejiang Shandong Guangxi Jiangxi Sichuan Yunnan Hunan Taiwan
Fusicladiumkaki Hori et Yosh. (Black star) Henan Shandong Jiangxi Guangxi Sichuan
Helminthosporium macrocarpum Grev Guangdong
Leptothyrium pomi (Mont. et Fr.) Sacc. (Black mold) Yunnan
Loranthus parasiticus (Linn.) Merr. (Mistletoe) Guangxi Guangdong
Macrophoma diospyri Earle (Leaf spot) Guangdong Guangxi
Monochaetia diospyri Yoshino (Leaf spot) Henan Hunan Guangdong
Mycosphaerella nawae Hiura et Ikata (Leaf spot) Hebei Henan Shanxi Shandong Guangxi Jiangxi Sichuan
Neocapnodium tanakae (Shiari et Hara) Yamam. (Black mildew) Taiwan
Pestalotia diospyri Syd. (leaf spot) Henan Jiangsu Hunan Guangdong Guangxi Jiangxi Yunnan
Phaeosaccardinula javanica (Zimm.) Yamam. (Black mildew) Taiwan
Phoma diospyri Sacc. (Brown leaf spot) Shanxi Yunnan
Phyllactinia kakicola Saw. (Powder mildew) Hebei Henan Anhui Fujian Guangxi Guizhou Yunnan Taiwan
Viscum angulatum Heyne (Mistletoe) Guangdong
Viscum orientale Willd. (Mistletoe) Guangxi Guangdong

***Diospyros lotus* L.**

Botrytis cinerea Pers. (Botrytis blight) Shanxi
Cercospora macClatchieana Sacc. et Syd. (Leaf spot) Huabei
Cercospora kaki Ell. et Ev. (Leaf spot) Liaoning Hebei Hubei Shandong Yunnan Hunan Shanxi
Colletotrichum gloeosporioides Penz. (Anthracnose) Hunan Guangdong

(=Gloeosporium kaki Hori)

Monochaetia diospyri Yoshii (Red leaf) Guangdong

Mycosphaerella nawae Hiura et Ikata (leaf spot) Hebei

Phyllactinia kakicola Saw. (Powder mildew) Hebei Henan Anhui Fujian Guangxi Taiwan Guizhou Yunnan

***Duabunaga grandiflora* (Roxb et DC.) Walp**

Meloidogyne incognita (Koforid et White) Chitwood (Root knot nematode) Guangdong

Elaeagnus spp.

Aecidium elaeagni Diet (Leaf rust) Hebei Shanxi Qinghai Jiangxi Fujian Sichuan Yunnan

Aecidium elaeagni-umbellatae Diet. (Leaf rust) Taiwan Yunnan

Cytospora elaeagni Allesch. (Canker) Xinjiang

Leveillula elaeagnacearum Golov. (Powder mildew) Xinjiang

Nectria cinnabarina (Tode) Fr. (Canker) Tibet

(Asexual : Tubercularina vulgaris Tode)

Phyllosticta argyrea Speg. (Leaf spot) Henan

Puccinia achroa Syd. (Leaf rust) Yunnan

***Elaeagnus angustifolia* L.**

Aecidium elaeagni Diet (Leaf rust) Shanxi Taiwan

Camarosporium caraganae Karst. (Dieback) Neimeng

Cytospora sp. (Canker) Xinjiang Neimeng

Cytospora elaegni Allesch. (Canker) Tibet

Fumago sp. (Black mildew) Ningxia

Leveillula leaegnacearum Golov. (Powder mildew) Xinjiang

Septoria argyrea Sacc. (Brown leaf spot) Liaoning Jilin Ningxia Xinjiang Neimeng Shanxi Gansu

***Elaegnus multiflora* Thunb.**

Aecidium elaeagni Diet (Leaf rust) Zhejiang

Aecidium elaeagni-umbellatae Diet. (Leaf rust) Henan Jiangsu

Phyllosticta argyrea Speg. (Leaf spot) Jiangsu Zhejiang

***Elaeagnus pungens* Thunb.**

Aecidium elaeagni Diet. (Leaf rust) Zhejiang

Irpex sp. (Decay) Shandong

Phellinus robustus (Karst.) Bourd. et Galz. (Sapwood brown rot) Shanxi

Phyllosticta argyrea Speg. (Leaf spot) Jiangsu Zhejiang

Septoria argyrea Sacc. (Brown leaf spot) Jiangsu

***Elaeis guineensis* Jacq.**

Pestalotia palmarum Cooke (Leaf spot) Fujian

***Enterolobium contortrisiliqum* Hauman**

Meloidogyne incognita (Koforid et White) Chitwood (Root knot nematode) Guangdong

Eucalyptus spp.

Cercospora epicoccoides Cooke et Mass. (Leaf spot) Taiwan

Cercospora eucalypti Ccoke et Mass. (Leaf spot) Guangdong

Colletotrichum sp. (Brown leaf spot)Guangdong

Coniothyrium kallangurense Sutton et Alcorn (Brown leaf spot) Guangdong

Corticium sp.

Cyclindroeladium sp(Brown leaf spot)Guangdong

Cylindrocladium(Brown leaf spot)Guangdong

Fusatium sp.(Damping off) Guangdong

Phoma eucalyptica (Thüm.) Sacc. (Canker) Guangdong

Phthium sp. (Damping off) Guangdong

Rhizoctonia sp. (Damping off) Guangdong

Sclerotium rolfsii Sacc. (Sclerotium) Guangdong
Septoria mortorlensis Penz. et Sacc. (Violet spot) Guangdong Guangxi
Thelephora sp. (Smother) Guangdong

***Eucalpytus exserta* F. Muell.**
Coniothyrium kallangurense Sutton et Alcorn (Brown leaf spot) Guangdong Guangxi
Phoma subnervisequa Desm. (Canker) Guangdong Guangxi
Septoria mortorlensis Penz. et Sacc. (Brown spot) Guangdong Guangxi

***Eucalpytus globulus* Labill.**
Cercospora epicoccoides Cooke et Mass. (Leaf spot) Taiwan
Cercospora eucalypti Cooke et Mass. (Leaf spot) Guangdong
Coniothyrium kallangurense Sutton et Alcorn (Leaf brown spot) Guangdong
Cryptostictis eucalpyti Pat. Yunnan
Pestalotia disseminata Thum. (Leaf spot) Yuannan

***Eucalyptus robusta* Smith**
Cephaleuros virescens Kunze (Algal) Guangdong
Cercospora epicoccoides Cooke et Mass. Taiwan
Coniothyrium kallangurense Sutton et Alcorn (Brown leaf spot) Guangdong Guangxi
Fusarium spp. (Root rot) Guangdong
Macrophomina phaseoli (Maubl.) Ashby. Guangdong
Phoma eucalytica (Thüm.) Sacc (Canker) Guangxi Guangdong
Virus (Witches' broom) Guangdong

***Eucalyptus tereticornis* Smith**
Agrobacterium tumefaciense (Smith et Towns) Conn. (Root gall canker) Guangdong
Cercospora sp. (Leaf spot) Guangdong
Coniothyrium kallangurense Sutton et Alcorn (Brown spot) Guangdong
Loranthus chinensis DC. (Mistletoe) Guangdong
Macrophomina phaseoli (Maubl.) Ashby (Stem canker)Guangdong
Septoria motorlensis Penz et Sacc.(Violet spot) Guangdong

***Eucommia ulmoides* Oliv.**
Cercospora sp.(Brown leaf spot) Henan Shanxi Hunan
Fusarium spp. (Damping off) Shanxi
Macrophomina phaseoli (Maubl.) Ashby. (Stem rot) Anhui Hunan
Myxosporium sp. (Dieback) Guizhou

Euonymus spp.
Cercosporellaeuonymi erikss. (Leaf spot) Hebei Henan Ningxia
Diplodia ramulicola Desm. (Dieback) Tibet
Macrophoma cylindrospora (Desm.) Berl. et Vogl. (Leaf spot) Henan
Phyllosticta aliena (Fr.) Sacc. (Leaf spot) Heilongjiang
Phyllosticta bolleana Sacc. (Leaf spot) Shanxi

***Euonymus bungeanus* Maxim**
Phyllosticta aliena (Fr.) Sacc. (Leaf spot) Liaoning Heilongjiang

***Euonymus japonica* Thunb.**
Cercospopra destractive Rav. (Brown Leaf spot) Henan Shandong Jiangsu Shanxi Hunan Sichuan
Cylindrosporium frigidum (Sacc.) Vass. Shandong
Fusicladium euonymi-japonici Hori (Fruit scab) Hunan
Macrophoma cylindrospora (Desm.) Berl. et Vogl. Shanxi
Oidium euonymi-japonicae (Arc.) Sacc. (Powder mildew) Sichuan
Pestalotia neglecta Thüm. (Leaf spot) Shanxi
Pestalotia planimi Vize (Leaf spot) Jiangsu
Sphaeropsis euonymi Desm. (Leaf spot) Jiangsu

***Euonymus japonica var. aureomarginiata* Nichols.**

Marmor euonymi Holmes. (Virus) Shanxi

***Euonymus sacrosancta* Koidz**

Pestalotia funerea Desm. (Brown leaf spot) Heilongjiang

***Euonymus venosus* Hemsl.**

Phoma subnervisequa Desm. (Leaf spot) Shanxi

***Eupatorium lindleyanum var. trifoliolatum* Makino**

Coleosporium eupatorii Arth. (Leaf rust) Liaoning

Ficus spp.

Irenina cheoi Hansf. (Black mildew) Guizhou
Irenopsis benguetensis Stev. et Rold. (Black mildew) Guangxi
Phakopsora fici-erectae Ito et Otani (Leaf rust) Jiangxi
Phakopsora nishidana Ito (Leaf rust) Jiangxi
Phyllachora ficuum Niessl (Black tar spot) Guangdong
Phyllactinia brossoentiae-kaempferi Saw. (Powder mildew) Zhejiang
Trabutia cinense Yates Zhejiang Guangdong
Trabutia elmeri Theiss. et Syd. Guangdong

***Ficus auriculata* Lour.**

Meliola microtricha Syd. (Black mildew) Guangdong
Septoria pirottae Tassi (Leaf spot) Yunnan

***Ficus benjamina* L.**

Trabutia elmeri Theiss. et Syd. Guangdong Yunnan

***Ficus retusa* L.**

Aithaloderma clavatispora Syd. (Black mildew) Taiwan
Coriolus versicolor (L. ex Fr.) Quél (Decay) Tibet
Neocapnodium tanakae (Shirai et Hara) Yamam. (Black mildew) Taiwan
Phaeosaccardinula javanica (Zimm.) Yamam. (Black mildew) Taiwan
Scorias communis Yamam. (Black mildew) Taiwan
Triposporiopsis sipinigera (Höhn.) Yamam. (Black mildew) Taiwan

***Firmiana simplex* (L.) W.F. Wight**

Cephaleuros virescenes Kunze (Algae)
Phyllactinia sinensis Yu (Powder mildew) Sichuan Fujian
(=Phyllactinia corylea (Pers.) Karst.)
Sclerotium rolfsii Sacc. (Sclerotium) Provinces of the South
Uncinula clintoniopsis Zheng et Chen (Powder mildew) Jiangsu Anhui Sichuan Guizhou

***Fontanesia fortunei* Carr.**

Cytospora sp. (Canker) Liaoning

Forsythia spp.

Pseudomonas savastanoi (E.F. Smith) Stev. (Stem gall) Guangxi

***Forsythia ovata* Nakai**

Phyllosticta forsythiae Sacc. (Leaf spot) Liaoning

Fraxinus spp.

Aecidium fraxini-bungeanae Diet. (Leaf rust) Jiangsu Guangxi
Coriolus versicolor (L. ex Fr.) Quél. (Decay) Hebei
Inonotus hispidus (Bull. ex Fr.) Karst. (Tip rot) Dongbei
Mycosphaerella fraxinea Peck (Leaf spot) Liaoning
Phyllactinia fraxini (DC.) Homma (Powder mildew) Sichuan
Pseudomonas savastanoi (E.F.Smith) Stev (Stem gall) Yunnan
Phyllosticta fraxini Ell. et Martin (Leaf spot) Hebei
Uncinula fraxini Miyabe (Powder mildew) Hebei Sichuan
Uncinula salmonii Syd (Powder mildew) Jiangsu

***Fraxinus americana* L.**

Cuscuta japonica Choisy. (Dodder) Xinjiang

Valsa fraxinina Peck (Canker) Xinjiang

***Fraxinus bungeana* DC.**

Cuscuta japonica Choisy (Dodder) Hebei Zhejiang

***Fraxinus chinensis* Roxb.**

Aecidium fraxini-bungeanae Diet. (Leaf rust) Jiangsu Zhejiang Henan Sichuan Guizhou

Loranthus parasiticus (Linn.) Merr. (Mistletoe) Guangxi

Phyllactinia Fraxini (DC.) Homma (Powder mildew) Jiangsu Zhejiang Sichuan

(=Erysiphe fraxini DC.)

(=Phyllactinia corylea (Pers.) Karst. em. Salm.)

(=Phyllactinia guttata (Wallr.) Vev.)

(=Phyllactinia suffulta (Rebent.) Sacc.)

(=Phyllactinia suffulta (Rebent.) Sacc. f. fraxini DC.)

Uncinula fraxini Miyabe (Powder mildew) Jiangsu Henan Guangxi Sichuan

***Fraxinus mandshurica* Rupr.**

Fomes fomentarius (L. ex Fr.) Kickx. (Decay) Heilongjiang

Inonotus hispidus (Bull. ex Fr.) Karst. (Leaf tip rot) Heilongjiang Jilin

Mycosphaerella fraxinea Peck (Brown leaf spot) Heilongjiang Dongbei

Septoria fraxini Desm. (Leaf spot) Liaoning

Uncinula salmonii Syd. (Powder mildew) Liaoning Dongbei

Valsa fraxinina peck (Canker) Xinjiang

***Fraxinus retusa* Champ.**

Phyllactinia fraxini (DC.) Homma (Powder mildew) Sichuan

***Fraxinus rhynchopylla* Hance**

Aecidium fraxini-bungeanae Diet. (Leaf rust) Jiangsu Zhejiang Anhui Jiangxi Sichuan

Coniothyrium fraxini Miura Jilin Liaoning

Uncinuia salmonii Syd. (Powder mildew) Jilin Heilongjiang

Uropyxis fraxini (Kom.) Magn. (Leaf rust) Jilin Liaoning

Ginkgo sp.

Coriolus versicolor (L. ex Fr.) Quél. (Decay) Zhejiang

***Ginkgo biloba* L.**

Discosia sp. (Canker) Henan

Fusarium sp. (Root rot) Shanxi

Glomerella cingulata (Stonem.) Spauld. et Schrenk (Anthracnose) Jiangsu

Macrophomina phaseoli (Maubl.) Ashby (Stem rot) Jiangsu Henan Anhui Guangdong Sichuan

(=Macrophoma phaseoli Maubl.)

Pestalotia sinensis Shen (Leaf blight spot) Jiangsu Henan Sichuan

Phyllosticta sp. (Leaf spot) Jiangxi

Rhioctonia solani Kühn (Root rot) Shanxi

***Gleditsia japonica* Miq.**

Macrophoma sp. (Dieback) Liaoning

***Gleditsia sinensis* Lam.**

Coriolus versicolor (L. ex Fr.) Quél. (Decay) Shandong

Microsphaera alni (Wallr.) Saslm. (Powder mildew) Shanxi

Phyllactinia caesalpiniae Yu (Powder mildew) Neimeng Sichuan

(=Phyllactinia corylea (Pers.) Karst.)

***Haloxylon ammodendron* (Mey.) Bunge**

Camarosporium paletzkii Sereb. (Dieback) Xinjiang

Leveillula saxaouli (Sorok) Golov (Powder mildew) Xinjiang Neimeng Gansu Ningxia

Stagonosporopsis haloxyli Syol (Black dry leaf) Xinjiang

Haloxylon persicum **Bunge**

Leveillula saxaouli (Sorok,) Golov. (Powder mildew) Ningxia Xinjiang

Hedysarum sp.

Uromyces hedysari-obcuri (DC.) Car. et Picc (Leaf rust) Shanxi

Hedysarum scoparium **Fisch. et Mey.**

Erysiphe sp. (Powder mildew) Ningxia

Trichocladia diffusa Jacz. (Powder mildew) Xinjiang

(=Microsphaera diffusa Cooke. et Peck)

Uromyces sp. (Leaf rust) Ningxia

Hippophae rhamnoides **L.**

Phellinus robustus (Karst.) Bourd et Galz. (Sapwood brown rot) Tibet Qinghai

Hypercium spp.

Melampsora kusanoi. Diet (Leaf rust) Hebei Qinghai Guangxi Sichuan Yunnan

Melampsora hypericorum Wint (Leaf rust) Jiangsu

Uromyces hyperici Curt. (Leaf rust) Guangxi Yunnan

Hypericum ascyron **L.**

Melampsora kusanoi Diet. (Leaf rust) Liaoning Shanxi Gansu

Ilex latifola **Thunb.**

Cephaleuros virescens Kunze (Algal) Guangdong Guangxi

Physalospora ilicella Teng Jiangsu

Illicium verum **Hook. f.**

Cephaleuros virescens Kunze. (Algal) Guangxi

Glomerella cingulata (Stonem.) Spauld. et Schrenk (Anthracnose) Guangxi

Loranthus sp. (Mistletoe) Guangxi

Juglans spp.

Antennaria pannosa Berk. (Black mildew) Yunnan

Botryodiplodia theobromae Pat. Hebei

Coriolus hirsutus (Wulf ex Fr.) Quél. (Decay) Hebei

Coriolus versicolor (L. ex Fr.) Quél. (Decay) Hebei

Cytospora juglandis (DC.) Sacc. (Canker) Neimeng Tibet Xinjiang

Diplodia juglandis Fr. (Dieback) Tibet

Fomitopsis pinicola (Sw. ex Fr.) Karst. (Brown cube rot) Hebei

Ganoderma applaudir (Pers.) Pat. (Decay) Hebei

Inonotus hispidus (Bull. ex Fr.) Karst. (Leaf tip rot) Hebei

Marssonina juglandis (Lib.) Magn. (Black leaf spot) Hebei

Melanconis juglandis (ell. et Ev.) Graves (Dieback) Jilin

Microsphaera yatagan (Salm.) Syd. (Powder mildew) Neimeng

Xanthomonas juglandis (Pierce) Dowson (Bacterial black leaf spot) Neimeng

Juglans cathayensis **Dode**

Melanconium juglandium Kunze (Dieback) Hebei Jiangsu Henan

Melanconium oblongum Berk. (Dieback) Jiangsu

Juglans mandshurica **Max.**

Cladosporium herbarum (Pers.) Link (Black mold) Jilin

Eriophyes sp. (Leaf mites) Hebei

Eriophyes tristesses-erineus Nal. (Leaf felt spot) Liaoning Heilongjiang

Haplosporella longpipes Ell. et Barth. (Dieback) Liaoning

Laetiporus sulphureus (Bull. ex Fr.) Bond. et Sing (Trunk brown rot) Jilin

Melanconium iuglandjum Kunze (Dieback)

(Sexual :Melanconis juglandis (ell. et Ev) Groves)

Liaoning Jilin Heilongjiang Shandong Hebei Jiangsu Zhejiang

Mealanconium oblongum Berk (Dieback) Liaoning
(Sexual :Melanconis juglandis (ell. et Ev.) Groves)
Marssonina juglandis (Lib.) Magn. (Brown leaf spot) Jilin
Microstroma juglandis (Bereng.) Sacc. (Frost leaf spot) Liaoning Jilin Heilongjiang
Microsphaera yatagan (Salm.) Syd. (Powder mildew) Liaoning
Phoma juglandis Sacc. (Dieback) Jilin
Phyllactinia juglandis-mandshuricae Yu (Powder mildew) Hebei
Phyllosticta juglandis (DC.) Sacc. (Leaf spot) Jilin Liaoning
Xanthomonas juglandis (pierce) Dowson (Bacterial black leaf spot) Liaoning Heilongjiang

***Juglans regia* L.**

Colletotrichum glucocorticoïdes Penz. (Anthracnose) Henan Shandong Hebei Shanxi Shanxi Jiangsu Xinjiang Liaoning
(=Gloeosporium rufomaculans (Berk.) Thüm.)
Cytospara juglandis (DC.) Sacc. (Canker) Xinjiang Gansu Shandong Henan Shanxi
Dothiorella gregaria Sacc. (Canker) Hebei Henan Shanxi Jiangsu
Eriophyes sp. (Leaf felt spot) Hebei Yunnan
Favolus squamosus (Huds. ex Fr.) Ames (Decay) Heilongjiang
Fomes fomentarius (L. ex Fr.) Kickx. (Decay) Xinjiang
Funalia hispida (Bagl.) Pat. (Decay) Sichuan
(=Trametes hispida Bagl.)
Fusarium avenaceum (Fr.) Sacc. Distribution not determined
Laetiporus sulphureus (Bull. ex. Fr) Bond. ex Sing. (Trunk brown rot) Henan Xinjiang
Loranthus parasiticus (Linn.) Merr. (Mistletoe) Yunnan
Marssonin juglandis (Lib.) Magn. (White star) Henan
Melanconium juglandium Kunze (Dieback) Hebei Henan Jiangsu Shanxi Zhejiang Dongbei
Melancoium oblongum Berk. (Dieback) Hebei Henan Shanxi Jiangsu Zhejiang Dongbei Yunnan
Microsphaera yatagan (Salm.) Syd. (Powder mildew) Hebei Henan Shanxi Jiangsu Neimeng Shanxi
Microstroma juglandis (Bereng.) Sacc. (Frost leaf spot) Henan Yunnan Shanxi Jilin
Phellinus igniarius (L. ex Fr.) Quél. (Heartwood white sponge rot) Xinjiang
Phellinus robustus (Karst.) Bourd. et Glaz. (Sapwood brown rot) Hebei
Phyllactinia juglandis Tao et Qin (Powder mildew) Shanxi Neimeng Sichuan Yunnan
(=Phyllactinia corylea (Pers.) Karst.)
Phyllosticta juglandis (DC.) Sacc. (Leaf spot) Hebei Henan Shanxi
Physalospora juglandis Syd. et Hara (Dieback) Henan Hunan Yunnan
Sclerotium rollier Sacc. (Root rot) Sichuan
Septobasidium bogoriense Pat. (Plaster) Zhejiang
Septobasidium tanakae (Miyabe) Boed. et Steinm (Plaster) Yunnan
Spongipellis lits-cages Lohw. (Heartwood white rot) Hebei
Trichothecium rose (Bull.) Link (Powder mold) Hebei Henan
Viscum album L. (Mistletoe) Yunnan
Xanthomonas juglandis (Pierce) Dowson (Bacterial black leaf spot) Hebei Henan Shandong Jiangsu Shanxi Guizhou Yunnan Ningxia Sichuan Neimeng

Juniperus spp.

Gymnosporangium japonicum Syd. (Leaf rust) Jiangxi
(=Roestelia photiniae P. Henn.)
Lophodermium juniperinum (Fr.) Rehm (Falling needle) Xinjiang
Spongipellis litschaueri Lohw. (Decay) Gansu

***Juniperus formosana* Hayata**

Gymnosporangium corniforme Saw. (Leaf rust) Taiwan

***Juniperus rigida* Sieb. & Zucc.**

Gymnosporangium haraenum Syd. (Leaf rust) Liaoning

***Juniperus sibirica* Burgsd.**

Gymnosporangium gaeumannii H. Zogg. (Leaf rust) Xinjiang

Lophodermium juniperinum (Fr.) Rehm (Needle cast) Xinjiang

***Kalopanax septemlobus* (Thunb.) Koidz.**

(=Acanthopanax ricinifolius Seem.)

Ascochyta acanthopanacis (Syd.) P.K.Chi (Leaf spot) Jiangsu Liaoning

Eriophyes sp. (Leaf mites felt spot) Gansu

Phyllactinia corylea (Pers.) Karst. (Powder mildew) Sichuan

***Keteleeria fortunei* (Murr.) Carr.**

Arceuthobium spp. (Drarf mistletoe) Yunnan Sichuan Tibet

Chrysomyxa sp. (Leaf rust) Yunnan

Koelreuteria spp.

Meliola koelreuteriae Hansf. (Black mildew) Zhejiang Guangxi

Nyssopora koelreuteriae (Syd.) Tranz. (Leaf rust) Zhejiang

***Koelreuteria bipinnata* Franch.**

Nyssopora formosana (Saw.) Lütj. (Leaf rust) Taiwan

Nyssopsora koelreuteriae (Syd.) Tranz. (Leaf rust) Zhejiang Hunan

Typhulochaeta koelreuteriae (Miyake) Tai (Powder mildew) Henan Hebei

(=Uncinula koelreuteriae Miyake)

(=Erysiphe koelreuteriae (Miyake) Tai)

***Koelreuteria henryi* Dummer**

Nyssopsora formosana (Saw.) Lütj. (Leaf rust) Taiwan

***Koelreuteria paniculata* Laxm.**

Bulbouncinula bulbosa (Taiet Wei) Zheng et chen (Powder mildew) Zhejiang

(=Uncinula clintonii Peck var. bulbosa Tai et Wei)

(=Uncinula bulbosa (Tai et Wei) Tai)

Eurotium herbariorum (Wigg.) Link Shanxi

Erysiphe sp. (Powder mildew) Jiangsu

Nyssopsora chinense (Tai et Cheo) Tai. (Leaf rust) Shanxi

Nyssopsora koelreuteriae (Syd.) Franz. (Leaf rust) Shanxi Zhejiang Shanxi Hunan

Typhulochaeta koelreuteriae (Miyake) Tai (Powder mildew) Hebei Jiangsu

Lagerstroemia indica

Uncinulieua australiana Zheng et Chen

Larix spp.

Armillariella mellea (Vahl. ex Fr.) Karst. (Root rot) Dongbei

Fomitopsis pinicola (Sw. Ex Fr.) Karst. (Brown cube rot) Heilongjiang

Guignardia laricina (Sawada) Yamamoto et K. Ito (Shoot blight) Liaoning Hebei Shandong

Laetiporus sulphureus (Bull. ex Fr.) Bond. et Sing. (Trunk brown cube rot) Heilongjiang Hebei Xinjiang Sichuan

Laricifomes officinalis (Vill. ex Fr.) Kotlaba ex Pouzar (Heartwood brown cube rot) Jilin Heilongjiang Hebei Shanxi Neimeng Xinjiang Sichuan

Melampsora larici-populina Kleb. (Leaf rust) Heilongjiang Qinghai Sichuan

Mycosphaerella larici-leptolepsis K. Ito et K. Sato (Falling leaf) Jilin Liaoning Heilongjiang Hebei Shandong Gansu Neimeng

Phaeolus schweinitzii (Fr.) Pat. (Trunk brown rot) Jilin Heilongjiang Hebei Xinjiang Sichuan Yunnan

Phellinus pini (Thore ex Fr.) Ames. (White pocket rot) Jilin Heilongjiang Hebei Shanxi Shanxi Gansu Xinjiang Ningxia Sichuan Yunnan Neimeng

Triphragmiopsis laricinum (Chou) Tai (Leaf rust) Jilin Liaoning Heilongjiang

(=Triphragmiopsis laricinum Chou.)

***Larix gmelini* (Rupr.) Rupr.**

Armillariella mellea (Vahl. ex Fr.) Karst. (Root rot) Dongbei
Caeoma laricis (West.) Hart. (Leaf rust) Heilongjiang Neimeng
Coriolus versicolor (L. ex Fr.) Quél. (Decay) Shandong
Cytospora abietis Sacc. (Canker) Heilongjiang Liaoning
Dothichiza sp. (Dieback) Heilongjiang
Fomitopsis pinicola (Sw. ex Fr.) Karst. (Brown cube rot) Heilongjiang Dongbei Xinjiang Neimeng
Fusarium spp. (Damping off) Neimeng
Fuscoporia weirii (Murr.) Aosh. (Decay) Neimeng
Ganoderma tsugae Murr. (Decay) Jilin Heilongjiang Shanxi
Gloephyllum abietinum (Bull ex Fr.) Karst. (Decay) Neimeng
Guignardia laricina (Sawada) Yamamot et K. Ito (Shoot blight) Jilin Heilongjiang Liaoning Hebei
(=Physalospora laricina Sawada)
Hypodermella laricis Tub. (Needle cast) Neimeng
Lachnellula wilkommii (Hart.) Dennis. (Canker) Heilongjiang
(=Trichoscyphella willkommii (Hart.) Nannf.)
Laetiporus sulphureus (Bull. ex Fr.) Bond. ex Sing. (Brown cube rot) Heilongjiang Neimeng
Laricifomes officinalis (Vill. ex Fr.) Kotlab ex Pouzar (Heartwood brown cube rot) Heilongjiang Hebei Neimeng
(=Fomitopsis officinalis (Vill. ex Fr.) Bond. et Sing.)
Lloydella abietina (Pers.) Ito (Decay) Neimeng
Melampsora larici-populina Kleb. (Leaf rust) Dongbei
Mycosphaerella larici-leptolepsis K. Ito et K. Sato (Falling leaf) Heilongjiang Liaoning Dongbei Shandong Jilin Neimeng
Nectria sp. (Dieback) Heilongjiang
Phaeolus schweintzii (Fr.) Pat. (Trunk brown cube rot) Neimeng
Phellinus pini (Thore ex Fr.) Ames (White pocket rot) Heilongjiang Dongbei Neimeng
(=Fomes pini (Thore) Karst.)
Phoma laricis Lév. (Sprout stem swell) Heilongjiang
Rhizoctonia solani Kühn (Damping off) Liaoning Heilongjiang Neimeng
Thelephora terrestris (Ehrh.) Fr. (Decay) Jilin Liaoning Dongbei
Triphragmiopsis laricinum (Chou) Tai (Leaf rust) Liaoning Jilin Heilongjiang
Tyromyces anceps (Peck) Murr. (Decay) Heilongjiang
Physical (Sunburn) Neimeng

***Larix griffithiana* (Lindl. et Gord.) Hort. ex Carr.**

Fomitopsis pinicola (Sw. ex Fr.) Karst. (Brown cube rot) Tibet

***Larix kaempferi* (Lamb.) Carr.**

(=Larix leptolepis (Sieb. et Zucc.) Gord.)
Botrytis cinerea Pers. (Damping off) Liaoning
Caeoma laricis (Westend.) Hart. (Leaf rust) Liaoning
Diaporthe conorum (Desm) Niessl (Stem canker) Liaoning
(Asexual: Phomopsis occulta Teav)
Fusarium spp. (Damping off) Liaoning
Guignardia laricina (Sawada) Yamamoto et K. Ito (Shoot blight) Dongbei
Melampsora larici-populina Kleb. (Leaf rust) Dongbei
Mycosphaerella larici-leptolepis K. Ito et. Sato (Falling leaf) Jilin Dongbei
Nectria cinnabarina (Tode) Fr. (Dieback) Liaoning
Phellinus pini (Thore ex Fr.) Ames (White pocket rot) Dongbei

Pythium spp. (Damping off) Liaoning
Rhizoctonia solani Kühn (Damping off) Dongbei
Thelephora terrestris (Ehrh.) Fr. (Decay) Heilongjiang
Triphragmiopsis laricinum (Chou) Tai (Leaf rust) Jilin Liaoning Heilongjiang

***Larix olgensis* Henry**

Botrytis cinerea Pers. (Damping off) Liaoning
Caeoma laricis (Westend.) Hart. (Leaf rust) Liaoning
Fusarium spp. (Damping off) Liaoning
Ganoderma tsugae Murr. (Decay) Jilin
Guignardia laricina (Sawada) Yamamoto et K. Ito (Shoot blight) Jilin Heilongjiang Liaoning Shandong
Mycosphaerella larici-leptolepis K. Ito et K. Stao (Falling leaf) Jilin Liaoning Heilongjiang Hebei Shandong Gansu
Nectria cinnabarina (Tode) Fr. (Dieback) Liaoning
Phythium spp. (Damping off) Liaoning
Rhizoctonia solani Kühn (Damping off) Liaoning
Sclerotium sp. (Larix blight) Jilin
Triphragmiopsis laricinum (Chou) Tai (Leaf rust) Jilin Liaoning Heilongjiang

***Larix potaninii* Batalin**

Fomitopsis pinicola (Sw. ex Fr.) Karst. (Heartwood brown cube rot) Sichuan Yunnan
Phaeolus schweinitzii (Fr.) Pat. (Trunk brown cube rot) Sichuan
Phellinus pini (Thore ex Fr.) Ames (White pocket rot) Sichuan Yunnan

***Larix principis-ruprechtii* Mayr**

Guignardia laricina (Sawada) Yamamoto et K. Ito (Shoot blight) Jilin Heilongjiang Liaoning Shandong
Phellinus pini (Thore ex Fr.) Ames (White rot) Hebei

***Larix sibirica* Ledeb.**

Alternaria tenuis Nees. Xinjiang
Armillariella mellea (Vhal. ex Fr.) Karst. (Root rot) Xinjiang
Botrytis cinerea Pers. (Damping off) Xinjiang
Cytospora abietis Sacc. (Canker) Xinjiang
Fomitopsis pinicola (Sw. ex Fr.) Karst. (Brown cube rot) Xinjiang
Fusarium spp. (Damping off) Xinjiang
Ganoderma tsugae Murr. (Decay) Xinjiang
Hypodermella laricis Tub. (Larch needle and shoot blight) Xinjiang
Laetiporus sulphureus (Bull. ex Fr.) Bond. ex Sing (Decay) Xinjiang
Laricifomes officinalis (Vill. ex Fr.) Kotlaba ex Pouzar (Heartwood brown cube rot) Xinjiang
Phacidium infestans Karst. (Leaf spot) Xinjiang
Phaeolus schweinitzii (Fr.) Pat. (Decay) Xinjiang
Phellinus pini (Thore ex Fr.) Ames (Heartwood white rot) Xinjiang
Pratylenchus sp. (Meadow nematode) Xinjiang
Rhizoctonia solani Kühn. (Damping off) Xinjiang

Lespedeza spp.

Cercospora latens Ell. et Ev. (Leaf spot) Jiangsu
Endoxylina sp. (Dieback) Liaoning
Erysiphe polygoni DC.
(Asexual :Oidium erysiphoides Fr.) (Powder mildew) Hebei
Phyllachora lespedezae (Schw.) Sacc. (Black mole) Yunnan
Uromyes lespedezae-bicoloris Tai et Cheo (Leaf rust) Jiangsu
Uromyces lespedezae-macrocarpae Liou et Wang (Leaf rust) Hebei Yunnan

Uromyces lespedezae-procumbentis (Schw.) Curt. (Leaf rust) Hebei Shanxi Shandong Jiangsu Anhui Zhejiang Jiangxi Henan Hunan Guangxi Sichuan Guizhou

Uromyces rugulosus Pat. (Leaf rust) Hebei Hunan

***Lespedeza bicolor* Turcz.**

Cercospora latens Ell. et Ev. (Leaf spot) Jiangsu

Erysiphe polygoni DC. (Powder mildew) Zhejiang Taiwan
(=Microsphaera polygoni (DC.) Saw.)

Pleospora lespedezae Miyake Hebei

Uromyces lespedezae-bicoloris Tai et Oheo (Leaf rust) Hebei Shanxi Jilin Shanxi Shandong Hubei

Uromyces lespedezae-macrocarpae Liou et Wang (Leaf rust) Hebei Yunnan

Uromyces lespedezae-procumbentis (Schw.) Curt. (Leaf rust) Jilin Liaoning Heilongjiang Hebei Shanxi Shandong Jiangsu Zhejiang Fujian Taiwan Henan Hunan Sichuan Yunnan Shanxi

Uromyces rugulosus Pat. (Leaf rust) Shanxi Hunan Sichuan

Xanthomonas lespedezae (Azers, Lefebre and Johnson) Starr (Bacterial leaf spot) Shanxi

***Lespedeza buergeri* Miq.**

Uromyces rugulosus Pat. (Leaf rust) Shanxi Shanxi

***Lespedeza chinensis* G. Don**

Uromyces lespedezae-procumbentis (Schw.) Curt. (Leaf rust) Liaoning Taiwan

***Lespedeza cuneata* (Dum. cours) G. Don**

Erysiphe polygoni DC. (Powder mildew) Jilin

Uromyces lespedezae-procumbentis (Schw.) Curt. (Leaf rust) Jilin Liaoning Jiangsu Henan

***Lespedeza cyrtobotrya* Miz.**

Cercospora latens Ell. et Ev. (Leaf spot) Taiwan

Uromyces lespedezae-procumbentis (Schw.) Curt. (Leaf rust) Liaoning Hubei

***Lespedeza davidii* Franch.**

Uromyces lespedezae-procumbentis (Schw.) Curt. (Leaf rust) Zhejiang Jiangxi

***Lespedeza dahurica* (Laxm.) Schindi.**

Erysiphe polygoni DC. (Powder mildew) Hebei

Uromyces lespedezae-procumbentis (Schw.) Curt. (Leaf rust) Jilin Liaoning Hebei Shanxi Shanxi Shandong Jiangxi Guizhou

***Lespedeza floribunda* Bunge**

Uromyces lespedezae-procumbentis (Schw.) Curt. (Leaf rust) Hebei Shanxi Sichuan

***Lespedeza formosa* (Vog.) Koeh.**

Uromyces lespedezae-procumbentis (Schw.) Curt. (Leaf rust) Jiangsu Henan Shanxi Guangxi Jiangxi

Uromyces rugulosus Pat. (Leaf rust) Shanxi

***Lespedeza homoloba* Nakai**

Uromyces lespedezae-procumbentis (Schw.) Curt. (Leaf rust) Taiwan

***Lespedeza juncea* (L. f.) Perr.**

Uromyces lespedezae-procumbentis (Schw.) Curt. (Leaf rust) Liaoning Hebei Shandong Jiangxi Henan Jiangsu Zhejiang Anhui Yunnan

***Lespedeza tomentosa* (Thunb.) Sieb.**

Cercospora latens Ell. et Ev. (Leaf spot) Shanxi

Erysiphe polygoni DC. (Powder mildew) Liaoning Hebei

Uromyces lespedezae-procumbentis (Schw.) Curt. (Leaf rust) Jiangsu Anhui

***Lespedeza virgata* (Thunb.) DC.**

Uromyces lespedezae-procumbentis (Schw.) Curt. (Leaf rust) Sichuan

Ligustrum spp.

Aecidium klugkistianum Diet. (Leaf rust) Hebei Henan Jiangsu Hunan Yunnan

Aecidium ligustricola Cumm. (Leaf rust) Liaoning Hebei Shanxi Xinjiang Guangxi
Coriolus hirsutus (Wulf ex Fr.) Quél. (Decay) Yunnan
Phyllosticta ligustrina Sacc. (Leaf spot) Jiangsu
Septobasidium bogoriense Pat. (Plaster) Jiangsu Zhejiang Guangxi

***Ligustrum japonicum* Thunb.**
Phaeosaccardinula javanica (Zimm.) Yamam. (Black mildew) Taiwan
Phyllosticta ligustri Sacc. (Leaf spot) Jilin

***Ligustrum lucidium* Ait**
Aecidium klugkistianum Diet. (Leaf rust) Jiangsu Hubei Hunan Guangxi
Cuscuta japonica Choisy (Cuseata high plant mistletoe) Henan
Cercospora ligustricola Tai (Leaf spot) Sichuan
Loranthus parsiticus (Linn.) Merr. (Mistletoe) Guangxi
Septobasidium bogoriense Pat. (Plaster) Zhejiang Hunan

***Ligustrum obtusifolium* Sieb. et Zucc.**
Aecidium klugkistianum Diet. (Leaf rust) Anhui
Cercospora ligustri Roum. (Leaf spot) Jiangsu
Phyllosticta ligustrina Sacc. (Leaf spot) Liaoning

***Ligustrum quithoui* Carr.**
Aecidium klugkistianum Diet. (Leaf rust) Jiangsu Henan Hubei

***Ligustrum sinense* Lour.**
Aecidium klugkistianum Diet. (Leaf rust) Jiangsu
Cercospora ligustricola Tai (Leaf spot) Sichuan

***Liquidambar formosana* Hance**
Cercospora liquidambaris Cooke et Ell. (Leaf spot) Taiwan
Cercospora tuberculans Ell. et Ev. (Leaf spot) Hunan
Cuscuta japonica Choisy (Cuseata high plant mistletoe) Zhejiang
Didymosporium liquidambaris Teng. (White leaf spot) Jiangsu
Elytranthe fordii (Hance) Merr. (Maple mistletoe) Guangxi Jiangsu
Loranthus parasiticus (Linn.) Merr. (Mistletoe) Guangxi
Loranthus yadoriki Sieb. et zucc. (Camphor mistletoe) Guangxi
Macrophomina phaseoli (Maubl.) Ashby. (Stem rot)
Uncinula liquidambaris Zheng et Chen (Powder mildew) Jiangsu Anhui Guangxi
Uncinula variabilis Zheng et Chen (Powder mildew) Hubei
Viscum articulatum Burm f. (Leafless maple mistletoe) Guangxi

***Livistona chinensis* R. Br.**
Anthostomella livistonae Girz. Guangdong
Botryodiplodia palmarum (Cooke) Petr. et Syd. Guangdong
Stylina disticha (Ehrenb.) Syd. Guangdong

***Livistona subglobosa* (Hassk.) Mart.**
Phaeosaccardinula javanica (Zimm.) Yamam. (Black mildew) Taiwan

Lonicera spp.
Fomes calcitratus Berk. et Curt. (Decay) Yunnan
Hirschioporus paragamenus (Fr.) Bond. et Sing. (Decay) Liaoning
Irenina lonicerae Yamam. (Black mildew) Guizhou
Kabatia latemarensis Bub. (Leaf spot) Liaoning Xinjiang
Melasmia lonicerae Jacz. (Black mole) Sichuan
(: Rhytisma lonicerae P. Henn.) Lonicera maackii (Rupr.) Maxim.
Microsphaera lonicerae (DC.) Wint. (Powder mildew) Xinjiang
Phellinus setulosus (Lloyd) Imaz. (Decay) Sichuan
Puccinia festucae Plowr. (Leaf rust) Xinjiang
Puccinia longirostris Kom. (Leaf rust) Xinjiang

Rhytisma lonicericola P. Henn. (Black mole) Hebei Xinjiang
Septoria lonicerae-maackii Miura (Leaf spot) Liaoning Xinjiang

***Lonicera chrysantha* Turcz.**
Rhytisma loniceriola P. Henn. (Black mole) Shanxi

***Lonicera coerulea* L.**
Coleosporium lonicerae Wang et S. X. Wei (Leaf rust) Tibet
Puccinia festucae Plowr. (Leaf rust) Sichuan

***Lonicera cyanocarpa* Franch.**
Rhabdospora decipiens (Berk. et Cooke) Sacc. Yunnan

***Lonicera japonica* Thunb**
Ascochyta tenerrima Sacc. et Roum. (Leaf spot) Jilin
Cercospora lonicericola Yamam. (Leaf spot) Taiwan
Cercospora periclymeni Wint. (Leaf spot) Anhui
Irenina lonicerae Yamam. (Black mildew) Taiwan
Puccinia festucae Plowr. (Leaf rust) Jiangsu Gansu
Melasmia lonicerae Jacz. (Black mole) Jilin
Puccinia festucae Plowr. (Leaf rust) Jiangsu
Rhytisma lonicericola P. Henn. (Black mole) Jilin Liaoning Heilongjiang Henan Sichuan
Septoria lonicerae-maackii Miura (Leaf spot) Liaoning

***Lonicera maximowiczii* Waxim.**
Leptostroma lonicericolum Rabenh. Shanxi

***Lonicera modesta* Rehd.**
Melasmia lonicerae Jacz. (Black mole) Jiangxi
Rhytisma lonicericola P. Henn. (Black Tar) Jiangxi

***Lonicera myrtillus* Hook f. et Thoms.**
Puccinia festucae Plowr. (Leaf rust) Tibet

***Lonicera nervosa* Maxim.**
Puccinia festucae Plowr. (Leaf rust) Sichuan

***Lonicera orientalis* Lam.**
Leptostroma lonicericolum Rabenh. Shanxi

***Lonicera stephanocarpa* Franch.**
Phyllachora xylostei (Fr.) Fuck. (Black tar) Shanxi

***Lonicera trichosantha* Bur. et Franch.**
Rhytisma lonicericola P. Henn. (Black tar) Yunnan

***Lonicera vesicaria* Kom.**
Phyllosticta caprifolii (Opiz) Sacc. (Leaf spot) Liaoning

***Loropetalum chinense* (R. Br.) Oliv.**
Loranthus maclurei Merr. (Mistletoe) Fujian Guangdong Guangxi Guizhou

***Lycium barbarum* L.**
Eriophyes sp. (Leaf mites) Ningxia Gansu Qinghai Neimeng Shanxi
Glomerella cingulata (Stonem.) Spauld. et Scherenk. (Anthracnose) Hebei Shanxi Shandong Ningxia

***Lycium chinense* Mill.**
Alternaria dauci (Kühn) Groves et Skolko f. sp. solani (Ell. et Mart.) Neerg Jilin
Alternaria solani (Ell. et Mart.) Jones et Grout (Lycium early blight) Neimeng Xinjiang
Cercospora chengtuensis Tai (Leaf spot) Sichuan
Cercospora lycii Ell. et Halst. (Leaf spot) Jilin
Eriophyes sp. (Leaf mites) Neimeng
Erysiphe cichoracearum DC. (Powder mildew) Hebei Xinjiang
(=Arthrocladiella mougeotii (Lév.) Vassilk.)
Fusarium solani (Mart.) App. et Wollenw. (Wilt) Qinghai

Microsphaera mougeotii Lév. (Powder mildew) Gansu
Peronospora lycii Ling et M. C. Tai (Frost mold) Sichuan

***Maackia amurensis* Rupr. et Max.**
Ascochyta sp. (Leaf spot) Dongbei
Cercospora cladrastidis Jacz. (Leaf spot) Jilin Liaoning Heilongjiang
Phyllosticta sp. (White leaf spot) Liaoning
Uromyces amurensis Kom. (Leaf rust) Jilin Liaoning Heilongjiang

Machilus spp.
Aecidium machili P. Henn. (Leaf rust) Sichuan
Aegerita webberi Fawcett Guangxi
Aschersonia placenta Berk. et Br. Guangxi
Fomes conchotus (Pers.) Gill. (Brown pocket rot) Hainandao
Melanopsichium inouyei (P. Henn. et Shirai) Ling (Black powder) Sichuan
Puccinia machili Cumm. (Leaf rust) Guangxi
Puccinia machilicola Cumm. (Leaf rust) Guizhou
Septobasidium bogoriense Pat. (Felt fungus) Taiwan

***Machilus bournei* Hemsl.**
Aecidium machili P. Henn. (Leaf rust) Sichuan

***Machilus thunbergii* Sieb. et Zucc.**
Aecidium machili P. Henn. (Leaf rust) Taiwan
Meliola machili Yamam. (Black mildew) Taiwan
Septobasidium albidum Pat. (Plaster) Taiwan

Magnolia spp.
Cephaleuros virescens Kunze (Algae) Fujian Guangdong Guangxi Taiwan Yunnan
Loranthus parasiticus (Linn.) Merr. (Mistletoe) Fujian Taiwan Hunan Guangdong Guangxi Yunnan Guizhou
Macrosporium trichellum Arc. et Sacc. Jiangsu
Phyllactinia magnoliae Yu et Lai (Powder mildew) Jiangxi

***Magnolia denudata* Desr.**
Colletotrichum gloeosporioides Penz. (Anthracnose) Jilin
(=Gloeosporioides magnoliae Sacc.)

***Magnolia liliflora* Desr.**
Phyllosticta magnoliae Sacc. (Leaf spot) Jilin

***Magnolia officinalis* Rehd. et Wils. var biloba Rehd et Wils.**
Coriolus versicolor (L. ex Fr.) Quél. (Decay) Jiangxi

Mahonia spp.
Gambleola cornuta Mass. (Leaf rust) Jiangxi Fujian Sichuan
Puccinia graminis Pers. (Leaf rust) Sichuan

***Mahonia bealei* (Fort.) Carr.**
Asterina mahoniae Keissl. Hunan
Chlorocyphella aeruginascens (Karst.) Keissl. Hunan
Gambleola cornuta Mass. (Leaf rust) Shanxi
Phyllosticta berberidis Rabenh. (Leaf spot) Jilin

***Mahonia fortunei* Mouillef**
Puccinia graminis Pers. (Leaf rust) Sichuan

***Mahonia llomariifolia* Takeda**
Gambleola cornuta Mass. (Leaf rust) Taiwan

***Mahonia sheridaniana* Schneid.**
Gambleola cornuta Mass. (Leaf rust) Guangxi

Melia spp.
Cercospora meliae Ell. et Ev. (Leaf spot) Jiangxi

Cercospora subsessilis H. et P. Syd. (Leaf spot) Jiangxi
Cerrena unicolor (Bull. ex Fr.) Murr. (Decay) Guizhou

***Melia azedarach* L.**

Cercospora meliae Ell. et Ev. (Brown leaf spot) Jiangsu Henan Hunan Guangdong Guangxi Fujian Taiwan Yunnan Sichuan
Cercospora subsessilis H. et P. Syd. (Brown leaf spot) Jiangsu Henan Hubei Guangdong Hunan Fujian Shanxi Taiwan Sichuan Yunnan Guangxi
Cuscuta japonica Choisy (Cuseata high plant mistletoe) Henan
Melanconium meliae Teng (Dieback) Jiangsu
MLO, BLO (Witches' broom) Zhejiang Guangdong Guangxi Hunan Hubei Jiangsu
Phellinus torulosus (pers.) Bourd. et Galz. (Decay) Guangdong
Phellinus williamsii (Murr.) Pat. (Decay) Taiwan
Phyllactinia toonae Yu et Lai (Powder mildew) Sichuan

***Metasequoia glyptostroboides* Hu et Cheng**

Macrophomina phaseoli (Maubl.) Ashby (Stem rot) Jiangsu
Monochaetia sp. (Leaf spot) Shanxi
Pestalotia funereal Desm. (Needle cast) Shanxi Hunan
Rhizoctonia solani Kühn (Damping off) Jiangsu Zhejiang
Physical (Physical drying) Shanxi

***Michelia macclurei var. sublanea* Dandy**

Cephaleuros parsiticus Karst (Algal) Guangxi
Cephaleuros virescens Kunze (Algal) Guangxi

***Michelia tenuipes* Dandy**

Cercospora sp. (Early leaf blight) Guangxi
Cephaleuros parasiticus Karst. (Algal) Guangxi

***Morina alba* Hand-Mazz.**

Puccinia heterocolor M. M. Chen (Leaf rust) Tibet

***Morus mongolica* Schneid.**

Aecidium mori Bracl. (Leaf rust) Liaoning
Clasterosporium mori Syd. (Leaf mold) Liaoning
Diplodia mori Westend. (Dieback) Heilongjiang
Nothopatella chinensis Miyake (Dieback) Liaoning Heilongjiang
Phyllactinia moricola (P. Henn.) Homma. (Powder mildew) Liaoning Heilongjiang
Wilt Heilongjiang

***Mytilaria laosensis* Lec.**

Cassytha filiformis L. (Cassytha mistletoe) Guangxi
Cercospora sp. (Leaf spot) Guangxi
Eriophyes sp. (Leaf felt spot) Guangxi

Olea spp.

Cercospora cladosporioides Sacc. (Leaf spot) Guangdong

***Olea europaea* L.**

Cercospora sp. (Leaf spot) Sichuan
Colletotrichum gloeosporioides Penz. (Anthracnose) Sichuan Yunnan Guangxi Hubei Jiangsu Shanxi
(=Gloeosporium olivae (Petri.) Foschi.)
Fusarium sp. (Damping off) Sichuan
Meloidogyne acrita (Chitwood) Esser, Perry and Toylor (Root-Knot nematode) Jiangxi Fujian Guangdong Guangxi
Meloidogyne arenaria (Neal) Chitwood (Root-knot nematode) Jiangxi Fujian Guangdong Guangxi

Meloidogyne incognita (Koforid & White) Chitwood (Root-knot nematode) Jiangxi Fujian Guangdong Guangxi

Meloidogyne javanica (Trenb) Chitwood (Root-knot nematode) Jiangxi Fujian Guangdong Guangxi

Pseudomonas savastanoi (E.F. Smith) Stev. (Root gall) Guangxi Sichuan Hubei Guizhou Yunnan Jiangxi

Pseudomonas solanacearum E.F. Smith (Root gall) Guangxi Hubei Jiangxi Fujian Sichuan Guangdong Hunan

Sclerotium sp. (Canker) Sichuan

Spilocaea oleaginea (Cast.) Hugh. (Peacock spot) Sichuan Yunnan
(=Cycloconium oleaginum Cast.)

***Ormosia formosana* Kaneh.**

Lembosia ormosiae Yamam. Taiwan

***Ormosia pinnata* (Lour.) Merr.**

Meliola bataanensis Syd. (Black mildew) Guangdong

Meliola franciscana Hansf. (Black mildew) Guangdong

***Paconia lactiflora* Pall.**

Aecidium paeoniae Kom. (Leaf rust) Heilongjiang

Cronartium flaccidum (Alb. et Schw.) Wint. (Autoecious host Needle rust) Liaoning Jilin Dongbei Jiangsu Zhejiang

Paliurus ramosissimus Poir.

Phakopsora zizyphi-vulgaris (P. Henn.) Diet. (Leaf rust) Hebei Henan Shandong

Paulownia spp.

Aleumtiolis

Candidatus Phytoplasma asteri (Witches' broom) Hebei Henan

Cercospora paulowniae Hori (Leaf spot) Henan

Colletotrichum gloeosporioides Penz. (Anthracnose) Henan Shanxi Huabei Huadong Huanan Hebei

Funalia hispida (Bagl.) Pat. (Decay) Shandong

MLO (Whiches' broom) Shandong Shanxi Jiangsu Anhui Jiangxi Henan Sichuan Shanxi Gansu

Phyllostica sp. (Leaf spot) Gansu

Sphaceloma paulowniae Hara (Paulownia scab) Henan Shandong

Valsa sp. (Canker) Shandong

Maerophoma phomopsis Colletotrichum sp.

Physical: (Sunburn) Henan

***Paulownia elongata* S. Y. Hu**

Meloidogyne incognita (Koforid et White) Chitwood (Root-knot nematode) Henan

MLO (Whiches' broom) Hubei Hunan Henan

Sphaceloma paulowniae Hara (Paulownia scab)Henan Shanxi Shandong Hunan

Valsa paulowniae Miyabe et Hemmi (Canker) Henan Shandong

***Paulownia fargesii* Franch.**

Phyllactinia salmonii Blum. (Powder mildew) Sichuan

***Paulownia fortunei* (Seew.) Hemsl.**

Asochyta paulowniae Sacc. et Brun (Leaf spot) Jilin

Colletotrichum gloeosporioides Penz. (Anthracnose) Huabei Huadong Huanan Xibei Shandong Shanxi Hunan Hubei
(=Colletotrichum kawakamii (Miyabe) Sawada)

Helicobasidium purpureum (Tul.) Pat. (root rot) Shanxi

Loranthus parasiticus (Linn.) Merr. (Mistletoe) Guangxi

Loranthus yadoriki Sieb. et Zucc (Mistletoe) Guangxi

MLO (Whiches' broom) Sichuan Henan Zhejiang Hebei Hubei Shanxi Shandong Guangxi Jiangsu Fujian Jiangxi Hebei Anhui
Phylllactinia salmonii Blum. (Powder mildew) Sichuan
Sclerotium rolfsii Sacc. (Sclerotium) Henan Hunan Guangxi
Sphaceloma paulowniae Hara (Paulownia scab) Henan Shandong Hunan Shanxi Hubei

***Paulownia Kawakamii* Ito**
(=Paulownia thyrosidea Rehd.)
Cercospora paulowniae Hori (Leaf spot) Taiwan

***Paulownia tomentosa* (Thunb) Steud.**
Cercospora paulowniae Hori (Leaf spot) Henan
Colletotrichum gloeosporioides Penz. (Anthracnose) Liaoning Henan Shandong
MLO (Whiches' broom) Henan
Mycosphaerella corylea (Pers.) Karst. (Leaf spot) Henan
Mycosphaerella paulowniae Shirai et Hara (Leaf spot) Henan
Phyllactinia paulowniae Yu (Powder mildew) Sichuan
Phyllactinia salmonii Blum. (Powder mildew) Sichuan
(=Phyllactinia corylea (Pers.) Karst.)
Septobasidium tanakae (Miyabe) Boed. et Stein (Plaster) Taiwan
(=Phyllactinia imperialis Miyabe)
Sphaceloma paulowniae Hara (Paulownia scab) Henan Shandong Shanxi Hunan
Uncinula clintonii Peck (Powder mildew) Sichuan

***Pedicularis resupinata* L.**
Cronartium ribicola J. C. Fischer ex Rabenhorst (Leaf rust) Dongbei
Puccinia clintonii Peck (Leaf rust) Shanxi

***Pedicularis resupinata var. ramose* Nakai**
Cronartium ribicola J.C. Fischer ex Robenhorst (Leaf rust) Dongbei

***Pedicularis spicata* pallas**
Cronartium ribicola J.C. Fischer ex Robenhorst (Leaf rust) Dongbei

***Phellodendron amurense* Rupr.**
Coleosporium phellodendri Kom. (Leaf rust) Liaoning Jilin Heilongjiang Shanxi
Phyllosticta phellodendri Allesch. (Leaf spot) Liaoning Heilongjiang

***Phellodendron sachalinense* Sarg.**
Ascochyta phellodendri Kab. et Bub. (Leaf spot) Jilin
Ascochyta pirina Pegl. (Leaf spot) Jilin
Coleosporium phellodendri Kom. (Leaf rust) Liaoning
Daldinia concentrica (Bolt.) Ces. et de Not. (Brown gall) Liaoning Heilongjiang
Schizophyllum commune Fr. (Decay)

***Philadelphus tenuifolius* Rupr. ex Maxim**
Phyllosticta vulgaris Desm. var. philadelphi Sacc. (Leaf spot) Jilin Heilongjiang

Phoebe sp.
Puccinia cinnamomi Tai (Leaf rust) Jiangxi

***Phoebe formosana* Hayata**
Aecidium machili P. Henn. (Leaf rust) Taiwan
Armata macrospora (Yamam.) Yamam. (Black mildew) Taiwan

***Phoebe nanmu* (Oliv.) Gamdle.**
Schlerotium rolfsii Sacc. (Schlerotium) Province not determined

Photinia spp.
Gymnosporangium japonicum Syd. (Leaf rust) Anhui Hunan Taiwan
Rhizoctonia solani Kuhn. (Damping off)

***Photinia serrulata* Lindl.**
Cercospora eriobotryae (Enj.) Saw. (Leaf spot) Jilin Henan

Pestalotia photiniae Thüm. (Leaf spot) Jiangsu
Roestelia wenshanensis (Tai) Tai (Leaf rust) Yunnan

***Photinia taiwanenesis* Hayata**

Neocapnodium tanakae (Shirai et Hara) Yamam. (Black mildew) Taiwan
Phaeosaccardinula javanica (Zimm.) Yamam. (Black mildew) Taiwan
Triposporiopsis spinigera (Höhn.) Yamam. (Black mildew) Taiwan

***Photinia villosa* (Thunb.) DC.**

Gymnosporangium haraenum Syd. (Leaf rust) Dongbei
Gymnosporangium japonicum Syd. (Leaf rust) Jiangsu Hunan

Phyllostachys spp.

Balansia take (Miyake) Hara (Whiches' broom) Jiangsu Zhejiang Henan Hunan Guizhou Hubei
Calonectria bambusae (Hara) Höhn. (Swell spot) Liaoning Jiangsu Guizhou
Coccostroma arundinariae (Hara) Teng Jiangsu Zhejiang Anhui Guangxi Sichuan
Coniosporium bambusae (Thöm. et Bolle) Sacc. (Leaf spot) Jiangsu Hunan Guangdong Guangxi Yunnan
Corticium bambusac Burt. (Damping off) Henan
Epichloë bambusae Pat. (Bamboo Epichloë disease) Jiangsu Zhejiang Guizhou
Helminthosporium foveolatum Pat. (Leaf spot) Province not determined
Homostegia fusispora Syd. Guangdong
Loculistroma bambusae Patters. et Charles Hubei
Meliola acristae Hansf. (Black mildew) Zhejiang
Meliola phyllostachydis Yamam. (Black mildew) Taiwan
Metashaeria deviata Syd. Zhejiang
Myriangium haraeanum Tai et Wei (Scale fungus) Jiangsu Zhejiang Hubei Guizhou Sichuan
Papularia arundinis (Corda) Fr. Zhejiang
Phyllachora orbicular Rehm (Black tar) Fujian Yunnan
Phyllachora phyllostachydis Hara (Black tar) Zhejiang
Phyllachora shiraiana Syd. (Black tar) Guangdong Sichuan Guizhou
Pseudolachnella scolecospora (Teng et Shen) Teng Zhejiang Hunan
Puccinia longicornis Pat. et Har. (Leaf rust) Jiangsu Zhejiang Anhui
Puccinia melanocephala Syd. (Leaf rust) Anhui
Puccinia phyllostachydis Kus. (Leaf rust) Jiangsu Zhejiang Henan Yunnan
Scolecotrichum phyllostachydis Teng Jiangsu
Shiraia bambusicola P. Henn. (Bamboo yellow gall) Jiangsu
Shiraiella phyllostachydis Hara (Small Bamboo yellow gall) Henan
Stereostratum corticioides (Berk. et Br.) Magn. (Leaf rust) Jiangsu Zhejiang Guangxi Sichuan Guizhou Hunan
Ustilago shiraiana P. Henn. (Bamboo smut) Jiangsu Jiangxi Fujian Henan Guizhou Taiwan

***Phyllostachys bambusoides* Sieb. Et Zucc.**

Balansia take (Miyake) Hara (Whiches' broom) Henan
Coniosporium saccardianum Teng (Awell l spot) Henan
Myriangium haraeanum Tai et Wei (Bamboo black rotund spot) Henan Jiangsu
Phyllachora phyllostachydis Hara (Bamboo black swell) Zhejiang
Phyllachora shiraiana Syd. (Black mole) Hunan
Phyllachora sinensis Sacc. (Black mole) Sichuan
Stereostratum corticioides (Berk. et Br.) Magn. (Leaf rust) Shandong Henan Hubei
Ustilago shiraiana P. Henn. (Black powder) Henan Fujian Hunan

***Phyllostachys congesta* Rendle**

Balansia take (Miyake) Hara (Witches' broom) Hunan
Shiraia bambusicola P. Henn. (Bamboo yellow gall) Henan Hunan

Stereostratum corticioides (Berk. et Br) Magn. (Stereostratum rust) Hunan
Ustilago shiraiana P. Henn. (Black powder) Jiangsu Henan Jiangxi Fujian Guizhou Taiwan Hunan

***Phyllostachys glauca* McCl.**
(=*Phyllostachys puberula (Miq.) Munro*)
Balansia take (Miyake) Hara (Witches' broom) Zhejiang Henan Hunan
Myriangium haraeanum Tai et Wei (Bamboo black rotund spot) Jiangsu Zhejiang Henan Hunan
Phyllachora phyllostachydis Haia (Black swell) Zhejiang
Puccinia melanocephala Syd. (Leaf rust) Guizhou
Puccinia phyllostachydis Kus. (Leaf rust) Yunnan
Stereostratum corticioides (Berk. et Br.) Magn. (Leaf rust) Jiangsu Zhejiang

***Phyllostachys makinoi* Hayata**
Chaetothyrium echinulatum Yamam. (Black mildew) Taiwan
Meliola phyllostachydis Yamam. (Black mildew) Taiwan
Neocapnodium tanakae (Shirai et Hara) Yamam. (Black mildew) Taiwan
Paranthostomella phyllostachydis Saw. Taiwan
Phaeosaccardinula javanica (Zimm.) Yamam. (Black mildew) Taiwan
Scorias communis Yamam. (Black mildew) Taiwan
Triposporiopsis spinigera (Hohn) Yamam. (Black mildew) Taiwan
Ustilago shiraiana P. Henn. (Black powder) Jiangsu Jiangxi Henan Fujian Guizhou Taiwan
Vermicularia nigronitentis Saw. (Anthracnose) Taiwan

***Phyllostachys pubescens* Mazel ex H. de Lehaie**
(=*Phyllostachys edulis A. et C. Riv.*)
Balansia taka (Miyake) Hara (Witches' broom) Guizhou Sichuan Hunan
Ceratosphaeria phyllostachydis (Shoot blight) Jiangsu Zhejiang Jiangxi Anhui Hubei Fujian
Fusarium sp. (Root rot) Sichuan Hunan
Fusarium moniliforme Sheld. (Trunk decay) Jiangsu Zhejiang Anhui
Helminthosporium foveolatum Pat. (Leaf dry spot) Hunan
Meliola phyllostachydis Yamam. (Black mildew) Taiwan Hunan
Rhizoctonia sp. (Root rot) Sichuan Hunan
Stereostratum corticioides (Berk. et Br.) Magn. (Leaf rust) Shandong Jiangsu Zhejiang Hunan

***Phyllostachys viridis* (Young) McClure**
Balansia take (Miyake) Hara (Witches' broom) Henan
Ustilago shiraiana P. Henn. (Black powder) Henan
Fusarium solani (Mart.) App. ex Wollenw. (Trunk brown rot) Jiangsu

Picea spp.
Arceuthobium chinense Lecomte (Dwraf mistletoe) Sichuan
Armillariella mellea (Vahl ex Fr.) Karst. (Root decay) Xinjiang
Cenangium ferruginosum Fr. ex Fr. (Dieback) Xinjiang
Chrysomyxa sp. (Leaf rust) Yunnan
Chrysomyxa deformans (Diet.) Jacz. (Leaf rust) Xinjiang
Chrysomyxa expansa Diet. (Leaf rust) Sichuan
Chrysomyxa ledi de Bary (Needle rust) Sichuan
Chrysomyxa pyrolae (DC.) Kostr. (Pine cone rust) Ningxia Qinghai Xinjiang Sichuan
Chrysomyxa rhododendri de Bary (Needle rust) Sichuan
Chrysomyxa succinea (Sacc.) tranz. (Leaf rust) Qinghai
Chrysomyxa weirii Jacks. (Needle rust) Xinjiang
Dothiorella sp. (Shoot blight) Sichuan
Fomitopsis pinicola (Sw. ex Fr. Karst.) (Brown cube rot) Xinjiang
Fomitopsis rosea (Alb. et Schw. ex Fr.) Karst. (Decay) Qinghai Xinjiang Sichuan Yunnan

Ganoderma applanatum (Pers.) Pat. (Decay) Xinjiang
Gloeophyllum abietinum (Bull. et Fr.) Karst. (Decay) Gansu Xinjiang
Gloeophyllum saepiarium (Wulf. ex Fr.) Karst. (Decay) Gansu Xinjiang
Gloeophyllum trabeum (Pers. ex Fr.) Murr. (Decay) Xinjiang
Heterobasidion annosus (Fr.) Bref. (Root rot) Xinjiang Sichuan Yunnan
Hirschioporus abietinus (Dicks. ex Fr.) Donk (Decay) Xinjiang
Hirschioporus laricinus (Karst.) teram. (Decay) Xinjiang
Hypoderma sp. (Falling needle) Sichuan
Inonotus dryadeus (Pers. ex Fr.) Murr. (Decay) Sichuan Yunnan
Laetiporus sulphureus (Bull. ex Fr.) Bond. et Sing. (Trunk rot) Qinghai
Lophodermium filiforme Dark. (Falling needle) Xinjiang
Lophodermium macrosporum (Hart.) Rehm (Leaf spot) Jilin Heilongjiang
Lophodermium picae (Fuck.) Hühn. (Falling needle) Xinjiang
Lophodermium pinastri (Schrad. ex Fr.) Chev. (Falling needle) Jilin Heilongjiang
Macrosporium helminthosporioides (Corda) Sacc. et Trav. Jiangsu
Melampsorella cerastii (Mart.) Wint. (Leaf rust) Sichuan
Phaeolus schweinitzii (Fr.) Pat. (Trunk brown cube rot) Dongbei Sichuan
Phellinus pini (Thore ex Fr.) Ames var. abietis Karst. (White pocket rot) Neimeng Jilin Heilongjiang Shanxi Hebei Xinjiang Sichuan Shanxi Gansu Ningxia Qinghai Tibet Yunnan
Phellinus hartigii (Allesch. et Schnabl) Imaz. (Brown rot) Sichuan
Phellinus yamanoi (Imaz.) Shaw. (Decay) Sichuan
(=Cryptoderma yamanoi Imaz.)
Poria odora (Peck) Sacc. (Decay) Yunnan
Poria purpurea (Fr.) Cooke (Decay) Xinjiang
Rhizoctonia solani Kühn (Damping off) Xinjiang
Stereum versicolor (Sw.) Fr. (Decay) Xinjiang
Thekopsora areolata (Fr.) Magn. (Pine cone rust) Jilin Heilongjiang Qinghai Xinjiang Sichuan Yunnan Tibet
(=Pucciniastrum padi Diet.)
Tyromyces guttulatus (Peck) Murr (Decay) Heilongjiang Hebei Yunnan
Hidden stem rot Xinjiang

***Picea asperata* Mast.**

Chrysomyxa deformans (Diet.) Jacz. (Leaf rust) Xinjiang
Chrysomyxa ledi de Bary (Leaf rust) Sichuan
Fomitopsis pinicola (Sw. ex Fr.) Karst. (Brown cube rot) Gansu
Fomitopsis rosea (Alb. et Schw. ex Fr.) Karst. (Decay) Gansu
(=Fomes roseus (Alb. et Schw. ex Fr.) Cooke)
Hirschioporus abietinus (Dicks. ex Fr.) Donk (White rot) Shanxi
Lophodermium filiforme Dark. (Long spot) Shanxi
Lopodermium macrosporum (hart.) Rehm (Leaf spot) Shanxi
Phellinus pini (Thore ex Fr.) Ames var. abietis karst. (White pocket rot) Shanxi Sichuan Tibet
Thekopsora areolata (Fr.) Magn. (Pine cone rust) Shanxi Gansu Xinjiang Sichuan Yunnan

***Picea brachytyla var. complanata* (Mast.) Cheng ex Rehd.**

Thekopsora areolata (Fr.) Magn. (Pine cone rust) Tibet

***Picea jezoensis Carr var. microsperma* (Linkdl.) Cheng et L.K. Fu**

Chrysomyxa pyrolae (DC.) Kostr. (Pine cone rust) Heilongjiang
Fomitopsis rosea (Alb. et Schw. ex Fr.) Karst. (Decay) Heilongjiang
Lophodermium pinastri (Schrad ex Fr.) Chev. (Falling needle) Dongbei
Phellinus pini (Thore ex Fr.) Ames var. abietis Karst. (White pocket rot) Heilongjiang Neimeng
Thekopsora areolata (Fr.) Magn. (Pine cone rust) Heilongjiang Neimeng

Tyromyces galactinus (Berk.) Bond. (Decay) Heilongjiang

Picea koraiensis **Nakai**

Chysomyxa pyrolae (DC.) Kostr. (Pine cone rust) Jilin Heilongjiang
Lophodermium pinastri (Schrad. ex Fr.) Chev. (Falling needle) Dongbei
Thekopsora areolata (Fr.) Magn. (Pine cone rust) Jilin Heilongjiang

Picea likiangensis **(Franch.) Pritz.**

Chrysomyxa ledi de Bary (Leaf rust) Sichuan Yunnan
Fomitopsis pinicola (Sw. ex Fr.) Karst. (Brown cube rot) Sichuan Yunnan Tibet
Heterobasidion annosus (Fr.) Bref. (Root rot) Sichuan Yunnan
Phellinus pini (Thore ex Fr.) Ames var. abietis Karst. (White pocket rot) Sichuan Yunnan Tibet
Thekopsora areolata (Fr.) Magn. (Pine cone rust) Sichuan Yunnan Tibet

Picea likiangensis var. balfourana **(Rehd. et Wils.) Hillier ex Slavin.**

Climacocystis borealis (Fr.) Kotlaba et Pouzar (Decay) Tibet Qinghai
(=Daedalea borealis (Fr.) Quel.)
Fomitopsis pinicola (Sw. ex Fr.) Karst. (Brown cube rot) Tibet Qinghai
Fomitopsis rosea (Alb. et Schw. ex Fr.) Karst. (Decay) Tibet Qinghai
Gloeophyllum saepiarium (Wulf. ex Fr.) Donk (Decay) Tibet Qinghai
Hirschioporus abietinus (Dicks. ex Fr.) Donk (Decay) Tibet Qinghai
Hirschioporus fusco-violaceus (Schrod. ex Fr.) Donk (Decay) Tibet
Oxyporus populinus (Schum. ex Fr.) Donk (Decay) Tibet

Picea likiangensis **(Franch.) Pritz var. linzhiensis Cheng et L. K. Fu.**

Coriolus versicolor (L. ex Fr.) Quél. (Decay) Tibet
Fomitopsis rosea(Alb. et Schw. ex Fr.) Karst. (Decay) Tibet
Phellinus pini (Thore ex Fr.) Ames var. abietis Karst. (White pocket rot) Tibet
Thekopsora sparsa (Wint.) Magn. (Pine cone rust) Tibet

Picea neoveitchii **Mast.**

Fomitopsis pinicola (Sw. ex Fr.) Karst. (Brown cube rot) Shanxi
Gloeophyllum subferrugineum (Berk.) Bond. et Sing. (Decay) Gansu
Gloeophyllum trabeum (Pers. ex Fr.) Murr. (Decay) Gansu
Hirschioporus abietinus (Dicks. ex Fr.) Donk (White rot) Shanxi
Irpex obliquus Fr. (Decay) Guizhou
Lophodermium macrosporum (Hart.) Rehm. (Leaf spot) Shanxi
Phellinus pini (Thore ex Fr.) Ames var. abietis Karst. (White pocket rot) Shanxi Gansu

Picea purpurea **Mast.**

Fomitopsis pinicola (Sw. ex Fr.) Karst. (Brown cube rot) Gansu
Phellinus pini (Thore ex Fr.) Ames var. abietis Karst. (White pocket rot) Gansu Sichuan
Thekopsora areolata (Fr.) Magn. (Pine cone rust) Gansu Sichuan

Picea schrenkiana **Fisch. et Mey.**

Chrysomyxa deformans (Diet.) Jacz. (Leaf rust) Xinjiang
Chrysomyxa pyrolae (DC.) Kostr. (Pine cone rust) Xinjiang
Chrysomyxa weirii Jacks. (Needle rust) Xinjiang
Fomes rufolaccatus Lloyd (Decay) Xinjiang
Fomitopsis pinicola (Sw. ex Fr.) Karst. (Brown cube rot) Xinjiang
Fomitopsis rosea (Alb. et Schw. ex Fr.) Karst. (Decay) Xinjiang
Fumago sp (Black mildew) Xinjiang
Heterobasidionannosus (Fr.) Bref. (Pine root rot) Xinjiang
Lophodermium piceae (Fuck.) Höhn. (Falling needle) Xinjiang
Megaloseptoria mirabilis Naum. Xinjiang
Pleurotus ostreatus (Jacq. ex Fr.) Quél. (Decay) Xinjiang
Pleurotus nidulans (Pers. ex Fr.) Gill. (Decay) Xinjiang
Rhizoctonia solani Kühn (Damping off) Xinjiang

Thekopsora areolata (Fr.) Magn. (Pine cone rust) Xinjiang

***Picea spinulosa* (Griff.) Henry.**

Thekopsora areolata (Fr.) Magn. (Pine cone rust) Tibet

Pinus spp.

Armillariella mellea (Vahl ex Fr.) Karst. (Root decay) Jilin Sichuan

Cercospora pini-densiflorae Horiet Nambu (Leaf spot) Ningxia Jiangsu Guangdong Fujian Hunan Zhejiang Guangxi Taiwan

Coleosporium solidaginis (Schw.) Thüm. (Leaf rust) Shanxi Yunnan Guizhou

Coriolus versicolor (L. ex Fr.) Quél. (Decay) Hunan Guangxi Yunnan

Cronartium quercuum (Berk.) Miyabe (Pine gall rust) Zhejiang Shanxi Anhui Hunan Henan

Cryptoporus volvatus (Peck) Hubb. (Decay) Hebei Jilin Hubei Guangdong Fujian Sichuan Yunnan

Daedalea biennis (Bull.) Fr. (Decay) Jiangsu

Fomitopsis pinicola (Sw. ex Fr.) Karst. (Brown cube rot) Dongbei Hebei Fujian Gansu

Fusarium oxysporum Schlecht. (Damping off) Shanxi Ningxia

Fusarium solani (Mart.) App. et Wollenw. (Damping off) Hebei

Fusarium spp. (Damping off) Shandong Ningxia Xinjiang Hunan Jilin

Gloeophyllum saepiarium (Wulf. ex Fr.) Karst. (Decay) Hebei Tibet

Gomphidius rutilus (Schaeff. ex Fr.) Lund. et Nannf. Hebei

Hypoderma desmazieri Duby (Red blight) Guangxi Sichuan Guizhou

Laricifomes officinalis (Vill. ex Fr.) Kotlaba ex Pouzar (Heartwood brown cube rot) Shanxi Hebei Heilongjiang Jilin Xinjiang Sichuan Yunnan Neimeng

Lophodermium sp. (Pine leaf spot) Yunnan

Lophodermium pinastri (Schrad ex Fr.) Chev. (Pine leaf spot) Distributed everywhere in our country

(Asexual :Leptostroma pinastri Desness)

Peridermium pini (Willd.) Kleb. (Dry rust) Heilongjiang Guizhou

Pestalotia funereal Desm. (Tip blight) Guangdong

Phaeolus schweinitzii (Fr.) pat. (Brown cube rot) Heilongjiang Jilin Hebei Xinjiang Sichuan Yunnan

Phellinus pini (Thore ex Fr.) Ames (White pocket rot) Heilongjiang Jilin Hebei Shanxi Shanxi Neimeng Gansu Xinjiang Ningxia Yunnan Sichuan

Phellinus robustus (Karst.) Bourd. et Galz. (Sapwood brown rot) Dongbei

Poria cocos (Fr.) Wolf (Decay) Henan Zhejiang Hubei Anhui Fujian Sichuan Yunnan Taiwan Guizhou

Pythium spp. (Damping off) Shandong Ningxia Xinjiang Hunan

Rhizoctania solani Kühn (Sprout damping off) Hebei Hunan Shandong Ningxia Jilin Xinjiang

Trametes cervina (Schw.) Bres. (Decay) Tibet

Tyromyces anceps (Peck) Murr. (Decay) Yunnan

***Pinus armandi* Franch.**

Armillariella tabescens (Scop. ex Fr.) Sing. (Armillaria root rot) Guizhou

Cercospora pini-densiflorae Hori et Nambu (Leaf spot) Anhui

Cladosporium nigrellum Ell. & Ev.

Cladosporium psoraleae M. B. Ellis

Coleosporium melampyri Tul. (Leaf rust) Shanxi

Coleosporium solidaginis (Schw.) Thüm. (Leaf rust) Sichuan Guizhou

Coleosporium senecionis (Pers.) Fr. (Leaf rust) Sichuan

Cronartium quercuum (Berk.) Miyabe (Pine gall rust) Henan Yunnan Tibet

Cronartium ribicola J.C. Fischer ex Rabenhorat (Five needle pine rust) Shanxi Shanxi Sichuan Hunan

Cryptocline conigena (Sacc. &Roum.) v. Arx

Cytospora pini Desm.
Eriophyes pini Nal. (leaf mites) Shanxi
Fusarium oxysporum Schlecht. (Damping off) Shanxi
Gloeophllum saepiarium (Wulf. ex Fr.)Karst. (Decay) Shanxi Gansu
Gloeophyllum subferrugineum (Berk.) Bond. et Sing. (Decay) Gansu
Hirschioporus abietinus (Dicks. ex Fr.) Donk (White rot) Shanxi
Hypoderma desmazieri Duby (Needle blight) Shanxi Sichuan
Lophodermium durilabrum (Scatter spot) Shanxi Hunan Guizhou
Lophodermiumpinastri (Scrad. ex Fr.) Chev. (Needle blight) Hebei Shanxi Sichuan Guizhou Hunan
Pestalotia funerea Desm. (Tip blight) Guizhou
Pestalotiopsis disseminata (Thuem.) Stey.
Phaeolus schweinitzii (Fr.) Pat. (Brown cube rot) Tibet
Phellinus pini (Thore ex Fr.) Ames (White pocket rot) Shanxi Gansu Sichuan Taiwan
Rhizoctonia solani Kühn (Damping off) Shanxi Hunan

***Pinus bungeana* Zucc. et Endl.**

Fusarium oxysporum Schlecht. (Damping off) Shanxi
Fusarium solani (Mart.) App. et Wollenw. (Damping off) Hebei
Gloeophyllum saepiarium (Wulf. ex Fr.) Karst. (Brown rot) Shanxi
Lophodermium pinastri (Schrad. ex Fr.) Chev. (Falling needle) Hebei Shanxi
Rhizoctonia solani Kühn. (Damping off) Hebei

***Pinus densata* Mast.**

Arceuthobium pini Hawksworth et Wiens. (Pine mistletoe) Tibet Sichuan
Fomitopsis rosea (Alb. et Schw. ex Fr.) Karst. (Decay) Tibet Sichuan
Gloeophyllum saepiarium (Wulf. ex Fr.) Karst. Tibet Sichuan
Lophodermium pinastri (Schrad. ex Fr.) Chev. (Falling needle) Sichuan Tibet
Phellinus pini (Thore ex Fr.) Ames (White pocket rot) Sichuan Tibet

***Pinus densiflora* Sieb. et Zucc.**

Bursaphelenchus xylophilus (Steiner et Buhrer) Nickle Jiangsu Anhui
Cronartium flaccidum (Alb et Schw.) Wint. (Fusiform rust) Dongbei Huabei
Fusarium oxysporum Schlecht. (Damping off) Dongbei
Lophodermium pinastri (Schrad. ex Fr.) Chev. (Needle blight) Jiangsu
Rhizoctonia solani Kühn (Damping off) Dongbei

***Pinus elliottic* Engelm**

Lecanosticta acicola (Thum) Sydow (Pine Needle brown spot) Fujian

***Pinus elliottii* Eng.**

Bursaphelenchus xylophilus (Steiner et Buhrer) Nickle Jiangsu Anhui
Diplodia pinea (Desm.) Kickx (Pine dry tip) Guangdong Hunan
(=Sphaeropsis ellisii Sacc.)
Fusarium oxysporum Schlecht. (Damping off) Hunan
Hypoderma desmazieri Duby (Leaf blight) Guizhou Sichuan Guangxi
Macrophomina phaseoli (Maubl.) Ashby. (Stem rot) Hunan
Pestalotia funereal Desm. (Pine tip blight) Guizhou Sichuan Guangdong Guangxi Hunan
Pythium aphanidermatum (Eds.) Fitzp. (Damping off) Hunan
Rhizoctonia solani Kühn (Damping off) Hunan
Septoria acicola (Thüm.) Sacc. (Brown leaf spot) Fujian Guangdong Hunan Sichuan
(=Scirrhia acicola (Dearn.) Siggers.)

***Pinus griffithii* McClelland.**

Arceuthobium chinense Lecomte (Pinus griffithii mistletoe) Tibet
Gloeophyllum subferrugineum (Berk.) Bond et Sing. (Decay) Tibet
Hirschioporus abietinus (Dicks. ex Fr.) Donk (Decay) Tibet

Lophodermium pinastri (Schrad. ex Fr.) Chev. (Falling needle) Tibet

Pinus koraiensis **Sieb. et Zucc.**

Armillariella mellea (Vahl ex Fr.) Karst. (Root decay) Dongbei Huabei Gansu Sichuan Yunnan
Cenangium ferrnginosum Fr. ex Fr. (Canker) Heilongjiang Hebei Liaoning Shandong Jiangsu
C. phellodendri
Coleosporium cacaliae Otth (Leaf rust) Heilongjiang
Coleosporium campanulae (Pers.) Tév. (Leaf rust) Liaoning
Coleosporium cimicifugatum Thüm. (Leaf rust) Liaoning Heilongjiang
Coleosporium ligulariae Thüm. (Leaf rust) Liaoning
Coleosporium solidaginis (Schw) Thüm. (Leaf rust) Dongbei Guizhou
Coriolus versicolor (L. ex Fr.) Quél. (Decay) Liaoning
Cronartium ribicola J.C. Fischer ex Rabenhorst (Fusiform rust) Liaoning Heilongjiang Jilin
Diplodia pinea (Desm.) Kickx. (Shoot blight) Heilongjiang Jilin
Fomitiopsis pinicola (Sw. ex Fr.) Karst. (Brown cube rot) Heilongjiang Jilin Neimeng
Fusarium oxysporum Schlecht. (Damping off) Dongbei
Fusarium solani (Mart.) App. et wollenw. (Damping off) Liaoning Heilongjiang
Gloeophyllum saepiarium (Wulf. ex Fr.) Karst. (Decay) Heilongjiang Jilin
Hirschioporus abietinus (Dicks. ex Fr.) Donk (Decay) Heilongjiang Jilin
Laetiporus sulphureus (Bull. ex Fr.) Bond et Sing. (Brown rot) Dongbei
Larcifomes officinalis (Vill. ex Fr.) Kotlaba ex pouzar (Heartwood brown cube rot) Dongbei
Lophodermium maximum B. Z. He. et Yang (Red pine needle loss) Liaoning
Lophodermium pinastri (Schrad. ex Fr.) Chev. (Falling needle) Liaoning Jilin Heilongjiang
(=Hyseterium pinastri Schrad.)
Nectria cinnabarina (Tode) Fr. (Canker) Dongbei
Peridermium pini-koraiensis Saw. (Leaf rust) Heilongjiang
Phaeolus schweinitzii (Fr.) Pat. (Brown cube rot) Heilongjiang Jilin Neimeng
Phellinus pini (Thore ex Fr.) Ames (White pocket rot) Dongbei
Pythium spp. (Damping off) Dongbei
Rhizoctonia solani Kühn (Damping off) Dongbei
Tympanis pithya (Fr.) Sacc. (Red pine canker) Liaoning
Valsa sp. (Canker) Liaoning Heilongjiang

Pinus massoniana **Lamb.**

Bursaphelenchus xylophilus (Steiner et Buhrer) Nickle Jiangsu Anhui
Cassytha filiformis L. (Cassytha mistletoe) Guangxi
Cercospora pini-densiflorae Hori et Nambu (Leaf spot) Jiangsu Henan Anhui Guangdong Taiwan Guangxi Hunan Guizhou
Coleosporium asterum (Diet.) Syd. (Leaf rust) Shanxi
Coleosporium senecionis (Pers.) Fr. (Leaf rust) Sichuan
Coleosporium solidaginis (Schw.) Thüm. (Leaf rust) Jiangsu Zhejiang Henan Anhui Jiangxi Fujian Hunan Sichuan Guizhou
Cronartium sp. (Stem rust) Anhui
Cronartium flaccidum (Alb. et Schw.) Wint. (Fusiform rust) Henan Sichuan Guizhou
Cronartium flaccidum (Alb. ex Schw.) Wint. f. sp. siphonostegium Jing et Wang, Acta Mycol. Sin.7(2):112, 1988
Cronartium quercuum (Berk.) Miyabe (Pine gall rust) Jiangsu Zhejiang Henan Shanxi Anhui Sichuan Yunnan Jiangxi Guizhou Hubei Guangdong
Cryptoporus volvatus (Peck) Hubb. (Decay) Fujian
Diplodia pinea (Desm.) Kickx (Shoot blight) Guangdong Hunan
Fomitopsis pinicola (Sw. ex Fr.) Karst. (Brown cube rot) Guangdong
Fusarium spp. (Damping off) Henan Sichuan
Fusarium lateritium Nees (Damping off) Shanxi

Fusarium scirpi Lamb. et Fautr (Damping off) Shanxi
Fusarium sambucinum Fuck. (Damping off) Shanxi
Gloeophyllum saepiarium (Wulf. ex Fr.) Karst. (Brown rot) Shanxi
Hirschioporus abietinus (Dicks. ex Fr.) Donk (Decay) Henan Fujian
Hypoderma desmazieri Duby (Needle cast) Shanxi Guangxi Guizhou Sichuan
Lophodermium durilabrum (Scatter spot) Shanxi
Lophodermium pinastri (Schrad. ex Fr.) Chev (Falling leaf) Jiangsu Zhejiang Henan Shanxi Guangdong Anhui Jiangxi Guizhou Taiwan Hubei Sichuan Hunan
Loranthus chinensis DC. (Mistletoe) Guangxi
Macrophomina phaseoli (Maubl.) Ashby (Stem rot) Henan Hunan
Peridermium pini (Willd.) Kleb. (Leaf rust) Shanxi
Pestalotia sp. (Needle blight) Hunan
Pestalotia funereal Desm. (Needle blight) Jiangsu Guangdong Guangxi Jiangxi Sichuan Guizhou Hunan Fujian
Pestalotia hartigii Tub. (Needle blight) Guangxi
Pestalotia zahlbruckneriana P. Henn. (Needle blight) Guangdong Sichuan Guizhou
Phellinus pini (Thore ex Fr.) Ames (White pocket rot) Sichuan
Poria cocos (Fr.) Wolf (Root decay) Henan
Pythium debaryanum Hesse (Damping off) Zhejiang Henan Sichuan
Rhizoctonia solani Kühn (Damping off) Shanxi Henan Guangdong
Sclerotium bataticola Traub. (Sprout stem rot) Henan

Pinus palustria Mill.

Pestalotia funereal Desm. (Needle blight) Jiangsu

***Pinus ponderosa* Dougl. ex Laws.**

Pestalotia funereal Desm. (Leaf rust) Jiangsu

***Pinus roxburghii* Sarg.**

Arceuthobium chinense Lecomte (Dwraf mistletoe) Tibet
Fomitopsis pinicola (Sw ex Fr.) Karst. (Brown cube rot) Tibet
Gloeophyllum saepiarium (Wulf. ex Fr.) Karst. (Decay) Tibet
Hirschioporus abietinus (Dicks. ex Fr.) Donk (Decay) Tibet

***Pinus sibirca* (Loud.) Mayr.**

Coleosporium sp. (Leaf rust) Xinjiang
Cronartium ribicola Fischer ex Rokenhorst (White pine blister rust) Xinjiang
Lophodermium pinastri (Schrad. ex Fr.) Chev. (Needle cast) Xinjiang
Phellinus pini (Thore ex Fr.) Ames (White pocket rot) Xinjiang
Rhizoctonia solani Kühn. (Damping off) Xinjiang

***Pinus sylvestris var. mongolica* Litvin.**

Coleosporium pulsatillae (Str.) Lév. (Leaf rust) Neimeng Heilongjiang
Cronartium flaccidum (Alb. et Schw.) Wint. (Leaf rust) Heilongjiang
Cronartium quercuum (Berk.) Miyabe (Pine gall rust) Heilongjiang Neimeng
Diplodia pinea (Desm) Kickx (Shoot blight) Heilongjiang Jilin
Fusarium oxysporum Schlecht. (Damping off) Dongbei
Lophodermium pinastri (Schrad. ex Fr.) Chev. (Falling needle) Liaoning Jilin Heilongjiang Neimeng Dongbei
Naemacyclus niveus (Pers.) Sacc. (Yellow spot dry needle) Heilongjiang
Peridermium pini (Willd.) Kleb. (Needle rust) Heilongjiang Neimeng
Phaeolus schweinitzii (Fr.) Pat. (Brown cube rot) Heilongjiang Neimeng
Phellinus pini (Thore ex Fr.) Ames (White pocket rot) Dongbei Neimeng
Rhizoctonia solani Kühn (Damping off) Dongbei
(Yellows) Neimeng

***Pinus tabulaeformis* Carr.**

Balanphora involucrate Hook. () Shanxi

Capnodium pini B. et C. (Black mildew) Neimeng
Cenangium sp. (Dieback) Shandong
Cenangium ferruginosum Fr. ex Fr. (Dieback) Liaoning Jilin Hebei Shanxi
Cercospora pini-densiflorae Hori et Nambu (Leaf spot) Henan
Coleosporium sp. (Leaf rust) Hebei
Coleosporium asterum (Diet.) Syd. (Leaf rust) Liaoning Sichuan
Coleosporium campanulae (Pers.) Lév. (Leaf rust) Henan Shanxi
Coleosporium melampyri Tul (Leaf rust) Shanxi
Coleosporium phellodendri Kom. (Leaf rust) Liaoning Jilin Shandong
Coleosporium senecionis (Pers.) Fr. (Leaf rust) Sichuan
Coleosporium tussilaginis Tub. (Leaf rust) Shanxi
Coniothyrium pini Oudemans. (Dry tip) Shanxi
Cronartium coleosporioides (D. et H.) Arth. (Leaf rust) Shanxi
Cronartium quercuum (Berk.) Miyabe (Pine gall rust) Zhejiang Henan Sichuan Shanxi
Cryptoporus volvatus (Peck) Hubb. (Decay) Hebei Yunnan
Cytospora sp. (Canker) Shandong
Fomitopsis pinicola (Sw. ex Fr.) Karst. (Brown cube rot) Shanxi
Fumago vagans Pers. (Black mildew) Liaoning
Fusarium spp. (Damping off) Dongbei Henan
Fusarium oxysporum Schlecht. (Damping off) Dongbei
Fusarium solani (Mart.) App. et Wellenw. var. martii (App. et Wollenw.) Wollenw. (Damping off) Hebei
Gloeophyllum saepiarium (Wulf. ex Fr.) Karst. (Brown rot) Shanxi
Lophodermium pinastri (Schrad. ex Fr.) Chev. (Falling needle) Jilin Heilongjiang Neimeng Hebei Dongbei Jiangsu Shanxi Shandong Shanxi Sichuan Ningxia Liaoning
Macrophomina phaseoli (Maubl.) Ashby Henan
Peridermium pini (Willd.) Kleb. (Leaf rust) Liaoning Jilin
Pestalotia funereal Desm. (blight) Henan
Phellinus pini (Thore ex Fr.) Ames (White pocket rot) Gansu Sichuan
Pythium spp. (Damping off) Dongbei Hebei Henan
Rhizoctonia solani Kühn (Damping off) Dongbei Hebei Henan
Sclerotium bataticola Traub. (Sclerotium) Dongbei Hebei Henan

Pinus taeda **Linnaeus**

Bursaphelenchus xylophilus (Steiner et Buhrer) Nickle Jiangsu Anhui
Diplodia pinea (Desm.) Kickx. (Pine dry tip) Shandong Hunan Guangdong Hebei Hubei
Fusarium oxysporum Schlecht. (Root rot) Hunan
Helicobasidium purpureum (Jul.) Pat. (Purple root)
Hypoderma desmazieri Duby (Needle blight) Guizhou Sichuan Guangxi
Macrophomina phaseoli (Maubl.) Ashby. (Stem rot) Hunan
Pestalotia funnerea Desm. (Pine blight) Sichuan Guangdong Guangxi Guizhou Hunan
Pythium aphanidermatum (Eds.) Fitzp. (Damping off) Hunan
Rhizoctonia solani Kühn (Damping off) Hunan
Septoria acicola (Thüm.) Sacc. (Brown leaf spot) Hunan Guangdong Fujian Sichuan

Pinus taiwanensis **Hay.**

Cercospora pini-densiflorse Hari et Nambu (Leaf spot) Taiwan
Coleosporium asterum (Diet.) Syd. (Leaf rust) Taiwan
Cronartium quercuum (Berk.) Miyabe (Pine gall rust) Zhejiang Anhui Jiangxi Guizhou
Lophodermium pinastri (Schrad. ex Fr.) Chev. (Falling needle) Anhui
Loranthus caloreus var. oblongifolius Lecente. (Long-leaf mistletoe) Anhui

Pinus thunbergii **Parl.**

Bursaphelenchus xylophilus (Steiner et Buhrer) Nickle Jiangsu Anhui
Cenangium ferruginosum Fr. ex Fr. (Dieback) Liaoning Jiangsu Shandong
(=Cenangium abietis var. japonica Henn.)
Cercospora pini-densiflorae Hori et Nambu (Leaf spot) Jiangsu Henan
Cronartium quercuum (Berk.) Miyabe (Pine gall rust) Henan
Lophodermium pinastri (Schrad. ex Fr.) Chev. (Falling needle) Jiangsu Shandong Taiwan
Pestalotia funereal Desm. (Leaf spot) Jiangsu
Pestalotia sydowiana Bres. (blight) Jiangsu

***Pinus yunnanensis* Franch.**
Arceuthobium pini (Mistletoe) Sichuan Yunnan
Cenangium ferruginosum Fr. (Canker) Sichuan
Coleosporium senecionis (Pers.) Fr. (Leaf rust) Sichuan Yunnan
Coleosporium solidaginis (Schw.) Thüm. (Leaf rust) Sichuan Yunnan Guizhou
Coriolus hirsutus (Wulf. ex Fr.) Quél. (Decay) Yunnan
Coriolus versicolor (L. ex Fr.) Quél. (Decay) Yunnan Tibet
Cronartium flaccidum (Alb. et Schw.) Wint. (Leaf rust) Sichuan Yunnan Tibet
Cronartium quercuum (Berk.) Miyabe (Pine gall rust) Sichuan Yunnan Guizhou Tibet
Cryptoporus volvatus (Peck) Hubb. (White rot) Sichuan Yunnan Tibet
Fomitopsis pinicola (Sw. ex Fr.) Karst. (Brown rot) Sichuan Yunnan Tibet
Gloeophyllum saepiarium (Wulf. ex Fr.) Karst. (Decay) Yunnan
Gloeophyllum subferrugineum (Berk.) Bond. et Sing. (Decay) Yunnan
Laetiporus sulphureus (Bull. ex Fr.) Bond. et Sing. (Brown rot) Sichuan Tibet
Laricifomes officinalis (Vill. ex Fr.) Kotlaba ex Pouzar (Heartwood brown cube rot) Sichuan Yunnan
Lenginus mellianus Lohw. Guangdong Huabei Yunnan Neimeng
Lophodermium pinastri (Schrad.) Chev. (Needle blight) Sichuan Yunnan Guizhou Tibet
Pestalotia funereal Desm. (Leaf spot) Yunnan Guizhou
Phellinus pini (Thore ex Fr.) Ames (White pocket rot) Sichuan Yunnan Tibet
Phellinus torulosus (Pers.) Bourd. et Galz. (Decay) Yunnan
Pleurotus ostreatus (Jacq. ex Fr.) Quél. (Decay) Yunnan
Poria cocos (Fr.) Wolf (Decay) Yunnan Tibet
Poria odora (Peck) Sacc. (Decay) Yunnan
Poria taxicola (Pers.) Bres (Decay) Yunnan

Pistacia spp.
Pileolaria pistaciae Tai et Wei (Leaf rust) Jiangsu Anhui
Pileolaria terebinthi (DC.) Cast. (Leaf rust) Anhui

***Pistacia chinensis* Bunge**
Cercospora pistaciae Chupp (Leaf spot) Taiwan
Microsphaera alni (Wallr.) Salm. (Powder mildew) Hebei
Pileolaria pistaciae Tai et Wei (Leaf rust) Jiangsu Zhejiang Shandong Anhui Jiangxi Taiwan
Pileolaria terebinthi (DC.) Cast. (Leaf rust) Henan
Uncinula verniciferae P. Henn. (Powder mildew) Jiangsu Henan

***Pistacia weinmarnnifolia* Poiss.**
Meliola rhoina Doidge (Black mildew) Yunnan
Pileolaria pistaciae Tai et Wei (Leaf rust) Yunnan
Pileolaria terebinthi (DC.) Cast. (Leaf rust) Yunnan

***Platanus acerifolia* L.**
(=Platanus hispanica Muenchh.)
Cercospora platanicola Ell. et Ev. (Mold spot) Jiangsu Henan Taiwan
(=Cercospora platanifolia Ell. et Ev.)
Gloeosporium nerviesquum (Fckl.) Sacc. (Anthracnose) Shanxi

***Platanus occidentalis* L.**

Cercospora platanicola Ell. et Ev. (Mold spot) Jiangsu Henan Taiwan

***Platanus orientalis* L.**

Cercospora platanicola Ell. et Ev. (Mold spot) Henan Jiangsu Taiwan
Cytospora personata Fr. (Dieback) Shanxi
Cytospera platani Fckl. (Canker) Shanxi Xinjiang
Phyllactinia corylea (Pers.) Karst. (Powder mildew) Shanxi Jiangsu
Stigmina platani (Fckl.) Sacc. (Mold spot) Henan

Platycarya spp.

Phylllactinia juglandis Tao et Qin (Powder mildew) Sichuan
Systremma natans (Tode) Theiss. et Syd. (Leaf spot) Anhui
Systremma sambuci (Pass. et Fr.) Mill. (Leaf spot) Anhui

***Platycarya strobilacea* Sieb. et Zucc.**

Coriulus versicolor (L. ex Fr.) Quél. (Decay) Guizhou
Phyllactinia juglandis Tao et Qin (Powder mildew) Jiangsu
(=Phyllactinia corylea (Pers.) Karst.)

***Platycladus orientalis* (L.) Franch**

Cladosporium herbarum (Pers.) Link Sichuan
Cuscuta japonica Choisy (Cuseata high plant mistletoe) Henan
Gloephyllum juniperinum (Teng et Ling) Teng (Decay) Hebei
Macrophomina phaseoli (Maubl.) Ashby. (Stem rot)
Rhizoctonia solani Kühn (Damping off) Henan

***Podocarpus macrophyllus* (Thunb.) D. Don**

Pestalotia foedans Sacc. et Ell. (Leaf spot) Taiwan

***Podocarpus macrophyllus var. maki* Endl.**

Pestalotia funereal Desm. (Leaf spot) Jiangsu

Populus spp.

Agrobacterium tumefaciens (Smith et Town) Conn. (Root cancer) Dongbei Huabei Xibei
(=Bacterium tumefaciens Smith et Town.)
Armillariella mellea (Vahl ex Fr.) Karst. (Root rot) Dongbei Xinjiang
Armillariella tabescens (Scop. ex Fr.) Sing. (Armillariella root rot) Hebei Shandong
Bjerkandera aadusta (Willd. ex Fr.) Karst. (Decay) Liaoning Xinjiang
Capnodium salicinum Mont. (Black mildew) Hebei Gansu
(Asexual:Fumago vagans Pers.)
Cenangium populneum (Pers.) Rehm (Dieback) Sichuan
Cercospora populina Ell. et Ev. (Populus blight) Ningxia
Cerrena unicolor (Bull. ex Fr.) Murr. (Decay) Hebei
Coniothyrium olivaceum Bon. Jiangsu
Coriolus hirsutus (Wulf ex Fr.) Quél. (Decay) Hebei
Cuscuta australis R. Br. (Cuseata high plant mistletoe) Hebei
Cuscuta japonica Choisy (Cuseata high plant mistletoe) Liaoning Heilongjiang Henan Shanxi
Cuscuta monogyna Vahl (Cuseata high plant mistletoe) Xinjiang
Cytospora chrysosperma (Pers.) Fr. (Canker)
(Sexual: Valsa sordida Nit.)
Heilongjiang Liaoning Huabei Sichuan Henan Hebei Xinjiang Neimeng Ningxia Shandong Shanxi Shanxi Qinghai
Daedaleopsis confragosa (Bolt. ex Fr.) Schröt. (Decay) Sichuan
Discosia artocreas (Tode) Fr. (Black leaf spot) Liaoning
Dothiorella gregaria Sacc. (Canker) Liaoning Henan Hebei Jiangsu Shanxi Shandong Shanxi
(Sexual: Botryosphaeria dothidea (Moug. ex Fr.) Ces. et den.)
Favolus squamosus (Huds. ex Fr.) Ames (White rot) Dongbei Sichuan Hebei Gansu

Fomes fomentarius (L. ex Fr.) Kickx. (Brown rot) Heilongjiang Tibet Gansu
Funalia trogii (Berk.) Bond. et Sing. (Decay) Liaoning Hebei
Fusicladium radiosum (Lib.) Lind. (Black star) Liaoning Xinjiang
Fusicladium tremulae Fr. (Black star) Gansu
Ganoderma applanatum (Pers.) Pat. (Decay) Heilongjiang Hebei Xinjiang Neimeng
Glomerella cingulata (Stonem.) Spauld. et Schr. (Anthracnose) Liaoning Henan Shanxi Ningxia
Helicobasidium purpureum (Tul.) Pat. (Purple root) Dongbei Henan Hebei Guangdong Sichuan
Anhui Jiangsu Shanxi Shandong
Hericium coralloides (Scop. ex Fr.) Pers. ex Gray (Decay) Xinjiang
Inonotus hispidus (Bull. ex Fr.) Karst. (Leaf tip rot) Xinjiang
Inonotus radiatus (Sow. ex Fr.) Karst. var. licentii Pilát. (Decay) Hebei
Inonotus rheades (Pers.) Pat. (Decay) Dongbei Heilongjiang Hebei Xinjiang Neimeng Ningxia
Irpex lacteus Fr. (Decay) Sichuan
Laetiporus sulphureus (Bull. ex Fr.) Bond. et Sing. (Trunk brown rot) Dongbei Xinan Xibei
Lenzites betulina (L.) Fr. (Decay) Liaoning
Leptothyrium sp. (Mold spot) Shanxi Henan Shanxi
Macrophoma tumeifaciens Shear (Branch gall) Henan Xinjiang Ningxia
Marssonina brunnea (Ell. et Ev.) Sacc. (Black leaf spot) Liaoning Jilin Heilongjiang Henan Hebei
Shanxi Neimeng Ningxia Shanxi Shandong
Marssonina populi (Lib.) Magn. (Black leaf spot) Sichuan Xinjiang Jiangsu Henan Shanxi
Shandong
Melampsora sp. (Leaf rust) Gansu
Melampsora larici-populina Kleb. (Leaf rust) Jilin Heilongjiang Sichuan Tibet Hebei Shanxi
Shanxi Yunnan Ningxia Neimeng Liaoning Gansu
Melampsora laricis Hart. (Leaf rust) Henan Heilongjiang
Melampsora magnusiana Wagn. (Leaf rust) Hubei Shanxi Xinjiang
Melampsora rostrupii Wagn. (Leaf rust) Yunnan Xinjiang
Mycosphaerella mandshurica Miura (Gray leaf spot) Liaoning Jilin Heilongjiang Hebei Shanxi
Ningxia Shanxi
(Asexual: *Coryneun populinum Bresad*)
Oxyporus populinus (Schum. ex Fr.) Donk (Decay) Sichuan
Phellinus igniarius (L. ex Fr.) Quél. (Heartwood white sponge rot) Jilin Henan Hebei Sichuan
Heilongjiang Shanxi Xinjiang Qinghai Yunnan Gansu Shanxi Neimeng Ningxia Tibet
Phellinus setulosus (Lloyd) Imaz. (Decay) Heilongjiang
Phellinus yucatenensis (Murr.) Imaz. (Decay) Yunnan
Pholiota adipose (Fr.) Quél. (Decay) Hebei Xinjiang Heilongjiang
Pholiota destruens (Brond.) Gill. (Decay) Jilin Hebei
Phyllactinia populi (Jacz.) Yu (Powder mildew) Liaoning Sichuan Hebei Shanxi Guizhou
Neimeng Ningxia Gansu Shanxi
(=Phyllactinia corylea (Pers.) Karst.)
(=Phyllactinia suffulta (Rebent.) Sacc. f. populi Jacz.)
Phyllosticta populina Sacc. (Leaf spot) Hebei Shanxi Ningxia
Physalospora populina Maubl. (Leaf spot) Henan Shanxi
Pleurotus calyptratus (Lindbl. in Fr.) Sacc (Decay) Henan Ningxia
Plenrotus carticatus (Fr.) Quél (Decay) Xinjiang
Pleurotus sapidus (Schwlz. ap. Kalchbr.) Sacc. (Decay) Xinjiang
Pycnoporus sanguineus (L. ex Fr.) Murr. (Decay) Xinjiang
(=Trametes sanguinea (L. ex Fr.) Lloyd)
Rhabdospora longispora Ferraris Jiangsu
Schizophyllum commune Fr. (Decay) Henan
Septoria populi Desm. (Brown leaf spot) Xinjiang Ningxia

(Sexual:Mycosphaerella populi (Auersw.) Kleb.)
Septoria populicola Peck (Brown leaf spot) Henan
(Sexual: Mycosphaerella populicola Thomp.)
Septotis populiperda (Moesz et Smarods)
Waterman et Cash (Big leaf spot) Shanxi
Stereum frustulosum (Fers.) Fr. (Decay) Heilongjiang
Taphrina populina Fr. (Leaf eviel)
Trametes suaveolens (L.) Fr. (Decay) Liaoning
Uncinula adunca (Wallr. ex Fr.) Lév. var. adunca (Powder mildew) Liaoning Sichuan Hebei
Xinjiang Yunnan Neimeng
Uncinula longispora Zheng et Chen (Powder mildew) Shanxi
Uredo tholopsora Cummis (Leaf rust) Anhui
Valsa populina Fuck. (Canker) Hebei Shandong
Virus (Virus) Shandong Henan Hebei Hubei Hunan
Viscum coloratum (Kom.) Nakai (True mistletoe) Heilongjiang Liaoning
Lack of iron: (Yellows) Neimeng Shanxi Xinjiang
Frostbite: (Willow rot) Henan Neimeng Shanxi Xinjiang Shandong Ningxia
Drought: (Flower tip) Neimeng
Physical and source of disease: (Red heart) Shanxi Henan Neimeng Ningxia Shandong
Sunburn: (Sunburn) Shanxi

***Populus adenopoda* Maxim.**

Marssonina populi (Lib.) Magn. (Black leaf spot) Shanxi Hunan
Melampsora larici-populina Kleb. (Leaf rust) Henan Guangxi
Melampsora rostrupii Wagn. (Leaf rust) Shanxi
Taphrina johansonii Sad. (Leaf eviel) Jiangsu
Uncinula adunca (Wallr. ex Fr.) Lév. var. adunca Jilin Anhui Yunnnan
(= Uncinula adunca (Wallr. ex Fr.) Lév.)
(=Uncinula salicis (DC.) Wint.)
Uncinula fragilis Zheng et Chen (Powder mildew) Sichuan

Populus adenopoda x tomentosa

Marssonina populi (Lib.) Magn. (Black leaf spot) Shanxi

***Populus alba* L.**

Alternaria alternata (Fr.) Keissler (Aspen dry leaf) Heilongjiang Jilin Liaoning Xinjiang Shanxi
Henan Hebei Shandong Beijing
Coryneum populinum Bres. (Gray leaf spot) Shanxi
Cuscuta japonica Choisy (Cuseata high plant mistletoe) Zhejiang
Cuscuta monogyna Vahl. (Cuseata high plant mistletoe) Xinjiang
Cytospora chrysosperma (Pers.) Fr. (Canker) Tibet
Gloeosporium populi-albae Desm. (Leaf spot) Hebei
Helicbasidium pupureum (Tul.) Pat. (Purple root) Dongbei Hebei
Laetiporus sulphureus (Bull. ex Fr.) Bond. et Sing. (Trunk brown rot) Xinjiang
Marssonina populi (Lib.) Magn. (Leaf spot) Xinjiang Shanxi
Melampsora larici-populina Kleb. (Leaf rust) Liaoning Henan
Melampsora magnusiana Wagn. (Leaf rust) Liaoning
Melampsora magnusiana Wagn. f. tomentosea Zhon. (Leaf rust) Henan
Melampsora rostrupii Wagn. (Leaf rust) Liaoning Xinjiang Jiangsu Shanxi
Phyllactinia populi (Jacz.) Yu (Powder mildew) Henan Hebei
Septotis populiperda (Moesz et Smarods)
Waterman et Cash (Big leaf spot) Shanxi

***Populus alba x tomentosa* (1)**

Coryneum populinum Bres. (Gray leaf spot) Shanxi

Marssonina populi (Lib.) Magn. (Black leaf spot) Shanxi

***Populus balsamifera* L.**

Coniothyrium populicola Miura (Leaf spot) Liaoning
Cytospora chrysosperma (Pers.) Fr. (Canker) Liaoning
(Sexual:Valsa sordida Nit.)
Venturia tremulae (Frank) Aderh. (Black star) Liaoning

Populus balsamifera x Pyramidalis

Marssonina brunnea (Ell. et Ev.) Sacc. (Black leaf spot) Shanxi

***Populus berolinensis* Dipp.**

Cytospora chysosperma (Pers.) Fr. (Canker) Xinjiang
Cryptosphaeria populina (Pers.) Wint. (Dieback) Dongbei
Dothiorella gregaria Sacc. (Canker) Shanxi
(=Dothiorella populina Thüm.)
Marssonina brunnea (Ell. et Ev.) Sacc. (Black leaf spot) Jilin Shanxi
(=Marssoninia populicola Miura)
Melampsora larici-populina Kleb. (Leaf rust) Liaoning Jilin Heilongjiang Dongbei Shanxi Neimeng

Populus berolinensis x simonii

Nelampsora larici-populina Kleb. (Leaf rust) Shanxi

Populus berolinensis x yunnanensis

Melampsora larici-populina Kleb. (Leaf rust) Shanxi
Phyllactinia populi (Jacz.) Yu (Powder mildew) Shanxi

***Populus bolleana* Lauche**

Capnodium salicinum Mont. (Black mildew) Xinjiang
Coryneum populinum Bres. (Gray leaf spot) Shanxi Shandong
Cusucta monogyna Vahl. (Cuseata high plant mistletoe) Xinjiang
Cytospora chrysosperma (Pers.) Fr. (Canker) Xinjiang
Fomes fomentarius (L. ex Fr.) Kickx. (Decay) Xinjiang
Funalia gallica (Fr.) Pat. (Decay) Xinjiang
Inonotus hispidus (Bull. ex Fr.) Karst. (Decay) Xinjiang
Inonotus rheades (Pers.) Pilát. (Decay) Xinjiang
Marssonina populi (Lib.) Magn. (Black leaf spot) Xinjiang Shanxi
Melampsora magnusiana Wagn. (Leaf rust) Xinjiang
Melampsora rostrupii Wagn. (Leaf rust) Xinjiang Shanxi
Pleurotus sapidus (Schwlz. ap. Kalchbr.) Sacc. (Decay) Xinjiang
Septotis populiperda (Moesz et Smarods) Waterman et Cash (Big leaf spot) Shanxi

***Populus Canadensis* Moench**

(=P. x euramericana (Dode) Guinier)
Agrobacterium tumefaciens (Smith et Towns.) Conn (Root cancer) Shanxi
Cercospora populina Ell. et Ev. (Leaf spot) Jiangsu
Cytospora chrysosperma (Pers.) Fr. (Canker) Jilin Henan Shanxi Jiangsu Neimeng
Dothiorella gregaria Sacc. (Canker) Jiangsu
Helicobasidium mompa (Tanaka) Jacz. (Purple root rot) Shanxi
Helicobasidium purpureum (Tul.) Pat. (Purple root rot) Henan
Leptothyrium sp. (Mold spot) Shanxi
Marssonina brunnea (Ell. et Ev.) Sacc. (Black leaf spot) Henan Shanxi
Melampsora larici-populina Kleb. (Leaf rust) Liaoning Heilongjiang Henan Shandong Neimeng
Melampsora magnusiana Wagn. (Leaf rust) Liaoning
Phyllactinia populi (Jacz.) Yu (Powder mildew) Henan Neimeng Shanxi Gansu
Sclerotinia sclerotiorum (Lib.) de Bary (Sclerotium) Henan

Populus canadensis x cathayana
Melampsora larici-populina Kleb. (Leaf rust) Shanxi
Phyllactinia populi (Jacz.) Yu. (Powder mildew) Shanxi
Populus canadensis x cathayana
Melampsora larici-populina Kleb. (Leaf rust) Shanxi
Phyllactinia populi (Jacz.) Yu (Powder mildew) Shanxi
Populus canadensis x cathayana
Melampsora larici-populina Kleb. (Leaf rust) Shanxi
Phyllactinia populi (Jacz.) Yu (Leaf rust) Shanxi
Populus Canadensis x nigra cv. 'Italica'
Marssonina brunnea (Ell. et Ev.) Sacc. (Black leaf spot) Shanxi
***Populus candicans* Ait.**
Cytospora chrysosperma (Pers.) Fr. (Canker) Xinjiang
Marssonina brunnea (Ell. et Ev.) Sacc. (Black leaf spot) Shanxi
Melampsora larici-populina Kleb. (Leaf rust) Shanxi
Melampsora rostrupii Wagn. (Leaf rust) Liaoning
Phyllactinia populi (Jacz.) Yu (Powder mildew) Xinjiang
***Populus canescens* Sm.**
Cytospora chrysosperma (Pers.) Fr. (Canker) Xinjiang
Dothiorella gregaria Sacc. (Canker) Shanxi
Fomes fomentarius (L. ex Fr.) Kickx. (Brown rot) Xinjiang
Melampsora magnusiana Wagn. (Leaf rust) Xinjiang
***Populus cathayana* Rehd.**
Coryneum populinum Bres. (Gray leaf spot) Shanxi Heilongjiang
Cytospora chrysosperma (Pers.) Fr. (Rot) Neimeng Heilongjiang
Fusicladium radiosum (Lib.) Lind. (Black star) Xinjiang Heilongjiang
(Sexual:Venturia populina (Viull.) Fabre)
Fusicladium tremulae Fr. (Black star) Henan Heilongjiang
Leptothyrium sp. (Mold spot) Shanxi
Marssonina brunnea (Ell. et Ev.) Sacc. (Black leaf spot) Shanxi Neimeng Jilin Heilongjiang
Melampsora larici-populina Kleb. (Leaf rust) Beijing Liaoning Jilin Heilongjiang Henan Shanxi
Shanxi Qinghai Yunnan Neimeng Hebei Xinjiang Gansu
Phyllactinia populi (Jacz.) Yu (Powder mildew) Neimeng Heilongjiang
Oxyporus populinus (Schum. ex Fr.) Donk (Decay) Shanxi
(=Fomes populinus (Schum. ex Fr.) Cooke)
Populus cathayana x pyramidalis
Marssonina brunnea (Ell. et Ev.) Sacc. (Black leaf spot) Shanxi
***Populus x dakuanensis* Hsu.**
Ascochyta tremulae Thuem (Cycle spot) Shanxi
Cercospora populina Ell. et Ev. (Leaf spot) Henan
Coniothyrium populicola Miura (Leaf spot) Shanxi
Coryneum populinum Bres. (Leaf spot) Shandong Shanxi
Dothiorella gregaria Sacc. (Canker) Henan
Fusicladium tremulae Fr. (Black star) Shanxi
Helicobsidium purpureum (Tul.) Pat. (Purple root rot) Henan
Leptothyrium sp. (Mold spot) Shanxi
Marssonina brunnea (Ell. et Ev.) Sacc. (Black leaf spot) Shanxi
Phyllactinia populi (Jacz.) Yu (Powder mildew) Henan Shanxi
Phyllosticta populina Sacc. (Leaf spot) Shanxi
Septotsi populiperda (Moesz et Smarods)
Waterman et Cash (Big leaf spot) Shanxi

***Populus davidiana* Dode**

Alternaria alternata (Fr.) Keissler (Aspen dry leaf) Heilongjiang Jilin Liaoning Xinjiang Shanxi Henan Hebei Shandong Beijing

Cercospora populina Ell. ex Ev. (Populus blight) Shanxi

Coryneum populinum Bres. (Leaf spot) Shanxi Heilongjiang

Eriophyes dispar Nal. (Leaf felt spot) Beijing Shanxi

Favolus squamosus (Huds. ex Fr.) Ames (Decay) Heilongjiang

Fomes fomentarius (L. ex Fr.) Kickx. (Brown rot) Gansu

Fumago vagans Pers. (Black mildew) Liaoning

Ganoderma applanatum (Pers.) Pat. (Decay) Gansu

Gloeosporium tremulae (Leb.) Pass. (Anthracnose) Heilongjiang Neimeng

Helicobasidium purpureum (Tul.) Pat. (Purple root rot) Ningxia

Inonotus rheades (Pers.) Pat. (Decay) Dongbei

Melampsora larici-populina Kleb. (Leaf rust) Hebei

Melampsora larici-tremulae Kirb. (Leaf rust) Tibet

Melampsora laricis Hart. (Leaf rust) Heilongjiang Liaoning Shanxi

Melampsora magnusiana Wagn. (Leaf rust) Heilongjiang Liaoning Sichuan Hebei Xinjiang

Oxyporus populins (Schu. ex Fr.) Donk (Decay) Shanxi

Phellinus ignaiarius (L. ex Fr.) Quél. (Heartwood white sponge rot) Jilin Heilongjiang Dongbei Sichuan Shanxi Xibei Neimeng Gansu

Phellinus igniarius (L. ex Fr.) Quél. f. tremulae (Heartwood white sponge rot) Tibet

Schizophyllum commune Fr. (Decay) Neimeng

Tectella calyptrate (Lindbl.) Sing. Ningxia

Trametes suaveolens (L.) Fr. (Decay) Hebei Shanxi Gansu

Uncinula adunca (Wallr. ex Fr.) Lev. var.adunca (Powder mildew) Liaoning Hebei Shanxi

Uncinula longispora Zheng et Chen var. minor Zheng et Chen (Powder mildew) Shanxi

Valsa sordida Nit. (Canker) Neimeng

Venturia populina (Vuill.) Fabr. (Black star) Sichuan

Viscum coloratum (Kom) Nakai (True mistletoe) Neimeng

Populus davidiana x adenopoda

Marssonina populi (Lib.) Magn. (Black leaf spot) Shanxi

***Populus deltoids var. missouriensis* Henry**

Marssonina brunnea (Ell. et Ev.) Sacc. (Black leaf spot) Shanxi

Melampsora larici-populina Kleb. (Leaf rust) Shanxi

Phyllactinia populi (Jacz.) Yu (Powder mildew) Shanxi

***Populus diversifolia* Schrenk.**

(=Populus euphratica Oliver)

Cytospora chrysosperma (Pers.) Fr. (Canker) Xinjiang

Melampsora larici-populina Kleb. (Leaf rust) Liaoning

Melampsora pruinosae Tranz. (Leaf rust) Xinjiang Neimeng

Septoris populi Desm. (Brown leaf spot) Xinjiang

Populus x euroamericana cv. 'Eugenei'

Ascochyta tremulae Thuem. (Leaf spot) Shanxi

Dothiorella gregaria Sacc. (Canker) Shanxi

Fusicoccum leucostomum Sacc. (Stem rot) Shanxi

Marssonina brunnea (Ell. et Ev.) Sacc. (Black leaf spot) Shanxi

Melampsora larici-populia Kleb. (Leaf rust) Shanxi

Phyllactinia populi (Jacz.) Yu (Powder mildew) Shanxi

Populus x euramericana cv. 'Gelrica'

Marssonina brunnea (Ell. et Ev.) Sacc. (Black leaf spot) Shanxi

Phyllactinia populi (Jacz.) Yu (Powder mildew) Shanxi

Populus x euramericana cv. 'Graupeaer-Selektionen Nr. 158'
Marssonina brunnca (Ell. et Ev.) Sacc. (Black leaf spot) Shanxi
Phyllactinia populi (Jacz.) Yu (Powder mildew) Shanxi

Populus x euramericana cv. 'Grandis'
Marssonina brunnea (Ell. et Ev.) Sacc. (Black leaf spot) Shanxi

Populus x euramericana cv. 'Leipzig'
Marssonina brunnea (Ell. et Ev.) Sacc. (Black leaf spot) Shanxi
Phyllactinia populi (Jacz.) Yu (Powder mildew) Shanxi

Populus x euramericana cv. 'Marilandica'
Marssonina brunnea (Ell. et Ev.) Sacc. (Black leaf spot) Shanxi

Populus x euramericana cv. 'Poska-15A'
Coniothyrium populicola Miura (Leaf spot) Shanxi
Coryneum populinum Bres. (Leaf spot) Shanxi
Dothiorella gregaria Sacc. (Canker) Shanxi
Fusicladium tremulae Fr. (Black star) Shanxi
Marssonina brunnea (Ell. et Ev.) Sacc. (Black leaf spot) Shanxi
Phyllactinia populi (Jacz.) Yu (Powder mildew) Shanxi
Phyllosticta populina Sacc. (Leaf spot) Shanxi
Septotis populiperda (Moesz et Smarods)
Waterman et Cash (Big leaf spot) Shanxi

Populus x euramericana cv. 'Regenerata'
(=Populus Canadensis Moench var. regenerata (Henry) Rehd.)
Melampsora larici-populina Kleb. (Leaf rust) Heilongjiang Neimeng Shanxi
Phyllactinia populi (Jacz.) Yu (Powder mildew) Shanxi

Popus euramericana cv. 'Robusta'
Dothiorella gregaria Sacc. (Canker) Shanxi
Marssonina brunnea (Ell. et Ev.) Sacc. (Black leaf spot) Shanxi
Melampsora larici-populina Kleb. (Leaf rust) Heilongjiang Neimeng Shanxi
Phyllactinia populi (Jacz.) Yu (Powder mildew) Shanxi

Populus x euramericana cv. 'Sacrau 79'
Alternaria alternata (Fr.) Keissler (Aspen dry leaf) Heilongjiang Jilin Liaoning Xinjiang Shanxi
Henan Hebei Shandong Beijing
Coryneum populinum Bres. (Leaf spot) Shanxi
Dothiorella gregaria Sacc. (Canker) Shanxi
Fumago vagans Pers. (Smoke mold) Henan
Marssonina brunnea (Ell. et Ev.) Sacc. (Black leaf spot) Shanxi
Phyllactinia populi (Jacz.) Yu (Powder mildew) Henan Shanxi
Phyllosticta populina Sacc. (Leaf spot) Shanxi
Septotis populiperda (Mcesz te Smardos) Waterman et Cash (Big leaf spot) Shanxi

Populus x euramericana cv. 'Sarce-Rouge'
Marssonina brunnea (Ell. et Ev.) Sacc. (Black leaf spot) Shanxi
Phyllactinia populi (Jacz.) Yu (Powder mildew) Shanxi

Populus x euramericana cv. 'Serotina'
Melampsora larici-populina Kleb. (Leaf rust) Shanxi
Phyllactinia populi (Jacz.) Yu (Powder mildew) Shanxi

Populus harbinensis Wang et Skurotzov.
Melampsora larici-populina Kleb. (Leaf rust) Heilongjiang Neimeng
Phyllactinia populi (Jacz.) Yu (Powder mildew) Shanxi

Populus harbinensis x pyramidalis
Marssonina brunnea (Ell. et Ev.) Sacc. (Black leaf spot) Shanxi
Melampsora larici-populina Kleb. (Leaf rust) Shanxi

Phyllactinia populi (Jacz.) Yu (Powder mildew) Shanxi

***Populus hoeiensis* Hu et Chow**

Cytospora chrysosperma (Pers.) Fr. (Canker) Neimeng
Marssonina brunnea (Ell. et Ev.) Sacc. (Black leaf spot) Shanxi
Marssonina populi (Lib.) Magn. (Black leaf spot) Shanxi
Melamspora sp. (Leaf rust) Ningxia
Melampsora larici-populina Kleb. (Leaf rust) Neimeng
Melampsora rostrupii Wagn. (Leaf rust) Shanxi

Populus x hybrida 275

Melampsora larici-populina Kleb. (Leaf rust) Shanxi

***Populus koreana* Rehd.**

Gloeosporium tremulae (Leb.) Pass. (Anthracnose) Liaoning
Melampsora larici-populina Kleb. (Leaf rust) Liaoning Heilongjiang Neimeng
Phellinus igniarius (L. ex Fr.) Quél. (Decay) Heilongjiang

Populus x kornik 5

Melampsora larici-populina Kleb. (Leaf rust) Shanxi

Populus x kornik 22

Melampsora larici-populina Kleb. (Leaf rust) Shanxi

***Populus laurifolia* Ledeb.**

Coniothyrium populicola Miura (Leaf spot) Liaoning Jilin
Cytospora chrysosperma (Pesr.) Fr. (Canker) Liaoning Jilin Xinjiang
Fusicladium radiosum (Lib.) Lind (Black star) Liaoning Xinjiang
(Sexual: *Venturia populina (Vuill.) Fabr.)*
Marssonina brunnea (Ell. et Ev.) Sacc. (Black leaf spot) Liaoning Jilin
Marssonina populi (Lib.) Magn. (Black leaf spot) Xinjiang
Macrophoma tumeifaciens Shear (Branch gall) Xinjiang
Melampsora larici-populina Kleb. (Leaf rust) Heilongjiang Xinjiang Neimeng
Mycosphaerella mandshurica Miura (Gray leaf spot) Liaoning Jilin
Phyllactinia populi (Jacz.) Yu (Powder mildew) Liaoning
Phyllosticta populina Sacc. (Leaf spot) Liaoning
Pleurotus ostreatus (Jacq. ex Fr.) Quél. Xinjiang
Septoria populi Desm. (Brown leaf spot) Xinjiang
Uncinula adunca (Wallr. ex Fr.) Lév. var. adunca (Powder mildew) Xinjiang
Uncinula adunca (Wallr. ex Fr.) Lév. var.
mandshurica (Miura) Zheng et Chen. (Powder mildew) Liaoning
Venturia tremulae (Frank) Aderh (Black star) Liaoning

Populus laurifolia x pyramid

Marssonina brunnea (Ell. et Ev.) Sacc. (Black leaf spot) Shanxi

***Populus mandshurica* Makai**

Melampsora larici-populina Kleb. (Leaf rust) Liaoning

***Populus maximowiczii* Henry**

Marssonina brunnea (Ell. et Ev.) Sacc (Black leaf spot) Dongbei Shanxi
Melampsora larici-populina Kleb. (Leaf rust) Shanxi
Phyllactinia populia (Jacz.) Yu (Powder mildew) Heilongjiang
Uncinula longispora Zheng et Chen (Powder mildew) Hebei

***Populus nigra* L.**

Alternaria alternata (Fr.) Keissler (Aspen dry leaf) Heilongjiang Jilin Liaoning Xinjiang Shanxi
Henan Hebei Shandong Beijing
Cytospora chrysosperma (Pers.) Fr. (Canker) Xinjiang
Cuscuta monogyna Vahl. (Cuseata high plant mistletoe) Xinjiang
Fomes fomentarius (L. ex Fr.) Kickx. (Brown rot) Xinjiang

Funalia gallica (Fr.) Pat. (Decay) Xinjiang
Melampsora larici-populina Kleb. (Leaf rust) Liaoning Heilongjiang Neimeng Xinjiang
Melampsora populina (Jcq.) Lev. (Leaf rust) Xinjiang
Phyllactinia populi (Jacz.) Yu (Powder mildew) Xinjiang
Rhabdospora longispora Ferraris Jiangsu
Trametes hirsute (Wulf. ex Fr.) Pilat. (Decay) Xinjiang
Uncinula adunca (Wallr ex Fr.) Lev. var. adunca (Powder mildew) Xinjiang
Uncinula tenuitunicata Zheng et Chen (Powder mildew) Xinjiang
Uredo tholopsora Cummis (Leaf rust) Guizhou

Populus nigra x laurifolia
Marssonina brunnea (Ell. et Ev. Sacc.) (Black leaf spot) Shanxi

***Populus nigra var. italica* (Muench) Koehne**
(=Populus pyramidalis)
Agrobacterium tumefaciens (Smith et Towns.) Conn. (Root cancer) Henan
Cercospora populina Ell. et Ev. (Leaf spot) Henan
Cytospora chrysosperma (Pers.) Fr. (Canker) Liaoning Henan Shanxi Xinjiang
Funalia gallica (Fr.) Pat. (Decay) Xinjiang
Funalia hispida (Bagl.) Pat. (Decay) Henan
Fusicladium radiosum (Lib.) Lind. (Black star) Xinjiang
Helicobasidium purpureum (Tul.) Pat. (Purple root rot) Henan
Inonotus hispidus (Bull. ex Fr.) Karst. (Decay) Xinjiang
Loranthus parasiticus (Linn.) Merr. (Mistletoe) Guangxi
Marssonina populi (Lib.) Magn. (Black leaf spot) Xinjiang Shanxi
Melampsora larici-populina Kleb. (Leaf rust) Heilongjiang Dongbei Hebei Neimeng Shanxi Xinjiang
Myxosporium tremulae Sacc. et Roum. Jiangsu
Pestalotia populi-nigrae Sawada et k. Ito (Dieback) Henan Jiangsu
Phoma populi-ngrae Allesch. (Leaf spot) Shanxi
Phyllactinia populi (Jacz.) Yu (Powder mildew) Henan
Phyllosticta populina Sacc. (Leaf spot) Shandong
Rhabdospora longispora Ferraris Jiangsu
Septoria populi Desm. (Brown leaf spot) Shanxi
Trametes hirsute (Wulf. ex Fr.) Karst. (Decay) Xinjiang
Venturia tremulae (Frank) Aderh. (Black star) Liaoning
(Sexual :Fusicaldium tremulae Fr.)

***Populus nigra var. italica (Muench.) Koehne x cathayana* Rehd.**
Coniothyrium populicola Miura (Leaf spot) Shanxi
Cornyeum populinum Bres. (Gray leaf spot) Shanxi Shandong
Cytospora chrysosperma (Pers.) Fr. (Rot) Xinjiang
Marssonina brunnea (Ell. et Ev.) Sacc. (Black leaf spot) Shanxi
Melampsora larici-populina Kleb. (Leaf rust) Heilongjiang Shanxi Neimeng
Phyllactinia populi (Jacz.) Yu (Leaf rust) Shanxi Xinjiang
Septotis populiperda (Moesz et Smarods) Waterman et Cash (Big leaf spot) Shanxi

***Populus nigra var. thevestina* (Dode) Bean**
Capnodium salicinum Mont. (Black mildew) Xinjiang
Coryneum populinum Bres. (Gray leaf spot) Shanxi Shandong
Cytospora chrysosperma (Pers.) Fr. (Canker) Neimeng Xinjiang
Dothiorella gregaria Sacc. (Canker) Henan
Marssonina brunnea (Ell. et Ev.) Sacc. (Black leaf spot) Neimeng
Marssonina populi (Lib.) Magn. (Black leaf spot) Xinjiang
Melampsora larici-populina Kleb. (Leaf rust) Shandong

Melampsora rostrupii Wagn. (Leaf rust) Xinjiang
Phyllactinia populi (Jacz.) Yu (Powder mildew) Gansu
Septoria populi Desm. (Brown leaf spot) Xinjiang
Virus (Virus) Ningxia

Populus nigra var. thevestina x nigra
Melampsora larici-populina Kleb. (Leaf rust) Shanxi
Phyllactinia populi (Jacz.) Yu (Powder mildew) Shanxi

Populus nigra var. thevestina x nigra cv. 'Italica'
Marssonina brunnea (Ell. et Ev.) Sacc. (Black leaf spot) Shanxi

Populus nigra var. thevestina x simonii
Marssonina brunnea (Ell. et Ev.) Sacc. (Black leaf spot) Shanxi
Melampsora larici-populina Kleb. (Leaf rust) Shanxi
Septoria populicola Peck (Brown leaf spot) Xinjiang

***Populus x opera* Hsu**
Coryneum populinum Bres. (Leaf spot) Shanxi
Dothichiza populea Sacc. et Br. (Skin canker)
Marssonina brunnea (Ell. et Ev.) Sacc. (Black leaf spot) Shanxi
Melampsora larici-populina Kleb. (Leaf rust) Shanxi
Phyllactinia populi (Jacz.) Yu (Powder mildew) Shanxi

***Populus pseudo-simonii* Kitag.**
Alternaria alternata (Fr.) Keissler (Aspen dry leaf) Heilongjiang Jilin Liaoning Xinjiang Shanxi Henan Hebei Shandong Beijing
Marssonina brunnea (Ell et Ev.) Sacc (Black leaf spot) Neimeng Jilin Heilongjiang
Melampsora larici-populina Kleb. (Leaf rust) Liaoning Heilongjiang Neimeng
Mycosphaerella mandshurica Miura (Gray leaf spot) Heilongjiang Shanxi
hyllactinia populi (Jacz.) Yu (Powder mildew) Neimeng

Populus pseudo – simonii x nigra
Coryneum populinum Bres. (Leaf spot) Shanxi Heilongjiang
Marssonina brunnea (Ell. e Ev.) Sacc. (Black leaf spot) Shanxi Heilongjiang

Populus pseudo-simonii x pyramidalis
Melampsora larici-populina Kleb. (Leaf rust) Shanxi
Phyllactinia populi (Jacz.) Yu (Powder mildew) Shanxi

***Populus purdomii* Reho.**
Ascochyta tremulae Theum. (Leaf spot) Shanxi
Cuscuta japonica Choisy (Cuseata high plant mistletoe) Shanxi
Eriophyes varius Nal. (Leaf felt spot) Shanxi
Fusicaldium tremulae Fr. (Black star) Shanxi
Melampsora larici-populina Kleb. (Leaf rust) Shanxi
Phyllactinia populi (Jacz.) Yu. (Powder mildew) Shanxi
Trametes Suaveolens (L.) Fr. (Decay) Shanxi

Populus pyramidalis x balsamifera
Marssonina brunnea (Ell. et Ev.) Sacc. (Black leaf spot) Shanxi

Populus pyramidalis x berolinensis
Marssonina brunnea (Ell. et Ev.) Sacc. (Black leaf spot) Shanxi
Septotis populiperda (Moesz et Smarods) waterman et Cash (Big leaf spot) Shanxi

Populus pyramidalis x 5
Marssonina brunnea (Ell. Et Ev) Sacc. (Black leaf spot) Shanxi
Melampsora larici-populina Kleb. (Leaf rust) Shanxi
Phyllactinia populi (Jacz.) Yu. (Powder mildew) Shanxi

Populus pyramidalis x 5 (59)
Melampsora larici-populina Kleb. (Leaf rust) Shanxi

Populus pyramidalis x 11
Marssonina brunnea (Ell. et Ev.) Sacc. Shanxi
Populus pyramidalis x 11 (53)
Melampsora larici-populina Kleb. (Leaf rust) Shanxi
Populus pyramidalis x 11 (55)
Melampsora larici-populina Kleb. (Leaf rust) Shanxi
Populus pyramidalis x 11 (58)
Melampsora larici-populina Kleb. (Leaf rust) Shanxi
Populus pyramidalis x Canadensis
Marssonina brunnea (Ell. et Ev.) Sacc (Black leaf spot) Shanxi
Populus pyramidalis x koreana
Melampsora larici-populina Kleb. (Leaf rust) Shanxi
Phyllactinia populi (Jacz.) Yu. (Powder mildew) Shanxi
Popuus pyramidalis x lasiocarpa
Melampsora larici-populina Kleb. (Leaf rust) Shanxi
Phyllactinia populi (Jacz.) Yu (Powder mildew) Shanxi
Populus pyramidalis x nigra
Melampsora larici-populina Kleb. (Leaf rust) Shanxi
Populus pyramidalis x purdomii
Melampora larici-populina Kleb. (Leaf rust) Shanxi
Phyllactinia populi (Jacz.) Yu. (Powder mildew) Shanxi
Populus pyramidalis x simonii
Ascochyta tremulae Thuem (Leaf spot) Neimeng
Marssonina brunnea (Ell. et Ev.) Sacc. (Black leaf spot) Shanxi
Melampsora larici-populina Kleb. (Leaf rust) Shanxi
Phyllactinia populi (Jacz.) Yu. (Powder mildew) Shanxi
Populus pyramidalis x simonii (45)
Melampsora larici-populina Kleb. (Leaf rust) Shanxi
Populus pyramidalis x simonii (47)
Melampsora larici-popuilna Kleb. (Leaf rust) Shanxi
Phyllactinia populi (Jacz.) Yu (Powder mildew) Shanxi
Populus pyramidalis x simonii (48)
Melampsora larici-populina Kleb. (Leaf rust) Shanxi
Populus pyramidalis x szechuanica
Marssonina brunnea (Ell. et Ev.) Sacc. (Black leaf spot) Shanxi
***Populus rotundifolia* Griff.**
Melampsora rostrupii Wagn. (Leaf rust) Yunnan
Populus serotina
Dothiorella gregaria Sacc. (Canker) Shanxi
Phyllactinia populi (Jacz.) Yu. (Powder mildew) Shanxi
***Populus simonii* Carr.**
Alternaria alternata (Fr.) Keissler (Aspen dry leaf) Heilongjiang Jilin Liaoning Xinjiang Shanxi Henan Hebei Shandong Beijing
Alternaria dauci (Kuhn) Groves et Skolko f. sp. solani (Ell. et Mart.) Neerg. Hebei
Capnodium salicinum Mont. (Black mildew) Neimeng
Coniothyrium populicola Miura (Leaf spot) Henan Shanxi
Cuscuta japonica Choisy (Cuseata high plant mistletoe) Shanxi Neimeng
Cytospora chrysosperma (Pers.) Fr. (Canker) Jilin Xinjiang Shanxi Neimeng
Dothichiza populea Sacc. et Br. (Canker) Jilin
Friophyes sp. (Leaf felt spot) Hebei
Fusicladium radiosum (Lib.) Lind (Black star) Shanxi

Fusicladium tremulae Fr. (Black star) Dongbei Henan Shandong Neimeng
Gloeosporium tremulae (Leb.) Pass. (Anthracnose) Heilongjiang
Leptothyrium sp. (Mold spot) Henan
Marssonina brunnea (Ell. et Ev.) Magn. (Black leaf spot) Jilin Liaoning Heilongjiang
Marssonina populi (Lib.) Magn. (Black leaf spot) Jilin Shanxi Neimeng Liaoning
Melampsora larici-populina Kleb. (Leaf rust) Liaoning Jilin Heilongjiang Henan Hebei Shandong Shanxi
Melampsora magnusiana wagn. (Leaf rust) Liaoning
Mycosphaella mandshurica Miura (Gray leaf spot) Jilin Heilongjiang Shanxi Neimeng
Myxosporium rimosum Fautr. Henan
Napicoladium asteroma Allesch. Jilin
Phyllactinia populi (Jacz.) Yu. (Powder mildew) Heilongjiang Hebei Henan Shanxi Neimeng Liaoning
Septoria populi Desm. (Brown leaf spot) Shanxi
Trametes suaveolens (L.) Fr. (Decay) Neimeng
Uncinula adunca 9Wallr. ex Fr.) Lév. var. adunca (Powder mildew) Beijing Henan Shanxi
Uncinula adunca (Wallr. ex Fr.) Lév. var. mandshurica (Miura) Zheng et Chen (Powder mildew) Dongbei Shanxi Neimeng Henan
(=Uncinula mandshurica Miura)
Venturia tremulae (Frank) Aderh. (Black star) Heilongjiang
Viscum album L.) (True mistletoe)Shanxi

Populus simonii x balsamifera
Marssonina brunnea (Ell. et Ev.) Sacc. (Black leaf spot) Shanxi

Populus simonii x nigra L.
Melampsora larici-populina. kleb. (Leaf rust) Shanxi Heilongjiang
Phyllactinia populi (Jacz.) Yu (Powder mildew) Shanxi Heilongjiang

***Populus simonii x nigra var. italica* (Muench.) Koehne**
Coryneum populinum Bres. (Gray leaf spot) Neimeng
Marssonina brunnea (Ell. et Ev.) Sacc. (Black leaf spot) Shanxi
Melampsora larici-populina Kleb. (Leaf rust) Shanxi
Phyllactinia populi (Jacz.) Yu (Powder mildew) Shanxi
Spetotis populiperda (Moesz et Smarods) Waterman et Cash (Big leaf spot) Shanxi

***Populus stalinetz* Jabl.**
Marssonina brunnea (Ell. et Ev.) Sacc. (Black leaf spot) Shanxi

***Populus suaveolens* Fisch.**
Daedaleopsis confragosa (Bolt. ex Fr.) Schröt. (Decay) Neimeng
Fomes fomentarius (L ex Fr.) Kickx. (Brown rot) Gansu
Ganodera applanatum (Pers.) Pat. (Decay) Neimeng
Melampsora larici-populina Kleb. (Leaf rust) Shanxi
Phyllactinia populi (Jacz.) Yu (Powder mildew) Hebei Neimeng

***Populus szechuanica* Schneid.**
Fomes fomentarius (L. ex Fr.) Kickx. (Decay) Gansu
Fusicaldium treumulae Fr. (Black star) Shanxi
Melampsora larici-populina Kleb. (Leaf rust) Neimeng
Phellinus igniarius (L. ex Fr.) Quél. (Decay) Gansu
Septoria populi Desm. (Brown leaf spot) Shanxi

***Populus talassica* Kom.**
(=Populus densa Kom.)
Armillariella mellea (Vahl ex Fr.) Karst. (Root decay) Xinjiang
Cuscuta monogyna Vahl. (Cuseata high plant mistletoe) Xinjiang
Cytospora chrysosperma (Pers.) Fr. (Canker) Xinjiang

Fomes fomentarius (L. ex Fr.) Kickx. (Brown rot) Xinjiang
Fusicladium radiosum 9Lib.) Lind (Black star)
(Sexual :Venturia populina (Vuill) Fabr.) Xinjiang
Ganoderma applanatum (pers.) pat. (Decay) Xinjiang
Ganodorma lobatum (Schw.) Atk. (Decay) Xinjiang
Hericium coralloides (Scop. ex Fr.) pers. ex Gray (Decay) Xinjiang
Marssonina populi (Lib.) Magn. (Black leaf spot) Xinjiang
Melampsora larici-populina Kleb. (Leaf rust) Xinjiang
Phellinus igniarius (L. ex Fr.) Quél. (Decay) Xinjiang
Phelilnus robustus (Karst.) Bound. et Galz. (Decay) Xinjiang
Phellinus tremulae Bond. et Sing. (Decay) Xinjiang
Pholiota adipose (Fr.) Quél. (Decay) Xinjiang
Phyllosticta cinerea Pass. (Leaf spot) Xinjiang
Pleurotus carticatus (Fr.) Quél.
Pleurotus sapidus (Schulz. ap Kalchbr.) Sacc.
Polyporus picipes Fr. (Decay) Xinjiang
Septoria populicola Peck (Brown leaf spot) Xinjiang
Uncinula adunca (Wallr ex Fr.) Lév. var. adunca (Powder mildew) Xinjiang

***Populus tomentosa* Carr.**

Agrobcterium tumefaciens (Smith et Towns.) Conn (Crow gall) Henan Hebei Shanxi Shanxi Shandong Liaoning
Alternaria alternata (Fr.) Keissler (Aspen dry leaf) Heilongjiang Jilin Liaoning Xinjiang Shanxi Henan Hebei Shandong Beijing
Ascochyta populi Delacr. (Leaf spot) Hebei
Ascochyta tremulae Thuem. (Cycle spot) Shanxi
Capnodium pelliculosum B et Rav. (Black mildew) Shanxi
Capnodium salicinum Mont. (Black mildew) Henan Hebei Shanxi Shandong
Cercospora populicola Tharp (Leaf spot) Shanxi
Cercospora populina Ell. et Ev. (Leaf spot) Shanxi
Coniothyrium populicola Miura (Leaf spot) Henan
Coryneum populinum Bres. (Leaf spot) Henan Shandong
Cytospora chrysosperma (Pers.) Fr. (Canker) Henan Shanxi
Cytospora harioti Briard. (Canker) Shanxi
Dothichiza populea Sacc. et Br. (Bark canker) Shanxi
Eriophyes dispar Nal. (Leaf felt spot) Beijing Shanxi Gansu
Funalia hispida (Bagl.) Pat (Decay) Henan Shanxi
Funalia trogii (Berk.) Bond. et Sing. (Decay) Shandong
Fusicoccum leucostomum Sacc. (Stem rot) Shanxi
Glomerella cingulata (Stonem.) Spauld. et Schr. (Anthracnose) Henan Shanxi
Helicobasidium purpureum (Tul.) Pat. (Violet root rot) Henan Jiangsu Shanxi
Inonotus hispidus (Bull. ex Fr.) Karst. (Leaf tip rot) Hebei
Marssonina brunnea (Ell. et Ev.) Sacc. (Black leaf spot) Shanxi Henan
Marssonina populi (Lib.) Magn. (Black leaf spot) Shanxi
Melampsora laricis Hart. (Leaf rust) Shanxi
Melampsora magnusiana Wagn. f. tomentosae Zhon (Leaf rust) Henan
Melampsora magnusiana Wagn. (Leaf rust) Liaoning Henan Shanxi Shanxi Shandong
Melampsora rostrupii Wagn. (Leaf rust) Henan Hebei Shanxi Shanxi Shandong Xinjiang
Myxosporium rimosum Fautr. (Dieback) Shanxi
Phyllactinia populi (Jacz) Yu (Powder mildew) Sichuan Hebei
Phyllosticta populina Sacc. (Leaf spot) Shanxi
Septoria populi Desm. (Brown leaf spot) Henan Jiangsu Shanxi

Septoria populicola Peck (Brown leaf spot) Shanxi Henan
Septotis populiperda (Moesz et Smarods) Waterman et cash (Big leaf spot) Shanxi
Taphrina aurea (Pers.) Fr. (Leaf blister) Shanxi
Uredo tholopsora Cummis (Leaf rust) Henan Hebei Shandong Shanxi Guangxi
Viscum album L. (Ture mistletoe) Henan

***Populus tremula* L.**

Coryneum populinum Bres. (Leaf spot) Xinjiang
Cuscuta japonica choisy (Cuseata high plant mistletoe) Xinjiang
Cytospora chrysosperma (Pers.) Fr. (Canker) Neimeng
Eriophyes dispar Nal. (Leaf felt spot) Beijing Shanxi
Fomes fomentarius (L. ex Fr.) Kickx. (Brown rot) Gansu
Fusicladium radiosum (Lib.) Lind (Black star) Xinjiang
Gloeosporium tremulae (Lib.) Pass. (Anthracnose) Neimeng
Marssonina populi (Lib.) Megn. (Black leaf spot) Xinjiang
Melampsora larici-populina kleb. (Leaf rust) Xinjiang Gansu
Melampsora larici-tremulae Kleb. (Leaf rust) Xinjiang
Melampsora rostrupii Wagn. (Leaf rust) Hebei
Phellinus igniarius (L. ex Fr.) Quél. (Decay) Heilongjiang Neimeng
Schizophyllum commune Fr. (Decay) Neimeng
Septotis populiperda (Moesz et Smarods) Waterman et Cash (Big leaf spot) Xinjiang
Uncinula adunca (Wallr. ex Fr.) Lév. var. adunca (Powder mildew) Xinjiang Yunnan
Viscum coloratum (Kom.) Nakai (Ture mistletoe) Neimeng

***Populus tremula var. villosa* Weam.**

Melampsora rostrupii Wagn (Leaf rust) Hebei

Populus usbekistanica cv. 'Afghanica'

Ascochyta tremulae Thuem. (Poplar leaf spot) Shanxi
Coryneum populinum Bres (Gray leaf spot) Shanxi
Cytospora chrysosperma (Pers.) Fr. (Canker) Shanxi
Dothiorella gregaria Sacc. (Canker) Shanxi
Eriophyes varius Nal. (Leaf felt spot) Shanxi
Glomerella cingulata (Stonem.) Spauld. et Schr. (Anthracnose) Shanxi
Macrophoma tumeifaciens Shear (Branch gall canker) Henan Xinjiang Ningxia
Marssonina brunnea (Ell. et Ev.) Sacc. (Black leaf spot) Shanxi
Melampsora larici-populina Kleb. (Leaf rust) Shanxi
Phyllactinia populi (Jacz.) Yu (Powder mildew) Shanxi
Phoma populi-nigrae Allesch. (Dieback) Shanxi
Septoria populi Desm. (Brown leaf spot) Shanxi
Septotis populiperda (Moesz et Smarods) Waterman et Cash (Big leaf spot) Shanxi

***Populus ussuriensis* Kom.**

Melampsora larici-populina Kleb. (Leaf rust) Heilongjiang Neimeng
Phellinus igniarius (L. ex Fr.) Quél. (Decay) Dongbei

***Populus velux* Hsu.**

Dothiorella gregaria Sacc. (Canker) Shanxi
Melampsora larici-populina Kleb. (Leaf rust) Shanxi
Phyllactinia populi (Jacz.) Yu (Powder mildew) Shanxi

***Populus yunnanensis* Dode**

Melampsora larici-populina Kleb. (Leaf rust) Yunnan
Uncinula pseudocedrelae Cheng et Chen (Powder mildew) Yunnan

Prunus spp.

Caeoma makinoi kus. (Leaf rust) Tibet
Cercospora circumscissa Sacc. (Leaf spot)

(Sexual :Mycosdhaerella cerasella Aderh.)
Jiangsu Guangdong Guangxi Sichuan
Clasterosporium carpophiolium (Lév.) Aderh Hubei
Coriolus hirsus (Wulf. ex Fr.) Quél. (Decay) Hebei
Coriolus versiolor (L. ex Fr.) Quel. (Decay) Hebei Zhejiang
Cylindrosporium padi Karst. Guangdong
Fomes fomentarius (L. ex Fr.) Kickx. (Brown rot) Xinjiang
Ganoderma applanatum (Pers.) Pat. (Decay) Hebei
Laetiporus sulphureus (Bull. ex Fr.) Bond et Sing (Trunk brown rot) Hebei
Leucotelium pruni-persicae (Hori) Tranz. Jiangsu Hunan Sichuan Guizhou
Monilinia laxa (Aderh. et Ruhl.) Honey (Brown rot) Sichuan Guizhou Liaoning Heilongjiang
(=Sclerotinia laxa (Ehrenb.) Aderh. et Ruhl.)
Napicladium brunaudii Sacc. (Black mildew) Hebei
Phellinus pomaceus (Pers. ex Gray) Quél. (Brown rot) Liaoning Hebei Shanxi Henan Ningxia Yunnan
Phoma enteroleuca Sacc. (Leaf spot) Province not determined
Phyllosticta persicae Sacc. (Leaf spot) Hebei
Polystigma ochraceum (Wahl.) Sacc. (Red leaf spot) Shanxi Hunan Heilongjiang
Poystigmina rubra Sacc. (Red leaf spot) Hebei Shanxi Sichuan Yunnan Liaoning Heilongjiang
Poria medulla-panis (Jacq. ex Fr.) Cooke (Decay) Zhejiang
Sphaeropsis malorum Peck Shandong
(:Physalopsora obtuse (Schw.) Cooke)
Taphrina cerasi (Fuck.) Sadeb. (Cherry witches' broom) Heilongjiang Liaoning
(=Exoascus cerasi Fuck.)
Taphrina deformans (Berk.) Tul. (Peach Leaf curl) Liaoning Shanxi Tibet
Taphrina pruni (Fuck.) Tul (Cherry leaf curl) Tibet
Thekopsora areolata (Fr.) Magn. (Leaf rust) Shanxi Heilongjiang Jilin Xinjiang
Truncospora truncatospora (Lloyd) Ito (Decay) Hebei

***Prunus amygdalus* Batsch**

Cylindrosporium padi Karst. Guangdong
Monilinia laxa (Aderh. et Ruhl.) Honey (Brown rot)
(Sexual :Manilia cinerea Bon.) Zhejiang Guangdong
Podosphaera tridactyla (Wallr.) de Bary (Powder mildew) Taiwan
Polystigmina rubra Sacc. (Red leaf spot) Xinjiang
Septobasidium bogoriense pat. (Felt fungus) Taiwan
Septobasidium tanakae (Miyabe) Boed. et Steinm () Taiwan
Taphrina deformans (Berk.) Tul. (Cherry leaf curl) Hubei

Pruus armeniaca var. ansu Maxim.

Botrytis cinerea Pers. (Brown rot) Sichuan
(Sexual :Sclerotinia fuckeliana (de Bary) Fuck.)
Monilinia laxa (Aderh. et Ruhl.) Honey (Brown rot) Jiangsu Heilongjiang
Polystigma deformans Syd. (Red leaf spot) Tibet Neimeng
Septobasidium tanakae (Miyabe) Boed. Et Steinm. (Felt fungus) Taiwan
Tubercularia vulgaris Tode (Canker) Tibet

***Prunus davidiana* Franch.**

Ascochyta prunicola P. K. Chi (Leaf spot) Jilin
Cercospora circumscissa Sacc. (Leaf hole) Liaoning
Graphium rhodophaeum Sacc. Et Trott. (Dieback) Liaoning
Phellinus pomaceus (pers. ex Gray) Quel. (Brown rot) Liaoning Hebei Tibet
Phyllosticta pirina Sacc. (Leaf spot) Jilin
Podosphaera tridactyla (Wallr.) de Bary (Powder mildew) Jilin Hebei

Polystigma deformans Syd. (Red leaf spot) Shanxi
Schizophyllum commune Fr. (Decay) Neimeng
Tranzschelia pruni-spinosae (Pers.) Diet. (Leaf rust) Liaoning Jilin Tibet

***Prunus humulis* Bunge**

Phyllosticta circumscissa Cooke (Leaf spot) Jilin
Podosphaera tridactyla (Wallr.) da Bary (Powder mildew) Jilin
Polystigmina rubra Sacc. (Red leaf spot) Hebei
Taphrina trunicola Kus. (Cherry leaf curl) Liaoning
(=Exoascus trunicola (Kusano) Shaw.)
Traqnzschelia pruni-spinosae (Pers.) Diet. (Leaf rust) Liaoning

***Prunus maackii* Rupr.**

Tranzschelia pruni-spinosae (Pers.) Diet. (Leaf rust) Liaoning

***Prunus padus* L.**

(=Padusracemosa Schneider)
Cuscuta japonica Choisy (Dodder) Liaoning
Cylindrosporium padi Karst. Sichuan
Eriophyes poderineus Nal. (Leaf felt spot) Xinjiang
Fomitopsis pinicola (Sw. ex Fr.) Karst. (Brown rot) Jilin
Nectria coccina (pers.) Fr. (Dieback) Xinjiang
Phellinus pomaceus (Pers. ex Gray) Quél. (Decay) Xinjiang
Podosphaera tridactyla (Wallr.) de Bary (Powder mildew) Xinjiang Heilongjiang
Polystigma ochrceum (Wahl) Sacc. (Red leaf spot) Xinjiang Heilongjiang
Thekopsora areolata (Fr.) Magn. (Pine cone rust) Xinjiang Heilongjiang
Xanthomonas pruni (Smith) Dowson (Bacterial spot of stone fruits) Heilongjiang Liaoning Neimeng Xinjiang

***Prunus padus var. pubescens* Rebel et Tiling**

(=Padus asiatica Kom.)
Podosphaera tridactyla (Wallr.) de Bary (Powder mildew) Jilin Heilongjiang Xinjiang
Polystigma ochraceum (Wahl.) Sacc (Red leaf spot) Heilongjiang Neimeng
Polystigmina rubra Sacc. Jilin
Taphrina pruni (Fuck.) Tul. (Bacterial spot of stone fruits) Liaoning
Taphrina pruni var. padi Jacz. (Bacterial spot of stone fruits) Heilongjiang
(=Exoascus pruni var. padi Jacz.)
Thekopsora areolata (Fr.) Magn. (Leaf rust) Heilongjiang Xinjiang
Viscum coloratum (Kom) Nakai (Turl mistletoe) Heilongjiang

***Prunus sibirica* L.**

Cercospora circumscissa Sacc. (Leaf hole) Liaoning
Polystigma deformans Syd. (Leaf spot) Liaoning

***Pseudolarix amabilis* (Nelson) Rehd.**

Discosia artocreas (Tode) Fr. (Leaf spot) Henan
Macrophomina phaseoli (Mnubl.) Ashby (Stem rot) Zhejiang Henan Anhui Guangdong Sichuan

***Pseudosasa japonica* (Sieb. et Zucc.) Makino**

(=Arundinaria japonica Sieb. et Zucc.)
Phyllachora orbicula Rehm (Black tar) Shandong

***Pseudoasa purpurascens* (Hack.)Makino**

Puccinia phyllostachydis Kus. (Leaf rust) Guangxi
Uredo ignava Arth. (Leaf rust) Guangxi

***Pterocarya stenoptera. C.* DC.**

Cuscuta japonica Choisy (Dodder) Zhejiang Henan
Loranthus parasiticus (Linn.) Merr. (Mistletoe) Guangxi

Loranthus yadoriki Sieb. et zucc (Mistletoe) Sichuan Zhejiang Fujian Guangdong Guangxi Hunan Hubei Shanxi
Melanconium juglandinum Kunze (Dieback) Jiangsu Hebei Henan Shanxi Liaoning
Melanconium oblongum Berk. (Dieback) Jiangsu Liaoning
Microstroma juglandis (Bereng.) Sacc. (Leaf spot) Jiangsu Henan Shandong Zhejiang Shandong Anhui Liaoning
Phyllactinia juglandis Tao et Qin. (Powder mildew) Jiangsu Henan Sichuan Anhui Guizhou
(=Phyllactinia corylea (Pers.) Karst.)
Phyllostica pterocaryai Thüm (Leaf spot) Hubei
Sclerotinia sclerotiorum (Lib.) de Bary (Sclerotinia) Henan
Viscum album L. (Ture mistletoe) Henan
Bacterial (Leaf spot) Gansu

Pteroceltis sp.

Phyllactiniapteroceltidis Yu et Han (Powder mildew) Anhui
(=Phyllactinia corylea (Pers.) karst.)

Quercus spp.

Armillariella mellea (Vahl ex Fr.) Karst. (Root decay) Sichuan
Aspergillus niger v. Tiegh. (Black mold) Henan
Cerrena unicolor (Bull. ex Fr.) Mull. (Decay) Hebei Heilongjiang
Ciboria pseudotuberosa Rhem (Resembling a sclerotium) Liaoning Heilongjiang
Coccdiscus quercicola P. Henn. (Dieback) Guangdong
Coriolus hirsutus (Wulf ex Fr.) Quél. (Decay) Henan Yunnan Heilongjiang
Coriolus versicolor (L. ex Fr.) Quél. (Decay) Hebei Gansu Fujian Yunnan Tibet Heilongjiang
Cronartium quercuum (Berk.) Miyable (Leaf rust) Jiangsu Zhejiang Anhui Hubei Guangxi Sichuan Jiangxi Heilongjiang Guizhou Yunnan
Cytospora microspora (Corda) Rabenh. (Dieback) Hebei
Daedaleopsis confragosa (Bolt. ex Fr.) Schröt. (Decay) Gansu
Endothia radicalis (Schw.) de Not. (Canker) Sichuan Heilongjiang
Erysiphe fagacearum Zheng et Chen (Powder mildew) Yunnan
Fomes fomentarius (L. ex Fr.) Kickx. (Decay) Jilin Heilongjiang Hebei Shanxi Gansu Sichuan Yunnan Tibet
Fomitopsis castaneus Imaz. (Decay) Jilin Heilongjiang
Fomitopsis pinicola (Sw. ex Fr.) Karst. (Decay) Hebei
Ganoderma applanatum (pers.) pat. (Decay) Hebei Henan Jiangsu Gansu Fujian
Ganoderma lucidum (Leyss. ex Fr.) Karst. (Decay) Tibet
Hericium erinaceus (Bull.) Pers. (Decay) Henan Sichuan Tibet
(=Hydnum erinaceus Fr.)
Hirschioporus paragamenus (Fr.) Bond. et Sing. (Decay) Sichuan
(=Polyporus paragamensu Fr.)
Inonotus dryadeus (Pers. ex Fr.) Murr. (Trunk brown cube rot) Guangxi Sichuan
(=Polyporus drydeus (Pers.) Fr.)
Inonotus gilvoides (Lloyd) Teng (Decay) Yunnan
Inonotus hispidus (Bull. ex Fr.) Karst. (Shoot tip rot) Hebei Henan
Inonotus radiatus (Sow. ex Fr.) Karst. var. licentii Pilát (Decay) Hebei Shanxi Zhejiang Anhui Jiangxi Guangxi Sichuan Yunnan
Inonotus rheades (Pers.) Pilát. (Decay) Heilongjiang
Irenina quercina Hansf. (Black mildew) Anhui
Laetioporus sulphureus (Bull. ex Fr.) Bond. et Sing (Brown rot) Sichuan Tibet Heilongjiang
Linospora conflicta (Cooke) Sacc. Zhejiang
Lloydella subpileata (Berk. et Curt.) Höhn. et Litsch. (Decay) Gansu
Loranthus europaeus Jacq (Mistletoe) Henan

Macrophoma fusispora Bub. (Dieback) Henan Shanxi Liaoning Heilongjiang
Marssonina martinii (Sacc. et Ell.) Magn. (Leaf spot) Hebei Jiangxi
Microstroma album (Desm.) Sacc. var. Japonicum P. Henn. (Dieback) Hunan
Microsphaera alphitoides Griff. et Maubl. (Powder mildew) Ningxia Hebei Dongbei Shanxi Taiwan Henan Hunan Sichuan
Mycosphaerella punctiformis (Pers.) Rabenh. (Leaf spot) Liaoning
Nectria cinnabarina (Tode) Fr. (Canker) Sichuan
Nectria coccinea (Pers.) Fr. (Canker) Sichuan
Nectria ditissima Tul. (Canker) Sichuan
Phellinns igniarius (L. ex Fr.) Quél (Heartwood white sponge rot) Jilin Heilongjiang Hebei Shanxi Shanxi Neimeng Gansu Ningxia Qinghai Xinjiang Sichuan Yunnan
Phellinus robustus (Karst.) Bourd. et Galz. (Brown rot) Heilongjiang Sichuan Tibet
Phellinus setulosus (Lloyd) Imaz. (Decay) Hubei Taiwan
Phellinus torulosus (pers.) Bourd. et Galz. (Decay) Jilin Heilongjiang Hebei Zhejiang Jiangxi Guangdong Guangxi Yunnan Taiwan
Phyllactinia roboris (Gachet) Blum. (Powder mildew) Jiangsu Anhui Sichuan
Phyllosticta quercus Sacc. et Speg. (White star) Liaoning
Poria lacerata Murr. (Decay) Hebei Jiangsu
Poria lurida Bres. (Decay) Zhejiang
Poria versipora (Pers.) Rom. (Decay) Hebei Yunnan
Pycnoporus cinnabarinus (Jacq.) Karst. (Decay) Liaoning Tibet
Pycnoporus sanguineus (L. ex Fr.) Mull. (Decay) Henan
Pyropolyporus pectinatus (Kl.) Murr. (Decay) Jiangsu
Sphaerotheca lanetris Harkn. (Powder mildew) Zhejiang Sichuan
Sphaerotheca writghtii (Berk. et Cart.) Höhn. (Powder mildew) Anhui
Spongipellis litschaueri Lohw. (Heartwood white rot) Jilin Heilongjiang Liaoning Hebei Shanxi Yunnan
Sterccherinum cirrhatum (Pers. ex Fr.) Teng Jilin
Stereum frustulosum (Pers.) Fr. (Decay) Heilongjiang
Stereum pubescens Burt (Decay) Liaoning
Stereum Purpureum (Pers.) Fr. (Decay) Sichuan
Trametes dickinsii Berk. (Decay) Henan
Trametes kusanoana Imaz. (Brown pocket rot) Hainandao
Trametes quercina (L.) pilát (Decay) Jilin Hubei Hebei
Trichothecium roseum (Bull.) Link (Powder mildew) Henan
Truncospora truncatospora (Lloyd) Ito (Decay) Hebei
Typhulochaeta japonica Ito et Hara (Powder mildew) Sichuan
Tyromyces amygdalinus (Berk. et Rav.) Teng. (Brown cube rot) Hainandao
Tyromyces fissilis (Berk. et Curt.) Murr. (Decay) Jilin Heilongjiang Hebei Shanxi

***Quercus acutissima* Carruth**

Cronartium quercuum (Berk.) Miyabe (Leaf rust) Henan Shanxi Anhui Guizhou Yunnan
Cytospora ep. (Canker) Shandong
Gnomonia setacea (pers.) Ces. et de Not. (Anthracnose) Jiangsu
Hymenochaete sp. (Decay) Shandong
Loranthus yadoriki Sieb. et zucc. (Mistletoe) Sichuan
Macrophomina phaseoli (Maubl.) Ashby (Trunk decay) Jiangsu
Microsphaera alphitoides Griff. et Maubl. (Powder mildew) Jiangsu Guizhou Sichuan
Nectria cinnabarina (Tode) Fr. (Dieback) Shandong Shanxi
Phyllactinia roboris (Gachet) Blum. (Powder mildew) Henan
Phyllosticta hranicensis Petr. (Leaf spot) Taiwan
Sphaerotheca lanestris Harkn. (Powder mildew) Anhui Guangdong Guangxi

Taphrina caerulescens (Desm. et Mont.) Tul (Oak leaf blister) Henan
Truncospora truncatospora (Lloyd) Ito (Decay) Hebei
Virus. (Leaf wrinkle) Shandong

***Quercus aliena* Bl.**

Colletotrichum gloeosporioides penz. (Anthracnose) Henan Shanxi
(=Gloeosporium quercinum West.)
Cronartium quercuum (Berk.) Miyabe (Leaf rust) Heilongjiang Shanxi Sichuan Henan Hubei Anhui Jiangsu Jiangxi Yunnan
Microsphaera alphitoides Griff. et Maubl. (Powder mildew) Hebei Sichuan Neimeng Henan
Mycosphaerella maculiformis (pers). Auersw. (Brown leaf spot) Henan
(Asexual Macrophoma fusispora Bub.)
Phyllactinia roboris (Gachet) Blum. (Powder mildew) Henan Hunan Guizhou
Taphrina caerulescens (Desm. et Mont.) Tul. (Oak leaf blister) Henan
Typhulochaeta japonica Ito et Hara (Powder mildew) Sichuan
(=Erysiphe japonica (Ito et Hara) Wei)

Quercus aliena Bl. var. acutidentata Maxim.

Ascochyta quercus Sacc. et Speg. (Gray leaf spot) Shanxi
Daldinia concentrica (Bolt.) Ces. et de Not. (Decay) Shanxi
Didymosphaeria atro-grisea Cooke. et Peck (Canker) Shanxi
Macrophoma fusispora Bub. (Brown leaf spot) Shanxi
Monochaetia pachyspora Bub. (Brown leaf spot) Shanxi

***Quercus bambusifolia* Hance**

Phellinus robustus (Karst.) Bourd. et Galz. (Sapwood brown rot) Hainandao
Pyropolyporus adamantinus (Berk.) Teng (Brown pocket rot) Hainandao
Pyropolyporus pusillus (Lloyd) Teng (Brown pocket rot) Hainandao
(=Fomes pusillus Lloyd)

***Quercus dentata* Thunb.**

Cronartium quercuum (Berk.) Miyabe (Pine oak gall rust) Henan Sichuan
Macrophoma fusispora Bub. (Leaf spot) Jilin
Microsphaera alphitoides Griff. et Maubl. (Powder mildew) Jilin Hebei Shandong Jiangsu Henan Gansu
Monochaetia kansensis (Ell. et Barth.) Sacc. (Anthracnose) Henan
Phyllactinia roboris (Gachet) Blum. (Powder mildew) Henan
Phyllosticta quercus Sacc. et Speg. (Leaf spot) Henan
Taphrina caerulscens (Desm. et Mont.) Tul. (Oak leaf blister)) Henan
Uncinula septata Salm. (Powder mildew) Henan

***Quercus fabri* Hance.**

Cronartium quercuum (Berk.) Miyabe (Pine oak gall rust) Jiangsu
Microsphaera alphitoides Griff. et Maubl. (Powder mildew) Jiangsu Hunan Henan
Phyllactinia roboria (Gachet) Blum. (Powder mildew) Hunan Guizhou Jiangxi
Typhulochaeta japonica Ito et Hara (Powder mildew) Jiangsu
Uncinula septata Salm. (Powder mildew) Jiangsu Hunan Guangxi

***Quercus glandulifera* Bl**

Cronartium quercuum (Berk.) Miyabe (Pine oak gall rust) Henan Jiangsu Zhejiang Sichuan
Leptothyrium quercinum (Lasch) Sacc. (Needle cast) Sichuan
Loranthus yadoriki Sieb. et zucc. (Mistletoe) Sichuan
Macrophoma fusispora Bub. (Quercus blight) Henan
Marssonina martinii (Sacc. et Ell.) Magn. (Black leaf spot) Jiangxi Sichuan
Microsphaera alphitoides Griff. et Maubl. (Powder mildew) Jiangsu Zhejiang Anhui Guangxi Sichuan Henan
Phyllactinia roboris (Gachet) Blum. (Powder mildew) Jiangsu

Scorias communis Yamam. Taiwan
Septobasidium bogoriense pat. (Felt fungus) Zhejiang
Sphaerotheca lanestris Harkn. (Powder mildew) Sichuan
Sphaerotheca wrightii (Berk. et curt.) Höhn. (Powder mildew) Henan
Triphragmium spinigera (Höhn.) Yamam. (Leaf rust) Taiwan
Typhulochaeta japonica Ito et Hara (Powder mildew) Sichuan

Quercus liaotungensis **Koidz**

Colletotrichum gloeosporioides penz. (Anthracnose) Shanxi Liaoning
Hericium caput-medusae (Bull. et Fr.) Pers. (Decay) Shanxi Liaoning
Loranthus europaeus Jacq (Europian mistletoe) Shanxi Liaoning
Phellinus igniarius (L. ex Fr.) Quél. (Heartwood white sponge rot) Shanxi Liaoning

Quercus mongolica **Fisch.**

Armillariellla mellea (Vahl ex Fr.) Karst. (Root decay) Jilin Neimeng
Ascochyta quercus Sacc. et Speg. (Leaf spot) Jilin
Ciboria pseudotuberosa Rehm (Resembling a sclerotium) Jilin Heilongjiang
Clithris quercina Rehm (Dieback) Heilongjiang
Coniothyrium quercinum Sacc. (Dieback) Jilin
Cronartium quercuum (Berk) Miyabe (Leaf rust) Heilongjiang Neimeng
Erysiphe polygoni DC. (Powder mildew) Taiwan
Fumago vagans pers. (Black mildew) Jilin
Fusicoccum quercus Oud (Canker) Heilongjiang
Ganoderma applanatum (pers.) pat (Decay) Dongbei Neimeng
Hericium erinaceus (Bull.) Pers. (Decay) Jilin Heilongjiang Neimeng
Inonotus hispidus (Bull. ex Fr.) karst. (Shoot rot) Dongbei
Inonotus krawtzewii (pilát) pilát (Decay) Neimeng
Inonotus rheades (pers.) pilát (Decay) Neimeng
Laetiporus sulphureus (Bull. ex Fr.) Bond. ex sing (Trunk brown rot) Heilongjiang Neimeng
Microsphaera alphitoides Griff. et Maubl. (Powder mildew) Dongbei Henan
Mycosphaeralla maculiformis (pesr.) Auersw. (Leaf spot) Jilin Liaoning Heilongjiang Henan
Phellinus igniarius (L. ex Fr.) Quél. (Heartwood white sponge rot) Dongbei Neimeng
Phellinus robustus (karst.) Bourd. et Galz. (Sapwood brown rot) Dongbei Neimeng
Spongipellis litschaueri Lohw. (Heartwood white rot) Dongbei
Steccherinum septentrionale (Fr.) Bank. Dongbei
Taphrina caerulescens (Desm. et Mont.) Tul. (Oak leaf blister) Dongbei
Trametes quercina (L.) Pilát (Decay) Heilongjiang
Typhulochaeta japonica Ito et Hara (Powder mildew) Jiangsu
Tyromyces fissilis (Berk. et Curt.) Donk. (Decay) Liaoning Heilongjiang

Quercus robur **L.**

Capnodium sp. (Black mildew) Xinjiang
Cytospora intermedia Sacc. (Canker) Xinjiang
Microsphaera hypophylla Nevod. em. Roll-Hans. (Powder mildew) Xinjiang

Quercus semicarpifolia **Smith**

Acantharia sinensis (Petr.) Arx Sichuan
Armillariella mellea (Vahl ex Fr.) Karst. (Root decay) Tibet
Daedaleopsis confragosa (Bolt. ex Fr.) Schröt (Decay) Tibet
Fomes fomentarius (L. ex Fr.) Kickx. (Brown rot) Tibet Yunnan Sichuan
Fomes marmoratus (Berk. et Curt.) Cooke (Decay) Tibet
Fomitopsis pinicola (Sw ex Fr.) karst. (Brown cube rot) Tibet
Ganoderma lobatum (Schw.) Atk. (Decay) Tibet
Ganoderma lucidum (Leyss. ex Fr.) Karst. (Decay) Tibet
Ganoderma oroflavum (Lloyd) Teng (Decay) Tibet

Ganoderma tsugae Murr. (Decay) Tibet
Hericium erinaceus (Bull. ex Fr.) Pers. (Decay) Tibet Sichuan
Hirschioporus sector (Ehrenb. ex Fr.) Teng (Decay) Tibet
Inonotus rheades (pers.) Pilát (Decay) Tibet
Irpex lacteus Fr. (Decay) Tibet
Laetiporus sulphureus (Bull. ex Fr.) Bond. et Sing (Trunk brown rot) Tibet Sichuan
Macrophoma suberis Prill. et Delacr. var. nigromaculata Keissl. (Dieback) Yunnan
Melanconis betulina Otth. Hedw. Tibet
Phellinus igniarius (L. ex Fr.) Quél. (Heartwood sponge rot) Tibet
Polyporellus picipes (Fr.) Karst. (Decay) Tibet
Polystictus meleagris (Berk.) Cooke (Decay) Tibet
Prillieuxina sinensis Petr. Yunnan
Pycnoporus cinnabarinus (Jacq.) Karst. (Decay) Tibet
Schizophyllum commune Fr. (Bark rot) Tibet
Stereum vellereum Berk. (Decay) Tibet
Trabutia sinensis Arx et Mull. Yunnan
Trametes quercina (L.) Pilát. (Decay) Tibet
(=Daedalea quercina (L.) Fr.)
Tyromyces pubescens (Schum. ex Fr.) Imaz. (Decay) Tibet

***Quereus variabilis* Bl.**

Balanophora nippanica Makino. (Balanophora mistletoe) Henan
Coriolus versicolor (L. ex Fr.) Quél. (Decay) Shandong
Cronartium quercuum (Berk.) Miyabe (Pine oak gall rust) Jiangsu Henan Shanxi Taiwan
Ganoderma applanatum (pers.) pat. (White rot) Shanxi
Ganoderma lobatum (Schw.) Atk. (White rot) Shanxi
Ganoderma lucidum (Leyss. ex Fr.) Karst. (Trunk decay) Shanxi
Macrophoma fusispora Bub. (Brown leaf spot) Shanxi
Microsphaera alpihtoides Griff. et Maubl. (Powder mildew) Jiangsu Anhui
Neocapnodium tanakae (Shirai et Hara) Yamam. (Black mildew) Taiwan
Phyllacitnia roboris (Gachet) Blum. (Powder mildew) Jiangsu Henan
Sphaerotheca lanestris Harkn. (Powder mildew) Shandong
Taphrina caerulescens (Desm. et Mont.) Tul. (Oak leaf blister) Henan
Trametes dickinsii Berk (Brown rot) Shanxi
Trametes quercina (L.) pilát (Brown rot) Shanxi
Tremella indurate Berk. et Br. (Skin rot) Yunnan
Triposporiopsis spinigera (Höhn) Yamam. (Black mildew) Taiwan
Truncospora truncatospora (Lloyd) Ito (Decay) Shanxi Yunnan
Uncinula septata Salm. (Powder mildew) Jiangsu

Rhamnus spp.

Aecidium alaternii Maire (Leaf rust) Hebei Shanxi
Coniothyrium rhamni Miyake (Leaf spot) Hebei Sichuan
Microsphaera divaricata (wallr.) Lev. (Powder mildew) Jilin Liaoning
(=Microsphaere alni (DC.) Wint.)
Puccinia coronata Corda (Leaf rust) Jilin Hebei Jiangsu Zhejiang Shanxi Anhui Henan Jiangxi Xinjiang Hubei Guangxi Sichuan Yunnan Heilongjiang
Septoria frangulae Guep. (Leaf spot) Jiangsu

***Rhamnus dahurica* Pall.**

Ascochyta rhamni Cooke et Shaw. (Leaf spot) Jilin
Cercospora rhamni Fuck (Leaf spot) Jilin
Coniothyrium dumeei Br. et Cav. (Leaf spot) Liaoning
Coniothyrium rhamni Miyake (Leaf spot) Jilin

Microsphaera divaricata (Wallr.) Lév. (Powder mildew) Heilongjiang Liaoning
Phyllosticta rhamnicola Desm. (Leaf spot) Jilin Liaoning
Puccinia poae-pratentsis Miura (Leaf rust) Jilin Liaoning Heilongjiang
Septobasidium borgoriense Pat. (Felt fungus) Liaoning Heilongjiang

***Rhamnus parvifolia* Bunge**
Septobasidium borgoriense Pat. (Felt fungus) Liaoning

***Rhamnus rugulosus* Hemsl.**
Microsphaera rhmnicola Yu (Powder mildew) Sichuan Gansu

***Rhamnus virgata* Roxburgh**
Puccinia coronata Corda (Leaf rust) Tibet

***Rhodea japonica* Roth.**
Colletotrichum gloeosporioides Penz. (Anthracnose) Jiangsu Anhui
(=Colletotrichum montemartinii Togn. var. rhodeae Trav)
Sphaerulina rhodeae P. Henn. et Shirai Jiangsu Zhejiang Anhui

Rhododendron spp.
Chrysomyxa expansa Diet. (Leaf rust) Sichuan Yunnan
Chrysomyxa rhdodendri de Bary (Leaf rust) Hebei Dongbei Qinghai Yunnan
Cordyceps sinensis (Berk.) Sacc. (Leaf spot) Tibet
Exobasidium japonicum Shirai (Rhododendron leaf gall) Jiangxi Hunan Yunnan
Exobasidium pentasporium Shirai (Rhododendron leaf gall) Jiangxi
Hendersonia bicolor Pat Yunnan
Lophodermium rhododendri Ces. (Ne edle cast) Sichuan Yunnan
Melasmia rhododendri P. Henn. et Shirai (Black tar) J iangxi Guizhou
Phellinus igniarius (L. ex Fr.) Quél. (Decay) Gansu Sichuan Tibet
Sporocybe azaleae (Peck) Sacc. Sichuan
Torula rhododendri Ces. Yunnan
Valsa subclypeata Cooke et Peck. (Canker) Hebei

***Rhododendron dahuricum* L.**
Chrysomyxa rhododendri de Bary (Leaf rust) Liaoning
Exobasidium japonicum Shirai (Rhododendron leaf gall) Henan
(=Exobasidium rhododendra Cram.)
Rhytisma rhododendri Fr. (Black mole) Henan

***Rhododendron fulvum* Bulf. et W. W. Sm.**
Chrysomyxa stilbae Y. –z. Wang, M. –m. Chen & Guo (Leaf rust) Tibet Qinghai

***Rhododendron simsii* Planch.**
Aecidium rhododendri Barclay (Leaf rust) Tibet
Chrysomyxa rhododendri de Bary (Leaf rust) Tibet
Exobasidium hemisphaericum Shirai (Rhododendron leaf gall) Sichuan
Exobasidium japonicum Shirai (Rhododendron leaf gall) Tibet
Melasmia rhododendri P. Henn. et Shirai (Black tar) Jiangsu Henan Hunan Zhejiang
(Sexual : Rhytisma rhododendri Fr.)
Phellinus igniarius (L. ex Fr.) Quél. (Decay) Tibet

Rhus spp.
Pileolaria dieteliana Syd. (Leaf rust) Guangdong
Pileolaria klugkistiana (Diet.) Diet. (Leaf rust) Anhui Jiangxi Sichuan Guizhou
Pileolaria shiraiana (Diet. et Syd.) Ito (Leaf rust) Jiangsu Zhejiang Anhui Jiangxi Hunan
Guangdong Guangxi Guizhou
Helieobasidium parpureum (Violet root rot)

***Rhus chinensis* Mill.**
Cronaritum quercuum (Berk.) Miyabe (Leaf rust) Guangxi
Mycosphaerella fushinoki Miura (Gray leaf spot) Liaoning Henan

Phyllactinia rhoina Doidge (Powder mildew) Sichuan
(=Phyllactinia corylea (Pers.) Karst.)
Pileolaria klugkistiana (Diet.) Diet. (Leaf rust) Shanxi Gansu Jiangsu Jiangxi Fujian Taiwan Hunan Guangxi Guizhou
Pileolaria shiraiana (Diet. et Syd.) Ito (Leaf rust) Shanxi Jiangsu Fujian Guangxi
Septoria sp. (Leaf spot) Shanxi
Tubercularia phyllophila H. et P. Syd. Taiwan
Uncinula verniciferae P. Henn. (Powder mildew) Jiangsu Zhejiang Beijing

***Rhus succedanea* L.**
Pileolaria shiraiana (Diet. et Syd.) Ito (Leaf rust) Guangdong Taiwan Yunnan
Uncinula verniciferae P. Henn. (Powder mildew) Taiwan

***Rhus sylvestris* Sieb. et Zucc.**
Nectria cucurbitula (Tode) Fr. (Canker) Shanxi
Pileolaria shiraiana (Diet. et Syd.) Ito (Leaf rust) Zhejiang

***Rhus verniciflua* Stokes**
Roniothyrium olivaceum Bon. Henan
Herpotrichia sp. (Dieback) Guizhou
Haplosporella sp. (Dieback) Guizhou
Loranthus parasiticus (Linn.) Merr. (Mistletoe) Guangxi Guizhou
Pileolaria shiraiana (Diet. et Syd.) Ito (Leaf rust) Zhejiang
Septobasidium sp. (Felt fungus) Guizhou
Uncinula verniciferae P. Henn. (Powder mildew) Jiangsu Zhejiang Anhui

Ribes spp.
Cronartium ribicola J.C. Fischer ex Robenhorst. (Leaf rust) Xinjiang Shanxi Tibet Sichuan Dongbei
Melampsora ribesii-purpureae Kleb. (Leaf rust) Dongbei
Melampsora ribesii-viminalis Kleb. (Leaf rust) Xinjiang
Puccinia ribesii-caricis Kleb. (Leaf rust) Jiangsu Xinjiang
Puccinia ribis DC. (Leaf rust) Shanxi Xinjiang

***Ribes mandshuricum* (Maxim.) Komarov**
Cronartium ribicola J.C. Fischer ex Robenhorst (White pine blister rust) Heilongjiang

***Ribes maximowiczianum* Komarov**
Cronartium ribicola J.C. Fischer ex Robenhorst (White pine blister rust) Heilongjiang

***Ribes pauciflorum* Turcz.**
Cronartium ribicola J.C. Fischer ex Robenhorst (White pine blister rust) Heilongjiang

Robinia sp.
Favolus squamosus (Huds. ex Fr.) Ames (Decay)
Nectria cinnabarina (Tode) Fr. (Canker) Hebei

***Robinia pseudoacacia* L.**
Botryosphaeria abrupta Berk. et Curt. Hebei Henan Shanxi Jiangsu
Collybia velutipes (Curt. ex Fr.) Quél. (Trunk decay) Shanxi
Cuscuta japonica Choisy (Dodder) Henan
Erysiphe polygoni DC. (Powder mildew) Liaoning Jiangsu Henan Shanxi
Fumago sp. (Black mildew) Ningxia
Fusarium spp. (Dieback) Shandong
Haplosporella robiniae (Ell. et Barth.) Pet. et Syd. (Dieback) Hebei
Helicobasidium prupureum (Tul.) Pat. (Violet root rot) Hebei Dongbei Jiangsu Zhejiang Shanxi Shandong Anhui Sichuan Yunnan Guangdong Henan
Macrophomina phaseoli (Maubl.) Ashby (Stem rot) Xinjiang
Microsphaera baumleri Magn. (Powder mildew) Shanxi Sichuan
Microsphaera robiniae Tai (Powder mildew) Liaoning Jiangsu Gansu

Phellinus gilvus (Schw.) Pat. (Decay) Shanxi
Phellinus pomaceus (Pers. ex Gray) Quél. (Decay) Shanxi
Phomopsis oncostoma (Thöm.) Höhn. (Brown blight) Jiangsu
Phyllosticta robiniella Miura (Leaf spot) Liaoning
Phytophthora cinnamomi Rands (Robinia canker) Jiangsu Liaoning Shandong
Rhizoctonia solani Kühn (Young shoot damping off) Henan
Sclerotium rolfsii Sacc. (Sclerotium) Liaoning Henan
Trametes robiniophila Murr. (Decay) Hebei Shanxi Shandong
Yellows Ningxia
Physical belt disease Ningxia

Robinia pseudoacacia var. inermis DC.

Fusarium sp. (Dieback) Liaoning Jiangsu
Phellinus pomaceus (Pers. ex Gray) Quél. (Brown rot) Shanxi
Trametes robiniophila Murr. (Decay) Shanxi

Rosa spp.

Botrytis cinerea Pers. (Gray mold blight) Jiangsu
Caeoma warburgianum P. Henn. (Leaf rust) Zhejiang Fujian
Cercospora rosae (Fuck.) Höhn. (Leaf spot) Huabei Shandong
Cercospora rosicola Pass. (Leaf spot) Guangdong
Colpoma rosae (Teng) Teng Zhejiang Hunan Guangxi
Coryneum rosaecola Miura (Gray leaf spot) Sichuan
Diplocarpon rosae Wolf Heilongjiang Jiangsu Zhejiang Fujian Yunnan
Gerwasia rosae Tai (Leaf rust) Anhui Guangxi
Inonotus hispidus (Bull. ex Fr.) Karst. (Decay) Shandong
Kuehneola japonica Diet. (Leaf rust) Taiwan
Medusosphaera rosae Golov. et Gamal (Powder mildew) Xinjiang
Monochaetia seiridioides (Sacc.) Allesch. Jiangsu
Mycosphaerella rosigena (Ell. et Ev.) Lindau Henan Yunnan
Myxosporium rosae Fuck. Jiangsu
Phragmidium motivagum Arth
Phragmidium mucronatum (Pers.) Schlecht.
Phragmidium roasae-davuricae Miura
Phragmidium rosae-multiflorae Diet. Zhejiang Anhui Jiangxi Shanxi Xinjiang Guangdong
Phragmidium rosae-rugosae Kasai (Leaf rust) Guizhou
Phragmidium tuberculatum Müll. (Leaf rust) Yunnan
Phyllosticta rosarum Pass. (Leaf spot)
Sphaceloma rosarum (Pass.) Jenk. (Scar spot)
Sphaerotheca pannosa (Wallr.) Lév. Zhejiang Jiangxi Sichuan Yunnan
Teloconia kamtschatkae (Anders.) Hirats. Hebei Dongbei Xinjiang Fujian
Valsa coronata (Hoffm.) Fr. (Canker) Jiangsu

***Rosa dahurica* Pall.**

Actinonema rosae (Lib.) Fr. (Black leaf spot) Xinjiang
Cercospora puderi Ben Davis (Leaf spot) Jilin
Cercospora rosae (Fuck.) Höhn. (Leaf spot) Jilin
Coryneum rosaecola Miura (Gray leaf spot) Jilin
Diplocarpon rosae Wolf Jilin
Cymnoconia peckiana (Howe) Trott. (Leaf rust) Liaoning
Phragmidium montivagum Arth. (Leaf rust) Jilin Liaoning
Phragmidium rosae-davuricae Miura (Leaf rust) Jilin Liaoning Heilongjiang
Phragmidium rosae-multiflorae Diet. (Leaf rust) Jilin Hebei
Septoria rosae Desm. (Leaf spot) Jilin

Sphaerotheca humuli (DC.) Burr. (Powder mildew) Jilin
Sphaerotheca pannosa (Wallr.) Lév. (Powder mildew) Jilin
Sphaceloma rosarum (Pass.) Jenk. (Scar spot) Taiwan
Teloconia kamtschatkae (Anders.) Hirats. (Leaf rust) Jilin Liaoning Heilongjiang

***Rosa maximowicziana* Rebel**
Cercospora rosae (Fuck.) Höhn. (Leaf spot) Liaoning
Phragmidium rosae-multiflorae Diet. (Leaf rust) Liaoning Taiwan

***Rosa rubus* Lév. et Vant.**
Uncinuliella simulans (Salm.) Zheng et Chen var. rosae-rubi Zheng et Chen (Powder mildew) Zhejiang Guizhou
(=Uncinula simulans (Salm.) sensu Tai et Wei)

***Rosa rugosa* Thunb.**
Actinonema rosae (L.) Fr. (Black leaf spot) Jiangsu Henan Shandong Sichuan
Phragmium mucronatum (Pers.) Schlecht. (Leaf rust) Hebei Jiangsu Shanxi
Phragmidium rosae-multiflorae Diet. (Leaf rust) Henan
Phragmidium rosae-rugosae Kasai (Leaf rust) Jilin Liaoning Hebei Sichuan
Septogloeum sp. (Leaf spot) Liaoning
Sphaerotheca fuliginea (Schlecht.) Poll. (Powder mildew) Jiangsu
Sphaerotheca humuli (DC.) Burr. (Powder mildew) Henan Jiangsu
Sphaerotheca pannosa (Wallr.) Lév. (Powder mildew) Jilin Liaoning Jiangsu Sichuan
Teloconia kamtschatkae (Anders.) Hirats. (Leaf rust) Taiwan

***Rosa xanthina* Lindl.**
Coryneum rosaecola Miura (Gray leaf spot) Liaoning
Diplocarpon rosae Wolf Jilin Hebei
Monochaetia concentrica (Berk. et Br.) Sacc. Jilin
Monochaetia seiridioides (Sacc.) Allesch. (Leaf spot) Henan
Phragmidium mucronatum (Pers.) Schlecht (Leaf rust) Neimeng Shanxi Xinjiang
Phragmidium rosae-multiflorae Diet. (Leaf rust) Neimeng
Septoria rosae Desm. (Leaf spot) Jilin
Sphaerotheca pannosa (Wallr.) Lév. (Powder mildew) Neimeng

Rubus spp.
Appendiculella calostroma Desm. Guangxi
Aschersonia tamurai P. Henn. Guangxi
Caeoma cheoanum Cumm. (Leaf rust) Guangxi Guizhou
Gerwasia rubi Racib. (Leaf rust) Hunan Guangdong Guangxi Sichuan Guizhou
Hamaspora acutissima Syd. (Leaf rust) Shanxi Jiangxi Taiwan Hunan Guangdong Guangxi Hainandao Guizhou
Hamaspora benguetensis Syd. (Leaf rust) Guizhou
Hamaspora hashiokae Hirats. f. (Leaf rust) Zhejiang Guangxi Sichuan
Hamaspora sinica Tai et Cheo (Leaf rust) Hunan Guangxi Sichuan
Hamaspora tairai Hirats. (Leaf rust) Hubei
Hendersonia vulgaris Desm. Hubei
Meliola formosensis Yamam. (Black mildew) Hunan
Mycosphaerella fragariae (Tul.) Lindau (Leaf spot) Zhejiang
(Sexual :Ramularia fragariae Peck)
Phragmidium griseum Diet. (Leaf rust) Anhui
Phragmidium nambuanum Diet. (Leaf rust) Guizhou
Phragmidium okianum Hara (Leaf rust) Heilongjiang
Phragmidium pauciloculare (Diet.) Syd. (Leaf rust) Liaoning Jiangsu Anhui Jiangxi Sichuan Guizhou
Phragmidium rubi-thunbergii Kus. (Leaf rust) Guizhou

Phragmidium shengezieense M. M. Chen & Chen (Leaf rust) Tibet
Phragmidium violaceum (Schultz) Wint. (Leaf rust) Yunnan
Phragmidium yamadanum Hirats. f. (Leaf rust) Sichuan
Rhytidhysterium prosopidis Peck Guizhou

***Rubus alexterius* Facke.**
Phragmidium zamonense M. M. Chen & Chen (Leaf rust) Tibet

***Rubus crataegifolius* Bunge**
Phragmidium griseum Diet. (Leaf rust) Heilongjiang Hebei
Phragmidium sinicum Tai et Cheo (Leaf rust) Hebei
Septoria rubi Westend. (Leaf spot) Jilin

***Rubus idaeus* L.**
Botrytis cinerea Pers. (Gray mold blight) Jilin
Gloeosporium venetum Sacc. (Anthracnose) Jilin
Mycosphaerella rubi (Westend.) Roark. (Leaf spot) Jilin
(Asexual: Septoria rubi Westond.)
Phragmidium pauciloculare (Diet.) Syd. (leaf rust) Hunan
Phragmidium rubi (Pers.) Wint. (leaf rust) Xinjiang
Phragmidium shensianum Tai et Cheo (leaf rust) Xinjiang Shanxi
Septoria brevispora Darr (Leaf spot) Jilin
Verticillium albo-atrum Reinke et Barth (Wilt) Jilin

***Rubus pungens* Camb.**
Phragmidium formosanum Hirats. (leaf rust) Taiwan
Phragmidium sikangense Petr. (leaf rust) Sichuan
Phragmidium yamadanum Hirats. f. (leaf rust) Shanxi

***Rubus saxatilis* L.**
Gymnoconia interstitialis (Schlecht.) Lagn. (Leaf rust) Heilongjiang
Gymnoconia peckiana (Howe) Trott. (Leaf rust) Dongbei

***Rubus trianthus* Focke**
Acrothecium rubi Saw. Taiwan
Appendiculella calostroma Desm. Taiwan
Hamaspora sinica Tai et Cheo (Leaf rust) Sichuan
Phragmidium griseum Diet. (Leaf rust) Taiwan
Septoria rubi Westend. var. brevispora Sacc. (Leaf spot) Taiwan

Rubus triphyllis Thunb.
Phragmidium pauciloculare (Diet.) Syd. (Leaf rust) Liaoning Shandong Jiangsu Zhejiang Jiangxi Taiwan Hunan Guangxi Yunnan
Phragmidium shensianum Tai et Cheo (Leaf rust) Shanxi
Phragmidium sikangense Petr. (Leaf rust) Shanxi

***Sabina chinensis (L.)* Antoine**
(=Juniperus chinensis L.)
Diplodia juniperi Westend (Dieback) Henan
Gloeophyllum juniperinum (Teng et Ling) Teng (Decay) Hebei
Gymnosporangium haraeanum Syd. (Leaf rust) Liaoning Hebei Jiangsu Zhejiang Henan Shandong Anhui Jiangxi Fujian Guangdong Guangxi Hubei Tibet Sichuan Shanxi
Gymnosporangium japonicum Syd. (Leaf rust) Hebei Jiangsu Shanxi Guangdong
Gymnosporangium yamadai Miaybe (Leaf rust) Liaoning Jilin Dongbei Hebei Henan Jiangsu Shanxi Sichuan
Pithya cupressi (Batsch et Fr.) Rehm Taiwan

***Sabina squnamata* (Buch. –Ham.) Ant.**
Gymnosporangium formosanum Hirats. f. et Hash. (Leaf rust) Taiwan

***Sabina tibetica* Kom.**

Arceuthobium oxycedri (DC.) M. Bieb. (Dwarf mistletoe) Tibet
Gymnosporangium japonicum Syd. (Leaf rust) Tibet
Hexagonia sp. (Decay) Tibet

***Sabina virginiana* (I.) Ant.**

Gymnosporangium haraeanum Syd. (Leaf rust) Hunan

***Sabina vulgaris* Antoine**

(=Juniperus Sabina L.)

Gymnosporangium confusm plowr. (Leaf rust) Xinjiang
Gymnosporangium fusisporum Ed. Fischer. (Leaf rust) Xinjiang
Gymnosporangium juniperinum (L.) Mart. (Leaf rust) Xinjiang
(=Gymnosporangium juniperi Link)
Gymnosporangium yamadai Miyabe (Leaf rust) Xinjiang
Lophodermium juniperinum (Fr.) Rehm (Falling needle) Xinjiang
Stigmatea juniperi Went (Leaf spot) Xinjiang

Salix spp.

Agrobacterium tumefaciens (Smith et Towns.) Conn. (Root crown gall) Dongbei Xibei
Apiosporium salicinum (Pers.) Kunze Henan Xinjiang
Armillariella mellea (Vahl ex Fr.) Karst. (Root decay) Gansu
Armillariella tabescens (Scop. ex Fr.) Sing. (Armillariella root rot) Hebei Shandong
Bjerkandera adusta (Willd. ex Fr.) Karst. (Decay) Liaoning
Capnodium salicinum Mont. (Black mildew) Liaoning Xinjiang Neimeng Ningxia
(sexual :Fumago vagans Pers.)
Cercospora populicola Tharp (Leaf spot) Jilin
Cerrena unicolor (Bull. ex Fr.) Murr. (Decay) Hebei Shanxi Yunnan
Cladosporium herbarum (Pers.) Link (Black mold) Shanxi
Coriolus hirsutus (Wulf ex Fr.) Quél. (Decay) Hebei Shanxi Sichuan Xinjiang
Coriolus versicolor (L. ex Fr.) Quél. (Decay) Hebei Jiangsu
Cuscuta japonica Choisy (Dodder) Liaoning Zhejiang
Cuscuta majory Choisy (CuseDodder) Henan
Cuscuta monogyna Vahl. (Dodder) Xinjiang
Cytospora ambiens Sacc. (Canker) Jiangsu Neimeng
Cytospora chrysosperma (Pers.) Fr. (Canker) Liaoning Heilongjiang Jilin Hebei Henan Huabei Neimeng Xinjiang Ningxia Sichuan Shandong
Daedaleopsis confragosa (Bolt. ex Fr.) Schröt (Decay) Hebei Jiangsu Zhejiang Heilongjiang Gansu Sichuan Yunnan
Diplodia salicina Lév. (Brown leaf spot) Jiangsu
Discella carbonacea (Fr.) Berk. et Br. Jiangsu
Discula microsperma (Berk. et Br.) Sacc. Hebei
Erinellina tomentella (Penz. et Sacc.) Tai Yunnan
Eriophyes sp. (Leaf mites) Neimeng Ningxia
Favolus squamosus (Huds. ex Fr.) Ames (Decay) Dongbei Henan
Fomes fomentarius (L. ex Fr.) Kickx. (Brown rot) Xinjiang
Fomitopsis pinicola (Sw. ex Fr.) Karst. (Brown cube rot) Hebei Sichuan
Funalia hispida (Bagl.) Pat. (Decay) Hebei Henan
Funalia trogii (Berk.) Bond. et Sing. (Decay) Hebei Dongbei Ningxia Liaoning Xinjiang Neimeng
Ganoderma applanatum (Pers.) Pat. (Decay) Dongbei Hebei Yunnan
Helicobasidium purpureum (Tul.) Pat. (Violet root rot) Dongbei Hebei Henan Jiangsu Zhejiang Anhui Guangdong Shanxi Sichuan Shandong
Inonotus hispidus (Bull. ex Fr.) Karst. (Leaf tip rot) Hebei Xinjiang
Inonotus rheades (Pers.) Pilát. (Decay) Xinjiang

Laetiporus sulphureus (Bull. ex Fr.) Bond. et Sing. (Trunk brown rot) Xinjiang
Macrophoma salicina Sacc. (Leaf spot) Jiangsu
Marssonina dispersu Nannf. (Leaf spot) Xinjiang
Marssonina salicigena (Bub. et Vieug) Nannf. (Leaf spot) Xinjiang
Marssonina salicis-purpureae Jaap (Leaf spot) Yunnan Xinjiang
(=Marssonina kriegeriana Bres.)
Melampsora capraearum Thüm. (Leaf rust) Xinjiang
Melampsora coleosporioides Diet. (Leaf rust) Liaoning Jiangsu Shanxi Neimeng Gansu Yunnan
Melampsora epiphylla Diet. (Leaf rust) Heilongjiang
Melampsora epitea Thüm. (Leaf rust) Xinjiang
Melampsora hartigii Thüm. (Leaf rust) Hebei
Melampsora larici-capracearum Kleb. (Leaf rust) Xinjiang
Melampsora larici-epitea Kleb. (Leaf rust) Jilin Heilongjiang Anhui Shandong Guizhou Yunnan Taiwan Sichuan
Melampsora ribesii-purpureae Kleb. (Leaf rust) Shanxi Shanxi
Melampsora salicis-albae Kleb. (Leaf rust) Gansu Yunnan Sichuan Xinjiang
Melampsora salicis-cupularis Wang (Leaf rust) Shanxi Qinghai
Metasphaeria spiculata (Wallr.) Sacc. Hebei
Phellinus igniarius (L. ex Fr.) Quél. (Heartwood white sponge rot) Heilongjiang Jilin Hebei Henan Shanxi Shanxi Gansu Ningxia Qinghai Xinjiang Neimeng Sichuan Yunnan
Phellinus robustus (Karst.) Bourd. et Galz. (Sapwood brown rot) Xinjiang
Phellinus adipose (Fr.) Quél. (Decay) Hebei
Phyllosticta translucens Bubak. et Kab. (Leaf spot) Liaoning Heilongjiang Gansu Neimeng
Pleurotus spathulatus (Fr.) Peck (Decay) Hebei
Poria mucida (Pers.) Fr. (Decay) Jiangsu
Ramularia rosea (Fckl) Sacc. (Leaf spot) Xinjiang
Rhytisma salicinum Fr. (Black mole) Jilin Hebei Henan Dongbei Neimeng Sichuan Yunnan Xinjiang
Septoria salicicola Sacc. (Gray leaf spot) Liaoning Jilin Heilongjiang Neimeng Xinjiang
Sphaerella salicina Ell. ex Ev. Jiangsu
Sphaeropsis salicicola Pass. Jiangsu
Trametes hirsute (Wulf. ex Fr.) Pilát. (Decay) Xinjiang
Trametes suaveolens (L.) Fr. (White rot) Liaoning Dongbei Shanxi Shanxi Henan
Tyromyces pubescens (Schum. ex Fr.) Imaz. (Decay)
Uncinula adunca (Wallr. ex Fr.) Lév. var. adunca (Powder mildew)
Uncinula adunca (Wallr. ex Fr.) Lév. var. mandshurica (Miura) Zheng et Chen (Powder mildew)
Valsa populina Fuck (Canker) Guizhou Sichuan Shandong
Virus (Willow yellow) Ningxia
Physical frostbite (Broken stomach) Ningxia
Physical (Yellow leaf) Ningxia

***Saliz babylonica* L.**

Apiosporium salicinuma (Pers.) Kunze Sichuan
Armillariella tabescens (Scop. ex Fr.) Sing. (Weeping willow root rot) Beijing
Cercospora salicina Ell. et Ev. (Leaf spot) Henan Taiwan Sichuan
(=Cercospora salicicola Saw.)
Cuscuta japonica Choisy (Dodder) Zhejiang
Cytospora chrysosperma (Pers.) Fr. (Canker) Henan Neimeng Beijing Hebei
Fumago vagans Pers. (Black mildew) Jiangsu
Laetiporus sulphureus (Bull. ex Fr.) Bond. et Sing. (Trunk brown rot) Xinjiang
Loranthus parasiticus (Linn.) Merr. (Mistletoe) Guangxi
Loranthus yadoriki Sieb. et Zucc. (Camphor mistletoe) Tibet

Melampspra allii-fragills Kleb. (Leaf rust) Sichuan
Melampsora coleosporioides Dieb. (Leaf rust) Hebei Henan Shanxi Jiangsu Taiwan Sichuan
Melampsora salicis-albae Kleb. (Leaf rust) Sichuan
Pleurotus ostreatus (Jacq. ex Fr.) Quél. (Decay) Dongbei Sichuan
Rhytisma salicinum Fr. (Black tar) Sichuan Tibet
Septoria salicicola Sacc. (Leaf spot) Jiangsu
Uncinula adunca (Wallr. ex Fr.) Lév. Var. adunca (Powder mildew) Hebei Henan Anhui Sichuan Tibet

***Salix caprea* L.**
Cytospora translucens Sacc. (Canker) Xinjiang
Melamspora larici-capraearum Kleb. (Leaf rust) Xinjiang Shanxi
Pleochaeta salicicola Zheng et Chen (Powder mildew) Shanxi
Rhytisma salicinum Fr. (Black tar) Henan Shanxi Xinjiang
Uncinula regularis Zheng et Chen (Powder mildew) Sichuan

***Salix chaenomeloides* Kimura**
Melampsora coleosporioides Diet. (Leaf rust) Hebei Shanxi Taiwan
Uncinula adunca (Wallr. ex Fr.) Lev. Var. adunca (Powder mildew) Henan

***Salix cheilophila* Schneid**
(=Salix mongolica Siuzev)
Agrobaceterium tumefaciens (Smith et Twons.) Conn. (Root cancer) Ningxia
Melampsora coleosporioides Diet. (Leaf rust) Henan
Rhytisma salicinum Fr. (Black tar) Ningxia
Septoria salicicola Sacc. (Leaf spot) Neimeng

***Salix cupularis* Rchd.**
Melampsora salicis-cupularis Wang (Leaf rust) Shanxi Neimeng

***Salix daphnoides* Vill.**
Uncinula adunca (Wallr. ex Fr.) Lév. Var. adunca (Powder mildew) Liaoning Heilongjiang

***Salix fragilis* L.**
Phyllosticta translucens Bubak et Kab. (Leaf spot) Neimeng
Septoria salicicola Sacc. (Leaf spot) Neimeng

***Salix lapponum* L.**
Rhytisma salicinum Fr. (Black tar) Xinjiang

***Salix lasiogyne* Seem.**
Apiosporium salicinum (Pers.) Kunze Jilin
Capnodium sp. (Black mildew) Liaoning

***Salix matsudana* Koidz**
Cercospora salicina Ell. et Ev. (Brown leaf spot) Henan
Cuscuta japonica Choisy (Dodder) Henan Shanxi
Cytospora ambiens Sacc. (Dieback) Neimeng
Coriolus hirsutus (Wulf. ex Fr.) Quél. (Decay) Tibet
Cytospora chrysosperma (Pers.) Fr. (Canker) Shanxi Hebei Neimeng Beijing Shanxi
Daedaleopsis confragosa (Bolt. et Fr.) Schröt. (White rot) Shanxi
Eriophyes sp. (Leaf mites) Xinjiang Neimeng
Favolus squamosus (Huds. ex Fr.) Ames (Decay) Tibet
Fuanlia trogii (Berk.) Bond. et Sing. (Decay) Neimeng Beijing
Loranthus parasiticus (Linn.) Merr. (Mistletoe) Guangxi
Melampsora sp. (Willow rust) Neimeng
Melamspora coleosporioides Diet. (Leaf rust) Hebei Henan Jiangsu Neimeng
Melampsora ribesii-purpureae Kleb. (Leaf rust) Shanxi
Melamspora salicis-albae Kleb. (Leaf rust) Gansu
Phellinus robustus (Karst.) Bourd. et Galz. (Sapwood brown rot) Shanxi

Pholiota adipose (Fr.) Quél. (Decay) Neimeng
Phyllosticta translucens Bubak et Kab. (Leaf spot) Neimeng
Rhytisma salicinum Fr. (Black mole) Shanxi Tibet
Schizophyllum commune Fr. (Bark rot) Neimeng
Septoria salicicola Sacc. (Canker) Shanxi
Septoria solanicola Ell. et Ev. (Gray leaf spot) Neimeng
Trametes suaveolens (L.) Fr. (Decay) Gansu Neimeng
Uncinula adunca (wallr. ex Fr.) Lev. var. adunca (Powder mildew) Hebei
Valsa ambiens (Pers. ex Fr.) Fr. (Canker) Tibet
Viscum album L. (Ture mistletoe) Henan

***Salix matsudana var. tortuosa* Vilm.**

Fumago vagans Pers. (Black mildew) Henan
Melampsora coleosporioides Diet. (Leaf rust) Henan
Valsa sordida Nits. (Canker) Henan Neimeng Beijing
(Sexual :Cytospora chrysosperma (Pers.) Fr.)

***Salix paraplesia* Schneid.**

Coriolus hirsutus (Wulf. ex Fr.) Quél. (Decay) Tibet
Funalia trogii (Berk.) Bond. et Sing. (Decay) Tibet
Phellinus salicinus (Fr.) Quél. (Decay) Tibet

***Salix purpurea* L.**

Eriophyes sp. (Leaf mites) Xinjiang
Marssonina salicis-purpureae Jacp (Leaf spot) Taiwan
Melampsora ribesii-purpureae Kleb. (Leaf rust) Hebei
Rhytisma salicinum Fr. (Black tar) Ningxia Neimeng
Septoria salicicola Sacc. (Leaf spot) Ningxia
Uncinula adunca (Wallr. ex Fr.) Lév. var. adunca (Powder mildew) Liaoning

***Salix purpurea var. stipularia* Franch.**

Agrobacterium tumefaciens (Smith et Towns.) Conn. (Root cancer) Henan
Cuscuta japonica Choisy (Dodder) Henan
Septoria salicicola Sacc. (Leaf spot) Henan

***Salix viminalis* L.**

Melampsora larici-epitea Kleb. (Leaf rust) Dongbei
Melampsora salicis-viminalis Wang et Guo (Leaf rust) Tibet
Uncinula adunca (Wallr. ex Fr.) Lév. var. adunca (Powder mildew) Dongbei

***Salix wilsonii* Seem.**

Melampsora coleosporioides Diet. (Leaf rust) Shanxi
Melampsora farinose Schröt. (Leaf rust) Hunan
Melampsora ribesii-purpureae Kleb. (Leaf rust) Jiangsu
Melampsora salicis-cupularis Wang (Leaf rust) Shanxi

***Salix xerophila* Floderus**

Rhytisma salicinum Fr. (Black mole) Heilongjiang

Sambucus sp.

Melanomma fuscidulum Sacc. Hebei

***Sambucus adnata* Wall.**

Erysiphe sambuci Ahmad var. crassitunicata Zheng et Chen (Powder mildew) Tibet
Eryisphe sambuci Ahmad var. sambuci (Powder mildew) Yunnan
(=Eryisphe polygoni DC.)

***Sambucus buergeriana* Bl.**

Ascochyta wisconsiana Davis (Leaf spot) Jilin
Cercospora depazeoides (Desm.) Sacc. (Leaf spot) Jilin
Septoria sambucina Peck (Leaf spot) Jilin

***Sambucus japonica* Thunb.**

Microsphaera grossulariae (Mallr.) Lév. (Powder mildew) Jiangsu

***Sambucus racemosa* L.**

Aecidium sambuci Schw. (Leaf rust) Heilongjiang

Microsphaera vanbrunetiana Ger. (Powder mildew) Heilongjiang

Puccinia bolleyana Sacc. (Leaf rust) Hebei Liaoning

***Sambucus wiliamsii* Hance**

Ascochyta wisconsiana Davis (Leaf spot) Jilin

Puccinia bolleyana Sacc. (Leaf rust) Hebei

Sapium sp.

Cercospora stillingiae Ell. et Ev. (Leaf spot) Jiangsu

***Sapium sebiferum (L.)* Roxb.**

Cercospora micromera Syd. (Leaf spot) Taiwan Guangdong

Cytospora sp. (Dieback) Jiangsu

Helminthosporium sapii Miyake (Leaf spot) Hunan

Macrophomina phaseoli (Maubl.) Ashby (Stem rot) Jiangsu Zhejiang Hunan

Meliola sp. (Black mildew) Guizhou

Phyllactinia corylea (Pers.) Karst. (Powder mildew) Taiwan

Sclerotium rolfsii Sacc. (Sclerotium) Zhejiang Jiangsu

***Sassafras tzuma* Hemsl.**

(=Pseudosassafras tzumu (Hemsl.) Lec.)

Fusicoccum sp. (Shoot blight) Guizhou

Phyllactinia linderae Yu et Lai (Powder mildew) Henan Jiangsu Anhui Sichuan Hunan Jiangxi

(=Phyllactinia corylea (Pers.) Karst.)

Septobasidium borgoriense Pat. (Plaster)

***Schefflera octophylla* (Lour.) Harms**

Meliola heteroseta Hönn (Black mildew) Guangxi

***Schisandra chinensis* (Turcz.) Baill.**

Phyllosticta sp. (Leaf spot) Liaoning

Sinarundinaria spp.

Puccinia melanocephala Syd. (Leaf rust) Sichuan

***Sinarundinaria nitida* (Mitf.) Nakai**

Phyllachora phyllostachydis Hara (Black tar) Zhejiang

Phyllachora sinensis Sacc. (Black tar) Sichuan

Stereostratum corticioides (Berk et Br.) Magn. Jiangsu Sichuan

Ustilago shiraiana P. Henn. (Black powder) Jiangsu Fujian

***Sinocalamus affinis* (Rendle) McClure**

Fusarium bambusicola Hara Sichuan

Melanconium shiraianum Syd. (Dieback) Sichuan

Munkiella shiraiana Miyake et Hara (Dieback) Sichuan

Phoma arundinacea (Berk.) Sacc. (Dieback) Sichuan

Phyllachora sinensis Sacc. (Black tar) Sichuan

Phyllosticta take Miyake et Hara (Leaf spot) Sichuan

***Sinocalamus latiflorus* (Munro) McClure**

Balansia take (Miyake) Hara (Witches' broom)

Corticum centrifugum (Lév.) Bres. (Fisheye fruit rot) Taiwan

Dastruella divina (Syd.) Mundk et Khesw. (Leaf rust) Taiwan

Dictyophora cinnabarina Lee Taiwan

Dimeriella dendrocalami Saw. et Yamma. (Black mildew) Taiwan

Scorias communis Yamam. (Black mildew) Taiwan

Triposporiopsis spinigera (Höhm.) Yamam. (Black mildew) Taiwan

Uredo dendrocalami Petch. (Leaf rust) Guangdong

***Siphonostegia chinensis* Benth**

Cronartium flaccidum (Alb. et Schw.) Wint. f. sp. siphonostegium Jing et Wang, Acta Mycol.Sin. 7(2):112, 1988

Sophora spp

Coriolus hirsutus (Wulf ex Fr.) Quél. (Decay) Shanxi
Cytospora sophorae Bres. (Canker) Shandong Neimeng
Diplodia sophorae Speg. et Sacc. (leaf spot) Yunnan
Inonotus hispidus (Bull. ex Fr.) Karst. (Leaf tip rot) Shanxi
Parodiella perisporioides (Berk. et Curt.) Speg. Sichuan
Uromyces truncicola P. Henn. et Shirai (Leaf rust) Anhui

***Sophora flavescens* Ait.**

Botrytis cinerea Pers. (Gray mold blight) Sichuan
Cercospora sophorae Saw. et Kats. (Leaf spot) Taiwan
Erysiphe sp. (Powder mildew) Neimeng
Phyllosticta sp. (Leaf spot) Neimeng
Phyllosticta sophoricola Hollos (Leaf spot) Jilin
Sclerotinia sclerotiorum (Lib.) de Bary (Sclerotinia) Sichuan
Uromyces sophorae-flavescentis Kus. (Leaf rust) Dongbei Hebei Shandong Guangdong

***Sophora japonica* L.**

Cytospora sophorae Bres. (Canker) Jiangsu Xinjiang
Diplodia sophorae Speg. et Sacc. (Dieback) Neimeng Shanxi
Dothiorella sp. (Canker) Hebei Henan Jiangsu Shanxi
Eutypella deusta Ell. et Ev. Hebei
Favolus squamosus (Huds. ex Fr.) Ames (Decay) Hebei
Fracchiaea heterogenea Sacc. Hebei
Fumago sp. (Black mildew) Ningxia
Fusarium sp. (Canker) Hebei Henan Jiangsu Shanxi
Haplosporella ailanthi Ell. et Ev. Hebei
Inonotus hispidus (Bull. ex Fr.) Karst. (Shoot rot) Hebei
Loranthus parasiticus (Linn.) Merr. (Mistletoe) Guangxi
Macrophoma sophorae Miyake (Leaf spot) Hebei Henan Shandong
Macrophoma sophoricola Teng. (Leaf spot) Jiangsu
Macrophomina phaseoli (Maubl.) Ashby (Trunk decay) Anhui
Nectria cinnabarina (Tode) Fr. (Dieback) Jiangsu
Phellinus rimosus (Berk.) Pilát. (Decay) Shanxi
(=Fomes rimosus (Berk.) Cooke)
Phomopsis sophorae (Sacc) Trav (Leaf spot) Jiangsu
Phyllosticta sophoricola Hollos (Leaf spot) Neimeng Beijing
Poria lurida Bres. (Decay) Hebei
Trametes robiniophila Murr. (Decay) Hebei
Trichothecium roseum (Bull) Link (Powder mold) Hebei
Tyromyces galactinus (Berk.) Bond. (Decay) Hebei
Uncinula sinensis Tai et Wei (Powder mildew) Henan Guizhou
Uromyces truncicola P. Henn. et Shirai (Leaf rust) Jiangsu Zhejiang Henan Anhui Jiangxi Shandong Sichuan Tibet
Physical belt disease Ningxia Beijing

***Sophora japonica var. pendula* Loud.**

Dothiorella sp. (Canker) Hebei Henan Jiangsu Shanxi
Fusarium sp. (Canker) Hebei Henan Jiangsu Shanxi

Sorbus spp.

Cerrena unicolor (Bull. ex Fr.) Mull. (Decay) Hebei Xinjiang
Pestalotia sorbi Pat. (Leaf spot) Yunnan

***Sorbus alnifolia* (Sieb. et Zucc. K. Koch.)**
Gymnosporangium nipponicum Yamada (Leaf rust) Guizhou
Monochaetia unicornis (Cooke et Ell.) Sacc. (Anthracnose) Jilin

***Sorbus amurensis* Koeh.**
Eriophyes sp. (Leaf felt spot) Xinjiang
Gymnosporangium sp. (Leaf spot) Heilongjiang

***Sorbus prattii* Koeh.**
Gymnosporangium clavariiforme (Jacq.) DC. (Leaf rust) Tibet

***Sorbus tianschanica* Rupr.**
Alternaria sp. (Cycle spot) Xinjiang
Cytospora sp. (Canker) Xinjiang
Entomosporium mesppili (DC.) Sacc. (Leaf spot) Xinjiang
Eriophyes pini Pegenst. (Leaf felt spot) Xinjiang
Fomes fomentarius (L. ex Fr.) Kickx. (Brown rot) Xinjiang
Gymnosporangium juniperinum (L.) Mart. (Leaf rust) Xinjiang
Gymnosporangium nipponicum Yamada (Leaf rust) Xinjiang
Nectria sp. (Dieback) Xinjiang
Phellinus igniarius (L. ex Fr.) Quél (White rot) Xinjiang
Podosphaera aucupariae Erikss. (Powder mildew) Xinjiang
Venturia inaequalis (Cooke) Wint. (Black scab) Xinjiang
(Asexual :Fusicladium dendriticum (Wallr.) Fuck.)

***Staphylea bumalda* DC.**
Aecidium staphyleae Miura (Leaf rust) Liaoning
Mycosphaerella staphyleae Miura (Leaf spot) Liaoning Dongbei

***Swietenia mahogany* Jacq**
Cassytha filiformis L. (Cassytha mistletoe) Guangxi
Glomerella cingulata (Stonem.) Spauld. et Schrenk (Anthracnose) Guangdong Guangxi

***Symplocos panciulata* (Thunb.) Miq.**
Erysiphe polygoni DC. (Powder mildew) Yunnan Taiwan
Microsphaera symploci Yu et Lai (Powder mildew) Hubei
Phyllactinia corylea (Pers.) Karst. (Powder mildew) Jiangsu
Septoria sydowii P. Henn. et Sacc. (Leaf spot) Liaoning

Syringa spp.
Fomitopsis pinicola (Sw. ex Fr.) Karst. (Brown cube rot) Jilin
Phellinus senex (Nees et Mont.) Imaz. (Decay) Hebei
Phyllosticta syringae West (Leaf spot) Neimeng

***Syringa amurensis* (Rupr.) Rupr.**
(=Ligustrina amurensis Reg.)
Cercospora macromaculans Heald. et Wolf. (Leaf spot) Heilongjiang
Hendersonia sarmentorum West. (Leaf spot) Heilongjiang
Phellinus setulosus (Lloyd) Imaz. (Decay) Heilongjiang
Phyllosticta syringae West. (Leaf spot) Heilongjiang
Septoria syringae Sacc. et Speg. (Leaf spot) Jilin

***Syringa oblata* Lindl.**
Microsphaera alni (Wallr.) Salm (Powder mildew) Jilin
Microsphaera syringae A. Jacz. (Powder mildew) Heilongjiang Liaoning Ningxia
Mycosphaerella syringae Bond. (Leaf spot) Heilongjiang Liaoning
Phellinus robustus (Karst.) Bourd. et Galz. (Sapwood brown rot) Ningxia
Pseudomonas syringae van Hall (Bacterial leaf spot) Heilongjiang Liaoning Ningxia

Septoria syringae Sacc. et Speg. (Leaf spot) Heilongjiang Liaoning
Verticillium albo-atrum Reinke et Berth (Wilt) Jilin
Virus (Mosaic) Heilongjiang Neimeng Ningxia

***Syringa pekinensis* Rupr.**
Phellinus stulosus (Lloyd) Imaz. (Decay) Hebei Beijing

***Syringa vulgaris* L.**
Microsphaera alni (Wallr.) Salm. (Powder mildew) Jilin
Septoria syringae Sacc. et Speg. (Leaf spot) Jilin
Verticillium albo-strum Reinke et Berth (Wilt) Jilin

Tamarix spp.
Ascochyta tamaricis Golov. (Leaf spot) Xinjiang
Inonotus rheades (Pers.) Pilát (Decay) Hebei Shanxi Neimeng
Inonotus tamaricius (Pat.) Maire (Decay) Xinjiang

***Tamarix chinensis* Lour.**
Cytospora tamaricella Syd. (Dieback) Neimeng Xinjiang
Eriophyes sp. (Leaf felt) Ningxia
Physical belt disease Neimeng

***Taxius cuspidate* Sieb. et Zucc.**
Fumago vagans Pers. (Black mildew) Liaoning
Macrophoma taxi (Berk.) Berl. et Vogl. (Leaf spot) Jiangsu

***Tectona grandis* L. f.**
Cercospora tectoniae Stev. (Leaf spot) Taiwan
Meloidogyne incognita (Kofoid et White) Chitwood. (Tecton root-rot nematode) Guangdong
Olivea tectonae Thirum. (Leaf rust) Taiwan Guangdong Guangxi Yunnan
Pseudomonas solanacearum E.F. Smith (Southern bacterial wilt) Guangdong
Uredo tectonae Racib (Leaf rust) Guangxi

Tilia spp.
Armillariella mellea (Vahl ex Fr.) Karst. (Root decay) Dongbei Huabei
Bjerkandera adusta (Willd. ex Fr.) Karst. (Decay) Liaoning
Cercospora microsora Sacc. (Linden leaf blight) Liaoning
Eriophyes tillae liosoma Nal. (Leaf felt spot) Liaoning
Fomes fomentarius (L. ex Fr.) Kickx. (Brown rot) Heilongjiang
Hericium erinaceus (Bull.) Pers. (Decay) Heilongjiang
Inonotus hispidus (Bull. ex Fr.) Karst. (Shoot tip rot) Heilongjiang
Irpex lacteus Fr. (Decay) Liaoning
Lenzites betulina (L.) Fr. (Decay) Liaoning
Phyllosticta vogelii (Syd.) Died. (Brown leaf spot) Liaoning
Pleurotus ostreatus (Jacq. ex Fr.) Quél. (Decay) Liaoning
Pycnoporus cinnabarinus (Jacq.) Karst. (Decay) Liaoning
Septoria tiliae Westend. (Leaf spot) Henan
Steccherinum septentrionale (Fr.) Bank. Jilin Heilongjiang Hebei

***Tilia amurensis* Rupr.**
Capnodium sp. (Black mildew) Heilongjiang
Cercospora microsora Sacc. (Linden leaf bligh) Heilongjiang
Eryoyhyes tiliae-liosoma Nal. (Leaf felt spot) Heilongjiang
Fomes fomentarius (L. ex Fr.) Kickx. (Brown rot) Dongbei
Fumago vagans Pers. (Black mildew) Liaoning
Nummularia bulliardi Tul. (Dieback) Liaoning
Pholiota adiposa (Fr.) Quél. (Decay) Dongbei
Phyllosticta vogelii (Syd.) Diet. (Brown rot) Liaoning
Pucciniastrum tiliae Miyabe et Hirats. (Leaf rust) Jilin Heilongjiang

Physical: Shoot belt disease Heilongjiang
Physical: Leaf mosaic Heilongjiang

***Tilia chinensis* Maxim.**
Cercopsora microsora Sacc. (Leaf spot) Shanxi
Coriolus versicolor (L. ex Fr.) Quél. (Decay) Gansu

***Tilia cordata* Mill.**
Coniothyrium tiliae Miyake (Leaf spot) Hebei

***Tilia mandshurica* Rupr. et maxim.**
Pucciniastrum tiliac Miyabe et Hirats. (Leaf rust) Heilongjiang

***Tilia tuan* Szysz.**
Pucciniastrum tiliae Miyabe et Hirats. (Leaf rust) Guizhou
Uncinula oleosa Zheng et Chen (Powder mildew) Sichuan

Toona spp.
Nyssopsora cedrelae (Hori) Tranz. (Leaf rust) Anhui Jiangxi
Phyllactinia toonae Yu et Lai (Powder mildew) Guangxi Hunan Hubei Guangdong Shandong Taiwan

***Toona microcarpa* (A. DC.) Harms**
Phyllactinia toonae Yu et Lai (Powder mildew) Sichuan

***Toona sinensis.* (A. Juss) Roem**
Botrytis cinerea Pers. (Gray mold blight) Shanxi
Helicobasidium purpureum (Tul.) Pat. (Violet root rot) Shanxi
Laetiporus sulphureus (Bull. ex Fr.) Bond. et Sing. (Trunk brown rot) Shanxi
Nyssopsora cedrelae (Hori) Tranz. (Leaf rust) Shandong Guangdong Guangxi Hunan Hubei Anhui Jiangxi Taiwan Guizhou
Phakopsora cheoana Cumm. (Leaf rust) Guizhou
Phyllactinia toonae Yu et Lai (Powder mildew) Hebei Jiangsu Shanxi Anhui Guangxi Yunnan Guizhou Hunan
Rhizoctonia solani Kühn (Damping off) Hunan
Uncinula cedrelae Tai (Powder mildew) Yunnan Hebei
(=Uncinula delavayi Pat. var. cedrelae (Tai) Tai)
Uncinula cedrelae Tai var. nodulosa Tai (Powder mildew) Hebei
(=Uncinula delavayi Pat. var. nodulosa (Tai) Tai)

Torreya grandis cv. 'Merrillii'
Macrophomina phaseoli (Maubl.) Ashby (Trunk decay) Jiangsu Zhejiang Hunan Anhui
Sclerotium rolfsii Sacc. (Sclerotium) Jiangsu Zhejiang Hunan Anhui

***Trachycarpus fortunei* (Hook. f.) H. Wendl.**
Leoptosphaeria trachycarpi Hara (Cane blight) Sichuan
Phytophthora palmivora (Butl.) Butl. (Plum bud rot) Hunan

***Tsoongiodendron odorum* Chun**
Colletotrichum sp. (Anthracnose) Guangxi
Macrophomina phaseoli (Maubl.) Ashby. (Stem rot) Guangxi
Pseudomonas sp. (Dry sprout) Guangxi

Tsuga spp.
Chrysomyxa tsugae-yunnanensis Teng (Leaf rust) Yunnan
(=Chrysomyxa tsugae Teng)
Fomitopsis pinicola (Sw. ex Fr.) Karst. (Brown cube rot) Taiwan
Phellinus robustus (Karst.) Bourd. et Galz. (Sapwood brown rot) Sichuan

***Tsuga chinensis* (Franch.) Pritz.**
Fomitopsis pinicola (Sw. ex Fr.) Karst. (Decay) Sichuan Tibet
Ganoderma tsugae Murr. (Decay) Gansu
Inonotus triqueter (Alb. et Schw. ex Fr.) Karst. (Decay) Tibet

Laetioporus sulphureus (Bull. ex Fr.) Bond. et Sing. (Trunk brown rot) Tibet
Loranthus caloreus var. oblongifolius Lecente (Long leaf mistletoe) Anhui
Phaeolus schweintzii (Fr.) Pat. (Brown cube rot) Tibet
Pucciniastrum sp. (Leaf rust) Shanxi

***Tsuga dumosa* (D. Don) Eichler**
(=Tsuga yunnanensis (Franch.) Mast.)
Chrysomyxa tsugae-yunnanensis Teng (Leaf rust) Sichuan

Ulmus spp.
Apiosporium salicinum (Pers.) Kunze (Black mildew) Heilongjiang
Capnodium sp. (Black mildew) Liaoning Heilongjiang Neimeng
Coriolus hirsutus (Wulf. ex Fr.) Quél. (Decay) Hebei Shanxi
Cerrena unicolor (Bull. ex Fr.) Murr. (Decay) Hebei
Cytospora sp. (Canker) Dongbei Xibei Neimeng
Favolus alveolaris (Bosc. ex Fr.) Quél. (Dieback) Liaoning
Favolus squamosus (Huds. ex Fr.) Ames (Decay) Heilongjiang Dongbei
Fomitopsis ulmaria (Sow. ex Fr.) Bond. et Sing. (Decay) Ningxia
Fumago sp. (Black mildew) Ningxia
Ganoderma applanatum (Pers.) Pat. (Decay) Liaoning
Gnomonia ahoarana Nishik. et Matsum. (Anthracnose) Jiangsu
Gnomonia ulmea (Sacc.) Thüm. (Anthracnose) Heilongjiang Dongbei Neimeng Liaoning
(=Gloesporium ulmeum Miles.)
Inonotus hispidus (Bull. ex Fr.) Karst. (Leaf tip rot) Jilin Hebei
Loranthus europaeus Jacq. (European mistletoe) Hebei Henan Shanxi Shanxi Sichuan Gansu Xinjiang
Melasmia ulmicola Berk. et Curt. (Black tar) Jiangsu Henan Hunan
Nectria cinnabarina (Tode) Fr. (Dieback) Liaoning
Phyllosticta ulmicola Sacc. (Gray leaf spot) Jilin Liaoning Heilongjiang
Pleurotus ostreatus (Jacq. ex Fr.) Quél. (Decay) Liaoning
Pleurotus ulmarius (Bull. ex Fr.) Quél. (Decay) Jilin Dongbei Qinghai
Poria mucida (Pers.) Fr. (Decay) Hebei
Septoria ulmi Hara (Leaf spot) Henan
Systremma ulmi (Duv. ex Fr.) Theiss. et Syd. (Swollen spot) Jilin
Uncinula clandestine (Biv.) Schröt. var clandestine (Powder mildew) Xinjiang
Uncinula kenjiana Homma (Powder mildew) Hebei Jiangsu Xinjiang
Valsa sordida Nit. (Canker) Dongbei Xibei Huabei
Viscum coloratum (Kom.) Nakai(ture mistletoe) Neimeng
Physical: (Brokenwounded trunk) Ningxia
Physical: (elm thick bark) Ningxia

***Ulmus carpinifolia* Gleditsch**
(=Ulmus campestris var. laevis Spach)
Mealsmia ulmicola Berk. et Curt. (Laequer spot) Dongbei
Nectia cinnabarina (Tode) Fr. (Dieback) Heilongjiang
(Asexual :Tubercularia vulgaris Tode)
Septoria yokokawai Hara (White leaf spot) Heilongjiang
Systremma ulmi (Duv. ex Fr.) Theiss. et Syd. (Swollen spot) Dongbei

***Ulmus davidiana* Planch**
stegophora aemula Syd Hebei

***Ulmus densa* Litw**
Cytospora sp. (Canker) Xinjiang

***Ulmus laevis* pall.**
Cytospora sp. (Canker) Xinjiang

***Ulmus macrocarpa* Hance**

Phyllosticta ulmicola Sacc. (Leaf spot) Jilin

Uncinula clandestine (Biv. Bern.) Schrot. var. ulmi-foliacea (Dzhaf.) Zheng et Chen (Powder mildew)

(=Uncinula clandestine (Biv. Bern.) Schröt. f. ulmi-folicea Dzhaf.) Zhejiang

***Ulmus parvifolia* Jacq.**

Coryneum intermedium Sacc. (Gray leaf spot) Jiangsu

Gnomonia oharana Nishik. et Matsum. (Anthracnose) Jiangsu Shanxi

Gnomonia ulmea (Sacc.) Thüm. (Anthracnose) Shanxi

Loranthus maclurei Merr. (Mistletoe) Fujian Guangxi Guangdong Guizhou

Melasmia ulmicola Berk. et Curt. (Black tar) Jiangsu

Uncinula clandestine (Biv.) Schröt. var. clandestine (Powder mildew) Jiangsu Hunan

***Ulmus propinqua* Koidz.**

Phyllosticta ulmicola Sacc. (Leaf spot) Liaoning

Septoria yokokawai Hara (Gray leaf spot) Heilongjiang

Septoria ulmi Hara (Dry leaf spot) Heilongjiang

Taphrina ulmi Johnas (Elm leaf curl) Heilongjiang

Ulmus propinqua. var. suberosa Miyabe

Phyllosticta bellunensis Martin (Leaf spot) Liaoning

***Ulmus pumila* L.**

Apiosporium salicinum (Pers.) Kunze (Black mildew) Jilin

Ascochyta ulmi (West.) Kleb. (Leaf spot) Jilin

Cercospora sphaeriiformias Cooke (Leaf spot) Jiangsu

Cuscuta monogyna Vahl (Dodder) Xinjiang

Cylindrosporium ulmi (Fr.) Vass. (Leaf spot) Jilin

Cytospora sp. (Canker) Xinjiang Neimeng

Favolus arcularius (Batsch ex Fr.) Ames (Decay) Liaoning

Favolus squamosus (Huds. ex Fr.) Ames (Decay) Heilongjiang Hebei

Fomes fomentarius (L. ex Fr.) Kickx. (Brown rot) Xinjiang

Gnomonia oharana Nishik. et Matsum. (Anthracnose) Henan Jiangsu Neimeng

Gnomonia ulmea (Sacc.) Thüm. (Anthracnose) Jilin Liaoning Heilongjiang Jiangsu Xinjiang

Melasmia ulmicola Berk. et Curt. (Black mole) Liaoning Dongbei

Nectria cinnabarina (Tode) Fr. (Dieback) Heilongjiang Jiangsu Jilin

Phyllosticta ulmicola Sacc. (Leaf spot) Liaoning

Ramulispora sp. (Mold spot) Liaoning

Septroai ulmi Hara (White leaf spot) Liaoning Jilin Heilongjiang Ningxia Neimeng

Systremma ulmi (Duv. ex Fr.) Theiss. et Syd. (Swollen spot) Dongbei

Taphrina ulmi Johans. (Leaf eviel) Henan

Tyromyces galactinus (Berk.) Bond. (Decay) Hebei Jiangsu

Uncinula clandestine (Biv.) Schröt. var. clandestine (Powder mildew) Jilin Liaoning Neimeng Anhui Heilongjiang

Uncinula kenjiana Homma (Powder mildew) Jilin Hebei Henan Shanxi Xinjiang

Viscum coloratum (Kom.) Nakai (ture mistletoe) Liaoning

***Vaccinium vitisidaea* L.**

Calyptospora goeppertiana J. Kuehn (Stem rust) Heilongjiang Xinjiang

Cercospora penicillata (Ces.) Fres. (Leaf spot)

***Vatica astrotricha* Hance**

Ganoderma applanatum (Pers. ex Gray) Pat. var. gibbosum (Bl. & Ness) Teng (Vatica white pocket rot) Hainandao

Hymenochaete rubiginosa (Dicks. ex Fr.) Lév. (White pocket rot) Hainandao

Phellinus rimosus (Berk.) Pilát (Yellow sponge rot) Hainandao

Phellinus yucatanensis (Murr.) Imaz. (Yellow-white sponge rot) Hainandao

***Viburnum opulus* L.**

Cercospora tinea Sacc. (Leaf spot) Hebei
Microsphaera viburni (Duby) Blum. (Powder mildew) Xinjiang

Viburnum sargentii Koehne

Aecidium viburni P. Henn. et Shirai (Leaf rust) Liaoning

***Xanthoceras sorbifolia* Bunge**

Erysiphe sp. (Powder mildew) Xinjiang
Pestalotia sp. (Cycle brown spot) Liaoning

Zanthoxylum spp.

Coleosporium zanthoxyli Diet. et Sxd. (Leaf rust) Jiangsu Zhejiang Anhui Jiangxi Hunan Guangxi Sichuan Guizhou Yunnan
Cuscuta japohica Cheisy (Dodder) Zhejiang
Rhytidhysterium scortechinii Sacc. et Berl. (Bark rot) Guizhou

***Zanthoxylum acanthopodium* DC.**

Coleosporium zanthoxyli Diet. et Syd. (Leaf rust) Hunan Sichuan Guizhou Yunnan

***Zanthoxylum alianthoides* Sieb. et Zucc.**

Cercospora fagarae Yamam. (Leaf spot) Taiwan
Coleosporium zanthoxyli Diet. et Syd. (Leaf rust) Taiwan

***Zanthoxylum alatum* Roxb.**

Aecidium zanthoxyli-schinifolii Diet. (Leaf rust) Yunnan
Coleosporium zanthoxyli Diet. ex Syd. (Leaf rust) Shanxi Jiangsu Henan Hubei Hunan Guizhou Yunnan

***Zanthoxylum armatum* DC.**

(=Zanthoxylum planispinum Sieb. et Zucc.)
Aecidium zanthoxyli-schinifolii Diet. (Leaf rust) Zhejiang
Coleosporium zanthoxyli Diet. et Syd. (Leaf rust) Taiwan

***Zanthoxylum bungeanum* Maxim**

Coleosporium zanthoxyli Diet. et Syd. (Leaf rust) Henan Anhui
Cuscuta chinensis Lam. (Dodder) Henan
Phellinus robustus (Karst.) Bourd. et Galz. (Sapwood brown rot) Shanxi

***Zanthoxylum nitidum* (Roxb.) DC.**

Calothyrium fagarae Saw. et Yamam. Taiwan
Cercospora zanthoxyli Cooke (Leaf spot) Taiwan
Coleosporium zanthoxyli Diet. et Syd. (Leaf rust) Taiwan
Meliola fagaricola Yamam. (Black mildew) Taiwan
Meliola macropoda Syd. (Black mildew) Guangdong

***Zanthoxylum piasezkii* Maxim.**

(=Zanthoxylum piperatum (L.) DC.)
Coleosporium zahthoxyli Diet. ex Syd. (Leaf rust) Shanxi Henan

***Zanthoxylum simulans* Hance.**

Aecidium zanthoxyli-schinifolii Diet. (Leaf rust) Jiangsu Shanxi
Asterina zanthoxyli Yamam. Taiwan
Coleospcrium zanthoxyli Diet. (Leaf rust) Hebei Jiangsu Shanxi Hunan Taiwan Guizhou Sichuan
Septobasidium bogoriense Pat. (Felt fungus) Sichuan Yunnan

***Zanthoxylum stenophyllum* Hemsl.**

Coleosporium zanthoxyli Diet. et Syd. (Leaf rust) Yunnan

***Zelkova serrata (Thunb.)* Makino**

Uncinula clintonii Peck (Powder mildew) Jiangsu
Valsa ambiens (Pers. ex Fr.) Fr. (Dieback) Jiangsu

Zizyphus spp.

Inonotus hispidus (Bull. ex Fr.) Karst. (Shoot tip rot) Shandong
Phakopsora zizyphi-vulgaris (P. Henn.) Diet. (Leaf rust) Hebei Jiangsu Guangxi Shanxi

Zizyphus jujuba **Mill.**
(=Zizyphus sativa Gaertn.)
Alternaria tenuis (Jujube fruit black spot) Shanxi
Ascoehyta zizyphi Hara (Brown leaf spot) Yunnan
Aspergillus niger v. Tiegh. (Stem rotted) Sichuan
Coniothyrium fuckelii Sacc. (Decay) Liaoning
Coniothyrium olivaceum Bon. (White rot) Henan
Cuscuta japonica Choisy (Dodder) Henan
Haplosporella ailanthi Ell. et Ev. (Leaf spot) Hebei
Loranthus parasiticus (Linn.) Merr. (Mistletoe) Guangxi
Macrophoma kawatsukai Hara (Leaf spot) Sichuan
MLO. (Witches' broom) Hebei Shanxi Henan Shandong Shanxi Ningxia Xinjiang
Nyssopsora koelreuteriae (Syd.) Tranz. (Leaf rust) Hunan
Neocapnodium tanakae (Shirai et Hara) Yamam. (Black mildew) Taiwan
Phakopsora zizyphi-vulgaris (P. Henn.) Diet. (Leaf rust) Taiwan Yunnan Sichuan Liaoning Dongbei Hebei Shanxi Henan Hubei Shandong Anhui Fujian Guizhou Jiangsu
Phoma sp. (Jujube fruit black spot) Shanxi
Rhizopus stolonifer (Ehrenb. ex Fr.) Vuill. (Soft rot) Sichuan

Zizyphus jujuba Mill. var. inermis **(Bunge) Rehd.**
(=Zizyphus sativa Gaertn var. inermis (Bumge) Rehd.)
MLO (Witches' broom) Shandong Hebei Henan Shanxi
Phakopsora zizyphi-vulgaris (P. Henn.) Diet. Hebei Shanxi Taiwan Guangxi Sichuan Guizhou

Zizyphus jujuba Mill. var. spinosus **Hu.**
(=Zizyphus sativa Mill. spinosus (Bunge) Schneid.)
Phakopsora zizyphi-vulgaris (P. Henn.) Diet. (Leaf rust) Hebei Shanxi Liaoning Dongbei Shandong Jiangsu

References

<<Forest and Fruit Tree Diseases Protection>>Writing Team 1974 Forest and Fruit Tree Diseases Protection Neimenggu People's Publishing House Agriculture College, Huhehaote. 5 pp

Aspen Root Cancer Disease Source Determination – Xijin Zhang 1980 Forestry Science Correspondence (3) 22-24

Bai, Jinkai etc. 1981 Leguminosae Fungi (*Petrakomyces caraganicola Pai et Hou*) New Species Plant Pathology Journal. (1) 61-62

Beijing Forestry Academy Editing 1979 Forest Disease Chinese Advanced Forestry Academy Trial Teaching Guide

Beijing Forestry Academy Journal Editing Office 1979 Beijing Forestry Academy Journal 1st Edition

Cao, Zhimin and Li, Zhenqi 1999 Rust Fungi of Qinling Mountains China Forestry Publishing House

Cen, Bingzhan and Xu, Shengjie 1979 Chinese fir Secondary Disease Source National Forest Disease Academic Symposium Information

Chen, Jituan 1980 Bamboo Trunk Decay Research Preliminary Report Forestry Science Correspondence (1) 28-32

Chen, Mo-mei 1979 Index of Tibet Forest Diseases

Chen, Mo-mei 1979 Some Fungi New Species and New Records of the China – Himalaya Region Plant Disease Journal 9(1) 25-30

Chen, Mo-mei 2001. Some studies of white pine blister rust flora of North America and East Asia PP. 10

Chen, Mo-mei 2001. Western Gall Rust Life Cycle in California

Chen, Mo-mei 2002 The Forest Ecology of Taiga fungi, Fairbanks, Alaska.PP. 45
Chen, Mo-mei 2002. The Fungi Resources of Tanoak (Lithocarpus densiflora (H. & A, Rehd.) in California PP.40
Chen, Mo-mei Jiuneng Zhang 1965 Catalog of Neimeng Zhelimu Protection Forest Disease Sources
Chen, Mo-mei Jiuneng Zhang Lizhuo Chen 1965 Catalog of Helan Mountain Forest Plant Disease Sources
Chen, Mo-mei Jiuneng Zhang etc. 1965 Catalog of Neimeng Man-made Forest Diseases Neimeng Academy of Agriculture Office of Forest and Fruit Tree Diseases Collective Information 1st Edition
Chen, Mo-mei and E.B. Smalley . 1988. Morphology of isolates of Ceratocystis ulmi from Wisconsin. Mycological Society of America Newsletter 39(I):24
Chen, Mo-mei and F.W. Cobb, Jr. 1985. White pine blister rust in young sugar pine plantations in the mid-elevation Sierra Nevada. Phytopathology
Chen, Mo-mei and F.W. Cobb, Jr.. 1988. Epidemiology of white pine blister rust in young sugar pine plantations in the central Sierra Nevada of California. 1988 Beijing International Symposium of Plant Pathology. pp. 3.
Chen, Mo-mei and H.J. McBeath. 1989. Index of Taiga forest fungi. Annual report to University of Alaska, Fairbanks. pp. 4
Chen, Mo-mei and H.J. McBeath. 1990. Ecology of Taiga fungi, Fairbanks, Alaska. Annual report to University of Alaska, Fairbanks. pp. 5 (draft)
Chen, Mo-mei, E.B. Smalley and R.P. Guries. 1988. Comparative resistance of Chinese sources of Ulnus pumila to North American and Eurasian strain of Ceratocystis (= Ophiostoma) ulmi. 1988 Beijing International Symposium on Plant Pathology. pp. 10.
Chen, Mo-mei, E.B. Smalley, and R.P. Guries. 1984. Disease resistance screening of elm species and cultivars. (manuscript)
Chen, Mo-mei, et al. 1988. Development of greenhouse inoculation techniques for western gall rust on radiata pine with some early results. Phytopathology 78(12):1603
Chen, Mo-mei, F.W. Cobb, Jr., W.J. Libby and D.R. Vogler. 1989. Evidence of variability in nomorphic population of western gall rust. Proceedings of the 36th Annual Western International Forest Disease Work Conference. pp. 95-106.
Chen, Mo-mei. , et al 1956. Ips typographus-caused spruce mortality in Heilongjiang. Report of Chinese Ministry of Forestry. 14 pp.
Chen, Mo-mei. , et al 1962. The seed fungus diseases of trees and shrubs. Fujean Forestry College. Namping Publishing House. 32 pp. (Translated from Russian to Chinese.)
Chen, Mo-mei. , et al. 1983. Forest tree diseases of China. Publishing House of Forestry. p. 245, Fig. 120
Chen, Mo-mei. 1955. Studies of white pocket rot (Phellinus pini) on larch (Larix gmelinii). Journal of Forestry Design 7:20-22.
Chen, Mo-mei. 1958. Investigation of forest diseases and insects (e.g., Phellinus pini, Polyporus betulinus) in Jin-Po Lake, Heilongjiang, on Pinus, Picea, Abies, Quercus forests (350,000 ha). Report of Chinese Ministry of Forestry. 32 pp.
Chen, Mo-mei. 1963. Studies of pine-oak rust (Cronartium quercuum) on var. Scotch pine. Heilongjiang Institute of Forestry Design. 5 pp.
Chen, Mo-mei. 1966. Studies of the host-pathogen interaction of needle rust (Coleosporium pulstillae) on var. scotch pine (Pinus sylestrix var. mongolica). Heilongjiang Institute of Forestry. 4 pp.
Chen, Mo-mei. 1971. Fruit tree diseases and their control. JOURNAL of Inner Mongolia College, Huhehaote I:45; II:46-91.
Chen, Mo-mei. 1972. Field studies of Scolytus schergwei in Inner Mongolia. Inner Mongolia College. 3 pp.
Chen, Mo-mei. 1973. Identification of forest tree pathogens of the Great Hsingan Mountain region. Inner Mongolia College of Agriculture, Huhehaote. 3 pp.
Chen, Mo-mei. 1973. Identification of spruce tree pathogens of the Helan Mountain area. Inner Mongolia

Chen, Mo-mei. 1974. Identification of shade tree and planting tree pathogens in central Inner Mongolia.
Chen, Mo-mei. 1974. Studies of poplar canker (Cytospora chrysosperma) and poplar resistance to canker. Inner
Chen, Mo-mei. 1975. Forest and fruit tree insects and diseases and their control. Inner Mongolia People's Publishing House, Huhehaote. p. 187 and Fig. 40.
Chen, Mo-mei. 1975. Studies of Ganoderma in the Sino-Himalaya. Symposium on Qinghai-Xizang Plateau. 4 pp.
Chen, Mo-mei. 1976. Studies of Tetropium and Dendroctanus on spruce. Commission for Integrated Survey of Natural Resources. 5 pp.
Chen, Mo-mei. 1979. Introduction of a serious tree quarantine. Disease Dutch Elm Disease. Journal of Chinese Forestry Science and Technology, Beijing. pp. 33-34
Chen, Mo-mei. 1979. Preliminary analysis of rust flora in the forests of Tibet. Scientia Silvae Sinicae, Beijing 79(1):1-8
Chen, Mo-mei. 1979. The management of larch forest decay (Phellinus pini) under the clear cutting system. Chinese Society of Forestry, 4 th Annual Symposium, p. 12
Chen, Mo-mei. 1980. The pore fungi flora of the Sino-Himalaya. Symposium on Qinghai-Xizang Plateau. pp. 146-147.
Chen, Mo-mei. 1980. The problems of diseases and insects in the Himalayan forest ecosystem. Symposium on Qinghai-Xizang Plateau. pp. 302-303.
Chen, Mo-mei. 1980. The rust flora of the Sino-Himalaya. Symposium on Qinghai-Xizang Plateau. pp. 144-145.
Chen, Mo-mei. 1981. Studies of pitch canker on Korea pine. Journal of Chinese Forest Tree Quar., 85:2.
Chen, Mo-mei and Liu Hui Zhen 1981. The forest tree disease pathogens in Northern China. (manuscript)
Chen, Mo-mei. 1981. The Polyporaceae flora of the Sino-Himalaya. pp. 1173-1178 In: Geological and Ecological Studies of Qinghai-Xizhang Plateau, Vol. II (Lieu, D-S., ed.) Science Press, Beijing.
Chen, Mo-mei. 1982. New species and new record of fungi in the Sino-Himalaya (II). Acta Phytopathologica Sinica. XII(1):27-32
Chen, Mo-mei. 1982. Review of epidemiological history of blister rust (Cronartium ribicola) in North America and East Asia. Journal Chinese Forest Tree Quar., p. 107
Chen, Mo-mei. 1983. The fungi of Tibet. In: Fungi of Xizang (Tibet) (Anonymous, ed.). Sinica Press. pp. 32-61
Chen, Mo-mei. 1984. A new genus, Stilbechrysomyxa Chen Gen. Nov. Scientia Silvae Sinicae 20(3):268-270
Chen, Mo-mei. 1984. The guidebook on laboratory identification of Polyporaceae. Inner Mongolia College of Agriculture, Huhehaote. (Translated from Russian to Chinese)
Chen, Mo-mei. 1985. Analysis of the distribution of forest pathogens in the United States and China: two case studies. Phytopathology 75(11):1322
Chen, Mo-mei. 1985. The Bio-geography of forest diseases in China. 13 pp. (manuscript)
Chen, Mo-mei. 1985. Tibetan forest diseases and insects. Chapter 6 In: The Forest of Tibet. pp. 266-309.
Chen, Mo-mei. 1986. Medicinal fungi in China. Proceedings of the 1st International Symposium and Workshop of the Society of Chinese Bioscientists in America. p. 1. (abstract)
Chen, Mo-mei. 1986. Studies on bark beetle isolation and Dutch elm disease in Xinjiang, China and in Wisconsin. 12 pp. (manuscript)
Chen, Mo-mei. 1988. Biogeography of Uredinales on Sino-Himalaya. Proceedings of the Symposium and Workshop of the Society of Chinese Bioscientists in America. p. 156 (abstract)
Chen, Mo-mei. 1988. Forest pathology research and management in China. Phytopathology 78(12):1538
Chen, Mo-mei. 1989. A study of the rust flora of Sino-Himalayan forests. Canadian Journal of Botany 67(3):827-833
Chen, Mo-mei. 1990. A list of timber pathogen names and literature from USSR ship logs from Siberia to California. pp. 3.(Unpublished)

Chen, Mo-mei. 1991. Coniferous forest nematodes of USSR, North America, and Far East Asia. (unpublished)
Chen, Mo-mei. 1991. Lists, citations and some slides for decay and insects of Birch. (unpublished)
Chen, Mo-mei. 1991. Lists, Citations of Quarantine Pest of USSR Logs Importation to China. USDA Forest Service Misc. Pub. No. 1495.
Chen, Mo-mei. 1991. Quarantine pest lists for inner China including common name, scientific name, distribution, province, and forest plant parts and products. (unpublished)
Chen, Mo-mei. 1991. Scotch pine disease of the Soviet Far East. USDA Forest Service Misc. Pub. No. 1495.
Chen, Mo-mei. 1991. Stem insects in larch and pine and their location in the log. USDA Forest Service Misc. Pub. No. 1495.
Chen, Mo-mei. 1991. The biogeography of *Monochamus* genus. (unpublished)
Chen, Mo-mei. 1991. Three major species of wood wasps on *Larix* and *Pinus* in Siberia. USDA Forest Service Misc. Pub. No. 1495.
Chen, Mo-mei. 1991.et.al. Pest-Risk Assessment of the Importation of larch from Siberia and Soviet Far East. USDA Forest Service Misc. Pub. No. 1495.
Chen, Mo-mei. Yuan Ge et al 1982. Studies of canker *(Phomesis acculata)* on Liaoning larch plantation. Journal of Science and Technology, Beijing. (8) 25-28
Chen, Mo-mei. Zhang Jiu Nan, Chen Li Zho 1964. The Index of the Great Hsingan mountain region.
Chen, Mo-mei., 1963. Studies of anthracnose *(Colletotrichum kawakamii)* on *Paulownia.*. Shannxi Institute of Forestry Design. 3 pp.
Chen, Mo-mei., et al 1956. Investigation of red pine *(Pinus koraiensis)* diseases (e.g., *Phellinus pini, Polyporus schweinitzii, Monochamus urussoviu, M. sutor)* in Langxiang County, Heilongjiang. Report of Chinese Ministry of Forestry. 21 pp.
Chen, Mo-mei., et al 1957. Investigation of forest diseases and insects in Yilan Mountain, Heilongjiang (240,000 ha). Report of Chinese Ministry of Forestry. 32 pp.
Chen, Mo-mei., et al 1958. Investigation of forest diseases and insects of *Pinus, Larix, Populus, Betula, Quercus,* etc. of Inner Mongolia, Hu-Ma region. Report of Chinese Ministry of Forestry. 28 pp.
Chen, Mo-mei., et al 1956. Investigation of Dahailing's forest diseases and insects *(Ips typographus)*. Report of Chinese Ministry of Forestry. 25 pp.
Chen, Mo-mei., et al 1956. Investigations of Dailing Experimental Forestry Station, Heilongjiang, on forest insects and diseases (e.g., *Cronartium ribicola)*. Chinese Ministry of Forestry. 44 pp.
Chen, Mo-mei., et al 1973. Identification of pathogens of desert plants in the western Inner Mongolia area. Inner Mongolia Agriculture College, Huhehaote. 4 pp.
Chen, Mo-mei., et al. 1954. Investigation of the Great Hsingan Region Forest (200,000 ha) diseases and insects. Chinese Ministry of Forestry. 305 pp.
Chen, Mo-mei., et al. 1955. Report of the Western Sichuan Forest (450,000 ha) diseases *(Phellinus pini* var. *abienis)* and insects *(Ips sexdentatus, I. acuminatus* on *Pinus yunnanensis)*. Chinese Ministry of Forestry. 52 pp.
Chen, Mo-mei., et al. 1963. Infection of leaf rust *(Melamprsora rustropii)* on Chinese white poplar. Journal of Plant Protection 1963 (2):16-20.
Chen, Mo-mei., et al. 1981. Literature review of *Cytospora* flora in China, (manuscript)
Chen, Shouchang 1979 Fir Decay Research National Forest Disease Academic Symposium Information
Chen, Shouchang 1981 Sichuan Tall Mountain Forest Regions Polyporous Fungi Dongbei Forestry Academy Journal (1) 29-34
Chen, Shouchang 2001 Chen, Shouchang Collective Works - Research on Forest Pathology Sichuan University Publishing House
Chen, Shouchang etc. 1964 mophological character, Frequency of Appearance, Pathogenicity and it's condition for Appearance of Oil tea's Anthracnose Fungi Plant Disease Journal 7(1): 45-52
Chen, Shouchang etc. 1981 Olive Peacock Spot Research I Source Fungi Physical Characteristics Plant Pathology Journal (3) 37-42

Chen, Wenshu and Meng, Meiqiong 1965 Tung oil Wilt Forestry Science 10(4): 362-363
Chinese Academy of Forestry Science Guangxi Daqing Mountain Experiment Office, Comprehensive Research Team Forest Disease Research Team 1980 Guangxi Daqing Mountain Disease Research
Chinese Academy of Medicinal Science Medicine Research Firm 1980 Chinese wolfberry (Lycium chinense)Black Fruit rot Preliminary Research Plant Protection (2) 24-26
Chinese Forestry Science Research Firm Edited 1982 Chinese Forest Diseases (Forestry Publishing House, Soon to be Published)
Dai, Fanglan Xiang, Wangnian and Zheng, Ruyong 1958 Catalog of Sources of Diseases of Chinese Economical Plants Science Publishing House
Dai, Fanglan 1979 Chinese Fungi Summary Science Publishing House
Deng, Huiqing and Nan, Yusheng 1979 Two of Strains of Olive and Peanut Pseudomonas solanacearum Research First Report Forestry Science Correspondence (10) 19-22
Department of Forestry Comprehensive Research Team 1957 1956-1957 "Xinjiang Resources Research" Three North slope of Tianshan
Department of Forestry Comprehensive Research Team 1957 1956-1957 "Xinjiang Forest Resources Research" Six South slope of Aertai Mountain
Fang, Huilan and Lian, Yueyan etc. 1979 Zhejiang Chinese fir Cause and Prevention National Forest Disease Academic Symposium Information
forest trees pathogen, The Institute of Forestry, Chinese Academy of Forestry, Beijing P.273
Fujian Forestry Academy Forestry Department 1979 Cedar Bacterial Leaf Spot Fungi Determination and Prevention
Gao, Ya 1981 Melampsora larici-populina Kleb Infection Life Cycle and Rust Resistance Tree Types Selection Dongbei Forestry Academy Journal (3) 10-18
Gao, Zhaowei 1980 Preliminary Research on Cypress Trunk and Branch Rust Occurrences of Our County Zhejiang Forestry Science (1) 13-14
Gong, Mingqin and Nie, Xilin 1981 Mytilaria laosensis Felt Gall Disease Preliminary Research Forestry Science Correspondence (6) 28-29
Guizhou Forestry Science Research Firm etc. 1979 Massonia Pine Needle cast Occurring Pattern and Losses Caused
Guo, Guanglin 1965 Catalog of Shanxi Province Forest Main Diseases Xibei Academy of Forestry Department of Forestry Forest Protection Team
He, Pingxun and Wang, Yongmin 1979 Deciduous Pine shoot blight and Its Prevention Research Jilin Forestry Science (2) 102-111
Heilongjiang Shoot blight Research Team 1979 Deciduous Pine shoot blight Research Dongbei Forestry Academy Journal (1) 32-42
Heilongjiang White Pine Blister Rust Research Team 1979 Red Pine White pine Blister Rust Research Forestry Science 15(2) 119-124
Henan Agriculture and Forestry Department Forestry Station etc. 1976 Henan Garden Diseases and Pests Summary
Hu, Jingjiang, Cao, Zhimin and Wen, Jianlei 2000 The Host Resistance on the Larch-Poplar Rust Forest Research 13: 114-118
Huang, Feilong 1979 Olive Gall Occurrence and Prevention Forestry Science Correspondence (9) 21-22
Huang, Tianzhang 1977 A New Species of Botryosphaeriaceae Microbiology Journal 17(4) 303-305
Huang, Tianzhang etc. 1981 Chinese fir Crown Blight Research 1. Symptoms and Disease Pathogen Fungi
Inner Mongolia Agriculture College, Huhehaote. 8 pp.
Jia, Yun 2000 A Study on the Biomass Loss of Korean Pine Needle Cast Forest Research 13: 130-135
Jiang, Jinsheng and Zhang, Mingzhong 1979 Chinese fir Yellows Observation and Research
Jin, Kaixuan 1963 Research Report on Wood Decay Fungi of Jianfengling forest region of Hainandao

Jin, Kaixuan and Hu, Qiongling 1963 Catalog of Chinese Forest Plants Disease Sources Chinese Forestry Academy Disease Office
Jin, Kaixuan and Cai, Xizhuo etc. 1982 Melia azedarach L.Witches' Broom Bacteria (BLO), Mycoplasma (MLO) Electron Microscope Observations Forestry Science 18 (4): 422-425
Jin, Kaixuan and Fu, Changsheng etc. 1978 Research on witches' broom of Paulownia Disease pathogen and channel of Infection Chinese Forestry Science (4) 1-4
Ju, Guozhu and Xu, Suqing etc. 1965 Research on Aspen Gray Leaf Disease Forestry Science 10(4): 325-330
Ju, Guozhu etc. 1979 Red Pine Root Decay Research Dongbei Forestry Academy Journal (2) 49-56
Ju, Longhe, Gao, Jie and Hu, Mingye 2000 Weeping Willow Root Rot and Its Control Forest Research 13: 13-17
Kang, Wentong, Wu, Tongxing, Lin, Zhiwei and Zhang, Rongsheng 2000 A Study on the Disease Characteristics of *Lecanosticta acicola* (Thum) Sydow in Fujian Forest pathogenicity among isolates from an enzymatically mo Research 13: 184-187
Lan, Ying 1980 Bamboo Stem Rot Research The Journal of Nanjing Academy of Forestry Products(1) 87-94
Li, Chuandao and Zhu, Xiqiao 1979 Pine Needle Brown Spot National Forest Disease Academic Symposium Information
Li, Chuandao and Zhu, Xiqiao etc. 1980 Cedar Anthracnose Research II Disease Source Fungi Determination Nanjing Academy of Forestry Industry Journal (3) 28-34
Li, Haiyan, Wang, Yuyan, Shu, Chaoran and Guo, Zhihong 2000 Effect of Moisture Content on pH-Value in Healthy and Pine Wilt Diseased Woods Forest Research 13: 58-62
Li, Wanbao, Wang, Zhiming, Jing, Tianzhong, Liu, Guorong and Li, Xiue 2000 A Study on Injury Index and Control Threstold of Larch Needle Brown Rust Forest Research 13: 136-139
Li, W-H, Y.F. Han and Chen, Mo-mei. 1985. The Forest of Tibet. Chinese 375 pp.
Liang, Zichao and Qi, Huifang 1979 Massonia Pine shoot blight Research National Forest Disease Academic Symposium Information
Linhai County Forestry Specialty etc. 1979 *Melia azedarach* Preliminary Research National Forest Disease Academic Symposium Information
Liu Shi Ji, Chen, Mo-mei., et al. 1982. Studies on *Cronartium.* Journal of Chinese Ministry of Forestry, Beijing (translated from Roger Peterson's English paper to Chinese), published by the Tottori Mycological Institute. The Japan Tokyo Research Center. Totter, Japan August 1973. pp. 203-223.
Liu, Huizhen 1982 Catalog of Main Chinese Tree Diseases Chinese Forestry Science Academy Forestry Research Firm
Liu, Xijing etc. 1981 The Studies of Tea-oil Rot Fungi Microbiology 21 (2) 154-163
Liu, Zengrong and Li, Wensheng etc. 1980 Jujube Rust Research Preliminary Report Forestry Science Correspondence (7) 20-23
Liu, Zhenkun 1978 Xinjiang Main Forest Plant Diseases Catalog
Liu, Zhenrong 1981 Wolf berry Wilt Research First Report Forestry Science Correspondence (5) 19-20
Liu, Zhenyu, Zhao, Lanyong, Ning, Fangyu and Li, Guilin 2000 Study on the Source of Primary Infection of *Euonymus* Brown Spot Forest Research 13: 23-28
Mongolia Agriculture College, Huhehaote. 5 pp.
Nanjing Academy of Forestry Forest Disease Research Group 1975 Research on Aspen Canker Nanlin Science (3)27-33
Nanjing Forestry Industry Academy etc. 1977 Chinese fir Leaf Spot Disease Source Bacteria – Chinese fir Microbiology Journal 17(3) 179-182
National Chinese fir Research Team 1978 Chinese fir yellows Preliminary Research Chinese Forestry Science (2) 42-46
Ningxia Academy of Agriculture Forestry Department Forest Protection Group 1975 Ningxia Garden Diseases Catalog

Parmeter, J.R. and Chen, Mo-mei. 1988. Occlusion of sapwood fungi carried by *Dendroctonus* species. 1988 Forest Pathology Symposium in China. p. 1. (abstract)
Parmeter, J.R., G.W. Slaughter, Chen, Mo-mei and D. Wood. 1989. Single and mixed inoculations of ponderosa pine with fungal associates of *Dendractonus* spp. Phytopathology 79(7):768-772.
Qin, Guofu, Zhao, Jun, Liu, Xiaoyong, Zhao, Jie and Tian, Shumin 2000 Detection of Latent Infection of Phytoplasmas in the Paulownia Plantation of Phytoplasma-Free Seedling Forest Research 13: 97-105
Qiu, Dexun and Tan, Songbo etc. 1980 Massonia Pine Needle Cast Preliminary Research Forestry Science (3) 203-207
Qiu, Fugeng and Zhang, Xijing etc. 1976 Research on Bamboo Shoot blight Forestry Science(3) 63-68
Rong, Hanquan 1980 Ningxia Sand Region Sand Plant Disease Research Preliminary Report Forestry Science Correspondence (5) 29-31
Rong, Hanquan and Tian, Naixiang 1979 Willow and Apple Rot Research First Report National Forest Disease Academic Symposium Information
Shandong Academy of Agriculture Forestry Department etc. 1979 Pine Rot Research First Report National Forest Disease Academic Symposium Information
Shandong Academy of Agriculture Forestry Department Forest Disease and Pest Research Office 1979 Walnut Anthracnose Research First Report National Forest Disease Academic Symposium Information
Shandong Forests Main Disease and Pest Picture Book Composing Team 1978 Shandong Forests Main Disease and Pest Picture book (Forest Plant Edition)
Shandong Fruit Trees Research Firm 1979 Chinese date (*Zizyphus jujuba*) Disease Source Research (1971-1978) Fruit Trees Science Research Special Topic Report Thirteen
Shanxi Agriculture and Forestry Academy Forest Research Institute 1977 Shanxi Forest Disease and Pest Book (First Edition)
Shao, Liping 1959 Tree Diseases of Dongbei Dongbei Academy of Forestry Forestry Resources No. 9
Shao, Liping and He, Bingzhang etc. 1979 Xinganling Deciduous Pine Cancer Swell Research First Report Dongbei Forestry Academy Journal (1) 27-31
Shao, Liping and Ju, Guozhu 1979 Scanning Microscope Observation of Three Chinese Pine Rust Fungi *Cronartium Spp.* Spore Surface Appearance Dongbei Forestry Academy Journal (2) 57-61
Shao, Liping and Pan, Xueren etc. 1979 Deciduous Pine Man-made Forest Heartwood Color Change and Decay Research
Shao, Liping and Shen, Ruixiang etc. 1981 Fungi Systematic Dongbei Forestry Academy
Shao, Liping etc. 1980 Red Pine White Pine Blister Rust Fungi Determination Forestry Science 16 (4) 279-282
Shen, Wensheng and Cao, Jidan etc. 1979 *Aleurites Montana* tree New Disease - *Aleurites Montana* Anthracnose (*Colletotrichum sp*) National Forest Disease Academic Symposium Information
Sheng, Shifa, Li, Nan and Yu, Liyun 2000 The Study of the Identification of the Fungi on the Cones of Declined *Pinus armandi* Forest Research 13: 1-7
Shi, Quanchang and Zheng, Wenfeng 1978 Olive Anthracnose Preliminary Observation
Shu, Ningbi 1981 Dalizhou Common Walnut Pest Types Yunnan Forestry Science (1) 85-86
Sichuan Forestry Science Research Firm 1979 Fire Pine and Wetland Pine Brown Spot Occurrence Observation First Report National Forest Disease Academic Symposium Information
Sichuan Forestry Science Research Institute 1979 Preliminary Catalog of Sichuan Forest Diseases
Song, Xiaobin, Zhang, Xingyao, Wang, Peixin, Zhang, Xuewu and Ma, Songtao 2000 Studies on Pathogen of the Bacterial Canker of Kiwifruit in Shanxi Forest Research 13:42-46
Su, Hai, Wang, Jun, Wu, Huixiong and Zhang, Jingning 2000 Variation in Pathogenicity of *Botrydiplodia theobromae* to *Cinnamomum cassia* Forest Research 13: 91-96
Su, Xiaohua, Zhang, Qiweng, Shen, Ruixiang, Yang, Wang, Gui, Fu and Wang, Yanmin 1998 Research on the Variation of Pupulus deltoides X P. cathayana F2 Generation against Alternaria alternata Forest Research 11(6): 565-568

Tan, Songshan 1980 The Index of Main Tree Diseases
Tan, Songshan 1980 Three Foreign Pine Diseases Observation Hunan Forestry Science (2) 25-28
Tan, Songshan and Deng, Lianhe etc. 1981 Armandi Pine Disease Preliminary Research Hunan Forestry Science (2) 22-25
Tao, Jiafeng and Qing, Jiazhong etc. 1977 Two New Species of Powder mildew Fungi Microbiology Journal 17(4) 293-296
Tian, Guizhong, Zhang, Xijin, Luo, Fei and Zhu, Suifang 1999 Responses of Resistant and Susceptible Clones of *In Vitro* Cultured Paulownia to the Graft Inoculation with Phytoplasmas Forestry Science 35(4): 31-39
Tian, Shumin, Qin, Guofu, Zhao, Jun, Zhao, Riliang and Zhang, Xizhi 2000 Antagonistic Effect of Wood Decay and Mycorrhizal Fungi on *Armillaria ostoyae* Forest Research 13: 75-80
Tian, Zhejun and Zhao, Dingquan etc. 1979 Preliminary Research on Western Sichuan Mountain Forest Regions Man-made New Forests Rust Fungi Chinese Forest Disease Academic Symposium Information
Tong, Benqun 1979 Scanning Microscope Observation of Chinese Pine Stem Rust Fungi Spore Surface Appearance Forestry Science 15(4) 241-243
Wan, Jianyu 1979 Deciduous Pine shoot blight Research – Disease Fungi Appearance and Physiology National Forest Disease Academic Symposium Information
Wang, Jun 1996 Several Factors Affecting the Determination of Disease Resistance in *Casuarina* Clones to *Pseudomonas Solanacearum* Forestry Science 32(3): 225-229
Wang, Laifa, Yang, Baojun and Li, Chuandao 1999 Evaluation of Pathogenicity of Parasitic Fungi to Root – Knot Nematodes Forestry Science 35 (5): 41-47
Wang, Qikai etc. 1981 Jujube witches' Broom Research Plant Pathology Journal (1) 15-18
Wang, Yunzhang and Chen, Mo-mei etc. 1980 Western China Rust Fungi New Species Microbiology Journal 20 (1) 16-27
Wang, Y-Z. and Chen, Mo-mei., et al 1981. Rust fungi of western China. Acta Mycologia Sinica 20(1):16-28
Wei, Anjing and Qi, Yinghe 1980 Tea-oil Anthracnose Fungi Sexual Stage Is Found In Our Province Zhejiang Tea-oil Science (1) 22
Wong, Yuexia 1979 Oil tea Anthracnose National Forest Disease Academic Symposium Information
Wu, Guangjin and Lin, Xuejian 1980 Tung oil Wilt Research Zhongnan Forestry Science (3) 30-41
Wu, Yousan etc. 1963 Research on Pine Sprout Damping Off Disease 2. Distribution, Loss, Symptom, Type, Source of Disease, and Effect on Plantation Plant Protection Journal 2(2): 179-185
Wu, Yubin and Zhang, Ronghuan etc. 1979 Red Pine Canker Occurring Pattern and Prevention Preliminary Research National Forest Disease Academic Symposium Information
Xiang, Cunti and Wang, Yongmin etc. 1964 Research on Black Leaf Spot Disease of Aspen Sprouts Plant Protection Journal 3(3): 201-207
Xiang, Cunti etc. 1981 Mongolia Pine Shoot Blight Research Dongbei Acade my of F orestry Journal (2) 1-9
Xiang, Yuying 1981 A Dangerous Poplar Disease Forestry Disease And Pest Protection 1st Edition
Xiang, Yuying and Xu, Meiqing 1980 Aspen Purple Root Occurrence and Prevention Research Plant Pathology Journal 10 (2) 125-131
Xiang, Yuying, Hua, Xiaomei and Zhao, Jingzhou 1979 Determination of the Disease Fungi of Aspen Canker Microbiology Journal 19(1) 57-63
Xibei Academy of Agriculture Forestry Department 1976 Shanxi Aspen Disease Catalog
Xinjiang Comprehensive Research Team 1958 Summary of Wild Fungi Research of South Xinjiang Regions Science Publishing House
Xu, Shiduo and Pan, Enqi etc. 1981 Chinese fir Sprout Blight Research First Report Forestry Science Correspondence (10) 24-2
Yang, Baojun and Zhong, Xiaowu 1980 Identification of The Olive Root Knot Nematode Disease Pathogen Forestry Science 16 (4) 264-265

Yang, Wang and Han, Guangming etc. 1980 Research on Chestnut Blight Diseases Of China Preliminary Report Forest Diseases and Pests Correspondence (2)
Yu ,Yongnian and Lai, Yiqi 1981 Chinese Microsphaera Systematic Research I Corylaceae, Menispermaceae, Hypericaceae, and Crassulaceae New Species, Names and Combinations Microbiology Journal 21 (1) 7-14
Yu, Yongnian and Zhao, Zhenyu 1981 Chinese Microsphaera Systematic Research IV Berberidaceae New Categories and Existing Ones Microbiology Journal 21 (2) 143-151
Yu, Yongnian etc. 1978 Systematic Research on Chinese Phyllctinia I Systematic of Types Microbiology Journal 18(2): 102-117
Yu, Yongnian etc. 1981 Chinese Erysiphales New Systematic Research Yunnan Plant Research 3(1): 33-42
Yu,Yongnian etc. 1979 China Phyllactinia Research Systematic Research III Long- appendage Type Microbiology Journal 19(2): 131-145
Yu,Yongnian etc. 1979 China Phyllactinia Systematic Research II Short-appendage new Type Microbiology Journal 19(1): 11-23
Yu,Yongnian etc. 1979 China Typhulochaeta Systematic Research Microbiology Journal 19(4): 379-382
Yu,Yongnian etc. 1981 Chinese Microsphaera Systematic Research II Aristolochiaceae, Celastraceae, Rhamnaceae, and Symplocaceae New Species Microbiology Journal 21 (1): 15-22
Yuan, Jianyu 1981 Deciduous Pine Brown Rust Research Plant Pathology Journal (2) 57-60
Yuan, Jianyu 1981 Deciduous Pine Dry Disease Research Dongbei Forestry Academy Journal (3) 5-9
Yuan, Jianyu 1981 Deciduous Pine Skin Rot Research First Report Forestry Science Correspondence (7) 29-31
Zeng, Sihai and He, Xueyou 2000 A Study on the Relationship between Wilting of Casuarina and Microbe in Its Xylem Forest Research 13: 18-22
Zhang, Hanwen etc. 1960 Catalog of Xinjiang Economical Plant Ranch Diseases Bayi Academy of Forestry Scientific Research Office
Zhang, Hesheng 1980 * Needle Rust Research Forestry Science Correspondence (2) 30-31
Zhang, Hesheng and Yang, Jinghua 1964 Preliminary Research on Forest Diseases of Liaoning Province
Zhang, Jingning 1980 *Melia azedarach* Shoot Witches' broom Pathogen of Disease Research Forestry Science Additional Edition 92-94
Zhang, Suxuan 1982 Bamboo Shoot Blight Fungi (*Ceratosphaeria*) New Species Nanjing Forestry Industry Academy Journal (1) 154-158
Zhang, Xingyao 1998 Review of Studies on the Fungus Genus *Exobasidium* Forestry Science 34 (1): 113-120
Zhao, Zhenyu 1963 Xinjiang Forest Diseases Xinjiang People's Publishing House
Zhao, Yunqin and Zhang,Ying etc. 1980 Chestnut Blight Occurrence and Prevention Research *Endothia Parasitica (Murr) P. J. et H. W Anders.* Plant Protection (1) 13-16
Zhao, Zhenyu Xinjiang The Desert Plant Mistletoe Fungi The Journal of Xinjiang Bayi Academy of Agriculture (3) 60-70
Zhao, Zhenyu 1959 The Falling Needle Disease of Xinjiang Needle Trees Xinjiang Agriculture and Science Communications 1959 (11)
Zhao, Zhenyu 1973 Leaf Rust Fungi of Xinjiang
Zhao, Zhenyu 1975 Xinjiang Forest Disease Catalog
Zhao, Zhenyu 1979 Xinjiang Aspen Diseases Xinjiang Bayi Forestry Academy Journal (2) 10-21
Zhao, Zhenyu 1979 Xinjiang Powder Mildew Fungi Journal Xinjiang People's Publishing House
Zhao, Zhenyu 1981 Chinese Erysiphales New Systematic Research V Bignoniaceae New Species Dongbei Forestry Academy Journal 2: 28-30
Zhao, Zhenyu etc. 1962 Research Report of Tree Disease Types of Tulufan Region

Zhao, Zijun, Wang, Lizhong, Zhao, Xiaojun, Lin, Zhongmin and Huo, Luyuan 2000 The Study on Chinese Jujube *Alternaria* and *Phome* Blight I. The Isolation and Identification of Pathogen Fungi Forest Research 13: 8-12

Zheng, Ruyong and Chen, Guiqing 1977 Systematic Research on Chinese Uncinula I Discussion of One Type of Chinese Uncinula Microbiology Journal 17(3) 189-197

Zheng, Ruyong and Chen, Guiqing 1977 Systematic Research on Chinese Uncinula II New Species and Combinations of Saliceae Microbiology Journal 17(3) 198-210

Zheng, Ruyong and Chen, Guiqing 1977 Systematic Research on Chinese Uncinula III New Species and Variety of Coriariaceae, Euphorbiaceae, Oleaceae, Chinese Sterculiaceae and Tiliaceae Microbiology Journal 17(4) 281-292

Zheng, Ruyong and Chen, Guiqing 1978 Categorizing Research on Chinese New Powder mildew genus II Asexual and sexual Phases of new genus: New Powder mildew Fungi Microbiology Journal 18(3) 181-188

Zheng, Ruyong and Chen, Guiqing 1978 Systematic Research on Chinese Uncinula IV New Species, Hybrids, and Combinations of Hamamelidaceae, Lauraceae, Monotropaceae and Ulmaceae Microbilogy Journal 18(1) 11-22

Zheng, Ruyong and Chen, Guiqing 1978 Systematic Research on Chinese New Powder mildew genus New Species of Salicaceae: 18(2) 118-121

Zheng, Ruyong and Chen, Guiqing 1979 China Systematic Research I Founding of Uncinula Microbology Journal 19(4) 375-378

Zheng, Ruyong and Chen, Guiqing 1979 China Uncinula Systematic Research I Founding of * and Determination of Chinese and Japanese Types 19(3) 280-291

Zheng, Ruyong and Chen, Guiqing 1980 Chinese Erysiphe Fungi Systematic Research I Caprifoliaceae New Species and New Variety Microbiology Journal 20(1) 45-49

Zheng, Ruyong and Chen, Guiqing 1980 Chinese Microsphaera Systematic Research I Microsphaera Determination and Aceraceae andHippocastanaceae New Species and Combinations Microbiology Journal 20 (1) 35-44

Zheng, Ruyong and Chen, Guiqing 1981 Chinese Erysiphe Systematic Research III Convolvulaceae, Coriariaceae, and Fagaceae New Species and Hybrids Microbiology Journal 21 (1) 23-30

Zheng, Ruyong and Chen, Guiqing 1981 Chinese Uncinula Systematic Research V Hamamelidaceae, Papilionaceae, Rutaceae, and Salicaceae New Species and Hybrids Microbiology Journal 21 (3): 298-307

Zhong, Zhaokang 1979 *Robinia pseudoacacia* Dieback Research First Report National Forest Disease Academic Symposium Information

Index of Host (China)

Index of Pathogen

Stories of Mo-Mei Chen's Scientific Expeditions in Tibet in the 1970s

By Wendy Helms and Mo-Mei Chen

In 1966 my husband and I were living in Ho He Hao Te, the capital city of Inner Mongolia. We were both employed in teaching and research in the Department of Forestry at the University of Inner Mongolia. In the spring of 1966 the University sent me, with a group of other teachers, to Chifeng in Liao Ning Province, to work on producing a teaching textbook on management of forest nurseries. At that time we began to hear more about the Cultural Revolution that was sweepint China, just like a proverb "The whole town is swept by wind and rain". It seemed to be some sort of power struggle that was happening far away and we did not think that it would have much effect on our lives. However, in May, our team had a telegram from the University Chancellor ordering any faculty members and students who were out in the field to return, as soon as possible, to campus to join the revolutionary movement. We returned immediately to Ho He Hao Te, full of fear and trepidation. We did not know what was going to happen and we felt like "birds startled by the mere twang of a bow string".

We were required to attend endless political meetings and meaningless debates, and much valuable time was wasted in group studies of Mao's Little Red Book. Open criticism was not permitted so it was necessary to find covert ways of resisting the suppression of intellectual activity. For example, I owned two copies of the Little Red Book, both of them being identical in appearance to the official Chinese edition; however, one was an English translation and the other was a Russian edition so that, in those tedious study groups, I was at least able to work on improving my language skills. I did not like politics. I was criticized for avoiding politics and blamed for my "bad" family background. The fact that my father was a professor and was in Taiwan was held against me as was the lack of workers, farmers and soldiers in my family, with the exception of one of my great-grandfathers who had been a carpenter. How ironic it was that my family's intellectual achievements, which had been such a source of pride, were now regarded as shameful! It seemed to me so unfair and irrational to blame people for their family history. I hated being criticized and I resented the invasion of my privacy, attracting further criticism for my "resistance" and "bad attitude.

There are many sad memories of that time. For two years our lives were turned upside down, though on the whole we suffered less and for a shorter time than many people in other parts of China. My husband and I had three strategies for survival during that chaotic time: we tried to keep quiet and avoid politics as much as possible, we immersed ourselves in study, research and writing and we sought to strengthen our own family support system. By 1969 the worst was over for us, we were allowed to resume our teaching duties and life became relatively "normal" again.

Since then I, my husband professor Ang-He Zhang, my six-year-old daughter Lily, with my pregnant May, we suffer so many Red Guards' humiliations, many devoid of gratitude stories, fortunately, in 1968 summer I get permitted to join return classroom group of teachers and students, it's proverb: "Said of people who fall into the same group because they are of a mind (as educator). Since then my best wishes for myself are "as a escape from politics movement!"

It was in 1972, while I was visiting my father in Beijing Normal University that I heard a radio program describing one of the Chinese Academy of Sciences' research expeditions to the upper reaches of the Yangtze River, which had its source in the high plateau of Tibet. I had previously imagined Tibet to be nothing but a remote, unpopulated, undeveloped area, cold, wild, and empty, and so I was amazed to hear of its great scenic beauty and rich diversity of plant and animal life. That radio program stimulated my imagination so much that I felt a great desire to see Tibet for myself, although I could not imagine how I might find a way to travel there. However, when I spoke to my father about the radio program and my desire to see Tibet, he was surprisingly encouraging. "If you want to go so much, then dream yourself there," he said, and it was not long before my dream came true.

In 1972/73, the Natural Resource Commission of the Chinese Academy of Sciences proposed an ambitious scientific survey of the Qinghai-Tibet Plateau. Ten small research teams had done some preliminary studies in Tibet the 1950s and 1960s, but this new survey was designed to involve 400 scientists, representing 30 different disciplines, in an intensive, systematic investigation of the physical geography and ecology of that unique area. Plate tectonic theory, first proposed in the 1920s and proven in the 1960s, had given a new understanding of the way in which the subduction of the Indian-Australian Plate under the Eurasian Plate had caused the gradual up-thrusting of the Himalayan mountains over a period of 60 million years, creating the Tibetan Plateau. During World War II, Premier Zhou En Lai had first proposed an expedition to the area to study the ecological effects of the raising of the Himalayas and the ways in which human life in the region had adapted. By the 1970s, China had additional strong incentives for arranging a geo-physical survey to assess the natural resources of Tibet, especially with regard to minerals and energy in the form of oil or hydro-electric potential which was badly needed to reduce China's dependency on coal and natural gas. Plans for the survey of the Tibetan Plateau were spearheaded by Mr. Sun Hong Lie, Director of the Natural Resource Research Institute of the Chinese Academy of Sciences. Mr. Sun came from an adventurous family. His father was a geologist who had made an expedition, on camel-back, to the deserts of Western China to look for oil in the 1920s, a successful venture commemorated by a statue of him in Yumeng Park, Gansu Province.

I was excited when I first heard about the Tibetan Plateau project in 1974. I was in Peking at the time, spending a sorrowful Chinese New Year with my family following the recent death of my mother. We happened to hear an evening radio program about "The Yangtze River of Southern Tibet," reporting on a research expedition to that area which had involved 200 scientists. I was immediately captivated by the story and decided to apply for a position on a research team, though I feared that both my sex and my age (I was then 43) would count against me. Indeed, Mr. Sun was at first unwilling to recruit women scientists for the Tibetan project in the belief that primitive field conditions at high elevations would be too harsh for "delicate females;" however, after talking with one of my microbiology classmates and learning of my extensive experience doing field work in remote areas, my good spirit, and capacity for hard work under difficult conditions, Mr. Wu finally offered me a position as both a researcher and a collector. I was proud to be one of only four women among 400 scientists, and I had full confidence in my own abilities, remembering my past success as a researcher and expedition team leader in virgin forests of the Great Xiang An Mountains and the Heng Duan Mountains.

In addition to my desire to see Tibet, further my professional career, and escape the Cultural Revolution, I had another reason for wanting to get away to the wilderness. My mother's death in January 1974 was not unexpected after her two-year battle with cancer of the stomach and uterus; however, I was deeply upset by her loss, which caused me to contemplate the sadness of her life. My mother had been an old-style woman, her life constricted in the way suggested by the ancient Chinese ideogram symbolizing "female" (nu) with a slanting line meaning "leaning" or "dependent" above two crossed lines suggesting "bound up." By contrast, the Chinese character for "male" (nan) uses a rectangle divided into four, to symbolize "field," above strong, upright lines denoting "energy." These ideograms, which originated in feudalistic times, have perpetuated in Chinese society over the centuries a deep-seated assumption that females are inferior to males and necessarily dependent on them, though during my mother's lifetime, considerable progress had been made, at least on paper, towards equality. Article 53 of China's constitution stated that "Women enjoy equal rights with men in all spheres of political, economic, cultural, social, and family life, Men and women enjoy equal pay for equal work.."

My mother shared little in common with my father who had, in any case, spent much of their married life living and working in cities far removed from the family. Like countless millions of women in China, my mother had devoted her life to ensuring the survival of her children through years of poverty, starvation, sickness, and danger, especially during the Japanese invasion. Yet, despite all the distress and upheaval in her life, she had managed to pass on to me, her daughter, a sense of hope and confidence, which I felt that I must work hard to justify. The expedition to Tibet became an opportunity for me to affirm my mother's hopes for improvement in the status of women in China. Years later, when I received from the Chinese Academy of Sciences a special award in honor of my distinguished participation in the Tibetan project, I felt like dedicating that award to my mother.

PREPARATION FOR THE EXPEDITION TO TIBET

Early in 1975, after a preparatory meeting in Peking, I went to Changdu to attend a two-week training session for scientists participating in the Tibetan expedition. Then we were sent for further orientation to the city of Xi Ming, at an altitude of 2,500 meters, both in order to get accustomed to high elevation and to learn about the geology, geography, flora, and fauna of the areas we were to survey. The information we received excited my imagination and made me eager to see for myself the natural wonders Tibet of the Quinghai

Because photographs of Tibet often focus on dramatic mountain panoramas of areas above 12,000 feet where climatic conditions are harsh, Tibet is often thought to be an arid land where agriculture is limited and vegetation sparse. In fact, the pleasant North Temperate zonal climate below 12,000 feet, combined with a high level of ultraviolet radiation, has provided ideal conditions for plant growth and high seed production, contributing, over the ages, to the evolution of an unusual abundance and diversity of plant life. To date, botanists have identified, in Tibet, 5,766 species of plants representing 2,008 families and 1,258 genera. Of the 5,766 species, 43% are specific to Tibet and occur naturally nowhere else in the world, a result of the unique and relatively recent geological history of the Tibetan Plateau. Another 37% of plant species found in Tibet are also common in East Asia, but some, like the 245 species of Tibetan rhododendron, can be identified as migrants from the ancient Mediterranean where they first evolved. The *Ericaceae* family, to which the rhododendron belong, is one of the five dominant families of Tibetan flora, the others families being *Leguminosae, Rosaceae, Gaminaceae*, and the highly evolved *Compositae.*

Over 1,300 of the plant species found in Tibet are tall shrubs and trees among which are many rapid-growing conifers of great economic value. Some of the conifers such as Yunnan pine, alpine pine, armand pine, west Sichuan spruce and yellow-cone Likiang spruce(*Picea linzhiensis* (Cheng et L. K.) Cheng et. K. Fu.) are also found in other provinces of China, but there are 15 species of conifer which are found only in Tibet. The most common of these native conifers are the Himalayan pine, long-leaf pine, long-leaf spruce, Himalayan fir, Himalayan larch and Himalayan cypress. Most of these species are found in the Tibet Forest Zone where magnificent virgin forests occur in southeastern Tibet, in the mountains of the southern Hengduan Range and the warm, wet, lower valleys of the Yarlung Zangbo River. In the lowest elevations of the Plateau such as the area around Yadong, subtropical evergreen forests are found with many species of valuable broadleaf hardwood trees and a richly varied understory that includes a great variety of fungi.

As a plant pathologist specializing in pathogens which affect the health of forest trees, and as a mycologist with a passion for mushrooms of every sort, it was very exciting for me to know that I was about to have the opportunity to enter areas that had never before been surveyed by modern scientists. I was especially thrilled by the very real possibility that I might discover a new species that would be named in my honor. This was a long-standing dream of mine, to discover a new species and join the ranks of those pioneer scientists and teachers who had been my professional mentors, those who had inspired me and passed on the knowledge and expertise that would enable me to follow in their footsteps.

It was in the mid-18th century that China was first introduced to international systems of organizing biological data. A French scientist, P.M. Cibot, traveled to China in 1759, later publishing papers on new species that he had found there and encouraging Chinese students to study overseas.

It is no small thing to identify a new species. The process is not one of creation or invention. The investigator needs to have a deep base of knowledge and experience and sufficient familiarity with a certain genus and the numbers of species in that genus to be able to recognize when key characteristics of a particular specimen do not fit within the recorded parameters of the genus. After nearly 20 years of fieldwork — in the course of which I had collected and identified large numbers of fungus specimens — I felt that I was ready to meet this challenge and I could hardly wait to set out on my first expedition to the Tibetan Plateau.

EXPEDITION TO THE TIBETAN PLATEAU (1975)

At last, in May 1975, when we were considered to be adequately prepared, we returned to Changdu and boarded a plane for Lhasa. Everybody was nervous as we approached Lhasa airport because there were steep mountain ranges on either side and we could see no place to land. We all breathed a sigh of relief when our plane was safely on the ground, but there was more discomfort ahead. The airport was 90 km from the city, a bumpy three-hour journey by truck, and by the time we arrived in Lhasa many of our group were already experiencing headaches and nausea in reaction to the 3,380 meter elevation. I was less affected than most because of my extensive experience working in mountain areas and I was one of the few who had an appetite for the good dinner, which has been prepared for us. We were lodged in a military hostel, which also housed mountaineers on their way to attempt the ascent of Qomolangma (Mt. Everest) since there were, at that time, no other accommodations in the city.

It happened that among our fellow residents were members of a Chinese climbing expedition returning from Everest, including a Tibetan woman, Pan Do, who had recently become the first Asian woman to reach the top of the highest mountain on Earth. Our group had followed the exploits of this team through tuning in to daily radio reports and we were full of admiration for their success and thrilled to meet them in person. Pan Do was a great inspiration to me. At 37, she was a good deal older than the other members of the team and yet she had succeeded where younger men had failed because both her body and her mind were strong. During one of our many talks, I told Pan Do, "I am proud of you and your accomplishment. I am only on your foothill." And she surprised me by saying "And I am also proud of you for being among the first scientists to come and study my homeland of Tibet." After hearing about the courageous way in which she successfully overcame many difficulties and tolerated the effects of elevations more than twice that of Lhasa, I felt challenged to live up to her high standard and determined to make the most of less the opportunities ahead.

Walking around Lhasa, I found the ancient streets to be bare, drab, windy and full of bad smells although these were easily forgotten with the magnificent sight of the Potala, the greatest of all the Tibetan Buddhist monasteries, dominating the landscape. I liked the city very much, especially at night when the combination of high elevation and clear atmosphere made the moon and stars seem very close and bright against an unusually black background. The high, clear air exposes the Tibetan Plateau to intense levels of ultra-violet radiation and causes a high incidence of skin cancer in the local population but is also responsible for Lhasa being called "Sunshine City." Vegetation around the city was rather limited and apart from patches of willow and poplar trees just coming into spring leaf, there were only a few stunted cypress trees to be seen. During our orientation meetings in Chengdu we had been told that there existed, in Lhasa, a 500-year-old cypress tree which had been planted by Wen Chang, a Chinese Tang Dynasty (618-907 AD) princess from Xian who had been married to the King of Tibet. We were told that this marriage forged the original union between China and Tibet, which was considered partial justification for China's action in forcibly reestablishing that union in 1951. Our instructors had impressed upon us the importance of respecting local customs.

We spent 10 days in Lhasa organizing the supplies and transportation necessary for 40 teams of investigators to spend the next four months in the field. The Chinese military had provided us with cooking equipment and military food rations, salt, oil, rice, flour, noodles, dried seaweed, candy, chocolate and a variety of canned goods. Our camping gear consisted of heavy canvas army tents and U.S.-made down sleeping bags which the Natural Resource Commission had somehow obtained as surplus equipment after the Korean War. Instead of a mattress or cot, each person was provided with a deerskin to use as a sleeping pad, an ancient practice that gave excellent protection against dampness. Personal equipment had to be chosen carefully, keeping in mind the fact that vehicles were restricted to the limited, primitive road system, and we would have to carry on our backs whatever was needed in field camps. For me, in addition to personal clothing, essentials included my precious microscope with slide preparation materials, key reference books and a hand lens so that I could identify specimens in the field, a plant press and a metal vasculum, a special container in which to preserve collected specimens.

I had been assigned to a forestry group as one of five scientists with six assistants (young Tibetans, recruited locally, to provide labor and act as interpreters) and two drivers for the truck and jeep which provided our transportation. When everything was finally packed and it was time to leave Lhasa, I climbed into the front of the truck together with a young woman scientist from Sichuan Forest College, while our team leader, Mr. Li Wen Hua and Mr. Han Yu-Fang, both foresters, and mycologist Mr. Zong Yu-Cheng rode in the back of the truck, perched uncomfortably on supply boxes. The rest of the party traveled in the jeep. Unfortunately, our truck driver was not pleasant or cooperative. Mr. Ge had retired from the People's Liberation Army and, true to the teachings of the Cultural Revolution, he regarded intellectuals as parasites, vastly inferior to soldiers and workers. He was unable to understand the reason for our need to travel around on the Tibetan Plateau and, because he resented having to drive for us, he treated us with angry contempt, which somewhat dampened the spirits of our group until we learned to ignore his ill humor.

That first day we took the main road south from Lhasa, crossing the Yarlung Zangbo, the highest river in the world, to camp at Yamzho Yumco Lake at an elevation of 4,200 meters. I was very impressed by this many-armed body of water, covering over 600 square kilometers, which was a beautiful, deep turquoise blue, serenely reflecting the sky. The lake edge near our rather barren campsite was so crowded with a type of small catfish that it was possible to scoop out five or six at a time with a bowl. Those fish would probably have made a delicious dinner but we threw them all back into the lake for we were afraid to eat them. We knew of the ancient Tibetan custom of "sky burial," which involved leaving dead bodies at exposed sites to be eaten by vultures. Those birds, flying over the lake, may have dropped carrion into the water. How were we to know that the fish had not fed on human flesh? We were not willing to take the chance, so we made do with a meal of half-cooked rice.

That first evening in camp we discovered that our normal method of cooking was going to be a problem. At that high elevation, the low boiling temperature of water meant that, even with the aid of a pressure cooker, rice took a very long time to cook. Our Tibetan assistants suggested that it would be better for us to eat the traditional local staple food, which they called *tsampa* (*zhang ba*, in Chinese). This was a mixture of roasted barley flour and bean flour which would be placed in a small pottery bowl then moistened with tea and kneaded into a dough which was more tasty and nutritious than rice and easier to prepare. We soon decided that we needed to carry the supplies for making *tsampa*, and we also learned the interesting traditional Tibetan method of making tea, which we called *shu yu cha*. While water was heated in a big pot over a fire fueled with yak dung, a portion of tea (broken from a compressed "brick" which was the usual way of packaging dried tea leaves), salt, and yak butter were placed in a long tube made of larch wood. Hot water was poured into the tube, and it was closed by a plug through which a wooden plunger was used to agitate the mixture. The liquid was poured back into the kettle to re-heat then returned to the wooden tube to be stirred again, and when it was ready, the first of the tea was used to moisten the waiting *zhang ba* flour before the rest was served to be sipped out of small bowls. We also learned that travelling Tibetans often carried potatoes in their clothing. They would throw these into a camp fire to roast, often accompanied by hot green peppers, and with *zhang ba* and *shu yu cha*, this made an easy, economical and nutritious meal. However, *zhang ba* is an acquired taste and most of us in the forestry group preferred to eat our familiar Chinese rations even though they were more time-consuming to prepare.

We continued southwards on the second day of our expedition following a road that climbed even higher into the mountains and traversed two 5,000-meter passes. I will never forget a most unusual plant we found growing at that great elevation. It was the snow lotus *(Saussurea),* one of many alpine plants which have evolved effective protection against extreme cold and strong solar radiation. The stems and leaves are covered with soft, fine hairy fibers which mat together to form insulating chambers preventing rapid evaporation and enabling the plant to produce a huge, creamy lotus-like flower which is amazing to see in such an environment. Our appreciation of the magnificent scenery was somewhat spoilt by concerns about altitude sickness, intense cold, and the glare of reflected sunlight. Many members of our group were suffering from severe headaches, and so it was a relief when the road descended again onto a grassland plateau as we headed towards Pali (Phari, in Tibetan) where we planned to camp for two days.

The vegetation of the alpine plateau was severely limited by harsh climate and poor soils, but the grassland provided food for wild yak, which could be seen grazing among patches of snow. We were able to smell the wild, pungent odor of the yak as we drove past them. There were a few herbaceous plants — such as sagebrush and wild chrysanthemum — but there were no trees other than occasional stunted alpine birch and some small, contorted sabina cypress. The absence of trees and shrubs posed a problem for the women when we needed to take a bathroom break on the roadside. The young Tibetan girl assistants were especially modest and shy, so I improvised a screen for them by using a large umbrella to protect their privacy.

Because the high plateau region was sparsely populated and we seldom saw signs of habitation, we were very interested to come upon a woman walking down the road with three children. When we stopped and spoke with her through an interpreter, she told us that she was 29 years old and lived close by with her husband who raised yaks and sheep. Her husband was not at home, she said, because he was making a two-day trip, on horseback, to Pali to see a movie! The woman invited us to visit her nearby house. It was built of wood but had no windows and so it was dark inside and smelled very strongly of yak, which was not surprising when we noticed that there were two yak calves sharing the living quarters. A large bed was spread with a huge fur rug on which nestled twin baby girls. I was amazed that the young woman was able to raise five children in such primitive conditions, but she looked healthy and happy, her skin was smooth and shiny and she wore silver jewelry with her bulky traditional clothing. She wanted to make tea for us but unfortunately we did not have time to stay longer. We gave the older children candy when we left. They had never seen candy before.

The village of Pali at an elevation of 4,360 meters is one of the highest communities in the world and its Tibetan name, Phari, means "Pig Hill", referring to a nearby mountain resembling a pig. The village lies on the edge of the Tibetan Plateau on a road which, only 14 km further on, plunges abruptly into the gorge-like Chumbi valley and descends 1,600 meters down to the town of Yadong overlooking the border between Tibet and India. We stayed only two days in Pali because we were anxious to proceed down to the lower elevations in order to survey the rich sub-tropical forests of the Chumbi valley. However, it was in the area around Pali that I was delighted to be able to collect a rare and unusual fungus called *Cordyceps sinensis*. As the common name "winter worm, summer fungus" suggests, the fungal spore becomes attached to a caterpillar host in winter and feeds on it as a parasite until the caterpillar dies and becomes filled with fungal tissue until it is essentially transformed into a fungus by summer. "Winter worm, summer fungus" had long been valued in China as a folk medicine known to be particularly effective in relieving symptoms of menopause. In 1975, dried cordyceps could be purchased in the local Pali markets for the equivalent of one cent a piece, but now, 20 years later, the price has suddenly escalated to $600 per pound, largely on account of publicity given to the fact that a Chinese woman athlete who won a gold medal at the 1996 Olympic Games was said to have used *Cordyceps sinensis* as part of her training diet. Unfortunately, intense interest in this rare species, which grows only in a small region of Tibet, has threatened its existence and unless it can be grown artificially, *Cordyceps sinensis* may become extinct. However, there are about 100 other similar species of *Cordyceps* in other parts of the world which may also prove to have medicinal value.

During the next few weeks, we worked in areas that were especially rich in fungus species, which afforded me a number of interesting experiences. On the way from Pali to Yadong, we came across a group of students from the Tibetan/Chinese Traditional Medical College in Lhasa. They were on a field trip to collect medicinal herbs and when they learned that I was a mycologist, they told me about a great crop of *ling zhi* mushrooms they had discovered the previous day, growing on some dead plum trees near a distant village. *Ling zhi* have long been highly prized for their medicinal qualities and the group wanted me to go and inspect the site so that I could identify the species for their professor, for there are numerous species in the *Ganoderma* genus to which *ling zhi* belongs. In fact, a learned Chinese monograph was written more than a thousand years ago describing the many different-colored varieties of ling zhi and the particular ecological location of each variety. When I decided that I wanted to see the site for myself, the students provided me with a detailed map and the following day the jeep driver took me and an assistant 250 km to the place that had been described. I was very excited to see such a treasure trove of *Ganoderma*. I collected more than 250 specimens and eagerly looked forward to making a microscopic examination in order to discriminate between closely related species. There were no storage boxes, so when we returned to the camp I carefully wrapped my precious specimens in my bed clothes and left them in my tent. It happened that a group of 20 paleontologists had arrived at our camp that day and I remember that they had shot some wild pigeons to eat for dinner and we all enjoyed a social evening. Unfortunately, my assistant told many people about our collecting expedition and the next day, when I was out in the field, somebody searched my tent and took every single one of the *Ganoderma* samples. I don't often cry, but I cried when I found that my *ling zhi* were gone. I felt very sad and angry to have lost so many important research specimens for I knew that there would be no chance to go back and collect more.

A few weeks later, when we were working out of Yadong, I was able to collect specimens of other fungi with important medicinal qualities, which somewhat compensated for my disappointment over the loss of the *Ganoderma*. One day I found some honey mushrooms *(Armillaria mellea)* growing on a decayed hardwood log. I knew, from my studies, that there was a kind of tuber *(Gastrodia elata),* called *tian ma*, which was symbiotic with the honey mushroom and was highly valued for its medicinal effectiveness in treating epileptic seizures. I had never seen *Gastrodia* before but finding the honey mushrooms (which made a delicious addition to our diet) led me to discover quantities of the precious tuber growing on the *Armillaria* mycelium hidden deep within the rotting wood. On my day off, I showed my friends where to find the *Gastrodia*, and they each enthusiastically gathered as much as five pounds of tubers, which were worth a lot of money, to be dried and carried home as gifts. The tubers were treasured in Tibet where epilepsy was a common medical problem along with other disorders associated with genetic deterioration caused by intermarriage. Traveling through the countryside we frequently saw local people who appeared to be mentally disabled, which made me very sad.

Another example of a fungus with significant medicinal value was the cloud mushroom *(Polystictus versicolor),* often used for the treatment of stomach cancer. Numerous samples of this very beautiful, many-colored, saprophytic mushroom were found in the Yadong area growing on oak trees, which were decaying because of infestation with this fungus. I gathered large quantities of cloud mushrooms for two reasons. By removing the fruiting bodies and preventing distribution of spores, I hoped to control the spread of oak rot; and, in addition, I planned to send many samples to Peking Chinese Medical Hospital for further research into their unique chemistry.

The town of Yadong (elevation 2,865m) lies at the junction of three rivers — the Tangka, flowing from the northwest, the Khangphu from the north, and the Amo from the northeast. It is the main center of commerce for the rich, sub-tropical Chumbi valley where barley, buckwheat, wheat and potatoes are grown. We stayed in Yadong for three weeks, lodging in the town where we could use local facilities and create a base from which to make trips into the field. There was a large military presence in Yadong since the border between China and India lay nearby. We provided something of a diversion for the isolated soldiers who were curious about our work and anxious to socialize. The local Tibetan people were also friendly and eager to be of assistance. In fact, we appreciated their agility and ingenuity in helping us to collect specimens from steep and difficult terrain. They told us "if you want the moon, we can get it for you" and we jokingly suggested that they must have learned their nimbleness from the monkeys which were very common in that area. Although monkeys were abundant, they were not eaten by locals, since the meat was considered unclean. However, one day, some local farmers sent us a gift of fresh meat which they said was "lamb." It happened that we had heard gun shots in the forest the previous evening and that same afternoon, Au te gen and I had noticed, on a local dock, a number of monkey paws laid out in the sun to dry. We all knew that the nearest sheep were far away, and so it was not hard to guess what kind of meat the "lamb" was and many of our group declared themselves too tired to eat that night.

In the month of June, the weather in Yadong was warm and very wet with rain falling two or three times a day. Out in the forest, we were never dry and we were plagued by leeches which were impossible to keep off our bodies. Even though we bound our legs tightly with puttees made from strips of cotton, by the end of the day the fabric would be soaked with blood. However, the constant warm moisture made the vegetation brilliantly green and created ideal conditions for the growth of mushrooms which I was delighted to find in great variety and profusion. With my assistant I could fill a large bamboo basket in one hour, some of the fungi being carefully wrapped as scientific specimens, others being collected for food, for there were many varieties that I recognized as being edible.

I frequently gathered *Boletus adulis*, the delicious white beef liver mushroom, also called porcini, the pineapple mushroom *(Boletus ananas),* the monkeyhead *(Hydnum erinacaus)* and the tiger bark mushroom *(Boletus spp.)* all of which I knew to be perfectly safe to eat. I was very surprised when local people expressed concern about our consumption of mushrooms and warned us that they may be dangerous. They seldom ate mushrooms, we were told, and one old man I talked to explained that, in the Yadong area, certain parts of even the well-known, delicious varieties may be toxic, for example the cap may be safe to eat but not the stem, or the mushroom may be poisonous when eaten in combination with other foods. This information was new to me and I found it hard to believe, especially since we had already eaten many mushrooms without adverse effect. However, one morning when 12 soldiers had joined our forestry group for breakfast, we ate a soup made with tiger bark mushrooms and a kind of lettuce which had a milky sap in the leaf stems, a combination which turned out to be disastrous. Within four hours, 18 of the 26 people, including myself, had to be taken to the local military hospital emergency room, vomiting blood and suffering from severe diarrhea. Interestingly enough, a few of the Tibetan people in our group were not ill, even though they had eaten the soup. We were treated with an emetic of powdered charcoal and most of us took a whole week to recover our strength, so that our stay in Yadong turned out to be longer than planned.

THE THREE-TOED HIPPARION

More than half of the scientists involved in the Tibetan Plateau expedition were experts in geology and related disciplines. Specialists in geophysics studied the physical properties of earth, seismologists measured earth movements, geochemists sampled earth's chemical makeup, mineralogists surveyed mineral resources, geomorphologists studied landforms, hydrologists looked at surface and sub-surface waters and paleontologists searched for fossilized plant and animal remains. The latter were an especially important group including representatives of the Paleoanthropology Institute which formed a prestigious part of the Chinese Academy of Sciences in Peking. The discovery of Peking Man *(Sinanthropus pekinensis)* at Choukoutien in 1927 had spurred interest in the study of the evolution of human beings and other vertebrates in Asia, and there was excitement at the prospect that the Tibetan Plateau might be rich in fossils of dinosours and ancestors of modern mammals. Small expeditions had been made in the 1920s, 1930s and 1940s headed by young paleantologists who later became leaders in their field. It was unfortunate that, by 1975, these highly experienced, senior specialists were too old to join the grueling Tibetan Plateau expedition and so the paleontology teams were made up of newly graduated young people, full of enthusiasm but unskilled in field work techniques.

At a research site on the northern part of the Tibetan plateau, it happened that our forestry group was camping with one of the geology teams close to a small village which supplied us with manual labor needed for such tasks as felling trees, digging soil pits and excavating fossils. The village was at an elevation of 4,300 meters and after out two teams had breakfasted together on *zhang ba* and *shu yu cha*, the geologists would head uphill to their research site at 4,600 meters while our forestry group descended to 3,800 meters where spruce trees grew. One day, the geologists invited us to go with them to see their current excavation of the fossilized skeleton of a three-toed hipparion, an extinct mammal common in the Miocene and Pliocene epochs, which was once thought to be a direct ancestor of the modern horse but is now considered only a related genus. As I stood in that wild mountain site looking at the remains of an animal that had died more than 10 million years ago, I felt overwhelmed by the thought that it had lived in a period when the Himalayan mountains were young and the high plateau where I was standing had not yet been formed. It was amazing to think that those fossilized remains had slowly risen thousands of meters as the rocks in which they were embedded were thrust upwards by the inexorable mountain-building forces of tectonic plate subduction.

We watched the inexperienced junior paleontologists who were conducting the excavation with only the most primitive of tools. They were using geologists' hammers to free the fossils from surrounding rock and, even as we watched, one young man made a careless blow with his hammer and broke a fossilized tooth. Since teeth are considered to be of primary importance in identifying species, the young scientist was severely criticized. Other members of the team berated him, and there was much angry shouting. "Don't use your hammer on rocks," he was told, "use it on your own head."

A few days later, another tooth of the three-toed hipparion was the cause of more drama. Local villagers had expressed much curiosity about the excavated fossils. They were amazed to think that those ancient remains had been in the ground under their feet all the time without their knowing it. So the geologists set up a small exhibition of fossils samples displayed on a table for inspection and the exhibit was visited, over a period of days, not only by villagers but by members of other research teams and by soldiers of the People's Liberation Army who were camped in the area, providing security for both scientists and villagers. Unfortunately, the exhibit was unsupervised and there was great consternation when it was discovered that a precious fossilized tooth was gone. Meetings were held, and everybody was questioned in an effort to find the missing treasure. When villagers reported that they had watched a large, military truck bring a group soldiers to the exhibit, the local military leader took steps to find out who had been part of that group, and after intensive questioning, it was found that the culprit was the officer who had been in charge of the visiting soldiers. Asked about his motivation for taking the tooth, which he produced from the pocket of his uniform, the officer described how he had heard about the importance of fossil teeth in identifying species and how he took the tooth because he wanted to learn more about fossils. He had picked it up from the table and impulsively dropped it into the basket of a Tibetan woman who was standing nearby, but there was a hole in the bottom of the basket through which the tooth had fallen to the ground, so the officer picked it up a second time and missed his chance for redemption when, instead of replacing it on the table, he put it in his pocket. He was punished with a long period of restricted activity and suffered severe criticism from his fellow soldiers, but his case turned out to be advantageous to the research teams for it demonstrated to everybody the vital importance of caring for scientific specimens collected in the field.

The quantity of specimens collected became something of a problem. Paleontologists dug up four tons of fossils, which took up all the room in their truck, leaving no space for other stores and personal possessions. However, the spirit of cooperation was essential for survival in those remote parts, so our forestry team helped out the other scientists by transferring some of their gear into our own truck, even though we had little room to spare and our unpleasant truck driver had yet another cause for complaint.

A Painful Ordeal

In June 1976, our team was working the in highlands of south-eastern Tibet looking for undisturbed, virgin stands of spruce trees *(Abies georger var. smithii).* Heavy rains, almost every day, had saturated our tents, bedding and clothing so that we thought we would never be dry again. On rare occasions when the sun appeared, we would spread out our arms in an effort to dry the clothes on our bodies, but our Tibetan assistants warned us against this for they held a superstitious belief that water evaporating from clothes could seep into the limbs and damage one's bones. Eventually, camping became so difficult that Mr. Li asked the local villagers if they had any spare room for us to sleep. Au te gen and I were delighted to be able to move into a house, although this turned out to be a mixed blessing. We did have a roof over our heads but we paid for protection from the elements by suffering other discomforts. The building where we lodged was a two-story affair built of wood, the lower level being completely open, without walls, to serve as a shelter for yaks. The upper story, which was supported on heavy wooden spruce log columns, had rough-hewn walls in which the only opening was one small aperture covered by a wooden shutter which let in no light. The floor was made of wooden slats with wide spaces between them allowing free passage of cold draughts from below which were heavily laden with the strong smell of animals and dung. We spread our sleeping bags on the floor next to piles of barley straw and in the darkness, all night long, we could feel mice running over our bodies. But worse than the mice were the fleas that the mice brought with them. If I turned on a flashlight I could see fleas jumping everywhere and we would both wake in the morning covered with bites. Au te gen was terribly tormented by the little pests and she eventually told me, one bleak morning, that she didn't think she could accompany me on another field trip, much as she valued the learning experience. The conditions were just too hard for her to stand, she said. I had become so accustomed to discomfort during my many field trips in wilderness areas that I thought I could withstand any hardship, but my own tolerance was soon to be put to a severe test when I suffered a painful injury.

The Forestry team had rented horses in a local village so that we could travel a long distance to reach undisturbed virgin spruce stands, since all forests close to Tibetan villages were heavily cut for both building materials and fuel. I had often ridden horses in Mongolia, but those had been quite small, stocky ponies suited to the open plains. The Tibetan horses were a very different breed, tall and long-legged to nimbly negotiate rough mountain paths and narrow logs over streams. I found them hard to mount and, from up in the saddle, it seemed to me a long way to the ground; however, my horse was quiet and the man in charge of the horses was full of good advice. We had been travelling for about four hours and were passing through a forested area on the edge of a steep canyon with a river far below.

I was leading the group when Mr. Li pointed ahead to a fine stand of spruce trees which was just what we had been looking for. I made straight for the site, in my enthusiasm failing to notice that my horse was passing under an over-hanging spruce branch. The next thing I knew, I was on the rocky ground, rolling fast towards the edge of the canyon. At the last moment, I was able to hook my foot around a bush and stop my fall, coming to rest on my back with my head almost hanging over the precipice. The heavy camera which I had been carrying over my shoulder was jammed under my back, its thick lenses smashed between my lower ribs and the sharp rocks beneath. I felt as though I was being stabbed in the back and when my companions pulled me up to safety, the pain was so great that I thought I would die. It was almost noon and we were four hours away from the village, with no medical supplies. I was in agony but I did not want to spoil the team's opportunity to investigate the spruce forest we had come so far to find. I knew that they had five hours work ahead of them so I told them to just let me lie on the ground until they had finished.

That afternoon was a nightmare in which the pain of every breath seemed more than I could bear, but there was worse to follow for I had to get back on my horse and endure four hours of torture on the trip back to the village. Even in the village there was no comfort or relief. I lay in the dark, smelly house on my mouse-and-flea-infested bed hearing the bones of my broken ribs scraping together and swallowing aspirin for the pain until our small supply of aspirin was all used up, which was just as well since my stomach could not have tolerated any more. I was ashamed to let others see me cry but in that agonizing situation I could not prevent the tears from falling.

Then I suffered the additional distress of hearing that my worried teammates wanted to send me back to Peking. For me, this would have been the ultimate humiliation, a sign of failure and a confirmation of Mr. Sun's initial fears that women scientists were not sufficiently strong and resilient to be included in the Tibetan Plateau expedition. I had worked so hard to prepare myself and to prove my worthiness. There was so much work yet to be done, so many more specimens to be collected and identified. It was unthinkable that I should have to give up! I knew that my injury was not life-threatening and I was sure that I could recover quickly if I could just obtain some relief from the overwhelming pain, so I planned to hold out and to prove that I was not a liability but fully deserved my place on the team. Our group had three days more work to do in the area after which they planned to make a one-day trek on horseback to the town of Chang du. My colleagues feared that I would not be capable of making the trip and suggested that they might obtain, in the town, some medication which could be sent back to enable me to travel to Chang du from where, they said, I could be transported to Peking for proper treatment. I finally managed to convince them that I was fit to travel with the team and I endured the painful journey in the hope that there would be a doctor in the town who could help me, for I was determined to stay with the expedition and finish my work.

Fortunately, we were able to locate a doctor of Chinese medicine who treated my back with hot poultices of herbal extracts and stimulated blood circulation by massaging my skin with pre-heated hands. He also used traditional small glass suction cups to draw excess blood and fluid from the bruised flesh. In this treatment method, a vacuum is created inside a glass cup by briefly flaming the inside with a burning cotton wad. The glass is applied to the skin, where it adheres firmly by suction until it is removed after about 20 minutes, leaving behind a bright red circle and significant relief of pain. I did my part by taking vitamin C and eating a high protein diet of milk, butter, cheese and meat provided by sympathetic Tibetan friends, and in only two weeks time I was well enough to return to work. I bear a permanent reminder of that ordeal in the form of a lump on my back where the broken ribs healed at a strange angle. That place tends to become painful in wet weather and the pain brings back distressing memories, but it also makes me proud to recall that I was not defeated by that harsh experience but survived one of the most difficult tests of my life.

七十年代青藏科学考察记

1972年，我在北京看望父亲时，听到一个广播节目中在讲中国科学院在西藏进行一次大规模学术考察。长江发源于青藏高原，我原来总觉得西藏只是一个遥远、没有人烟的落后地区，寒冷、荒芜、空旷。因此听到它如画的美景和各种各样的动植物时，我被深深地吸引了。那个广播节目激发了我的想象力，以至于我特别想亲眼看一看西藏，尽管我不知道怎样才能去那里。当我对父亲讲起那个广播节目和我的愿望时，父亲非常鼓励我，他说："青藏科学考察是个宏伟的项目，我赞成你争取参加！ 果真经多方奔走努力表达我的理想．不久中国科学院终于接收我作为一名科考队员，西藏高原深深吸引了我！

1974年，第一次听说赴青藏高原的探险计划时，我十分激动．我那会儿在北京，妈妈在不久前刚刚去世，我们家度过了一个悲伤的新年。一天晚上，我们碰巧听到了一个广播节目"西藏南部的长江"，报道一个去那里考察的、由两百名科学家组成的科学探险队。我立刻被那个故事迷住了，决定要申请加入这支考察队，尽管我担心自己的性别和年龄（我那时43岁了）也许会是两个不利因素。的确，孙鸿烈大队长一开始并不想在赴西藏的这个项目中招收女科学家，因为他觉得高海拔地区野外艰苦的条件会让"未经锻炼过的女性"也许吃不消。但是，当他与我一个学微生物的同学交谈，并了解到我已在偏远区进行野外工作所积累的大量经验、我在艰苦条件下做困难工作的良好精神和能力之后，孙鸿列最后同意让我既做一个研究人员，也做一个采集人员。能成为四百名科学家中仅有的四名女成员之一，我感到十分自豪，同时，想起自己在大兴安岭和横断山脉的原始林中作为一个研究人员和探险小队长的成功，这次看到集体考察队员精神力量，我也对自己的能力充满了信心。

1975年初，在北京开了一个准备会后，我去成都参加了一个对参加西藏探险的科学家进行的为期两周的训练。有时还会被送往位于海拔2500米的西宁，既为了适应高海拔，同时也为了学习我们将要考察的那一地区的地质地理、植物区系和动物区系。我们了解到的一些东西让我十分激动，跨过 15个 5000米大山和甘孜大地震区的艰难道路，我的理想是尽快亲身为建设青藏高原宝地贡献智慧。

1976年 6月，我们队在西藏东南的高地上寻找尚未有人到过的原始云杉群。每天几乎都有暴雨，帐篷、床和衣服全都湿透了．有时真的不好受 。在极少极少的、太阳出来的时候，我们就伸开胳膊，想把衣服在身上晾干。但藏族助手们却告诉我们不要这样，因为他们有一种迷信，认为从衣服上蒸发出来的水会渗入四肢和骨头中。 最后.雨太多，要野营已是很困难了，于是李文华队长就问当地的村民有没有空房给我们睡觉。我的蒙族学生敖特根和我很高兴能住到一间屋子里去，但这似乎并不是一件好事。我们的头上的确有了屋顶，可我们为这个保护却忍受了其他许多不适。我们住的是一栋两层的木头房子，底下一层四面都没有墙，用作牦牛棚。上面一层建在粗大的云杉木柱子上，有粗糙的木墙，墙上只有一个小孔，盖着木头窗板，一点光都透不进来。地上铺着木板，板与板之间空隙很大，下面牲畜难闻的气味就和冷风一起灌了进来。我们在一堆一堆的大麦杆旁边铺开睡袋，整夜里，我们在黑暗中都能感到老鼠在身上跑来跑去。但比老鼠更烦人的是老鼠带来的跳蚤。我如果打开手电筒，就能看见到处都有跳蚤在跳，而每天早上我们醒来时，身上都是被跳蚤咬过的痕迹。敖特根说她实在受够了这种小东西，终于在一个寒冷的早晨，她对我说，她不能和我一起进行下一次野外考察了，尽管她知道那是很有价值的，她说条件太艰苦了。而在我数次对野外地区进行的考察中，我已经逐渐习惯了各种不同环境，因此我觉得自己经历过许多艰苦的条件，但在一次痛苦的受伤中，我自己的忍耐能力也受到了一次严峻的考验。

72、73年，中国科学院的自然资源委员会准备对青藏高原进行一次较大规模的科学考察。五六十年代，已有十支考察队在西藏做了一些预备性的研究。但这次考察将由来自三十个不同学科的四百名科学家组成，要对那块特殊地区的自然地理和生态情况作一次集中系统的研究。1920年首次提出、并在1960年得到了证明的板块构造学说给出了一种新见解，认为印澳板块从欧亚板块中分离，从而导致了喜马拉雅山脉在六千年中逐渐隆起，并形成了青藏高原。二战期间，周恩来总理第一个提出，应当对这个地区进行一次科学探险，研究喜马拉雅山脉隆起所带来的生态影响，以及这一地区人们的生活方式。七十年代，中国又有了其他一些强烈的动机，要安排一次地球物理学的勘测行动，以估定西藏的自然资源。尤其考虑到当时极需石油和水力等矿产能源，来减少中国对煤炭和天然气的依赖。对青藏高原的考察计划由中国科学院自然资源研究所所长孙鸿列领头，孙鸿列来自一个冒险之家，他父亲是一个地理学家，曾在二十年代骑着骆驼在中国西部的沙漠中寻找石油。那次探险成功了，人们在甘肃的玉门公园里为他父亲立了一座雕像以资纪念。

除了想看看西藏、让自己的学术生涯能有所发展、以及逃开文化大革命之外，我之所以想要去野外，还有另一个原因。妈妈患胃癌和子宫癌已经两年了，因此她在1974年1月的去世并不让人感到惊讶，但我还是因她的离开而心烦意乱，不由得想起她一生的不幸。妈妈是一个旧社会的妇女，她一生的缩影正符合汉字“女”所体现的一种方式。“女”字下面是两笔交叉，意为“束缚”，交叉的上面一笔是个斜划，意为“依靠”。与此相对，汉字的“男”上面是一个被分成四份的长方形，象征着“田地”，下面则是一个笔画笔直有力的“力”。这些在封建时代形成的汉字，多少世纪以来将一种观念深植于人们心中，即，女子地位低于男子，必须依靠男子。在我妈妈生活的年代，在男女平等上已有了许多进步，在1949年以后，起码是在一些规定中是这样.“中华人民共和国宪法第五十三条中，在括号里写道：“妇女与男子在政治、经济、文化、社会和家庭生活等一切领域中享有平等权利，男女同工同酬。”父母结婚后，父亲经常在离家很远的城市中工作和生活，因此妈妈很少能和爸爸在一起。与中国无数妇女一样，妈妈的一生都忙于在多年的贫穷、饥饿、病痛和危险中将几个孩子带大。然而，尽管一生中有如此之多的不幸和动乱，她还是将一种充满希望和自信的态度传给了我，她的女儿。我觉得自己必须努力工作以证明这种今 天的妇女能力。去西藏的探险是一个机会，可以让我实现妈妈希望中国妇女的处境能有所改善的愿望。数年之后，当我因在赴西藏的项目中的出色表现而被中国科学院授予特殊奖时，我觉得应该把这个奖献给我妈妈。

赴西藏探险前的准备

由于大部分反映西藏的照片都集中于12000英尺以上雄伟的山景，那里的气候条件极其恶劣，因此人们总认为西藏是一片不毛之地，农业非常有限，植被稀疏。事实上，12000英尺以下，她有舒适的北温带气候，加上很强的紫外线辐射，为植物的生长提供了理想的条件，并且使植物结籽很多。所以经过多年的演进，这里植物的种类和数量都极为丰富。迄今为止，植物学者已经在西藏发现了属于2008个科和1258个属的5766种植物。在这5766种植物中，由于青藏高原近期独特的地质演变史，43%为西藏特有，在世界其他地方没有发现。另外的37%在东亚也比较常见，但有一些，比如说有245种西藏杜鹃，就是从古地中海迁徙过来的，它们最初是在那里被发现的。

杜鹃所属的 *Ericaceae* 科是西藏植物区系五大科之一，其他四个科是 *Leguminosae*, *Rosaceae*, *Gaminaceae* 以及高度进化了的 *Compositae*。

在西藏除了发现的 1300 种植物以外，还有高大的灌木和乔木，有许多经济价值很高的、生长迅速的针叶树。有些针叶树，比如云南松、高山松，喜马拉雅松，西四川云杉和黄果云杉等在中国其他省份也有发现，但有十五种针叶树只生长于西藏，它们中最常见的有喜马拉雅松、长叶松、喜马拉雅枞、喜马拉雅落叶松和喜马拉雅丝柏等。这些树种大多数都发现于西藏东南的森林带，即横断山脉的南部山区及雅鲁藏布江温暖潮湿的峡谷中的大片原始森林。在高原的低纬度区，例如亚东附近，有亚热带的常绿林，里面有许多宝贵的阔叶硬木树，和种类丰富的蕨，苔 藓，地衣，也包括许多真菌。

作为一个研究影响林木健康的病菌的植物病理学工作者，和一个对各种蘑菇感兴趣的菌类学工作者，当得知将有机会进入现代科学工作者从未勘察过的地区时，我激动极了。更让我激动的是，我很有可能发现一个新的物种，并以我的名字为它命名。这是我的一个夙愿，发现一个新物种，加入那些科学家和老师们的先锋行列。他们在学术上是我的指导者，他们激励我，向我传授知识和技能，从而使我也有可能踏着他们的脚步前进。

整理生物数据的国际体系是在十八世纪中期传到中国的。1759 年，一位法国科学家 P.M.

来到中国，不久以后发表了他在中国发现的新物种，这促使中国学生开始学习国外的知识。鉴别一个新物种可不是一件小事，其过程类同于发明创造。鉴别者必须有坚实的知识经验基础，并且对某一个属及那个属中不计其数的种都十分了解，这样，当某个标本的主要特征不符合那一属的已知参数时，他才能辨认出来。在二十年的野外工作中，我采集和鉴别了许许多多的真菌标本，觉得自己已经能应付这个挑战了。我急不可待地想要赶紧出发，开始我对青藏高原的第一次探险。

正式科学考察 (1975)

终于，在 1975 年 5 月，当觉得已经准备充分了时，我们由北京飞到成都，登上了一架飞往拉萨的飞机。快到拉萨机场时，每个人都十分紧张，因为两边都是崇山峻岭，我们看不到可以降落的地方。飞机安全着陆后，大家都松了一口气，但更令人不舒服的头疼又来了。机场距市区有 90 公里，我们坐卡车颠簸了三个小时，当到达拉萨时，好多人都因 3380 米的海拔而感到头疼和恶心。我因为经常在山区工作，所以比大多数人的反应都要轻微，是极少数几个能有胃口吃下为我们准备的美味晚餐的人之一。我们住在一个部队招待所里，这里也为那些要去攀登珠穆朗玛峰的登山者提供住宿，因为那时在拉萨还没有别的招待所。非常凑巧，与我们同住的有一支从珠峰回来的中国登山探险队，里面有一个藏族女队员潘多，她刚成为世界上第一个登上珠峰的女性。我们从广播中听到了这支登山队的成绩，特别敬仰他们的成功，很为能亲眼看到他们而激动。潘多对我是个很大的鼓舞，她三十七岁了，比队里其他一些年轻队员要大得多，然而在许多年轻人失败了的地方，她成功了，因为她的身体和意志都很坚强。我俩经常聊天，有一次我告诉潘多："我为你和你的成就而骄傲，我跟你比真是差远了。"让我惊讶的是，她说："我也为你而骄傲，因为你是第一批来到并研究我的家乡西藏的科学家之一。"听她讲了她是如何克服了许多困难，忍受了海拔比拉萨还要高一倍多的地方的高山反应之后，我觉得要想达到她的标准还很难，我决定，要充分利用前面的每一次机会。

我在拉萨市内走了走，发现这里古老的街道空荡荡的，十分单调，街上风很大，气味很难闻，但在看到雄伟的布达拉宫时，我就把这些都忘了。布达拉宫是西藏最宏伟的佛教寺庙，也是最美的景色。我非常喜欢这个城市，尤其是夜里，高海拔和干净的空气使得星星看起来格外近，在漆黑的天幕的衬托下显得格外明亮。稀薄干净的空气使青藏高原暴露在强烈的紫外线辐射之中，使得当地人很容易患上皮肤癌，但也为拉萨赢得了“日光城”的美名。城市周围的植被很有限，除了几小块刚刚吐绿的柳树和杨树外，就只看得到一些矮小的柏树了。我们在成都开计划会时，听说在拉萨有一棵五百年的古柏，是由唐朝从长安远嫁吐蕃的文成公主种下的。我们得知，这次联姻在中原与吐蕃之间建立起了最初的联盟，这被认为是 1951 年中国努力重建这一联盟的部分原因。我们的指导向我们强调，一定要尊重当地的风俗和宗教信仰。

我们在拉萨呆了十天，为四十个考察队将在野外度过的四个月准备必要的供给和交通工具。部队给我们提供了厨具和军用食品，盐、油、米、面粉、面条、干海藻、糖、巧克力和其他各种罐头。我们的野营装备有很重的军用帆布帐篷，和一些美国产的鸭绒睡袋，这是自然资源委员会设法在朝鲜战争后作为剩余物品弄到的。这次发给每个人的睡垫不是床垫或吊床，而是鹿皮，这是一种能有效防潮的古老方法。个人的装备必须严格挑选，我们得意识到，有限而简陋的交通使得车辆受到了限制，我们得自己背着野营的必需品。我的装备除了几件衣服之外，主要就是我那台贵重的防滑显微镜、重要的参考书、一个用来在野外鉴别标本的轻便透镜、一本植物杂志、一个植物采集箱和一个用来装标本的特殊容器。

我被分到一个林学组，我们组共有五名科学工作者，六个助手（从当地雇的藏族年轻人，既是劳力，也当翻译），还有两个司机，各开一辆卡车和一辆吉普车。所有的东西都装好了后，该出发了，我和另一个来自四川林学院的女队员爬进了卡车驾驶室，我们的队长李文华、另一名林学家韩育方和真菌学家宗毓臣则在卡车车斗里，很不舒服地坐在一堆箱子中间。其他人坐吉普车。很不走运，我们的卡车司机不太合作，他姓葛，是退役军人，他坚信文化大革命的那一套，认为知识分子都是寄生虫，比战士和工人要差远了。他无法理解为什么我们得在青藏高原上奔波。他不喜欢给我们开车，总是用一种轻蔑的态度对我们，这使得我们组的士气多多少少受到了一些影响，直到后来我们学会了对他的这种态度不加理会。

第一天我们沿公路从拉萨往南走，经过了世界上最高的河流----雅鲁藏布江，在海拔 4200 米的羊卓雍错边宿营。羊卓雍错给我留下了非常深刻的印象，它支流众多，占地 600 平方公里。湖特别美，倒映着蓝天，湛蓝湛蓝的。离我们几乎是荒芜的宿营地不远，就是湖岸，水里满是一种小鲶鱼，用碗一次就可以舀上来四五条。这些鱼本来可以成为我们一顿丰盛的晚餐的，但我们还是把它们都放回了湖里，因为我们不敢吃。我们听说过藏族传统的“天葬”，人们将死尸放在露天处，让秃鹰来吃。这些在湖面上空盘旋的鸟，说不定会把一些尸肉掉到湖水里。谁知道这些鱼是不是吃了人肉的呢？我们可不想试试，所以那天我们还是吃的一餐半生不熟的米饭。

野营的第一个晚上，我们就发现，通常煮饭的方法在这里不太适用了。在海拔很高的地方，水的沸点降低，这就意味着，即使是用压力锅，饭也需要很长时间才能熟。我们的藏族助手们建议我们最好是吃当地的传统主食----糌粑。糌粑是将炒大麦粉和豆粉放在一个陶碗中，然后用茶和成面团，它比米饭要好吃和有营养一些，也更容易做。我们很快就决定了，应该带上做糌粑要用的东西。我们还学会了很有意思的西藏传统的泡茶方法。他们的茶叫酥油茶，在牛粪烧的火堆上将水烧开后，把一块茶（从“茶砖”上掰下来的。“茶砖”是通常用来包装干茶叶的一种方法）、盐和牦牛油放入一根用松木做成的长管子中，把开水倒进去，再盖好盖子，有一个木头活塞穿过盖子，在里面搅拌。然后把管子里的水倒入水壶，再次加热后，又倒入木头管子进行搅拌。茶泡好后，第一道用来和糌粑，其余的就倒入小碗中喝。我们还听说，走远路的西藏人常常带着土豆，把它们和很辣的青椒一起扔到火堆里烤，再加上糌粑和酥油茶，这就是简单经济、营养丰富的一餐饭了。但不管怎么说，糌粑毕竟是一种学会吃的东西，我们组里的大多数人还是愿意吃熟悉的米饭，尽管它得花更多时间才能做好。

第二天我们继续往南，公路随山爬得很高，我们经过了两个海拔 5000 米的山口。我永远不会忘记我们看到的一种生长在那么高的海拔处的、最不寻常的植物----雪莲，它是高山植物的一种，高山植物都具有有效防寒和防强烈阳光照射的结构。雪莲的茎和叶上都长有绒毛，它们缠结在一起，形成了一个个防止水分迅速蒸发的隔离室，从而使得植株能开出像莲花一样的巨大奶油色花朵，它在这样一种环境中看起来格外迷人。但我们对这一美景的欣赏却被高山反应、极度的寒冷和阳光的反射所打扰。我们组的很多人都头疼得厉害，所以当公路又下到一处长满青草的高原上时，大家都轻松多了。我们继续向帕里前行，计划在那里呆两天。

由于恶劣的气候，加上土地贫瘠，高原的植被非常有限，但还是有能为野牦牛提供食物的草地。我们看见野牦牛在一块块残雪之间吃草，当车从它们身边驶过时，还能闻到一股刺鼻的气味。路边有一些草木，像山艾树和野菊花，但除了一些矮小的高山桦和扭曲的塞宾柏之外，就没有其他什么树了。没有树或是灌木丛，我们几个女性要在路边方便一下就成了问题。我们的几个藏族女助手都特别文静害羞，我就用一把大伞给她们做临时的帘子。

高原人口稀少，很少能看到人烟，因此当在路上碰到一个带着三个孩子的妇女时，我们觉得特别有意思。我们停下来，通过一个翻译和她聊天。她告诉我们，她二十九岁了，和丈夫一起住在附近，喂养牦牛和绵羊。她说她丈夫不在家，因为他骑马去帕里（得花两天）看电影了！这个妇女邀请我们去她不远处的家看看。房子是用木头做的，但没有窗户，所以里面很暗。我们看到屋子里还有两头牦牛犊，因此也就不奇怪屋里有一股强烈的牦牛味了。屋里有一张大床，铺着一大块毛毯，上面躺着一对双胞胎女孩。我感到十分惊讶，这个女人竟然能在如此艰苦的条件下养育五个孩子。但她穿着肥大的传统服装，戴着银首饰，皮肤平滑有光泽，看起来很健康，也很幸福。她想给我们烧茶喝，但是很遗憾，我们没有时间呆下去了。临走时，我们给了最大的那个孩子一些糖，他们从来没有见过糖。

帕里海拔 4360 米，是世界上最高的社区之一。它的名字在藏语里的意思是“猪山”，因为它附近有一座山形状像头猪。帕里位于西藏高原的边缘，村边有一条长十四公里的路，它直插入藏布河谷，路的那一头海拔下降了 1600 米，是位于中印边界的亚东。我们在帕里只呆了两天，因为我们迫不及待地想要去海拔较低的藏布河谷勘测那里丰富的亚热带森林。但就是在帕里附近，我非常高兴地采集到了一种稀有的菌类--虫草（*Cordyceps sinensis*），一般被称为“冬虫夏草”。正如它的俗名所显示的，冬天时，它的孢子附着在一种毛虫身上，并以之为食，直到毛虫死去,其体内也充满了菌类的组织，最后，到夏天时，它就长成了一棵菌类植物。“冬虫夏草”在中国一直是一种民间草药。1975 年时，晒干的冬虫夏草在帕里市场上能以每片一分的价格买到，而二十年后的今天，价格已经猛涨到了每磅六百美元。这主要是由于一件事实的公开，即，一位在 1996 年奥运会上获金牌的中国女运动员据说是在训练饮食中加入了冬虫夏草。很不幸，对这种只生长在西藏少数地区的稀有物种的强烈注意已经威胁到了它的生存，除非人们能对它进行人工种植，否则冬虫夏草有可能会灭绝。不过，在世界其他地方，还生长着虫草属的其他一百多个种类，它们也被证明是有药用价值的。

在接下来的几个星期我们所工作的地区中，生长有大量菌类，这使得我有了很多有趣的经历。在从帕里到亚东的路上，我们碰到了一队来自拉萨西藏中医学院的学生，他们正在进行野外的中草药采集。当他们得知我是一个真菌学者时，便告诉我头一天他们在一个较远的村子里发现的一大片生长在死了的李树上的灵芝。灵芝一直因其药用价值而十分珍贵。这些学生希望我能去那里看一看，帮他们的老师鉴定一下那些灵芝，因为灵芝属(*Ganoderma lucidium*)有许多不同种类。早在一千多年以前，中国就曾有一篇学术文章，描述了不同颜色的各种灵芝，及其特殊的生态位置。我决定要去那里看一看。学生们给我画出了一幅详细的地图，第二天，吉普车司机载着我和另一个助手去了250公里以外他们所描述的地方。看到这么多的无主灵芝，我十分激动，一口气采集了250多个标本，急切地想要用显微镜对它们进行观察，以寻找相近种类之间的区别。我没有装标本的盒子了，所以回到营地后，我小心谨慎地将这些珍贵的标本用床单包起来，放在帐篷里。正巧那天晚上有二十个古生物学者到了我们的营地，我记得他们打了野鸽子作晚餐，我们还开了一个联欢会。很不幸，我的助手向很多人讲到了我们的采集，结果第二天，当我在野外时，有人到了我的帐篷里，拿走了每一个灵芝的唯一标本。我不是个爱哭的人，可当我发现灵芝被偷时，我哭了。丢掉了这么多重要的研究标本，我又伤心又愤怒，因为我知道不会再有机会回去采集了。

几个星期后，当我们在亚东工作时，我采集到了其他一些有药用价值的菌类标本，这多多少少对我是一个补偿。有一天我在一段腐朽的硬木上发现了一些蜜菇 假蜜环菌 (*Armillaria mellea*)，根据我所学到的知识，有一种叫做天麻 (*Gastrodia elata*) 的块茎植物是与蜜菇共生的，它在治疗癫痫病方面有极高的药用价值。我从未见过天麻，但发现的蜜菇（我们美餐了一顿）使得我找到了大量在腐烂的木头深处、生长在蜜菇上的这种块茎植物。休息的那天，我告诉朋友们我是在哪里找到天麻的，他们每个人都赶快采了大概有五磅，准备晒干了带回家送人，这可值一大比钱呢。天麻在西藏十分珍贵，在这里，由于近亲结婚，天麻及其他一些病症极其普遍。我们在乡间经常能看到一些精神失常的当地人，看到他们，我觉得十分难过。

另一种具有药用价值的菌类是云芝(*Polystictus versicolor*)，经常用于治疗胃癌。在亚东地区，发现了大量这种色彩斑斓、非常美丽的腐生蘑菇，它们生长在橡树上，这些橡树由于云芝的生长而枯朽。我采集了大量的云芝，这有两个原因，一是砍掉长满了云芝的树枝，防止孢子扩散，我希望这样能控制橡树的腐烂；此外，我还想给北京中医院送一些样本，让他们可以做医学上的进一步研究。

亚东（海拔2865米）位于三条河的交汇处----从西北方流来的唐卡河、从北边流来的康普河，和从东南方流来的阿莫河。它是肥沃的亚热带藏布河谷的主要商业中心，河谷里的人们种植大麦、荞麦、小麦和土豆。我们在亚东呆了三个星期，住在镇上，我们可以使用当地的一些设备，并建立起一个基地，从那里出发去野外。由于亚东靠近中印边界，所以有一支驻军。这些与外界隔绝的战士们对我们和我们的工作感到新奇，很想和我们交往。当地的西藏人也都十分友好，很愿意做我们的助手。事实上，他们的敏捷灵巧能帮助我们从险要的地方采来标本。他们对我们说："如果你们要月亮，我们也可以弄来。"我们开玩笑说，这些人肯定是从当地众多的猴子那里学得这么灵活。虽然猴子很多，但当地人从不吃猴肉，因为他们觉得猴肉很脏。有一天，几个当地的农民送给我们一些新鲜肉，说是"羊肉"。而我们碰巧在头一天晚上在森林里听到了枪声，那天下午，敖特根和我看到在一个码头上晒着好几个猴爪子。我们都知道最近的羊群也离这里很远，因此不难猜出那些"羊肉"是什么。那天晚上，我们组的好几个人都说自己太累了，不想吃饭。

六月，亚东的天气温暖潮湿，几乎每天都要下一两场雨，我们呆在野外，身上就没有干过。更烦人的是蚂蝗，我们简直没办法躲开它们。哪怕我们用棉绑腿紧紧地缠在脚上，一天的工作结束时，绑腿还是会被血浸透了。但持续的温暖湿气却使得草木青翠欲滴，并为蘑菇的生长提供了理想的条件，而能看到这么多数量种类都极其繁多的蘑菇，是叫我格外高兴的。在助手的帮助下，我一个小时就可以采到满满一大篮蘑菇。有些被小心地包起来，用作标本，其他的则作为食物，因为我看到有些蘑菇是可食用的。我采集到的有美味牛肝菌（*Boletus adulis*），即味道鲜美的白色菇，也叫，还有菠萝菇（凤梨小牛肝菌，*Boletus ananas*），猴头菇（*Hydnum erinacaus*）和虎皮菇（*Boletus spp.*），这些都是可以吃的。我很奇怪，当地人对我们吃这些蘑菇表示担心，并警告我们它们可能会有毒。我们得知，他们很少吃蘑菇。一个老人告诉我说，在亚东地区，哪怕是一些众所周知的、味道鲜美的蘑菇，它的有些部分也会有毒，比如说，菌帽也许能吃，但菌柄就不能吃了，或者是当把某种蘑菇和其他食物一起吃的时候，它就有毒了。这些话我闻所未闻，觉得简直难以置信，尤其是我们已经吃过很多蘑菇了，也并没有出什么事。可后来有一天早上，有十二个战士和我们组一起吃早饭，我们喝了一盘用虎皮菇和一种叶茎里有奶液的莴苣做的汤，这盘汤最后被证明是灾难性的。四个小时后，二十六人中有十八个，包括我，又吐血又拉肚子，都被送到了当地军医院的急诊室。可奇怪的是，我们组里有几个西藏人虽然也喝了汤，却安然无恙。我们服下木炭粉以催吐，大多数人都花了整整一个星期才恢复过来，因此我们在亚东比原计划的要多呆了一阵子。

三趾马化石群发现的故事

在西藏高原考察的科学家中，有一多半的人是地质学或相关学科方面的专家。在那里，地球物理学者研究地表的自然特征，地震学者测量地壳运动，地球化学学者对地球的化学成份进行取样，矿物学者勘测矿物资源，地形学者研究地形，水文学者观察地表水和地下水，古生物学者则寻找动植物的化石。古生物学者组特别重要，这一组有来自古人类学学院的代表，而古人类学学院在中国科学院颇有名望。1927年周口店北京猿人的发现推动了对人类及其他脊椎动物进化过程的研究。而西藏高原很可能蕴藏着大量恐龙和现代哺乳动物祖先的化石。二十世纪二十到四十年代，由一些年轻的古生物学者领头，进行了几次小规模的考察，那些年轻人后来都成为了他们学科领域中的大家。但是很遗憾，1975年时，这些经验丰富的高级学者年纪都太大了，没法参加这次大工作量的考察活动。因此古生物学者组里都是些刚刚毕业的年轻人，热情有余，但经验不足。

在高原北部的一个考察点，我们林学组正好和一个地质学队一起在一个小山村附近宿营，村民们帮助我们干一些诸如砍树、挖土坑和挖化石之类的体力活。村子位于海拔4300米的地方，早上，我们两队人一起吃过早饭后，地质学者们就上山，到他们海拔4600米的考察点去，而我们林学组就下到生长着云杉的3800米的海拔处。有一天，地质学者们邀请我们去看看他们正在挖掘的一个三趾马的化石，三趾马是一种生活在中新世和上新世的哺乳动物，现在已经灭绝了，人们曾经认为它是现代马的祖先，但现在一般认为它只是一种与现代马较近的种类。我站在野外山地上，看着这个在一万多年以前死去的动物的遗骸，不禁想，在它生活的时代，喜马拉雅山还很年轻，而我脚下的这块高原还没有形成呢。想一想，构造板块分离时的巨大作用力使得高山形成，这些化石也随着它们所嵌于其中的岩石被推起而缓慢上升了几千米，这是多么奇妙的事呀！我们看着这些缺乏经验的年轻古生物学者用最简陋的工具进行着挖掘工作，他们用地质学者的锤子来将化石从周围的岩石中剥离出来。就在我们看着时，一个年轻人不小心用锤子敲断了化石的一颗牙齿。由于牙齿在鉴别物种上有着极其重要的作用，这个年轻人受到了严厉的批评。队里的其他人都责骂他，很多人都愤怒地大叫：“别拿锤子砸石头，砸你自己的脑袋吧！”

几天之后，这个三趾马的另一颗牙齿引起了更大的骚动。当地的村民对这个发掘出来的化石感到十分惊讶，他们很奇怪，这个远古的遗骸竟然在他们脚底下躺了这么久，而他们对此一无所知。因此地质学家们就在一张桌子上举办了一个小小的化石标本展，几天里，不但村民，其他队的队员、还有驻扎在这一地区、保卫科学工作者和村民安全的战士们也都来看展览。但展览没有人看守，结果人们惊讶地发现，有一颗珍贵的牙齿化石不翼而飞了。各方召开了紧急会议，每个人都被叫去问话，以寻找这个丢失的宝贝。村民们汇报说他们看到一辆军用大卡车载着一车战士来看展览，于是地方的部队领导赶快调查是哪些人。经过大量问话之后，终于找到了肇事者，就是这些来参观的战士们的负责人，他从军装口袋里掏出了那颗牙齿。在问到动机时，他说，他听到牙齿化石在鉴定物种上的重要性，所以他拿了这颗牙齿，想对化石多了解了解。当时他把它从桌子上拿起来，由于激动而将它掉到了旁边一个藏族妇女拎着的篮子里，可篮子底有个洞，牙齿化石又从洞里掉到了地上。他再次将它拿起来，但没有把这颗化石放回桌子上，而是将它放进了自己的口袋，他终于失去了一个补过的机会。他受到了惩罚，很长一段时间行动都受管制，战士们也都对他进行了严厉的批评。但这件事对考察队却有一点好处，就是，它向每个人说明了，看管好从野外采集到的标本是多么的重要。

我们所采集到的标本的数量渐渐地成为了一个问题。古生物学者们挖出了四吨重的化石，占去了他们卡车的所有空间，没有地方放其他物品了。在那种偏远的地方，要想生存，团结合作的精神就很重要，所以我们林学组向其他科学工作者伸出了援助之手，让他们把一些东西放到我们的卡车上，尽管我们也空不出多少地方来，而我们那位令人不快的司机就又有了抱怨的理由了。

一次落马的经历

由于西藏各山村附近的森林都因建筑和燃料而被大量砍伐，林学组就从一个村子里租来了几匹马，这样我们就可以走较长的路，去一些无人到过的原生林了。我在内蒙时经常骑马，但那都是些矮胖的小马，适于在平原上行走。可西藏的马大不一样，它们十分高大，腿很长，能够灵活安全地在山路上和独木桥上行走。我发现要骑上它们非常困难，坐在马鞍上看下去，我觉得自己离地面特别远。好在我的那匹马很安静，而且负责这些马的人也很有经验。那一天，我们已经走了四个小时了，正在穿过一处陡峭的峡谷边上的一个林区，峡谷下面就是河。我走在队伍前面，这时李队长指着前面一处特别好的、我们正在寻找的云杉群。我满心激动，直冲过去，却没有发觉我的马上方就是一个倒悬着的云杉枝。接下来我所知道的就是，我正在石头地面上迅速地滚向峡谷边缘。在最后一分钟，我终于用脚勾住了一株灌木，停止了滚动。我仰面朝天，脑袋几乎就是悬在悬崖边上。背着的相机被压在身下，厚厚的镜头在我的肋骨和岩石之间被压得粉碎。我觉得好象有什么东西戳在背上，当队友们将我拖到安全的地方时，我感到特别特别疼，以至于觉得自己都要死掉了。那会儿是中午，我们离村子有四小时的路程，手头也没有药物。我当时十分痛苦，但我们队大老远来这片云杉林考察，我不想让这次机会因为我而泡汤。我知道他们还得花五个小时，我就告诉他们，就让我躺在地上，等他们弄完了再说。

那个下午在我看来简直就是一场恶梦，每呼吸一次，疼痛似乎就让我更加无法忍受。可更难受的还在后面，我必须再回到马背上，忍受又一个四小时的煎熬，回到村子。在村里，我的疼痛也没有丝毫减轻。我在黑暗而臭气熏天的屋子里，躺在我那张老鼠和跳蚤肆虐的床上，听着我断裂的肋骨互相磨擦，不停地吞下阿司匹林止疼。结果我们的那一点阿司匹林都被我吃完了，正好那时我的胃也已经再也受不了了。我觉得让别人看见自己哭是一件很不好意思的事，但在那样一种痛苦之中，我的眼泪止不住地往下掉。然后我又听说我焦急的队友们想把我送回北京，我在疼痛之外又感到了沮丧。对我来说，这是最丢脸的事了，它标志着失败，同时又证实了孙鸿列的担心，认为女科学家不够强壮，不宜于参加对青藏高原的探险。我一直努力工作，准备好一切，并证明了自己的价值。还有很多很多工作要做，还有很多很多标本要采集和鉴定，可我却得放弃，这简直不可想像！我知道自己的伤并不是致命的，我相信只要疼痛能减轻，我很快就能恢复过来，因此我决定要坚持住，并证明我不是个麻烦，而是能胜任在队里的位置。我们组在这个地区还要呆三天，然后他们就准备骑马长途跋涉去昌都镇。同伴们担心我无法完成这次行程，他们建议说，他们可以从昌都给我弄一些药送回来，这样我就可以去昌都，然后从那里我就可以去北京了。我费了好大劲，最后终于让他们相信我能和队伍一起去。我忍受着行程中的痛苦，想着也许能在镇上找到一个能帮助我的医生，因为我已经下定决心要留下来，和探险队一起，完成我的工作。很幸运，我们找到了一个中医，他在我背上贴上烤热的膏药，并把手搓热了给我按摩，以促进血液循环。他还用传统的方法给我拔火罐，吸出瘀血。拔火罐就是先用烧着的棉花团在小罐内部形成真空，然后将小罐按到皮肤上，由于吸力，小罐会牢牢地吸附到皮肤上，大约二十分钟后将它取下来。皮肤上会留下一个红红的圆圈，而疼痛也大大减轻了。我自己则服用维生素 C，并吃一些含有大量蛋白质的东西，比如牛奶、黄油、奶酪和猪肉，这些都是好心的西藏朋友们送给我的。只用了两个星期，我就能重新工作了。但这次痛苦的经历却给我留下了一个永久的后遗症，断裂的肋骨成一个奇怪的角度自己长好了，在我的背上形成了一个肿块。当天气潮湿时，那个肿块就会疼，这疼痛又会勾起我痛苦的回忆，但它也让我骄傲地想起，我没有被那次艰苦的经历所打败，而是战胜了我一生中最困难的考验之一。

1976 年毛主席逝世以及随后四人帮倒台之后，政府对知识分子的态度有了变化。邓小平宣布说，以后，知识分子不再是腐朽的资产阶级的一部分，而是工人阶级的一部分，因为“科学技术是生产力的一部分”。作为这种变化的结果，科学家和教授们回到了北京，重新进行他们的教研工作。政府还努力恢复长期受到冷落的高等教育和科研中心，林业科学院也包括在内。由于要重新进行一些研究计划，所以急需一些有经验的工作人员。我和我的丈夫不仅是经验丰富的研究人员，而且在各自的学科中都有较高的威信，被认为是领头人。结果，在 1977 年，林业科学院联系到了在内蒙的我们，希望我们到北京工作。我将负责一个森林病理实验室，昂和则将继续研究他的专业森林经理学。我们觉得很光荣，很激动，但还有一些重要的问题需要解决。要回到北京，我们必须要有北京户口，这是每一个市民都必须有的。户口很难解决，因为这是政府用来控制北京人口的一个重要办法。昂和弄到了所有必需的文件，最后我们终于得到了北京户口。但到了北京后，我们面临着更多的复杂问题。工作是有了，但我们却被告知不仅必须自己找房子住，还得负责为进行工作的实验室选定一间办公室，因为林业科学院的房子还被部队占着呢。一九八十年后我们家因得知婆母消息及 University of Wisconsin/Madison 植物病理系邀请我合作研究荷兰榆病及锈病等国际流行病；这样，又开始了生命新的一页。(梅雪琴教授研究生小组　整理)

INDEX OF FOREST FUNGI OF PHYTOGEOGRAPHY

INDEX OF SUBJECT

INDEX OF HOST SCIENTIFIC NAME

INDEX OF FUNGI SCIENTIFIC NAME

INDEX OF OTHER PATHOGEN AND

INSECT SCIENTIFIC NAME

INDEX OF SUBJECT

INDEX OF HOST SCIENTIFIC NAME

INDEX OF FUNGI SCIENTIFIC NAME

Dawn Redwood

OTHER PATHOGENS

Bacteria

MLO

High plant Parasite

Nematode

Others Pathogen and physical reason

INDEX OF INSECT SCIENTIFIC NAME

Dendroctonus frontalis 89
Dendroctonus sp. 89
Hylestes techangensis 89
Hylurgops major 89
Cinara tibetapini 89
Cinara. paxilla 89
Matsucoccus sinensis 89
Crisicoccus pini 89
Selenephera lunigera 91
Ichneumonidae 92
Tetropium oreinum 98, 99
Tetropium castaneum 99
Dendroctonus micans 99
Dendroctonus armandi 99
Leptomias acutus 115
Leptomias semilircularis 115
Pulvinnaria targioni 115
Ceroplastes 116
Rosanococcus 116
Pulvinaria citricola 116
Pterocomma bailangense 117
Pemphigus chomoensis 117
Pemphigus tibetensis 117
Callaphis nepalensis 117
Cavariella thasana 117
Pterocomma tibetasalici 117s
Elatobium chomoense 117